Measurement of Residual and Applied Stress Using Neutron Diffraction

NATO ASI Series

Advanced Science Institutes Series

A Series presenting the results of activities sponsored by the NATO Science Committee, which aims at the dissemination of advanced scientific and technological knowledge, with a view to strengthening links between scientific communities.

The Series is published by an international board of publishers in conjunction with the NATO Scientific Affairs Division

A Life Sciences **B Physics**	Plenum Publishing Corporation London and New York
C Mathematical **and Physical Sciences** **D Behavioural and Social Sciences** **E Applied Sciences**	Kluwer Academic Publishers Dordrecht, Boston and London
F Computer and Systems Sciences **G Ecological Sciences** **H Cell Biology** **I Global Environmental Change**	Springer-Verlag Berlin, Heidelberg, New York, London, Paris and Tokyo

NATO-PCO-DATA BASE

The electronic index to the NATO ASI Series provides full bibliographical references (with keywords and/or abstracts) to more than 30000 contributions from international scientists published in all sections of the NATO ASI Series.
Access to the NATO-PCO-DATA BASE is possible in two ways:

– via online FILE 128 (NATO-PCO-DATA BASE) hosted by ESRIN,
Via Galileo Galilei, I-00044 Frascati, Italy.

– via CD-ROM "NATO-PCO-DATA BASE" with user-friendly retrieval software in English, French and German (© WTV GmbH and DATAWARE Technologies Inc. 1989).

The CD-ROM can be ordered through any member of the Board of Publishers or through NATO-PCO, Overijse, Belgium.

Series E: Applied Sciences - Vol. 216

Measurement of Residual and Applied Stress Using Neutron Diffraction

edited by

Michael T. Hutchings

National Non-Destructive Testing Centre,
AEA Technology, Harwell Laboratory
Didcot, Oxfordshire, U.K.

and

Aaron D. Krawitz

Department of Mechanical and Aerospace
Engineering and Research Reactor Center (MURR),
University of Missouri, Columbia, Missouri, U.S.A.

Springer Science+Business Media, B.V.

Proceedings of the NATO Advanced Research Workshop on
Measurement of Residual and Applied Stress Using Neutron Diffraction
Oxford, United Kingdom
March 18–22, 1991

Library of Congress Cataloging-in-Publication Data

Measurement of residual and applied stress using neutron diffraction /
    edited by Michael T. Hutchings and Aaron D. Krawitz.
        p.   cm. -- (NATO ASI series. Series E, Applied sciences ; no.
    216)
      Includes bibliographical references and index.
      ISBN 978-0-7923-1809-5   ISBN 978-94-011-2797-4 (eBook)
      DOI 10.1007/978-94-011-2797-4
      1. Neutron radiography--Congresses. 2. Residual stresses-
    -Measurement--Congresses.   I. Hutchings, Michael T. II. Krawitz,
    Aaron D.   III. Series.
    TA417.25.M43   1992
    620.1'124--dc20   92-16842

ISBN 978-0-7923-1809-5

Printed on acid-free paper

TABLE OF CONTENTS

8. COMPARISON OF NEUTRONS WITH X-RAYS AND OTHER STRESS PROBES

Note: Speaker is Underlined in multi-author papers.
            +denotes Poster Paper.

PREFACE

The accurate, absolute, and non-destructive measurement of residual stress fields within metallic, ceramic, and composite engineering components has been one of the major problems facing engineers for many years, and so the extension of X-ray methods to the use of neutrons represents a major advance. The technique utilizes the unique penetrating power of the neutron into most engineering materials, combined with the sensitivity of diffraction, to measure the separation of lattice planes within grains of polycrystalline engineering materials, thus providing an internal strain gauge. The strain is then converted to stress using calibrated elastic constants. It was just over ten years ago that the initial neutron diffraction measurements of residual stress were carried out, and during the ensuing decade measurements have commenced at most steady state reactors and pulsed sources around the world. So swift has been the development of the field that, in addition to fundamental scientific studies, commercial measurements have been made on industrial components for several years now. The use of neutrons is ideally suited to the determination of triaxial macrostress tensors, macrostress gradients, and microstresses in composites and multiphase alloys as well as deformed, plastically anisotropic metals and alloys. To date, it has been used to investigate welded and heat-treated industrial components, to characterize composites, to study the response of material under applied loads, to calibrate more portable methods such as ultrasonics, and to verify computer modelling calculations of residual and applied stress.

The small but active neutron stress measurement community convened in Oxford from March 18-22, 1991, for the first full week's meeting dedicated to discussing the problems, progress, and future potential of the method. Most of the current experts in the technique were able to participate, as well as representatives of related disciplines. This volume is the proceedings of that meeting, and as such represents a definitive statement of the current status of the field. The papers contained herein, all of which have been professionally refereed, represent a comprehensive, largely self-contained, source of information on the use of neutrons for residual stress analysis. Although it is probably best to let the contributors speak for themselves through these papers, a brief outline of the Workshop, and thus the status of the field today, may prove useful to readers of this volume.

The scope of the Workshop programme can be summarized as follows. An overview of the method was first presented, followed by background presentations on the industrial need for such measurements, examples of industrial X-ray applications, and various analytical perspectives. Fundamentals were then extensively addressed, including the extraction of stress tensors and the errors involved, separation of macro- and microstresses, stress-free reference standards, problems caused by anisotropy, large grains and texture, plasticity, and grain interaction and relaxation issues. In the consideration of these points the interaction of those measuring stress with experts in materials science and mechanical metallurgy is proving of increasing value to all concerned. Instrumentation was a major theme and included discussion of optimization of instruments, the use of position-sensitive detectors and soller collimators, microbeam methods, instrument and sample alignment methodology, Fourier techniques, peak fitting and analysis errors, and resolution in both strain and gauge volume sampled. The use of pulsed sources was explored and their virtues and limitations relative to steady state sources was a central theme. The study of composites and the measurement of microstresses was described and examples were shown. Examples of applications and problems in the measurement of bulk components were presented. Interesting results on stress gradients within one millimetre of a surface led to considerable debate about how close to a free surface one can measure, and whether or not probe volumes can be allowed to partially leave a sample. In contrast to these points of detail in strain measurement, an invited evening speaker recounted his experience using conventional strain gauges to deal with practical in situ problems occurring in industry. In all of this work the value of analytical, theoretical, and numerical support for diffraction measurements was apparent, as was, conversely, the requirement for validation of numerical and analytical models of stress fields. The concluding presentations dealt with a comparison of the neutron method with other stress probes. Three discussion sessions on topics of interest were held during the Workshop, and brief reports of their conclusions are included in this volume. Participants had an opportunity to make an afternoon visit to the ISIS spallation neutron source at the nearby Rutherford Appleton Laboratory.

Although these remarks hopefully capture the primary focus of the meeting, other points are worth noting. Though it was the charge of the Workshop to deal with problems in the field, and many of the papers therefore dwell on them, it should be emphasised that this volume is basically a positive statement of the considerable progress and success of the method. It was intentional that some of the participants had used only X-rays for the measurement of stress, since the neutron method is an outgrowth of decades of successful X-ray stress methods which will remain the primary diffraction tool for near-surface residual stress problems because they are more appropriate, cheaper and more readily accessible. A final point is that although the need for answers to practical industrial problems will probably be the main impetus to work in the field, the use of the method to give new unique information on the more fundamental aspects will undoubtedly continue to develop and, in return, to widen the application.

Finally, the Workshop participants expressed a desire to pursue a number of actions. It was agreed in principle that round robin testing of a standard specimen would be useful, and steps were taken to select an appropriate sample. It was also agreed to begin considering an ASTM standard for neutron stress measurements and, as a first step, the recently written X-ray procedure will be reviewed. Lastly, a search for a regular organ for communication within the field was initiated, probably through the new journal Neutron News.

ACKNOWLEDGEMENTS

The NATO Advanced Research Workshop Programme is gratefully acknowledged for financial support, under grant No ARW-900814, which made this most timely and successful meeting possible. We would also like to acknowledge the National NDT Centre of AEA Technology for their help with the organisation, and the staff of Jesus College, Oxford, for their hospitality and impeccable logistic support. We thank T Leffers and L Pintschovius for joining us on the Organising Committee of the Workshop.

M T Hutchings, Harwell Laboratory

A D Krawitz, University of Missouri

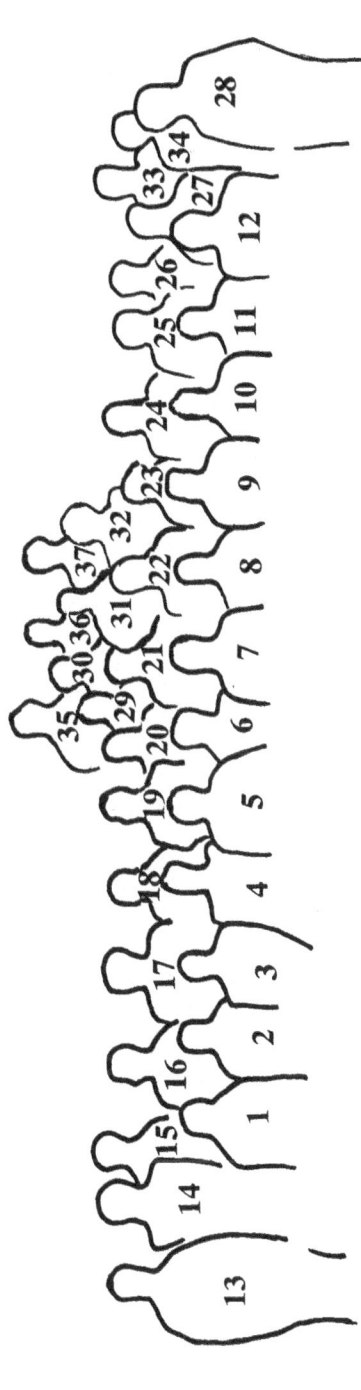

35 H PRIESMEYER, 36 T LEFFERS, 37 P WITHERS

30 A ALLEN, 31 T LORENTZEN, 32 A LODINI

29 A BOWEN

33 H PRASK, 34 P J WEBSTER

18 M KIJEK, 19 L EDWARDS, 20 S SJÖSTRÖM, 21 M HUTCHINGS, 22 A EZEILO, 23 C WINDSOR, 24 P BRAND

25 W REIMERS, 26 J RICHARDSON, 27 KOCSIS

15 M BOURKE, 16 T HOLDEN, 17 F RUSTICHELLI

3 P HOLDWAY, 4 M JOHNSON, 5 M JAMES, 6 R WINHOLTZ, 7 F MARGAÇA, 8 I NOYAN, 9 S HULL, 10 P S WEBSTER

28 X WANG

13 A KRAWITZ, 14 S SKOLIANOS

1 G MILLS, 2 J GOLDSTONE

11 L PINTSCHOVIUS, 12 E PLUYETTE

LIST OF PARTICIPANTS

Director:

M T HUTCHINGS,   *NNDTC, AEA Technology, B521.1, Harwell Laboratory, Didcot, OX11 0RA, UK.*

Committee:

A D KRAWITZ,   **(Deputy Director)** *Dept. of Mechanical and Aerospace Engineering and Research Reactor Center (MURR), Univ. of Missouri, Columbia, MO 65211, USA.*

L PINTSCHOVIUS,   *KFK, Institut fur Nukleare Festkorperphysic, Postfach 3640, D-7500 Karlsruhe, Germany.*

T LEFFERS,   *Materials Dept., Riso National Laboratory, P.O. Box 49, DK-4000 Roskilde, Denmark.*

USA:

M A M BOURKE,   *MS H805, P-LANSCE, Los Alamos National Laboratory, Los Alamos, New Mexico 87545, USA.*

J A GOLDSTONE,   *MS H805, P-LANSCE, Los Alamos National Laboratory, Los Alamos, New Mexico 87545, USA.*

M R JAMES,   *Rockwell International Science Center, 1049 Camino Dos Rios, P.O.Box 1085, Thousand Oaks, CA 91358, USA.*

I C NOYAN,   *IBM Research, T G Watson Research Center, P.O.Box 218, Yorktown Heights, NY 10598, USA.*

H J PRASK,   *Bldg. 235, NIST, Gaithersburg, MD 20899, USA.*

J W RICHARDSON, JR.,   *IPNS Div., Bldg.360, Argonne National Laboratory, 9700 S. Cass Avenue, Argonne, IL 60439, USA.*

R A WINHOLTZ,   *Dept. of Materials Science and Engineering, Northwestern Univ., Evanston, IL 60208, USA.* (Present address: *Dept. of Mechanical and Aerospace Engineering and Research Reactor Center (MURR), Univ. of Missouri, Columbia, MO 65211, USA.*)

Canada:

T M HOLDEN,   *Neutron and Solid State Physics Branch, Atomic Energy of Canada Ltd., Research Company, Chalk River Nuclear Laboratories, Chalk River, Ontario, Canada K0J 1J0.*

U.K.:

A J ALLEN,   *NNDTC, AEA Technology, B521.1, Harwell Laboratory, Didcot, OX11 0RA, UK.* (Present address: *Bldg. 223, NIST, Gaithersburg, MD 20899, USA.*)

A W BOWEN,   *Materials Structures Dept., RAE Farnborough, Hants. GU14 6TD, UK.*

L EDWARDS,   *Fracture Research Group, Faculty of Technology, The Open University, Milton Keynes, MK7 6AA, UK.*

A N EZEILO,   *Dept. Mechanical Engineering, Imperial College, Exhibition Road, London SW7 2BX, UK.*

P HOLDWAY,   *Materials Structures Dept., RAE Farnborough, Hants. GU14 6TD, UK*

S HULL,   *ISIS Facility, Bldg. R3, Rutherford Appleton Laboratory, Chilton, Didcot, OX11 0QX, UK.*

M W JOHNSON,   *ISIS Facility, Bldg. R3, Rutherford AppletonLaboratory, Chilton, Didcot OX11 0QX, UK.*

G MILLS,   *Dept. of Civil Engineering, Univ. of Salford, Salford, M5 4WT, UK.*

E PROCTER,   66, Kenilworth Road, Leamington Spa, Warwickshire, CV32 6JX, UK.

X WANG,   Dept. of Civil Engineering, Univ. of Salford, Salford, M5 4WT, UK.

G A WEBSTER,   Dept. Mechanical Engineering, Imperial College, Exhibition Road, London
                     SW7 2BX, UK.

P J WEBSTER,   Dept. of Civil Engineering, Univ. of Salford, Salford, M5 4WT, UK.

P S WEBSTER,   Rolls Royce plc., PO Box31, Derby, DE2 8BJ, UK.

C G WINDSOR,   NNDTC, AEA Technology, B521.1, Harwell Laboratory, Didcot, OX11
                     0RA, UK.

P J WITHERS,   Dept. of Materials Science and Metallurgy, Univ. of Cambridge, Pembroke
                     Street, Cambridge, CB2 3QZ, UK.

France:

M KOCSIS,   Institut Laue Langevin, BP 156 Centre de Tri, 38042 Grenoble Cedex,
                     France.

A LODINI,   Ecole Superieure d'Ingenieurs en Emballage et Conditionnement, 51100
                     Reims Cedex, France.

E PLUYETTE,   Laboratoire MecaSurf, ENSAM, 2 Cours des Arts et Metiers, 13617 Aix-en-
                     Province Cedex, France.

Netherlands:

P C BRAND,   ECN, PO Box 1, 1755 ZG Petten, The Netherlands. (Present address: Reactor
                     Radiation Division, NIST, Gaithersburg, MD 20899, USA.)

Germany:

H G PRIESMEYER,   Institut fur Reine und Kernphysik Universitat Kiel, c/o GKSS Research
                     Center, Box 1160, D-2054 Geesthacht, Germany.

W REIMERS,   Hahn-Meitner-Institut Berlin GmbH, Group N5, Glienicker Str. 100, W-1000
                     Berlin 39, Germany.

Denmark:

T LORENTZEN,   Materials Dept., Riso National Laboratory, P.O. Box 49, DK-4000 Roskilde,
                     Denmark.

Italy:

F RUSTICHELLI,   Instituto di Fisica Medici, Facolta di Medicina e Chirurgia, Universita
                     degli Studi, Via Ranieri- Monte d'Ago, 60131 Ancona, Italy.

Greece:

S M SKOLIANOS,   Dept. of Mechanical Engineering, Laboratory of Physical Metallurgy,
                     Aristotle University of Thessaloniki, P O Box 1552, 540 06 Thessaloniki,
                     Greece.

Portugal:

F M A MARGAÇA,   Dept. Fisica, ICEN, EN10, 2685 Sacavem, Portugal.

Sweden:

S SJÖSTRÖM,   Dept. of Mechanical Engineering, Linkoping University, S-58183 Linkoping,
                     Sweden.

Australia:

M KIJEK,   Dept. of Physics, Monash University, Clayton, Victoria 3168, Australia.

1. OVERVIEW

NEUTRON DIFFRACTION MEASUREMENT OF RESIDUAL STRESS FIELDS: OVERVIEW AND POINTS FOR DISCUSSION

MICHAEL T. HUTCHINGS
National NDT Centre
AEA Technology
Harwell Laboratory
Didcot
Oxfordshire OX11 0RA U.K.

ABSTRACT. Design and Production Engineers wish to know the magnitude of residual stresses throughout their fabricated components and weldments in order that they can estimate their maximum loading, fatigue life, etc. The technique of stress measurement using neutron diffraction is outlined, with its advantages and outstanding problems. The latter are emphasised as points for discussion. Its use in a number of cases, covering both practical and fundamental problems, is described.

1. Introduction

Inhomogeneous heat treatment, such as welding, and inhomogeneous plastic deformation during fabrication can leave strong residual stresses locked within components. These stresses can affect the component life in service as they may add to applied loads causing fatigue and failure. Consequently engineers have to allow for their presence, and in absence of measured values over-conservative estimates must usually be used with correspondingly increased costs of manufacture or construction. The engineer has therefore sought for many years an absolute, nondestructive, means of measurement of the internal residual stress field within weldments and fabricated components. He or she is principally concerned with macrostress which varies over distances which are relatively large compared to the grain size of the material. A simple small 'black-box', with a probe which could be placed on the surface of the component, with which, on dialling the depth and direction of stress value required, he or she could read off the stress level on a meter or display would do very nicely! Unfortunately this ideal is still some way off full realisation, although a wide range of techniques exist to give limited information on the stress field. The engineer also has recourse to computer calculations of stress fields using finite element methods or analytical models.

In this paper the neutron diffraction method of stress measurement is described and examples of its use and of its outstanding problems are given. To introduce topics for discussion at the Workshop the latter are emphasised by use of *italics*. Indeed the aim of the Workshop is to concentrate on those points which require further elucidation, both experimental and theoretical, to clear up uncertanties and to develop the range of the technique and the information it can provide. In exploring these finer points the technique is providing materials scientists with new

3

M. T. Hutchings and A. D. Krawitz (ed.),
Measurement of Residual and Applied Stress Using Neutron Diffraction, 3–18.
© 1992 UKAEA.

insights into the fundamental behaviour of materials under stress. It should however be clearly stated that in most practical applications the neutron diffraction technique can now be used with confidence to provide the engineer with the information he or she requires. Indeed it does perhaps come the closest of all the available techniques to fulfilling his or her dream of a black box, although falling a good deal short of the ideal. It enables macrostress tensors, macrostress gradients, stresses due to applied loading, and microstress to be measured. The technique was first developed in 1980, with parallel work being carried out at Harwell in the UK [1,2], in Germany [3] and in the U.S.A. [4]. It is now used worldwide as a nondestructive technique, and its range of application is increasing. In this paper only examples from work by the Harwell group will be given. Work carried out at other laboratories is described in many of the other papers in this volume.

Other techniques currently used to give information on the residual stress field range from totally destructive methods such as sectioning, partially destructive methods such as hole drilling and trepanning, to nondestructive methods such as ultrasonics, thermoelastic techniques and X-ray diffraction, and the magnetic techniques such as measurement of induction, permeability, anisotropy, flux leakage, Barkhausen and magneto-acoustic effects [5]. The neutron diffraction method closely parallels and complements the X-ray diffraction method. It enables in-depth measurements to be made on account of the penetration of the neutron into most materials, whereas the X-ray method examines near-surface stress fields to a maximum depth of ~100μm. The major drawback of the neutron diffraction method is its *need of intense neutron beams*, available only at a medium or high-flux reactor or at an accelerator-based, usually time-pulsed, neutron source. This limits its use to relatively small portable components, with typical maximum dimension ~50-100cm. However it can be used to *test computer codes* using model samples, and also to test assumptions made in the use of the portable, but limited, techniques, and to *calibrate these techniques* for use in the factory or field.

2. Principles of the Neutron Diffraction Method

2.1 BASIC PRINCIPLES

In this section the principles of the method, in which the lattice plane spacing of grains acts as an internal strain gauge, will be briefly outlined. Full details are given by Allen et al [2]. The diffraction method is illustrated in Figure 1, and a typical diffractometer on a steady reactor source is shown in Figure 2. Standard powder neutron diffractometers are often quite suitable for stress measurement, but *dedicated instruments* have recently been designed and constructed.

The white neutron beam from the reactor is first monochromated to a chosen wavelength λ by Bragg reflection from a large single crystal monochromator. This monochromated beam is defined in direction by a soller slit collimator, or by apertures, to pass over the 'sample axis' about which the detector rotates. The detector counts neutrons scattered through an angle φ, with the scattered beam again defined in direction by a soller slit assembly. Both the beam incident on the sample and the beam entering the detector are defined in area by horizontal and vertical apertures in a neutron absorbing mask, made of cadmium for example. The 'gauge volume' or 'volume sampled' by the diffractometer is defined by the intersection of the incident and scattered beam as shown in Figure 1. A sample placed wholly within this volume will have its average property measured, whereas a large sample may be moved through the gauge volume in order to obtain a profile of a property, such as strain.

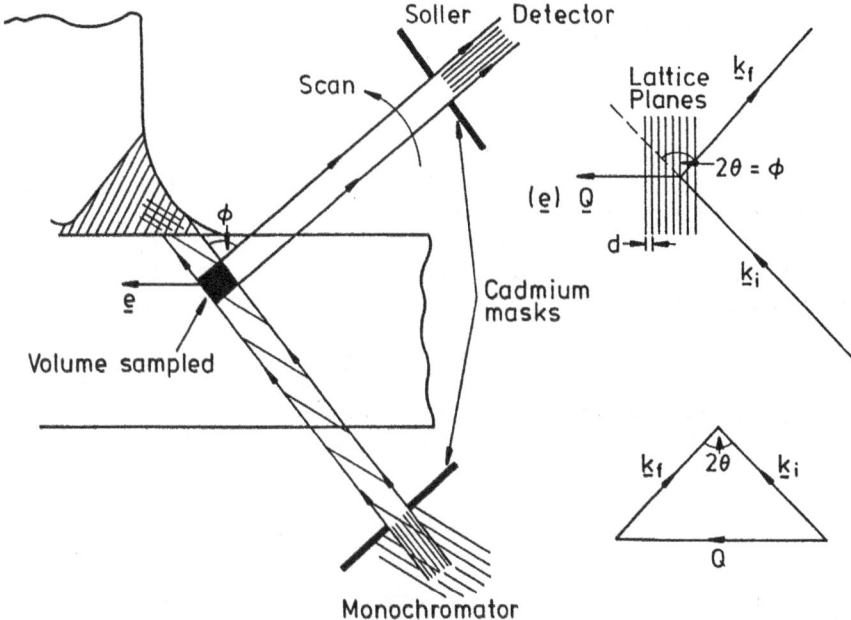

Figure 1. The principles of strain measurement, showing the definition of gauge volume sampled and the direction of strain measured.

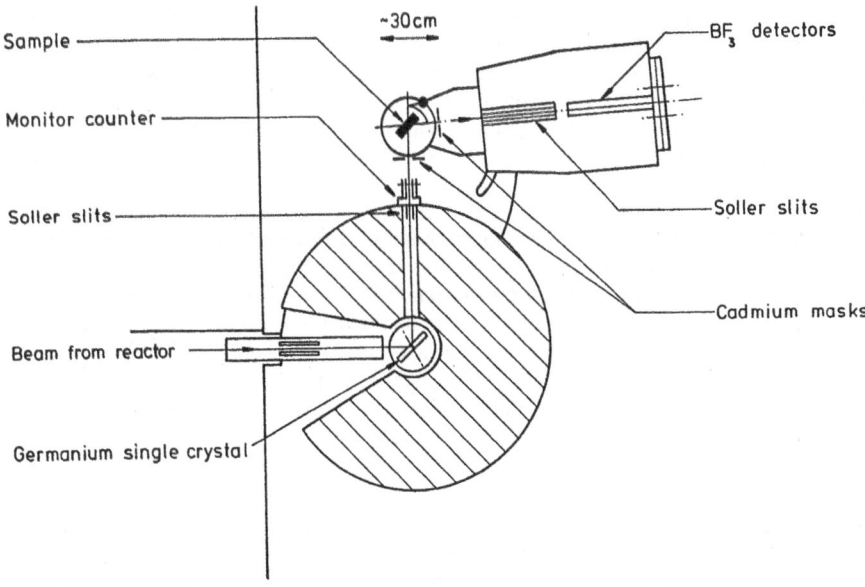

Figure 2. Outline of the PANDA diffractometer at the PLUTO reactor Harwell.

Polycrystalline materials give rise to cones of diffracted neutrons at angles $\phi_{hkl} = 2\theta_{hkl}$ about the incident beam on the sample. These angles are given by Bragg's Law:

$$2d_{hkl} \sin\theta_{hkl} = \lambda, \tag{1}$$

where d_{hkl} is the lattice spacing of planes with Miller indices hkl. The cone of scattering arises from just those crystallites or grains in the sample which satisfy Bragg's Law and which are oriented so that the planes are at an angle θ to the incident beam. It is usual to define a scattering vector $\mathbf{Q} = \mathbf{k_i} - \mathbf{k_f}$, where $\mathbf{k_i}$ and $\mathbf{k_f}$ are the incident and scattered neutron wavevector, of magnitude $|\mathbf{k}| = 2\pi/\lambda$, as shown in Figure 1. For Bragg reflection $\mathbf{Q}$ is normal to the planes giving the diffracted beam, and of magnitude equal to that of the reciprocal lattice vector τ of the planes, $|\mathbf{Q}| = |\tau_{hkl}|$, [6].

The detector is scanned through ϕ_{hkl} to determine the peak-count angle, which, in the case of a large sample, corresponds to the average d_{hkl} of the grains in the gauge volume sampled. Usually a fitting routine using least-squares minimisation, such as PKFIT, is used to determine the angle and angular width of the peak. The average lattice macrostrain in the volume sampled is then given by

$$e_{hkl} = (d_{hkl} - d_{ohkl})/d_{ohkl} = - (\cot \theta_{ohkl}).(\phi_{hkl} - \phi_{ohkl})/2, \tag{2}$$

where d_{ohkl} is the lattice spacing of a 'stress free' sample of the same material composition, and $\phi_{ohkl} = 2\theta_{ohkl}$ the corresponding diffraction angle. The direction in which the strain e_{hkl} is measured is that of the scattering vector $\mathbf{Q}$. We here use the term *macrostrain* to denote the strain component which is constant over many grains. The *microstrain* is strain which varies on a spatial scale of the order of the grain size, and the *mean lattice microstrain* in the volume sampled is related to the *angular peak width*.

As the lattice spacing can only change elastically it is an elastic strain which is always measured, but this may be a result of *intergranular strains* arising from a plastic deformation or thermal treatment of the sample. If an accurate value of d_o, or ϕ_o, can be measured, the strain determined is absolute. However obtaining a *true* ϕ_o may prove difficult in practice. A small annealed sample with measurements made in several orientations and averaged, or an extreme part of a component, may be taken to be in zero strain. Alternatively a *sum rule or boundary conditions*, after conversion to stress, may be used.

In order to determine the strain in different directions in the sample the sample must be *rotated accurately* about the centre of the gauge volume so that each direction lies along $\mathbf{Q}$. This is often quite *difficult to do in practice*, and requires very *careful alignment and centering* of the sample. Large samples may hit the spectrometer hardware, or path lengths of the beam in the sample may become excessive, preventing some orientations. It should be noted that the volume sampled is independent of the vertical height of the aperture in the scattered beam, so in order to increase intensity this can be as large as possible consistent with the definition in direction of e required. A vertical position sensitive detector (PSD) can be used with advantage. At $\phi = 90°$ the cone becomes a plane of scattering, and this is the optimum scattering angle for *definition of the gauge volume*. Although the resolution in strain increases with ϕ and is best at $\phi \sim 180°$, using $\phi \sim 90°$ scattering at both monochromator and sample (the focussing condition) is often a good *compromise*. However an angle of $\phi >$ or $< 90°$ may have the advantage that it can provide more ready access to the point in the sample at which the strain is to be measured. The *gauge volume shape and size must be chosen* with due consideration paid to the direction and magnitude of the strain gradient, the intensity of scattering and consequent measurement time, and the size of the grains in the sample. Typically, gauge volumes are between 1 and 27 mm^3.

The *use of a horizontal PSD* can speed up data collection by effectively performing the scan of ϕ at one setting. However the *definition of the gauge volume becomes difficult* as masks must be placed very close to the sample, and these may well prevent rotation of the sample to give strains in different directions. It should be noted that, in contrast to normal powder diffraction, higher wavelength orders, $\lambda/2$, $\lambda/3$, etc., from the monochromator can be used with advantage in strain measurement, as they increase the scattered intensity from higher order sample reflections 2h,2k,2l and 3h,3k,3l which measure the same strain, providing that these do not overlap other reflections h'k'l'. In order to obtain diffraction angles of $\phi\sim90°$ neutron wavelengths in the region $\lambda = 1.5\text{-}2.5$ Å must usually be used.

An alternative to a steady reactor source of neutrons is the *pulsed white beam* of neutrons given by a spallation source. In this case polycrystalline diffraction is observed at a fixed ϕ. On most standard instruments ϕ is usually $\sim150°$ to $180°$ to give very good resolution, and time of flight t is used to scan the different lattice planes. The strain is now given by $e = \Delta t/t = \Delta\lambda/\lambda$, where $t = L/v = (\lambda m/h)L$ is the time taken for a neutron with mass m and velocity v to travel a path length L, and h is Planck's constant. The advantage of this method is that strains may be measured from *many lattice planes* {hkl} at once. However Q is nearly along k_i in the back-scattering arrangement. As for the reactor instrument *counters at 90°* must be used to give measurement of the full strain tensor, and such instruments are currently under development. The *positioning of the sample is even more critical* here as it may effect L and hence t.

In most cases the strain and stress are triaxial and can be represented by a strain tensor $\underline{\varepsilon}$ and stress tensor $\underline{\sigma}$. If we define axes Oxyz within the sample, and define $\underline{\varepsilon}$ at a point with respect to these axes, then the strain measured in a direction specified by direction cosines (l'm'n') relative to Oxyz is given by:

$$e(l'm'n') = l'^2\varepsilon_{xx} + m'^2\varepsilon_{yy} + n'^2\varepsilon_{zz} + 2l'm'\varepsilon_{xy} + 2m'n'\varepsilon_{yz} + 2n'l'\varepsilon_{zx} . \qquad (3)$$

At least six measurements of e(l'm'n') are therefore necessary to determine the six terms in $\underline{\varepsilon}$, but clearly the *accuracy can be improved if more are made*. The simplest directions are those along, and at 45° to, the axes Oxyz. Having determined $\underline{\varepsilon}$, the principal strain axes OXYZ and the principal strains $\underline{\varepsilon}^D$ along them may be found by diagonalisation. The principal internal stresses along these axes are then related, *assuming an elastically isotropic model* with macroscopic Young's modulus E and Poisson's ratio v, by:

$$\begin{bmatrix} \sigma^D_{xx} \\ \ \sigma^D_{YY} \\ \ \ \sigma^D_{ZZ} \end{bmatrix} = \frac{E}{(1+v)} \begin{bmatrix} \varepsilon^D_{xx} \\ \ \varepsilon^D_{YY} \\ \ \ \varepsilon^D_{ZZ} \end{bmatrix} + \frac{vE}{((1 - 2v)(1 + v))} (\varepsilon^D_{xx} + \varepsilon^D_{YY} + \varepsilon^D_{ZZ}) \begin{bmatrix} 1 \\ \ 1 \\ \ \ 1 \end{bmatrix}$$

$$(4)$$

Plane strain or plane stress are special cases of these general expressions. Typically, for steel a shift in ϕ from that of a stress-free sample of 0.01° at 90° indicates a strain of $\sim100\mu\varepsilon$ or a stress of ~20 MPa. Accuracy in measurement of 0.005° is usually easily obtained by peak profile fitting using a computer.

Practical path lengths, that is of the total incident plus diffracted beams, in steel and in aluminium are typically 50 and 100mm respectively, for volumes sampled of $27mm^3$. This means that one can measure to a depth of ~2.5cm into steel from a flat surface. Longer paths in the material, for deeper penetration, can be realised by sampling larger volumes. In principle all

8

crystalline materials can be examined, but some materials, such as nickel, scatter neutrons incoherently, and others, such as titanium alloy, have small Bragg intensities making measurements at depth correspondingly more difficult. A few materials absorb neutrons strongly.

2.2 ELASTIC ANISOTROPY

In most materials, the values of E *and* v in equation (4) are *dependent on the lattice planes* (hkl) used as the 'strain gauge', since the elastic response of each grain to an applied stress is anisotropic. Indeed one needs to determine 'neutron elastic constants' by calibration, using for example a compact stress rig such as that shown in Figure 3 which can be accommodated on most neutron diffractometers. This rig allows a known uniaxial stress to be applied in tension or compression to a sample of material, and the strain to be measured either along the direction of stress (//) or perpendicular to it ($\perp$).

For a single crystal of ferritic steel the effective elastic modulus, stress/strain, can readily be calculated. For strain measured in the same direction as the applied stress, it is found to vary from 131 GPa in the [100] direction to 283 GPa in the [111] direction. These values are to be compared with the usual bulk value of Young's modulus for polycrystalline steel, measured with an external strain gauge, of 207 GPa. Clearly inclusion of the effects of elastic *anisotropy is of paramount importance* in interpreting neutron diffraction stress data. It should be

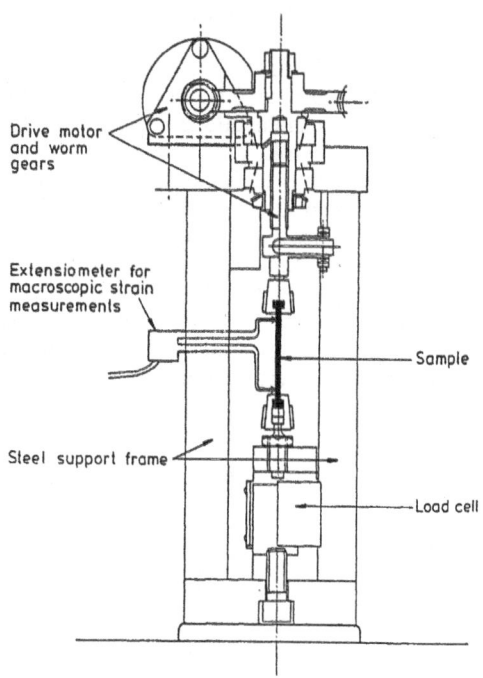

Figure 3. Compact stress rig designed for neutron diffraction measurements.

noted that the intensity of the reflection, related to its multiplicity and the effects of any texture present, has also to be considered when making a choice of hkl.

In practice the calculation of the *response of polycrystalline material to an applied uniaxial stress* is not straightforward, and assumptions as to continuity at grain boundaries must be made to simplify the procedure. The two simplest approximations give extreme bounds to the actual response. These are the Voigt [7] and Reuss [8] models in which strain or stress is respectively assumed continuous across grain boundaries. A more realistic model is given by Kröner [9] which approximates to the mean of the Voigt and Reuss bounds. The diffraction experiment samples a *subset of grains* in a polycrystalline specimen, that is it measures the average response of only those grains which are oriented with their plane normals along **Q** and which satisfy Bragg's Law. In measuring strain in a given direction in the sample using different reflections hkl one is sampling a different subset of grains, and in making measurements of strain in different directions, whether or not the same reflection is used, one is again largely sampling different grains.

In the Voigt limit all grains of a polycrystalline sample are taken to exhibit the same strain, whereas in the Reuss limit they are taken to experience the same stress. It is clear that for uniaxial stress the strain calculated along the stress direction in the Reuss limit is the same as that for a single crystal, whereas the strain perpendicular to an applied uniaxial stress must be determined by *averaging over all possible grain orientations* with chosen diffracting planes perpendicular to the direction of strain measurement. One finds [2]:

$$e_{//}{}^{poly} = \sigma[S_{11} - 2SA_{hkl}]$$

and

$$e_{\perp}{}^{poly} = \sigma[S_{12} + SA_{hkl}], \tag{5}$$

where $A_{hkl} = (h^2k^2 + k^2l^2 + l^2h^2)/(h^2 + k^2 + l^2)^2$ is an anisotropy factor for the particular plane. $S = (S_{11} - S_{12} - S_{44}/2)$, and the S_{ij} are the single crystal elastic compliance constants. The Poisson's ratios may be determined in this approximation from $v = - e_{\perp}{}^{poly}/e_{//}{}^{poly}$. In practice the measured anisotropy lies between the Voigt and Reuss limits. In textured samples $e_{//}$(Reuss) will be independent of the texture, but $e_{\perp}$(Reuss) will be weighted by the appropriate crystallite orientation distribution function [10]. It is possible in principle to measure the *texture at each point* in the sample, and then to use the above theory for the observed reflection hkl to calculate a *general expression* relating strain to stress corresponding to equation (4). In practice this is a lengthy procedure and approximations must be used.

The above considerations apply only to the elastic region. If stresses beyond the yield point are applied, *slip will occur in the grains but will be constrained by the surroundings*. The *theory for such effects is not yet fully developed* [11] and will not be discussed here. It is felt that results obtained from the measurements using the stress rig described in the section below will provide key data for a better *understanding of plastic strain at the grain interaction level*.

3. Examples of Strain and Stress Field Measurement

In this section a selection of examples of measurements are briefly described to illustrate the range of application of the technique. Full details can be found in the references given.

3.1 NEUTRON ELASTIC CONSTANTS AND ANISOTROPIC EFFECTS

3.1.1 *'As Received' Mild Steel.* The measurement of the lattice strain of the 211 reflection of ordinary mild steel using the stress rig shown in Figure 3, is shown in Figure 4, where it is compared with the overall macroscopic strain of the sample measured simultaneously with an extensiometer. The effective neutron Young's Modulus E is close to the macroscopic value, and the response is linear even in the plastic region indicating that it is the elastic component which is measured by the lattice for this reflection. On release of the applied stress there is no residual elastic lattice strain, whereas the extensiometer shows a residual plastic strain of ~0.04%.

The value of the anisotropy factor for the 211 reflection is $A_{211}= 0.25$, equal to that of the 110 and 321 reflections which are therefore expected to show similar response. These three reflections are the best to use for practical stress measurement. Other reflections do not exhibit a

10

linear response in the plastic regime, and do show a residual lattice strain. Thus the strain response of the 222 reflection (A_{222}= 0.33) tends to curve upwards exhibiting less strain in the plastic region, and the 200 reflection (A_{200}= 0.0) tends to show increased strain in the plastic region and a positive residual strain on release of the load. This behaviour is also exhibited for the ferritic (α) component of a duplex alloy of ~50:50 ferrite (α) and austenite (γ) shown in Figure 12. It suggests that by measurement of lattice strain from several reflections, not only can the stress state be determined but also some *information on stress history* can be obtained from samples which have been taken past the yield point.

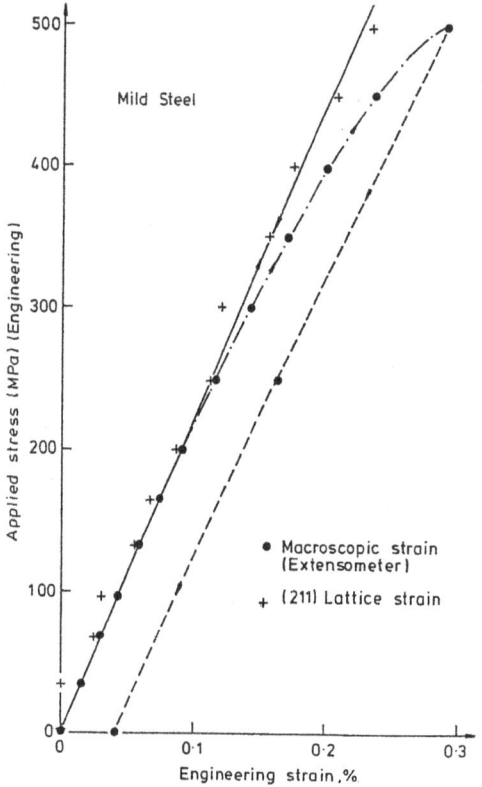

Figure 4. The (211) lattice macrostrain response parallel to applied stress levels. Extensiometer measurement of the overall specimen strain is also shown.

3.1.2 *Annealed Mild Steel.* Recent measurements made on annealed mild steel have revealed interesting behaviour of the lattice strain on entering the plastic regime as the stress exceeds the yield point [12]. The 211 Bragg peak measured by time of flight on the High Resolution Powder Diffraction Diffractometer (HRPD) at ISIS, Rutherford Appleton Laboratory, is shown in Figure 5. At low stresses the peak shape is limited by the instrumental resolution, and the strain increases linearly with stress. At ~250 MPa, the yield stress, there is significant *relaxation in the strain*, and above this stress the peak broadens rapidly showing *increase in microstrain*. This general behaviour is exemplified by Figure 6, which shows the lattice microstrain response both parallel ($e_{//}$) and perpendicular ($e_{\perp}$) to applied tensile and compressive stresses of the 310 reflection. Relaxation is observed in both $e_{//}$ and $e_{\perp}$, and the microstrain increases with stress in the plastic regime. *Different reflections are found to exhibit different relaxation*, which clearly depends on the amount of slip allowed by neighbouring grains, and on the anisotropy of the grains. Further experiments on other materials are underway to examine this behaviour more generally, and to help shed light on current possible theories of the plastic regime.

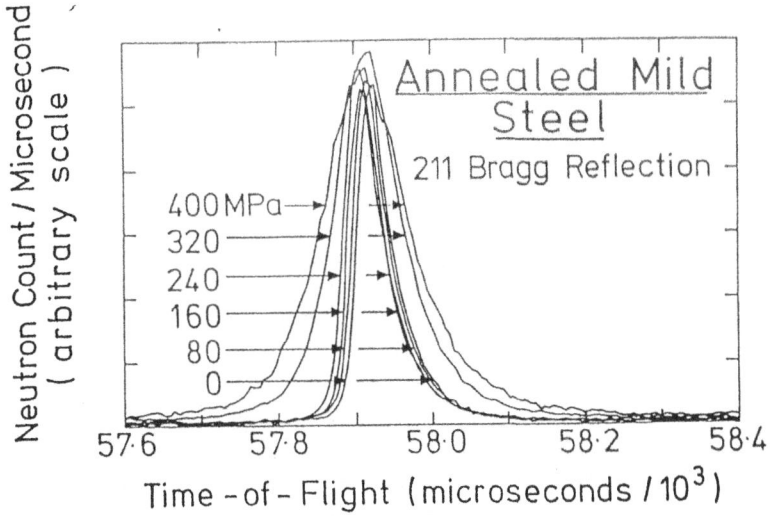

Figure 5. The 211 Bragg reflection observed by time-of-flight diffraction from an annealed mild steel sample subjected to successive stress levels. The strain is measured perpendicular to the applied stress. The time-of-flight is proportional to the plane spacing [12].

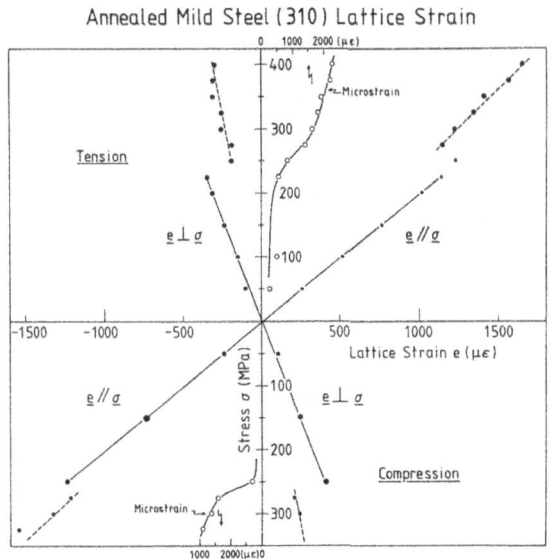

Figure 6. The lattice macrostrain response from the 310 Bragg reflections in annealed mild steel, measured parallel and perpendicular to applied tensile and compressive stress. The errors are given by the size of points. The microstrain parallel to the applied stress, as given by the full width at half maximum of the diffraction peak, is also shown [12].

3.1.3 *Elastic Constants.* The *elastic compliance constants* determined from the slopes of the stress-strain relation are plotted as a function of the anisotropy factor A_{hkl} in Figure 7 for both the 'as received' and annealed mild steel. The broken lines show the calculated Voigt [5] and Reuss variation, equation (5). The isotropic average Reuss value corresponds to A = 0.2. The solid line in Figure 7 is drawn through the measured data points for the parallel data, and has a slope of 44% of the full Reuss variation. It lies close to the variation expected from the Kröner model. The perpendicular data for the annealed mild steel lie close to the solid line which is deduced from the parallel variation multiplied by -0.3 (Poisson's ratio), whereas the 'as received' mild steel data follow more closely the Reuss bound. It is interesting to note that in the duplex alloy referred to below, a similar variation is observed [12].

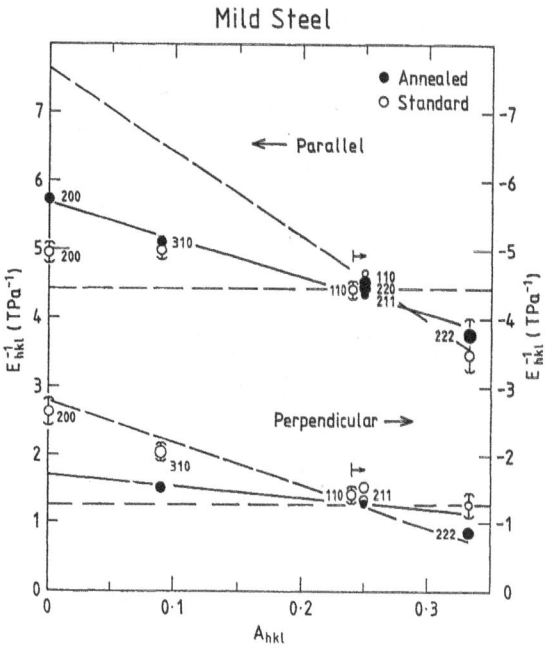

Figure 7. Anisotropic variation of elastic compliance constants parallel and perpendicular to applied tensile stress for annealed and unannealed mild steel. The solid lines through the data points are described in the text. The broken lines are calculated from the Reuss and Voigt bounds [12].

3.2 STRESS IN WELDMENTS

One of the key uses of the neutron diffraction method of stress measurement has been to investigate in-depth stresses in and around the weld in a variety of weldments. These measurements can actually test the assumptions and theories which had hitherto been used to estimate the stress levels.

3.2.1 *Double-V Weldment.* The strain and stress variation through a section cut from a double-V weldment, fabricated from 50D C-Mn steel as shown in the inset of Figure 8, were measured using the 211 Bragg reflection [2]. The dimensions of the section were 240mm long (Y), 42mm through the weld (Z), and 13.5mm thick (X). In this case plane-stress should be a good approximation. The gauge volume sampled was 3 x 3 x 3mm^3, and strain data taken at 3mm intervals were converted to stress using $E = 207$ GPa and $v = 0.28$. The resulting stress values are shown in Figure 8. The shaded area gives the range of data taken by a destructive sectioning method on adjacent sections, and there is overall good agreement with the diffraction data. The value of σ_z *does not go to zero* at the edges of the weld, although it is expected to since the stress normal to a free surface must be zero. This could be a result of an incorrect *'zero' stress reference* ϕ_o *value*, the plane stress approximation used, or the fact that data could *not be measured actually at the surface, a potential limitation of the neutron technique* except in special cases.

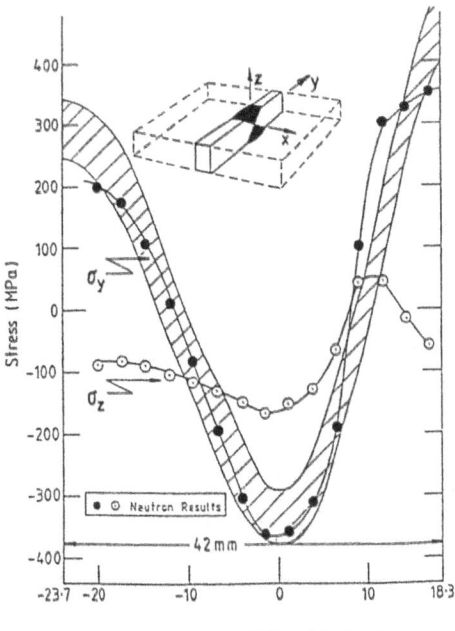

Figure 8. Variation of residual stress with position through a double-V weldment sample, cut as shown in inset. The points are neutron diffraction results and the shaded area denotes the range of sectioning technique measurements [2].

3.2.2 *T-Butt Weldments.* An extensive series of measurements have been made on two T-butt weldments as shown in the inset in Figure 9, made from 50D steel used in 'offshore' applications [13]. These concentrated on the strain and stress at and below the toe of the weld in the centre position, and again used the 211 Bragg reflection. Although the weldments were made in an identical manner the strain e_x as a function of depth below the toe was not exactly the same in the two weldments! This indicates the extent of *variation in residual stress which can occur due to small changes in the processing conditions.* The full stress tensor was measured at the toe position and involved *careful manipulation to orient the weldment* into six orientations. One weldment was subjected to fatigue cracking and the strain variation redetermined, the other to a post weld heat treatment typical of that given to offshore structures. The results of the latter on the strain variation is shown in Figure 9, where it is seen that the strain is essentially reduced to zero level by the treatment. This example illustrates the use of the nondestructive technique to examine the same sample before and after processing.

14

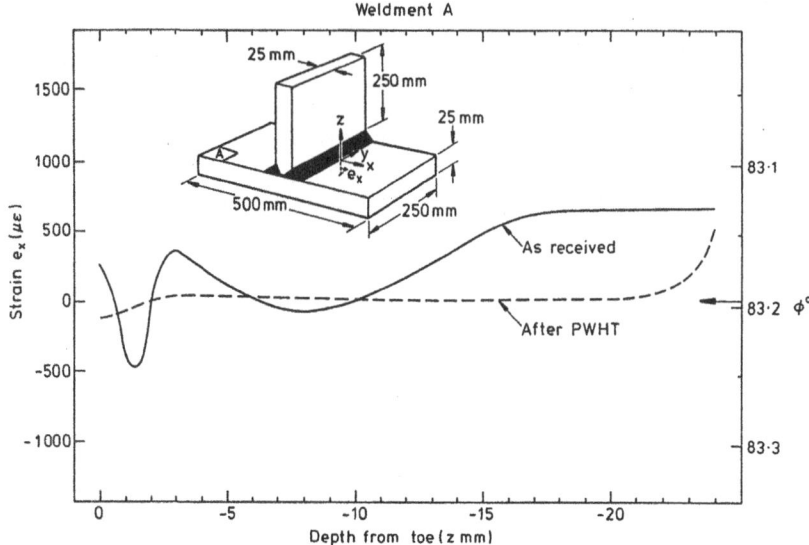

Figure 9. Variation of strain e_x through the plate beneath the toe of the weld in a T-butt weldment (inset) before and after post weld heat treatment (PWHT). The direction of strain measured is perpendicular to the weld line [13].

3.3 FATIGUE CYCLING OF A COMPACT TENSION SPECIMEN

A small ($31.5 \times 49.6 \times 13.1$mm^3) compact tension specimen of BS 4360 steel was subjected to 20,000 cycles of fatigue loading, with a stress intensity level varying between a minimum of K=3 and maximum of K=34 MPa mm$^{1/2}$, in order to produce a small crack of 3.6mm length. The triaxial stress levels along the line of cracking were measured at the maximum load by bolting at the peak level, and are shown in Figure 10. The data were taken with a gauge volume of only 1mm^3, and used the 211 Bragg reflection [14]. The stress field was found to be neither plane strain nor plane stress, states often assumed in theoretical calculations, and the peak stress levels are less than expected from theory assuming plane strain. *Instrumental volume resolution* could have contributed to this reduction. On removal of the bolt to give the minimum load situation, it was found that the peak level moved away from the crack tip with consequences for hydrogen or sulphur embrittlement.

3.4 LOAD SHARING IN COMPOSITES AND MULTIPHASE MATERIALS

3.4.1 *SiC/Al Composites.* The ability of the diffraction method to separate the strain (and stress) levels in each component of a composite material, by observation of the characteristic diffraction peak from each, has led to the measurement of residual stress levels induced by thermal processing in, for example, metal matrix composites (MMC) of silicon carbide in aluminium [15]. It also enables the load partition between the two components to be determined when a sample is subjected to known stress levels. An example of the latter is shown in Figure 11 for a composite sample with a 2014 aluminium alloy matrix reinforced with 20%vol. SiC particles. The lattice strain response measured by three reflections in each of the phases of the sample subjected to stress levels up to 500 MPa is shown. There is relatively little elastic anisotropy in either component. The SiC strain is much less than that of the aluminium, and the response of each lies within, but close to, the respective Reuss values, shown by the solid lines. These are calculated using E(Al) = 72 GPa and E(SiC) = 420 GPa. The strain response.

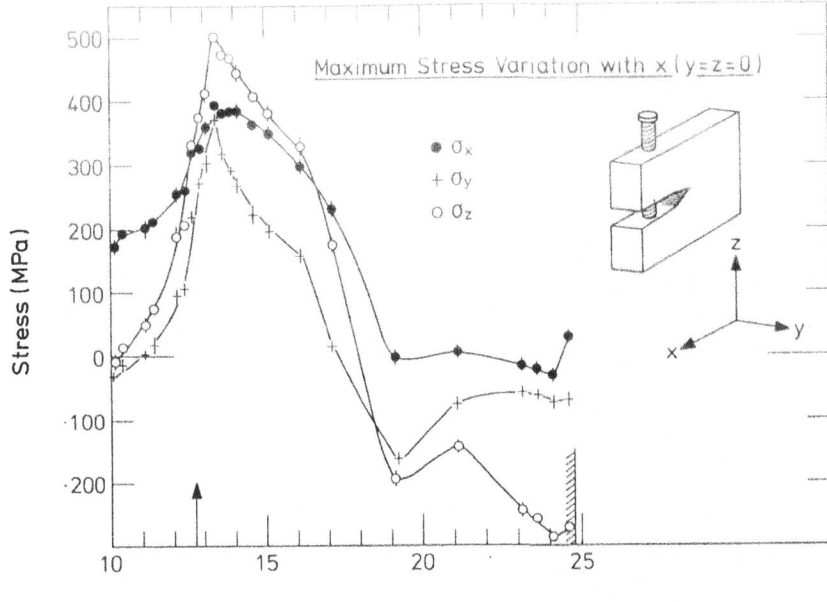

Figure 10. Triaxial stress variation with position x in a cracked fatigue test specimen bolted in the maximum crack-tip stress configuration. The arrow denotes the position of the crack tip on the specimen centre line, and the shaded line the outer edge of the specimen [14].

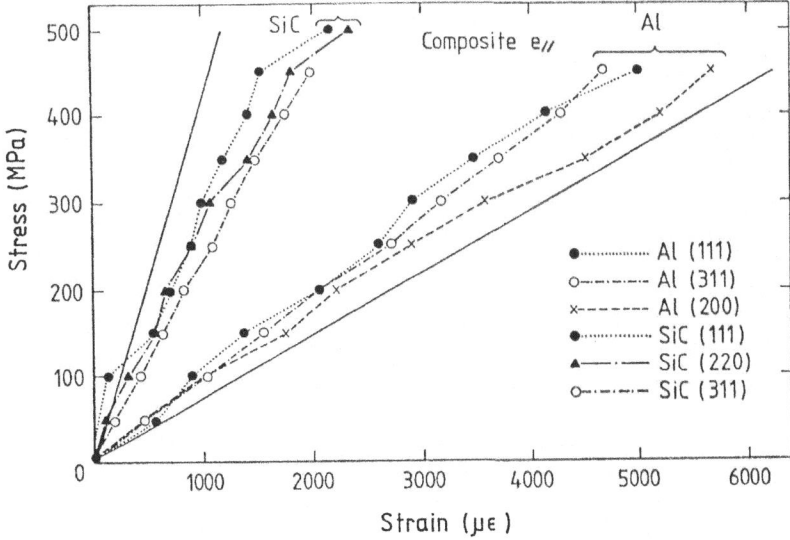

Figure 11. Al and SiC strain response parallel to stress applied to a MMC composite rod of 20% vol. SiC particulate in an Al (2014) matrix. The two solid lines are calculated for the simple individual response of each component subjected to the stress [15].

perpendicular to the applied stress is also closer to the Reuss bound than the Voigt, indicating that the latter is a poor approximation for calculating the response. A more detailed model, based on Eshelby's theory, is given by Allen et al. [16], and reproduces the data very well.

On removing the load, residual strains relative to the original state are observed perpendicular, but not parallel, to the applied stress direction. These differences are probably due to the effect of *relaxation*.

3.4.2 *Duplex Steel*. A second example is that of a 50:50 duplex alloy of ferritic (α) and austenitic (γ) steel with yield stress of~ 220 MPa, already referred to in earlier sections [12]. The response of the ferritic α component parallel to the applied stress is shown in Figure 12. This shows clearly the characteristic upturn of the (222) plane response, and the greater strain of the (002) planes, in the plastic regime. On reducing the stress, residual lattice strains relative to the initial state are observed and are shown on the abscissa.

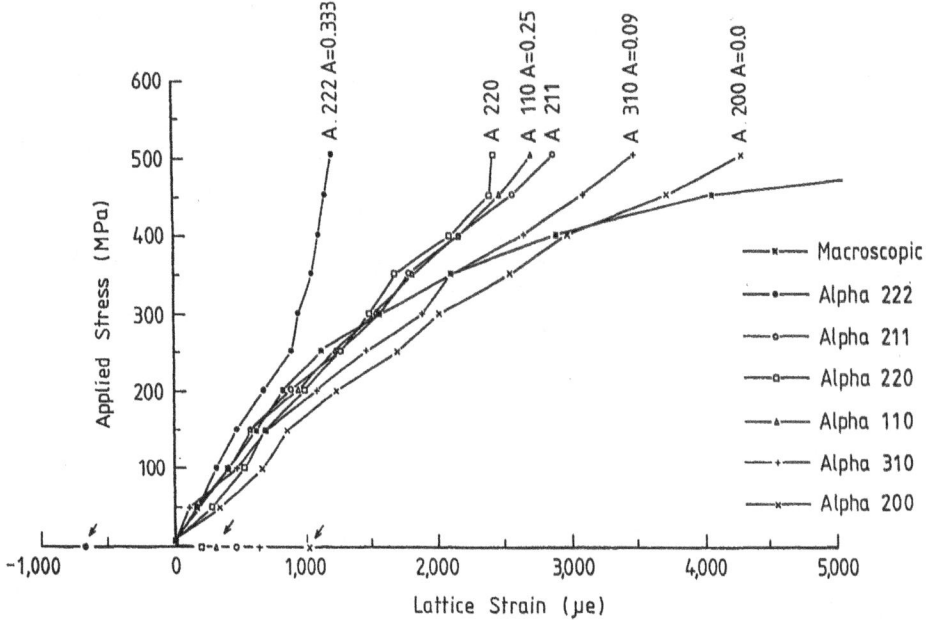

Figure 12. Lattice strain response from several Bragg reflections in the ferritic phase of a 50:50 duplex alloy measured parallel to the applied stress. Residual strains on relieving the stress are marked on the abscissa [12].

4. Conclusions

Neutron diffraction provides a unique, nondestructive, and absolute method of residual stress measurement at depths of up to ~2.5cm in steel and ~7cm in aluminium. However it falls well short of the Engineer's ideal in that a high-flux neutron source must be used, precluding its use for in situ measurements on plant and structures. It can be used for portable samples of less than 1m typical size, and for validation and calibration of several other portable but limited techniques. Unfortunately *portable neutron sources of sufficient intensity* for general use in the field are unlikely to become available in the foreseeable future.

The neutron method is also becoming increasingly valuable in the fundamental materials science field, in the understanding of the stress distribution in new materials to be used by Engineers such as metal matrix composites, and also in shedding light on the *strain response at the individual grain level of metals stressed into the plastic regime.*

There are a number of areas which need considerably more development. On the practical level these include the *improvement of volume resolution,* particularly *near surfaces,* preparation of *zero-stress reference samples,* and improved *data collection rate* by the use of powerful spallation source *time-of-flight techniques* and of *area detectors,* although the latter have limitations. There are a number of *refinements in data analysis* yet to be made, such as *corrections for absorption, extinction and multiple scattering,* and their effect on the mean position of strain measurement. Measurements on *large-grained samples,* such as cast iron or aluminium, need to be developed. *Theory needs to be developed* in line with experiment on the effects of *anisotropy, texture, and grain interaction stresses.*

All the above developments are currently being addressed (as the following papers at the Workshop testify), and in the next few years will provide increasing confidence in the use of this relatively new technique, and lead to the development of *standards of practice* to be used by the Engineers. The technique has now taken its place worldwide alongside the other more traditional methods of nondestructive testing.

Acknowledgements

The author wishes to thank his colleagues, A.J. Allen, C. Andreani, M. Bourke, W.I.F. David, S. Dawes, C. Hippsley, A.D. Krawitz, V. Rainey and C.G. Windsor for their collaboration in aspects of this work. The work was undertaken as part of the Corporate Research Programme of the UK Atomic Energy Authority. Some of it was supported by external contracts from customers, and has received support from the UK Department of Energy and The Welding Institute.

References

1. Allen, A.J., Andreani, C., Hutchings, M.T. and Windsor, C.G. (1981) 'Measurement of internal stress within bulk materials using neutron diffraction', NDT International, October, pp. 249-54.
2. Allen, A.J., Hutchings, M.T., Windsor, C.G. and Andreani, C. (1985) 'Neutron diffraction methods for the study of residual stress fields', Adv. in Phys. 34, 445-73.
3. Pintschovius, L., Jung, V., Macherauch, E., Schäfer, R. and Vöhringer, O. (1982) 'Determination of residual stress distribution in the interior of technical parts by means of neutron diffraction', in E. Kula and V. Weiss (eds.), Residual Stress and Stress Relaxation, Plenum, New York, pp 467-482.
4. Krawitz, A. D., Brune, J. E. and Schmank, M. J. (1982) 'Measurement of stress in the interior of solids with neutrons', in E. Kula and V. Weiss (eds.), Residual Stress and Stress Relaxation, Plenum, New York, pp. 139-155.
5. James, M.R. and Buck, O. (1980) 'Quantitative nondestructive measurements of residual stresses', Critical Reviews Solid State Mat. Sci. 9 (1), 61-105.

18

See also Allen, A. J. (1992) 'Calibration of portable techniques for residual stress measurement', this volume.

6. Kittel, C. (1986) Introduction to Solid State Physics, 6th Edition, Wiley, New York.

7. Voigt, W. (1928) Lehrbuch der Kristall Physik, Teubner, Leipzig.

8. Reuss, A. (1929) 'Calculation of flow limits of mixed crystals on basis of plasticity of single crystals', Z. Angew. Math. Mech. 9, 49-58.

9. Kröner, E. (1958) 'Berechnung der elastischen konstanten des vielkristalls aus den konstanten des einkrystalls', Z. Phys. 151, 504-18.

10. Sayers, C.M. (1984) 'The strain distribution in anisotropic polycrystalline aggregates subjected to an external stress field', Phil. Mag. A49, 243-62.

11. Leffers, T. (1981) 'Microstructures and mechanisms of polycrystal deformation at low temperatures', in N. Hansen, A. Horsewell, T. Leffers and H. Lilholt (eds.), Deformation of Polycrystals: Mechanisms and Microstructures, Riso Nat. Lab., Roskilde, pp. 55-71.

12. Allen, A.J., Bourke, M., David, W.I.F., Dawes, S., Hutchings, M.T., Krawitz, A.D. and Windsor, C.G. (1989) 'Effects of elastic anisotropy on the lattice strains in polycrystalline metals and composites measured by neutron diffraction', in G. Beck, S. Denis and A. Simon (eds.), International Conference on Residual Stress ICRS2, Elsevier, London, pp. 78-83.

13. Allen, A.J., Hutchings, M.T. and Rainey, V.S. (1988) 'Measurement of through-thickness residual stress in T-butt weldments of offshore steel by high resolution neutron diffraction', in J. M. Farley and R.W. Nicols (eds.), Non-Destructive Testing, Pergamon, Oxford, pp. 1808-17.

14. Hutchings, M.T., Hipplsey, C. and Rainey, V. (1990) 'Neutron diffraction measurement of the stress field during fatigue cycling of a cracked test specimen', Mat. Res. Soc. Symp. Proc. 166, 317-321.

15. Allen, A.J., Bourke, M., Hutchings, M.T., Krawitz, A.D. and Windsor, C.G. (1987) 'Neutron diffraction measurement of internal stress in bulk materials:- metal-matrix composites', in E. Macherauch and V. Hauk (eds.), Residual Stresses in Science and Technology, DGM Informationsgesellschaft Verlag, Oberursel, Vol. 1, pp. 151-7.

16. Allen, A.J., Bourke, M., Dawes, S., Hutchings, M.T. and Withers, P.J. (1992) 'The analysis of internal strains measured by neutron diffraction in Al/SiC metal matrix composites', (submitted to Acta metall.).

2. BACKGROUND

ROLE OF NEUTRON DIFFRACTION IN ENGINEERING STRESS ANALYSIS

G.A.WEBSTER
Department of Mechanical Engineering,
Imperial College, London, SW7 2BX

ABSTRACT. A method of measuring residual stresses accurately and non-destructively by neutron diffraction is described. Examples are presented of the residual stress fields that can be introduced by welding, shot-peening, wear, autofrettage and ahead of cracks. Data are included on the influence of residual stress on fatigue crack propagation and interpreted in terms of fracture mechanics concepts. It is shown how neutron diffraction results can be used to validate engineering stress analysis calculations and provide information on how fabrication processes can be adjusted to achieve improved resistance to failure in components.

1. INTRODUCTION

The presence of residual stresses in engineering components can significantly affect their load carrying capacity and resistance to fracture. Residual stresses can be introduced into components during fabrication and as a result of creep and plastic deformation during use. Examples of manufacturing processes which can lead to residual stresses include welding, forging, bending, and machining operations.

Residual stresses can be beneficial or detrimental to the resistance to failure of components depending on the magnitude and direction of the applied loading (Hermann and Reid, 1986 and Webster, 1989). Tensile surface residual stresses can be particularly harmful since they may aid the initiation and growth of fatigue cracks from surface defects and stress concentrations. Consequently heat-treatments are frequently employed to relax these stresses. Alternatively, manufacturing processes, such as shot-peening and autofrettage, which cause compressive surface residual stresses can be applied to critical components to inhibit crack initiation and growth and enhance fatigue lifetimes.

A quantitative assessment of the influence of residual stresses on the resistance to failure of components requires an accurate knowledge of their magnitude and distribution. Several methods are available for determining these stresses (Parlane, 1978). They include X-ray diffraction (Hughes, 1967), neutron diffraction (Stacey et al, 1985, Allen et al, 1985 and Webster, 1990), hole drilling (Beaney and Proctor, 1974) and slicing (Williams et al, 1981) or boring (Sachs, 1927) methods. However, only the neutron diffraction technique is capable of

21

M. T. Hutchings and A. D. Krawitz (ed.),
Measurement of Residual and Applied Stress Using Neutron Diffraction, 21–35.
© 1992 *Kluwer Academic Publishers.*

establishing residual stresses in the interior of metallic components non-destructively.

In this paper the application of neutron diffraction for measuring residual stress is outlined. A representative selection of stress distributions developed by a range of manufacturing processes is examined. Some comparisons are made with strain gauge determinations and finite element predictions. It will be shown how the results can be employed in engineering stress analysis calculations to make accurate assessments of fatigue crack growth and resistance to fracture in components containing residual stresses.

2. THE NEUTRON DIFFRACTION METHOD

With the neutron diffraction method components of strain are obtained directly from measurements of the separation between crystallographic planes. For a mono-chromatic beam of neutrons, a small change Δd in the lattice spacing d will result in a change $\Delta\theta$ in the angular position θ of the Bragg reflection so that the lattice strain ε in the direction of the scattering vector Q (Fig 1) is given by

$$\varepsilon = \Delta d/d = -\Delta\theta \cot \theta \qquad (1)$$

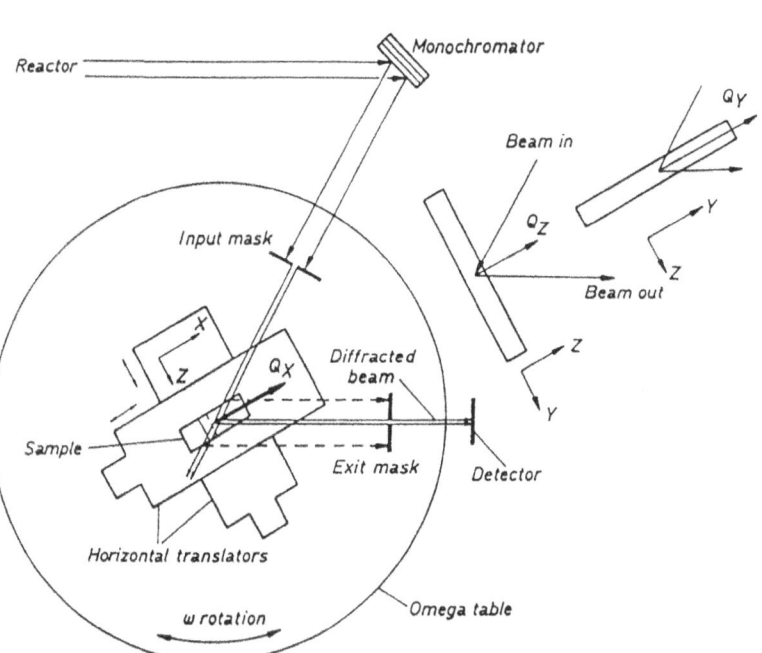

Fig 1 Sample in neutron beam showing orientations for strain measurements in x, y and z directions.

In general, to define the strain tensor at a point completely, measurements in six orientations are required. However, when the principal directions are known, three orientations will suffice. When the principal directions coincide with the coordinate directions x, y and z and the material is isotropic with a Young's modulus E and Poisson's ratio v, the principal stresses are obtained from

$$\sigma_x = \frac{E}{(1+v)(1-2v)} \left[(1-v)\varepsilon_x + v(\varepsilon_y + \varepsilon_z)\right] \tag{2}$$

$$\sigma_y = \frac{E}{(1+v)(1-2v)} \left[(1-v)\varepsilon_y + v(\varepsilon_x + \varepsilon_z)\right] \tag{3}$$

$$\sigma_z = \frac{E}{(1+v)(1-2v)} \left[(1-v)\varepsilon_z + v(\varepsilon_x + \varepsilon_y)\right] \tag{4}$$

Figure 1 shows a schematic of a manipulator, with three coordinate and two rotational axes, that is capable of locating specimens in a neutron beam to an accuracy of 0.1 mm. The sensitivity of the method depends upon the angular definition of the system and the volume of material sampled. The volume sampled is controlled by collimating and masking the beam. Measurements are obtained at any desired location by suitable positioning of the specimen in the beam. Further readings can be repeated at other locations by traversing the specimen through the beam. Strains in the x, y and z directions are determined by orientating the Q vector to coincide with the coordinate axes as indicated in the figure. With this facility, it is possible to determine stresses with a resolution of 15 MPa and identify stress gradients of 2000 MPa/mm using sampling volumes as small as 8 mm^3.

Some examples of the application of the neutron diffraction method for determining residual stress distributions in a selection of specimens and components will now be considered.

3. APPLICATIONS

Figure 2 shows the stresses that have been measured along the centre line through a double vee butt weld in an aluminium alloy (Smith et al, 1988). The weld was made in several passes as indicated in (a). The residual stress distribution generated due to differential contraction from the welding procedure is clearly apparent. Good agreement between the neutron diffraction technique and a destructive strain gauge layering method is achieved. The sensitivity to small changes in the zero strain reference diffraction angle θ_0 is also shown since this value is often difficult to determine accurately. The peak stresses recorded in this case are about a half the yield strength of the material. It is apparent that the tensile residual stresses developed during welding will assist fatigue crack initiation and growth from the sides.

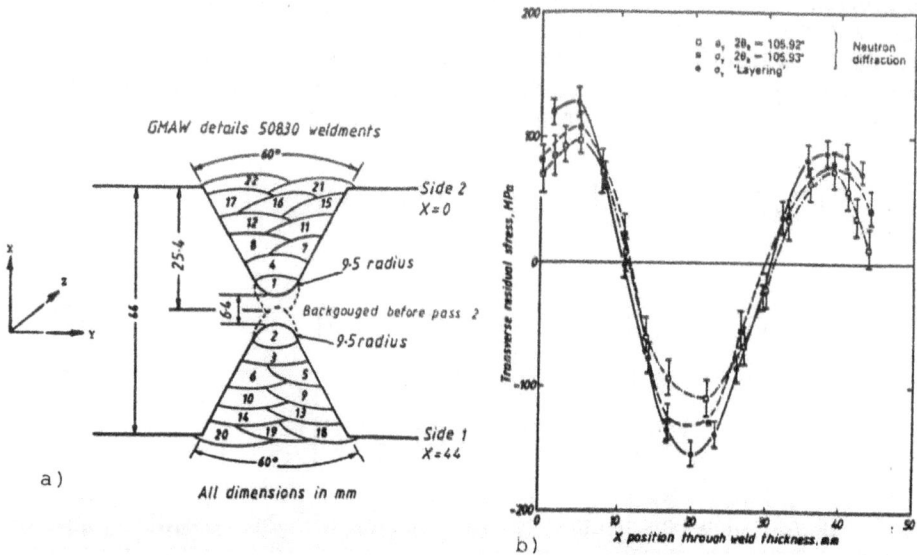

Fig. 2 Aluminium alloy butt weld showing a) pass sequence, b) comparison between neutron diffraction and layering determinations of transverse residual stress distribution.

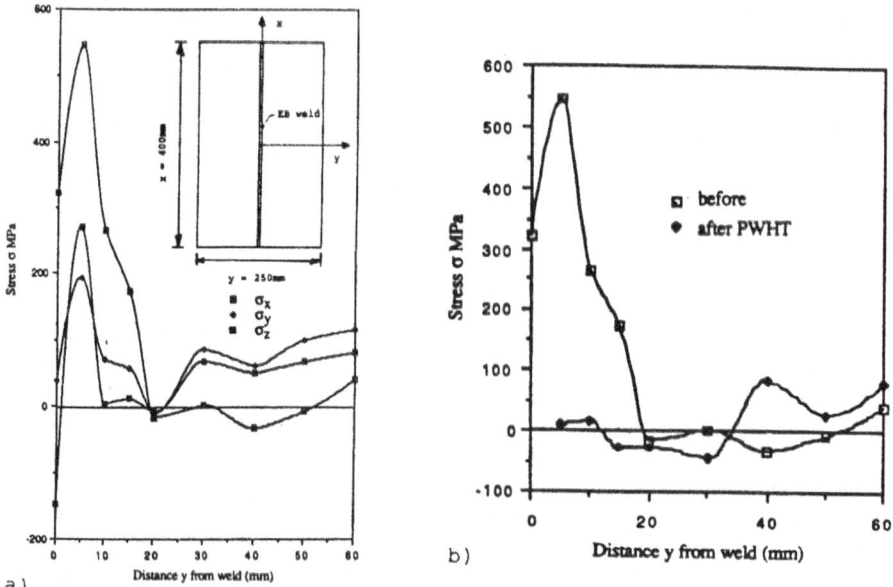

Fig. 3 Residual stress distributions across an electron beam weld in steel a) before, b) after post weld-treatment.

Corresponding results for the residual stresses measured across an electron beam weld in a 25 mm thick steel plate are depicted in Fig 3. Figure 3a) shows that although stresses approaching yield are observed, the region over which they are significant is much smaller than is expected in a normal fusion weld. An illustration of the relaxation that is achieved in the longitudinal residual stress by post weld heat-treatment is indicated in Fig. 3b). It is clear that the heat-treatment has caused almost complete annealing and elimination of the potentially detrimental tensile residual stress. This application demonstrates that neutron diffraction can be employed to investigate the effectiveness of stress relief heat-treatment processes.

Resistance to fracture of components can often be improved by shot-peening critical surface regions. The magnitude of the residual stresses generated is sensitive to the intensity and coverage of the process. An example of the influence of shot-peening the surface of an aluminium alloy plate is illustrated in Fig. 4 (MacGillivray et al, 1987, Bourke et al, 1987). This figure shows the in-plane compressive residual stress developed close to the surface as a function of depth z from the surface. This compression is balanced by tension sub-surface. The presence of the compressive residual stress field should give some protection against fatigue crack initiation and growth from the surface until the region of sub-surface tension is reached. It is apparent that neutron diffraction can be used to quantify the residual stress distribution produced by shot-peening.

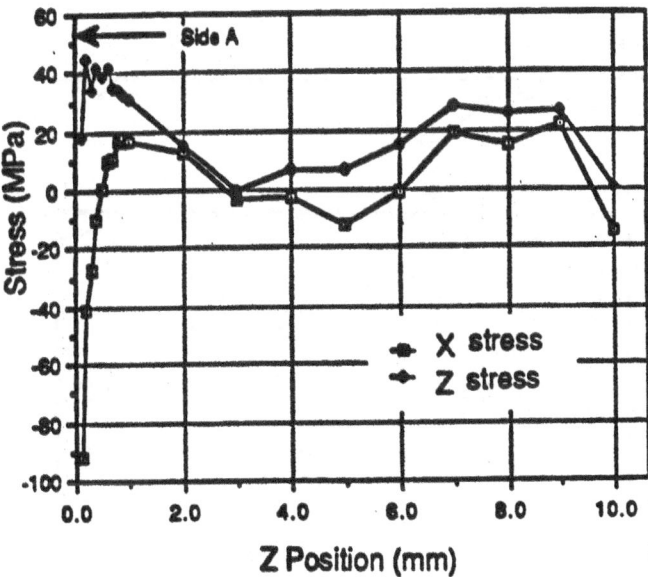

Fig. 4 Residual stress distributions with distance z from surface of a shot peened aluminium alloy plate assuming equal planar stresses in the x and y directions.

Figure 5 shows the results of residual stress measurements that have been made in the head of a slice of railway rail that was cut from a piece of used track (Webster et al, 1989, 1990). The stress contours were obtained by making horizontal and vertical scans of the neutron beam through the rail head in a series of parallel traverses. It is apparent that progressive wear due to wheel contact along the running-line (which for this rail was about 6mm off the centre-line) has caused an asymmetric stress pattern with steep gradients just below the surface. Such information, in combination with a knowledge of the applied stresses, should assist in explaining failures in track due to rolling contact fatigue and fracture.

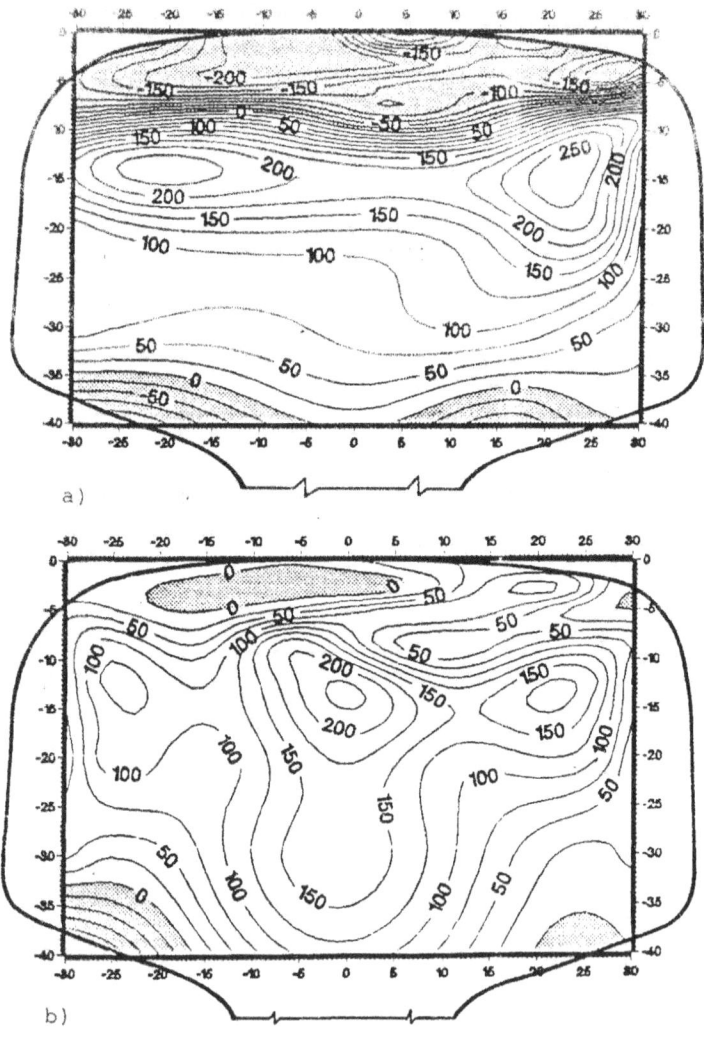

Fig. 5 a) Transverse and b) vertical residual stress contours in the head of a slice of used rail.

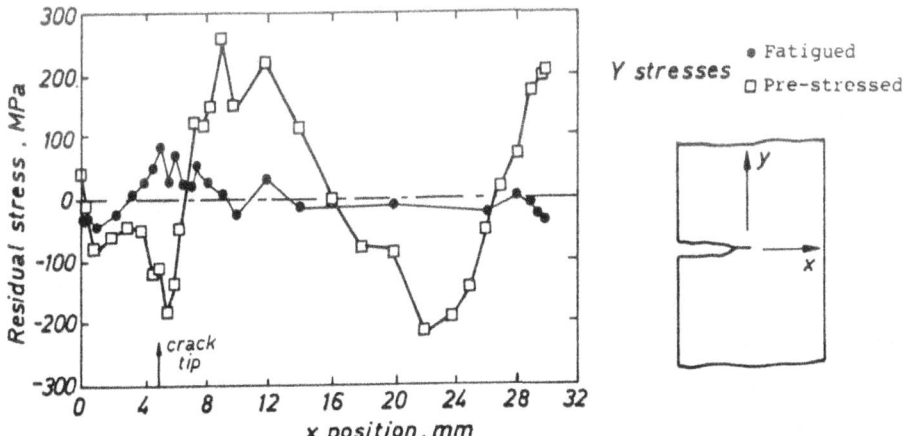

Fig. 6 Residual stress field ahead of a fatigue crack and after warm pre-stress in a pressure vessel steel

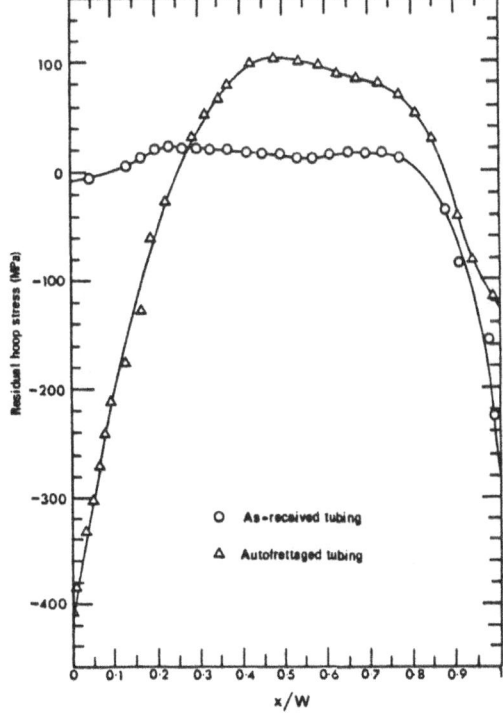

Fig. 7 Influence of autofrettage on residual hoop stress distribution in thick-walled tubing

Figure 6 indicates the residual stress distributions that can be generated ahead of a crack tip due to fatigue and warm pre-stress (Bourke et al, 1990). This process, when applied to steels, involves overloading in tension at a high enough temperature to be in the ductile region. Residual compression at the crack tip is then formed on cooling. It is anticipated that this compression will give improved resistance to further fatigue crack growth and protection against possible brittle fracture in subsequent low temperature use. The neutron diffraction data enable the magnitude of the compression developed by the warm pre-stress process to be estimated.

Figure 7 compares the residual stress distributions that have been measured with distance x from the bore in as-received and autofrettaged tubing of wall thickness W (Stacey and Webster, 1984, 1988a and 1988b). The autofrettage process involves applying internal pressure to the tubing to cause yielding in tension part-way through the wall so that depressurization leaves the residual stress pattern shown. The compression adjacent to the outer diameter is a result of the tube manufacturing process. It has been found (Stacey and Webster, 1988c) that an accurate knowledge of the tube material yielding behaviour is required in engineering stress analysis calculations for satisfactory predictions to be made of the influence of the autofrettage process.

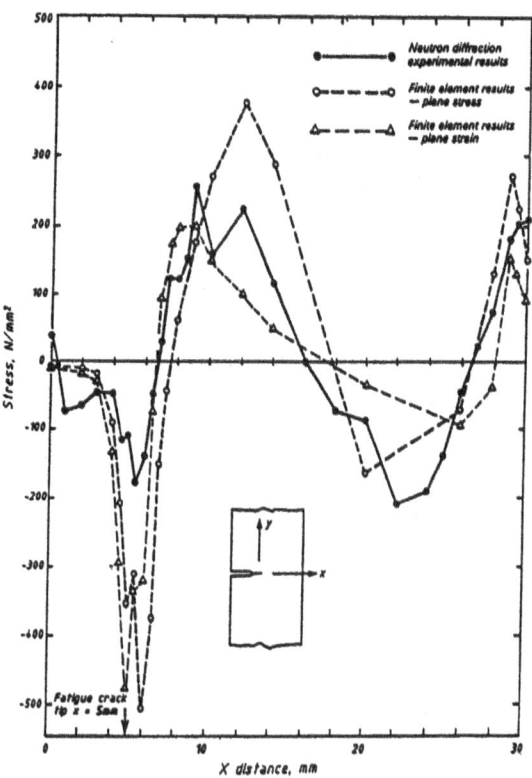

Fig. 8 Comparison between finite element prediction and neutron diffraction measurements of residual stress field ahead of a crack after warm pre-stress

4. COMPARISON WITH FINITE ELEMENT PREDICTIONS

To make predictions of residual stress distributions it is necessary to have an accurate model of the process by which the stresses were introduced and of the material properties. Although analytical solutions are sometimes possible, numerical methods are usually needed. When finite element procedures are adopted, unless a full three-dimensional calculation is performed, some knowledge of the state of stress in the component is required. Figure 8 shows a comparison between finite element estimates of the residual stresses generated by warm pre-stress ahead of a crack tip and those obtained by neutron diffraction in Fig. 6. It can be seen that the assumption of plane stress or plane strain gives satisfactory bounds to the experimental data except immediately adjacent to the crack tip where reverse yielding and a 'Bauschinger effect' may be expected. No allowance for the 'Bauschinger effect' was included in the calculations. Allowance for this effect was needed previously (Stacey and Webster 1988c) to achieve satisfactory predictions of the residual stresses generated by autofrettage. Its inclusion in this case would reduce the compression obtained on the crack face and improve correlation with the experimental measurements. It is apparent from this example that neutron diffraction data can be used to aid the development of more refined finite element calculation procedures.

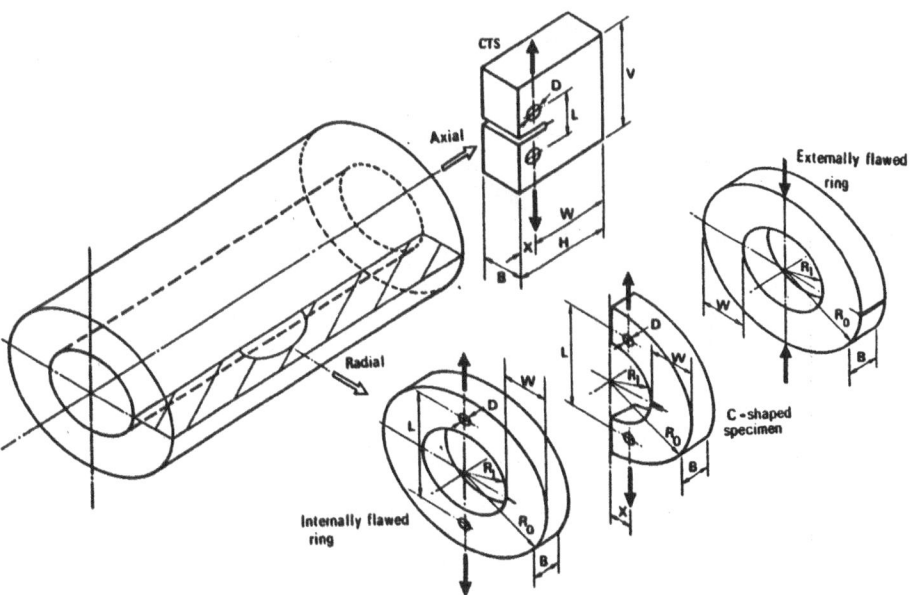

Fig. 9 Specimens for investigating fatigue crack growth in thick-walled tubing

5. PREDICTIONS OF FATIGUE CRACK GROWTH

The use of neutron diffraction data for quantifying the influence of residual stress on fatigue crack growth will now be examined.

Thick walled cylindrical pressure vessels subjected to pulsating internal pressure frequently experience failure by fatigue crack growth from the bore as illustrated in Fig. 9. Resistance to fatigue failure can be improved by generating compressive residual stresses at the bore by autofrettage prior to use as indicated in Fig. 7 (Stacey and Webster, 1988b). Fatigue cracks can be propagated through

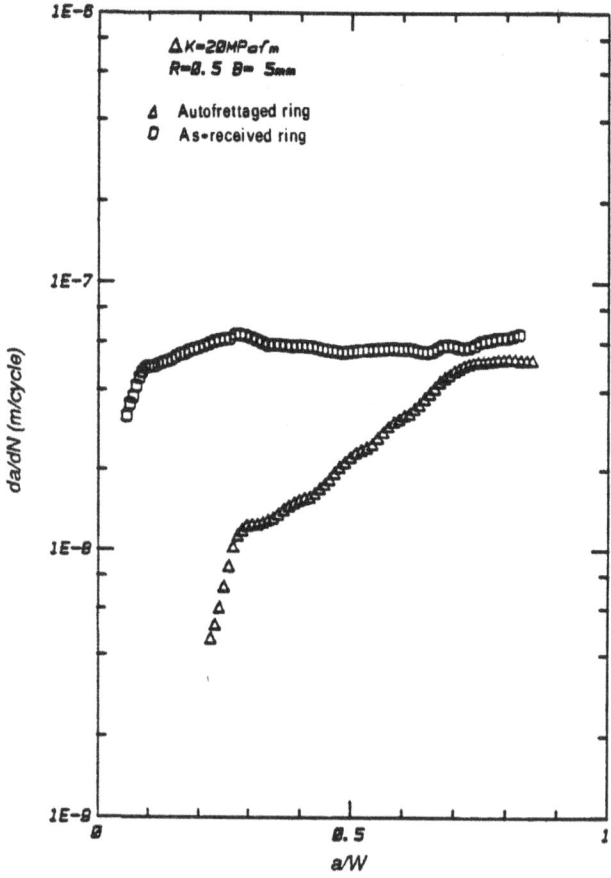

Fig. 10 Fatigue crack propagation at a constant ΔK of 20 MPa √m in internally cracked rings

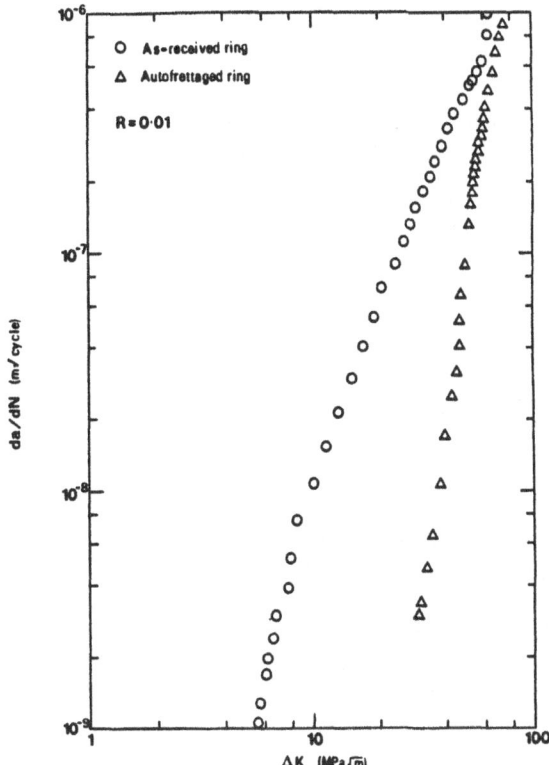

Fig. 11 Fatigue crack growth in internal cracked rings at R = 0.01

different residual stress fields by testing compact tension (CTS), C-shaped and ring specimens taken from sections of the as-received and autofrettaged tubing and subjecting them to fatigue loading as shown in Fig. 9. Results of tests performed on internally cracked ring specimens at a constant applied stress intensity factor range ΔK to demonstrate the influence of autofrettage are depicted in Fig. 10 (Stacey and Webster, 1984). The tests were conducted at a minimum to maximum load ratio R = 0.5. A constant crack growth/cycle, da/dN, as a function of crack depth through the wall, a/W, is observed for crack propagation from the bore of an as-received ring. This corresponds with the region where the residual hoop stress is very small (Fig. 7). In contrast, in the autofrettaged sample, an initially low crack growth/cycle is noticed which progressively approaches that of the as-received specimen with increase in crack depth. This is consistent with the crack propagating out of the compressive residual stress field adjacent to the bore (Fig. 7).

The dependence of crack growth/cycle on the applied stress intensity factor range for additional tests performed at a constant load amplitude and R = 0.01 on as-received and autofrettaged internally cracked ring samples is presented in Fig. 11. An increase in ΔK corresponds with an increase in crack length. It is apparent that a much larger stress intensity factor range must be applied to cause the same fatigue crack growth rate at short crack lengths close to the bore in the autofrettaged sample than in the as-received specimen. This is the region where the compressive residual hoop stress is greatest in the autofrettaged ring. As the crack in this sample propagates out of this region, the crack growth/cycle approaches that in the as-received ring in agreement with the behaviour observed in Fig. 10.

The influence of residual stress on the fatigue crack growth process can be rationalized by applying the principle of superposition to define an effective stress intensity factor K_{eff} such that

$$K_{eff} = K + K_{res} \tag{5}$$

where K is the stress intensity factor due to the applied loading and K_{res} is that due to the residual stress field. Solutions for both K and K_{res} for the testing conditions considered here have been determined by Stacey and Webster (1988a). All the values of K_{res} were found to be negative over the range of a/W of interest.

To determine the effect of residual stress on fatigue crack growth an effective stress intensity factor range ΔK_{eff} is defined which is responsible for the cracking process. It is supposed that no damage is incurred when compressive stresses are transmitted across the crack faces. Consequently, for cycling between minimum and maximum stress intensity factors of K_{min} and K_{max} respectively, when K_{min} >|K_{res}| the crack will remain open and $\Delta K_{eff} = \Delta K$ but R will change to

$$R_{eff} = \frac{K_{min} + K_{res}}{K_{max} + K_{res}} \tag{6}$$

When $K_{min} < $ |K_{res}| crack closure will occur so that $R_{eff} = 0$ and

$$\Delta K_{eff} = K_{max} + K_{res} \tag{7}$$

The data of Fig. 11 plus additional test results are shown plotted against ΔK_{eff} in Fig. 12. It is apparent that all the curves lie close to the essentially stress free internally cracked ring and CTS specimen data. Similar correspondence was also achieved for other geometries examined by Stacey and Webster, (1988b). As K_{res} was always negative for the tests reported, the influence of the residual stress fields is to reduce R towards zero and leave $\Delta K_{eff} \leq \Delta K$. Since little effect of R on fatigue crack growth was apparent for the tube material, plotting da/dN against ΔK_{eff} causes all the data to superimpose. From these results it is apparent that neutron diffraction can be employed in conjunction with fracture mechanics concepts to quantify the influence of residual stress on fatigue failure in engineering components.

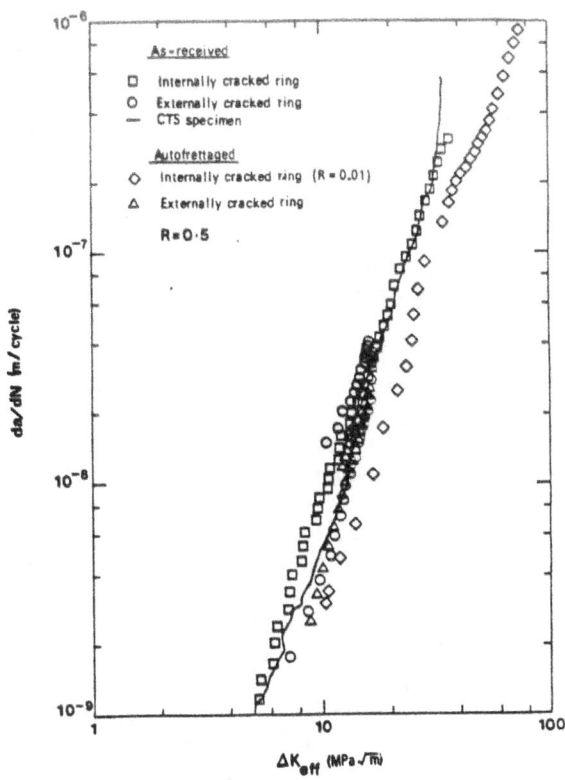

Fig. 12 Correlation of fatigue crack growth in autofrettaged ring specimens with effective stress intensity factor range.

6. CONCLUSIONS

The use of neutron diffraction for measuring residual stress accurately and non-destructively has been outlined. A representative selection of residual stress distributions has been presented. Some comparisons have been made with strain gauge and finite element determinations. Fatigue crack growth through regions of different residual stress has been examined and explained in terms of fracture mechanics concepts. It has been shown how neutron diffraction can be employed in engineering stress analysis to validate calculations and provide improved predictions of the residual stress fields that can be introduced into engineering components during manufacture.

34

7. ACKNOWLEDGEMENTS

The author wishes to acknowledge the assistance of his colleagues, M.A.M. Bourke, K.S. Low, H.J. MacGillivray, G. Mills and P.J. Webster with the neutron diffraction measurements and the SERC, Welding Institute and British Rail for the provision of specimens and resources.

8. REFERENCES

Allen, A.J., M.T. Hutchings, C.G. Windsor and C. Andreani, (1985). 'Neutron diffraction methods for the study of residual stress fields'. Advances in Physics, 34, 445-473.

Beaney, E.M. and E. Proctor (1974). 'A critical evaluation of the centre hole technique for the measurement of residual stresses'. Strain, 10, 7-14.

Bourke, M.A.M., H.J. MacGillivray, G.A. Webster, K.S. Low and P.J. Webster (1987). 'Improving the resolution of neutron diffraction residual stress measurements in engineering components'. IITT Int. Conf. on Fatigue and Stress, Paris.

Bourke, M.A.M., D.J. Smith, G.A. Webster & P.J. Webster (1990). 'Neutron diffraction measurements of residual stresses in plastically deformed cracked beams', 9th Conf on Experimental Mechanics, 3, Copenhagen, 1990, pp 1198-1206.

Hermann, R. and C.N. Reid (1986). 'Slow crack growth in the presence of tensile residual stresses'. Residual Stresses in Science and Technology, 2, E. Macherauch and V. Hauk (Eds), 759-766.

Hughes, H. (1967), 'X-ray techniques for residual stress measurements', Strain, 3, 26-31.

MacGillivray, H.J., M.A.M. Bourke, G.A. Webster and P.J. Webster (1987). 'Recent advances in neutron diffraction residual stress measurements in engineering components'. Int. Conf. Advanced Measurement Techniques, BSSM, London.

Parlane, A.J.A. (1978). 'The determination of residual stresses: a review of contemporary measurement techniques'. in Residual Stresses in Welded Construction and their Effect. Welding Inst. Cambridge, 63-78.

Sachs, G. (1927). 'Der nachweis immerer spannungen in stangen und rohren'. Zeit Met, 19, 352-357.

Smith, D.J., R.H. Leggatt, G.A. Webster, H.J. MacGillivray, P.J. Webster, and G. Mills (1988). 'Neutron diffraction measurements of residual stress and plastic deformation in an aluminium alloy weld'. J. Strain Analysis, 23, No 4, 201-211.

Stacey, A. and G.A. Webster (1984). 'Fatigue crack growth in autofrettaged, thick-walled high pressure tube material'. in High Pressure in Science and Technology, C. Homan, R.K. MacCrone and E. Whalley (Eds), Elsevier Publishing Co., New York, Part III, 215-219.

Stacey, A., H.J. MacGillivray, G.A. Webster, P.J. Webster and K.R.A. Ziebeck (1985). Measurement of residual stresses by neutron diffraction, J. Strain Analysis 20, 93-100.

Stacey, A. and G.A. Webster (1988a). 'Stress intensity factors due to residual stress fields in autofrettaged tubing', in Analytical and Experimental Methods for Residual Stress Effects in Fatigue, (Eds R.L. Champoux, J.H. Underwood and K.A. Kapp), ASTM STP 1004, Am. Soc. for Testing and Matls., 37-53.

Stacey, A. and G.A. Webster (1988b). 'Influence of residual stress on fatigue crack growth in thick-walled cylinders', ib id., 107-121.

Stacey, A. and G.A. Webster (1988c). 'Determinations of residual stress distributions in autofrettaged tubing'. Int. J. Pres. Ves and Piping, 31, 205-220.

Webster, G.A. (1989). 'Propagation of fatigue cracks through residual stress fields' in Fatigue and Stress (Ed. H.P. Lieurade), IITT International, Gournay-sur-Marne, France, 9-20.

Webster, G.A., M.A.M. Bourke, H.J. MacGillivray, P.J. Webster , K.S. Low, D.F. Cannon and R.J. Allen (1989). 'Measurement of residual stresses in railway rails', in Int Conf on Residual Stresses, (Eds G. Beck, S. Denis and A. Simon), Elsevier Applied Science, 203-208.

Webster, P.J., K.S. Low, G. Mills and G.A. Webster (1990). 'Neutron measurement of residual stresses in a used railway rail', (Eds S.M. Shapiro & others), Mat.Res. Soc. Symp. Proc., 166, 1990, pp 311-316.

Webster, P.J. (1990). 'The neutron strain scanner : a new analytical tool for engineers', Steel Times, 218, No 6, 321-323.

Williams, J.G., J.M. Hodgkinson and A. Gray (1981). 'The determination of residual stresses in plastic pipe and their role in fracture', Polymer Eng. Sci, 21, 816-821.

APPLICATION OF X-RAY RESIDUAL STRESS MEASUREMENTS IN INDUSTRIAL R&D

M. R. JAMES
Rockwell International Science Center
1049 Camino Dos Rios
Thousand Oaks, CA, 91360
USA

ABSTRACT. The measurement of residual stress by x-ray diffraction techniques is a mature and cost effective approach for obtaining relevant data in many industrial circumstances. This is made possible by the recent development of commercial portable equipment having appropriate software to acquire the raw data, calculate the measured stress and provide an estimate of the precision based on counting statistics, regression analysis and fluctuations in peak intensity, all at relatively high speed compared to ten years ago. Despite its potential, and the fact that few materials can be manufactured and assembled without introducing residual stresses, these hardware developments have not resulted in widespread use of the technique in industry. The integrity of the results are still highly dependant on the skills of the operator, and the utility of the results remains contingent on an understanding of the consequences of residual stress on material performance.

In an industrial research and development laboratory, we have become involved in two important uses of x-ray residual stress measurements: origin and control, and failure analysis. Two case studies of each are cited with the intention of showing how x-ray measurements provided a key element in identifying important aspects of the particular problem. The four studies originate on the aerospace side of Rockwell International where issues of fatigue, stress corrosion cracking and hydrogen embrittlement are important failure mechanisms and in which the integrity of the surface is paramount. Issues relating to dimensional stability and static strength where the bulk residual stress influence properties are less amenable to x-ray analysis because of the near surface nature of the technique.

1. INTRODUCTION

It is not without trepidation that a practitioner of one technique discusses his viewpoints in front of a group representing an alternative method. However, I do not feel apprehensive because x-ray and neutron diffraction should not, and must not, be considered as

37

M. T. Hutchings and A. D. Krawitz (ed.),
Measurement of Residual and Applied Stress Using Neutron Diffraction, 37–50.
© 1992 *Kluwer Academic Publishers.*

competing techniques. They are instead complementary techniques which, along with a host of other supplementary laboratory tools, can be used to elucidate the role and behavior of residual stresses in advanced components. My approach in this paper is to highlight the strengths and weaknesses of the x-ray diffraction (XRD) technique using examples which can be contrasted to the neutron diffraction technique. The studies also serve to illustrate typical problems in which measurement of the residual stress state was beneficial to improving a manufacturing process or solving a service problem.

Most processing sequences in the manufacture of a component such as forming, heat treating, machining, surface finishing, plating and assembly impart a residual stress state. Almost universally, a compressive residual stress state is desirable whereas a tensile state is deleterious. The residual stress state can effect 5 categories of primary concern to the performance of a part as stated in Fig. 1

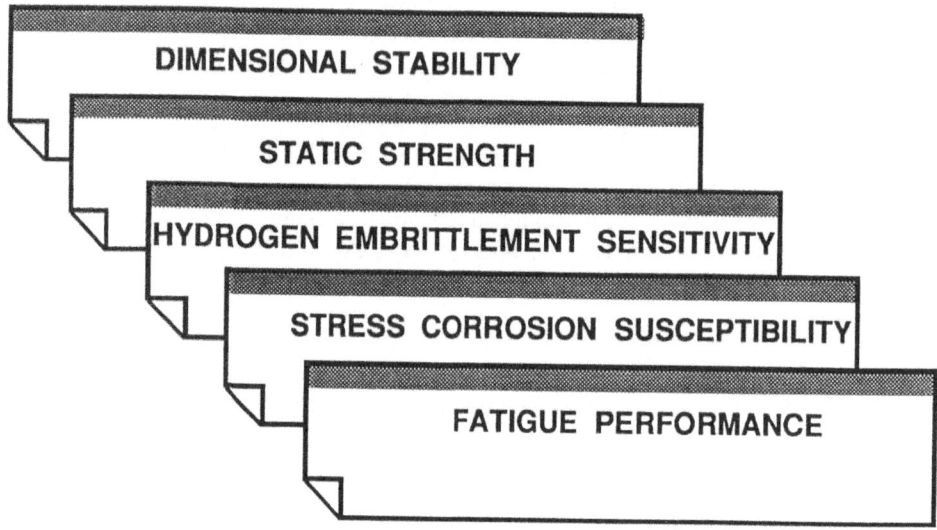

Fig. 1 Residual stress effects on performance

The bulk state of stress is important in the first two categories and, as such, problems in these categories do not lend themselves to efficient use of a surface sensitive technique such as XRD. Typical examples might be the stress state induced in a plate by rolling or extruding, in a cast structural component which experiences non-uniform cooling, or in a large welded structure. The hole drilling technique, and more recently, ultrasonic approaches have been used for these situations, as could neutron diffraction. Nonetheless, many performance issues are much more surface related. This is especially true in the aerospace field where the last three categories are pertinent. Many high performance aerospace components experience dynamic loading and/or aggressive environments in which inadequate knowledge or control of residual stresses may lead to deficient performance. Proper manufacturing should consist of adequate control of the residual stress state, but invariably scant attention is given until a component fails to meet expectations. This results from our inability to control the residual stress state during manufacturing, a consequence associated with a lack of modelling of the generation of residual stresses in all processing sequences. Two research programs recently carried out at the Science Center are described to illustrate where the measurement of residual stresses can be useful in developing appropriate predictive models of the generated residual stress state.

By far the most abundant applications of XRD stress measurements in our laboratory have been in areas associated with failure analysis, be it failure of a part in service, or failure of a processing sequence in the manufacture of a part. Numerous examples could be cited from the literature (1-5) as well as our own efforts, but two will be highlighted here to emphasize special aspects.

Before discussing these case studies, a few comments on the general techniques employed are warranted. Two computer controlled diffractometers are used for stress measurements in our laboratory. One has a liquid nitrogen cooled solid state detector which provides excellent energy resolution for use in obtaining diffraction peaks from highly fluorescing alloys (superalloys, copper alloys). This step scanning system is also used for routine phase determination, line broadening and texture analysis. The other diffractometer is equipped with a one-dimensional position sensitive detector for rapid acquisition of the diffraction profile. In addition to rapid XRD stress measurements, the system is equipped with a high temperature camera for phase equilibria studies. Both systems provide for complete computer control of data acquisition and calculation using either the biaxial or triaxial stress analysis methods. Multiple profile fitting algorithms are available ranging from a simple multiple point parabolic fit to a Pearson VII function to determine the peak location. The determining factor in their use is the degree of similarity in peak shape at all ψ tilts; greater consistency in peak shape allows for simpler profile fits.

2. EXAMPLES IN ORIGIN AND CONTROL OF RESIDUAL STRESSES

Many approaches have been used to deliberately control the residual stress state, from heat treating [6] to shot peening, and recently more exotic methods such as laser glazing [7]. Two studies are briefly reviewed which demonstrate cases where modelling the origin of the residual stress state was important so as to optimize those parameters which control the stress state.

2.1 Transformation Toughened Ceramics

The strength of ceramics or glasses can be increased by placing their surfaces into compression. Techniques include ion exchange, temperature glazing, surface chemical reactions and stress-induced phase transformation. The latter has also been used to dramatically toughen ceramics by enabling the phase transformation to take place in localized zones around cracks. Fig. 2 shows examples of these for zirconia-containing ceramics which undergo a martensitic tetragonal to monoclinic transformation with a concomitant volume expansion placing the transformed region into compression. The energetics to drive the stress induced phase transformation can be supplied by grinding the surface, or by the propagation of a crack. An optimal compressive state requires a narrow range of microstructures in which ZrO_2 grains or precipitates are on the verge of spontaneous transformation. We have conducted a number of studies in the Al_2O_3/ZrO_2 and MgO/ZrO_2 systems to determine the influence of different parameters on the residual stress state as shown in Fig. 3 [8-10].

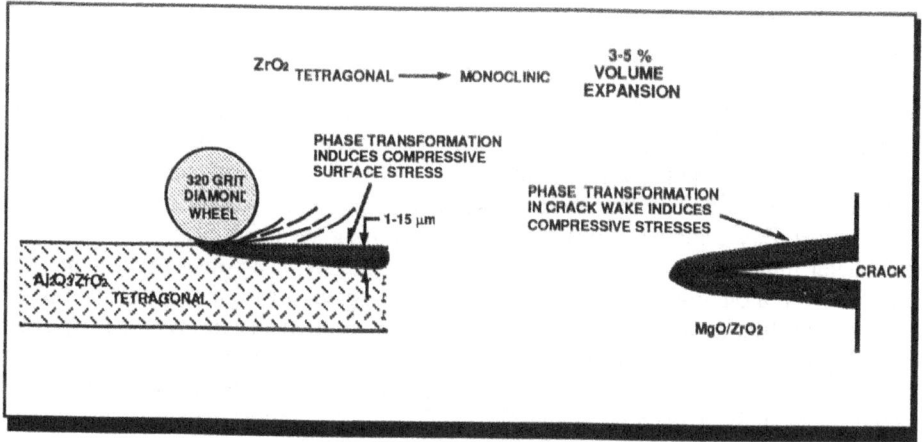

Fig. 2 Transformation toughening in ceramics.

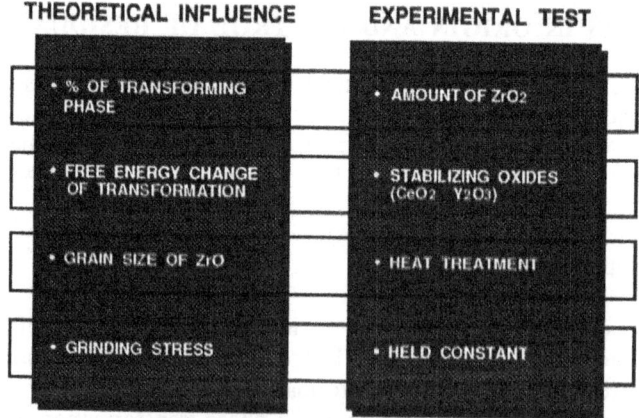

Fig. 3 Factors influencing degree of transformation.

We have shown that, as expected, the residual compressive stress increases with increasing volume percent ZrO_2 retained in its tetragonal structure in the sintered material (8). The compressive stresses can exceed 1 GPa with high volume percent ZrO_2. The energetics of the phase transformation depend on the ZrO_2 grain size and free energy change of the transformation. Fig. 4 shows the results of stress profile measurements carried out on the Al_2O_3-30 v/o ZrO_2 system. The larger grain size produces slightly greater compressive stress, but not to the degree predicted by a theoretical analysis of the free energy change associated with unconstrained transformation of inclusions [8]. More importantly, the figure shows that the compressive stress decreases with increasing distance from the initial ground surface. The material was removed by very slow automated polishing using fine diamond paste. A residual compressive stress of ~ 100 MPa still persisted after a large portion of the initial surface was removed, suggesting that the mechanical polishing operation may produce some transformation and or plastic deformation that results in a surface compressive stress. Other studies on polished ceramic surfaces also show a small compressive residual stress even on the most carefully prepared surfaces [11,12].

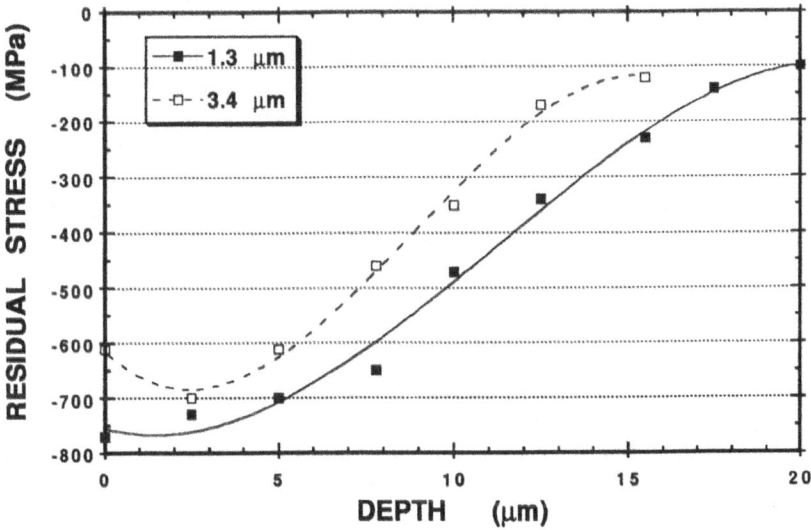

Fig. 4 Influence of grain size on residual stress.

The strengthening aspect of a compressive residual stress depends on the crack size and on the magnitude and, to a large degree, the depth of the compressive zone. It is important to understand the process variables, particularly if the compressive stress varies with depth. The near surface sensitivity and the ability to measure the surface stress with small amounts of layer removal (as shown in Fig. 4) contributed to the usefulness of the x-ray technique in this application.

2.2 Metal Matrix Composites

High temperature strength and stiffness and reduced weight demanded in advanced aerospace applications are being met by use of fiber and particle reinforced metal matrix composites. The significant difference in thermal expansion coefficient between the metal matrix and ceramic reinforcement generates large mismatch strains during cooldown from the fabrication temperature. Accurately predicting their magnitude is difficult because of the many unknowns that affect the fiber/matrix interface, the extent of plastic deformation in the matrix and the fiber distribution. Finite element methods (FEM) have been used to predict the full three dimensional stress state around each fiber, and show that for typical metal/continuous ceramic fiber systems, the longitudinal (parallel to fiber direction) and hoop stresses in the matrix are tensile, while the radial stress acting on the fiber is compressive.

Recent advances in triaxial stress analysis have made it possible to analyze the residual stress state in both the metal and reinforcement phase in particulate reinforced composites [13, 14]. For this application, the standard biaxial stress analysis procedure cannot be used because, as shown in Fig. 5, the reinforcement produces a normal stress within the penetration depth of the x-rays. For continuous fiber reinforced systems, however, the fiber diameter is large and considerable matrix is present between the fiber and the surface so that the normal component induced by the fiber is zero near the surface. We have shown this to be the case in two titanium matrix systems reinforced with 145 μm diameter SiC fibers [15]. Fig. 6 shows a comparison of measured values calculated by each analysis procedure using SiC/Ti-6Al-4V (12% vol. fraction fiber). Only within 50 μm of the fibers, which in this case lie 150 μm below the surface, is there any real difference between values calculated by the biaxial and triaxial stress analysis procedures. Considering that the triaxial procedure requires three times the amount of data, and has greater error because of the need to measure the stress free interplanar spacing, the biaxial analysis procedure was considered satisfactory for this application. Other techniques such as measurement of the residual stress in the fiber and determination of the crack opening profile at zero load were used to verify the value obtained by x-ray diffraction in the SiC/Ti-25Al-10Nb-3V-1Mo system [15].

The apparent slope in Fig. 6 is due to bending of the sample. Layer removal produces an asymmetric lay-up and changes the volume fraction of fiber in the cross-section. After correction for bending, the surface measurement is nearly identical to the value of the matrix stress in the fiber/matrix region for this composite [15]. This is not always the case as shown in Ref. 15 in that even after correction for bending, the residual stress depth profile still exhibited a gradient for the SiC/Ti-25Al-10Nb-3V-1Mo system.

Comparison of predicted and measured longitudinal surface residual stress can be used to assess or even calibrate a finite element model, as is done in Fig. 7 for four titanium matrices reinforced with SiC SCS-6 fibers. Using ABAQUS™ finite element solver, strains developed during cooldown from the consolidation temperature were calculated from constituent elastic-plastic temperature dependent material properties and appropriate boundary conditions. The resulting stress field was then calculated to obtain an average

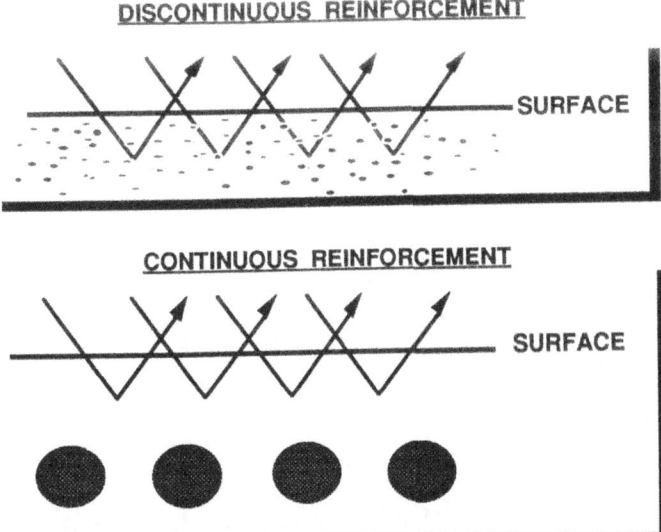

Fig. 5 X-ray analysis of composite matrices.

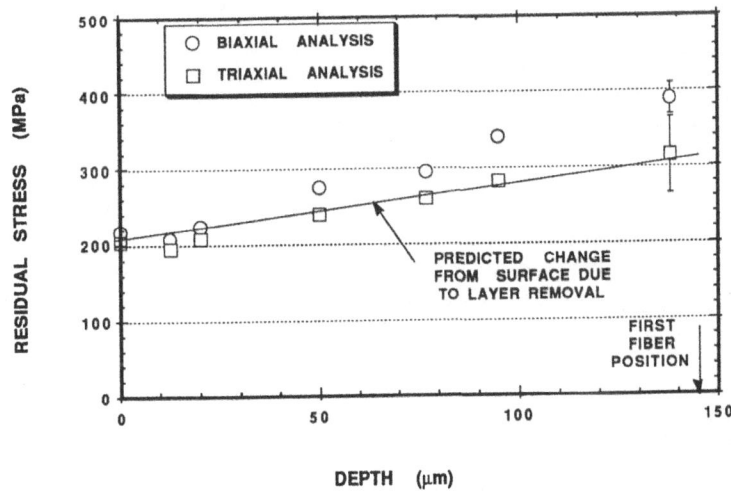

Fig. 6 Calculated residual stress using two analysis procedures.

longitudinal residual stress at the surface for comparison to the x-ray results   The fiber volume fraction was 35% except for the Ti-6-4/SiC system, as denoted in the figure. For the two ductile matrix systems, Ti-15Al-3Al-3Sn-3Cr and Ti-6Al-4V, the FEM calculation agrees quite well with the measured value. Considerable plastic flow is experienced in

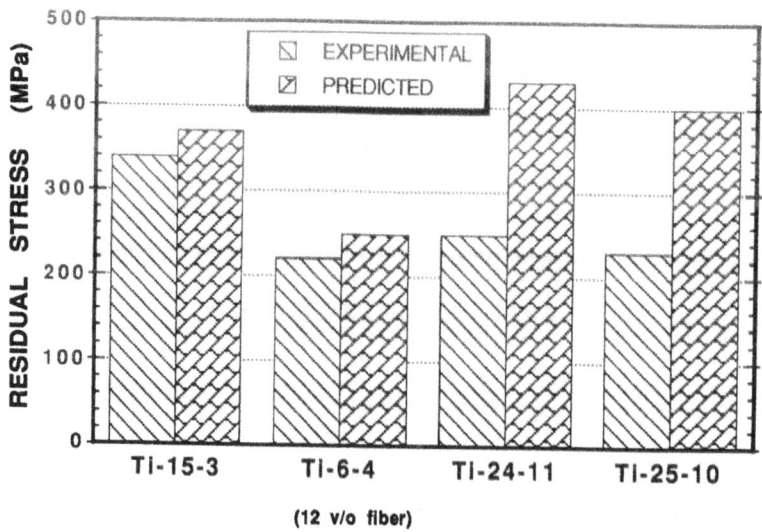

Fig. 7 Comparison of FEM and measured longitudinal residual stress.

these two systems during cooldown which is adequately modelled by FEM. The two intermetallic matrices have higher high temperature strength and do not undergo much plastic deformation. In these cases, the measured values are considerably less than predicted. X-ray depth profile measurements show a stress gradient in these two composites indicating that the surface value is not representative of the matrix residual stress state in the bulk. Thus, surface residual stress measurements alone are not enough to ascertain the residual stress state in the fiber/matrix core region. In this case, neutron diffraction would be a better method to measure the residual stress state for comparison to predicted values. Other aspects that have been studied by x-rays include the stability of the tensile matrix stress during thermal exposure, mechanical fatigue and thermal-mechanical fatigue (14).

These two examples show how measured depth profiles of the surface residual stress can be useful in developing models to predict the magnitude and distribution of residual stresses. In both cases the necessary depth resolution was on the scale of tens of microns.

3. EXAMPLES OF FAILURE ANALYSIS AND REMEDY RESOLUTION

Failure analyses often proceeds towards an inconclusive inquiry before the role of residual stresses is ever assessed. Then, *when all else fails, blame it on residual stresses!* Fortunately, with transportable equipment and a mature technology, we can now quantify the role of residual stresses in many cases and help verify any remedial procedures instituted to prevent the failure from recurring. We have experienced many such examples over the past decade [5], including numerous cases where residual stress were shown not to play a role in the deficient performance of a part. The two cases described here have been chosen to illustrate situations where XRD might not have been thought to be a good analysis technique, but where the ease and efficiency of measurement provided a quick and cheap means to verify the repair.

3.1 Turbine Blades

Turbine blades on both the high pressure fuel and high pressure oxygen turbopumps of the Space Shuttle main engines experienced significant crack formation early in the shuttle program. The blades were made from directionally solidified DS Mar-M-246, a cast nickel base superalloy. An illustration of the high pressure oxygen turbopump (HPOTP) blade in Fig. 8 shows that the root of the blade had extensive ground surfaces including the leading and trailing faces as well as the firtree. Because the DS blades are composed of between 5 and 15 grains running along the vertical axis, intuition would suggest that XRD stress measurements would not be successful on this component. However, the surfaces which experienced cracking had a significant subgrain structure induced during grinding. Measurements on the cracked surface gave tensile stresses of up to 250 MPa normal to the crack direction. Ground blanks with tensile surface stresses were found to develop cracks when exposed to high pressure hydrogen even without an applied load.

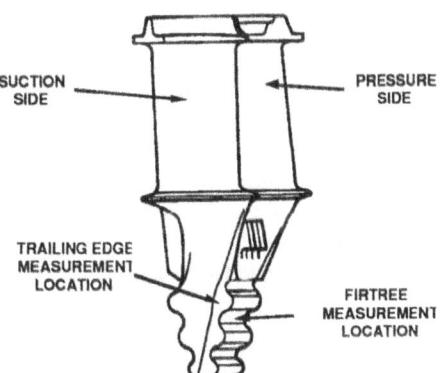

Fig. 8 HPOTP turbine blade

The high pressure hydrogen environment of the turbopumps was sufficient to induce hydrogen embrittlement on the ground surfaces that had a tensile residual stress state. While tensile stresses occurred only on a limited basis due to some improper grinding (momentary lack of coolant, worn abrasive wheel, etc), even heat treating after grinding could not guarantee that some tensile stresses would not be present. A rigorous program was undertaken to investigate the use of shot peening. A strain-to-crack-initiation test in high pressure hydrogen using tapered tensile specimens showed that, on average, crack initiation began at 0.8% strain on ground specimens, whereas over 2% strain could be achieved without crack initiation on shot peened specimens. The minimum value was raised from 0% (cracked without applied load) on the ground surfaces to 0.8% on the peened surfaces. However, because of large carbides in the microstructure, a peening intensity which did not crack the carbides (thereby immediately causing a site for hydrogen egress and crack initiation) had to be used. XRD and metallography were employed to identify the optimal intensity to produce a sufficiently compressive surface stress without cracking the 10-15 μm carbides. Measurements with Cr and Cu Kα radiation showed that the compressive layer was ~ 15 μm thick with a peak stress of ~ -800 MPa.

These tests on flat coupons demonstrated that shot peening could be a viable remedy. However, verifying that the entire ground surface of the small blades could be adequately peened was necessary before the approach could be implemented. The measurement locations shown in Fig. 8 required a beam size of 0.75 mm square. Only XRD with its small spatial resolution could have provided the necessary data. Using a turntable and appropriate inclinations of three nozzles, a reproducible shot peening procedure was developed. Along with certain design modifications, shot peening has completely eliminated crack formation on the ground surfaces, and blade life is no longer the limiting factor in the life of the turbopumps.

3.2   Laser Beam Welding of Titanium Aluminides

Compared to conventional aerospace materials, titanium aluminides of various compositions offer attractive properties for potential structural applications, including high temperature strength and creep resistance, improved high temperature environmental resistance and relatively low density. Unfortunately, they suffer from low ductility and moderate fracture resistance at ambient temperature. This greatly increases the difficulty associated with joining. Initial efforts at producing laser beam welds resulted in delayed crack formation normal to the weld bead as shown in Fig. 9. While the severe texture and large grain size of the weld bead precluded measurement of the residual stresses in the fusion zone, measurements were feasible in the heat affected zone and parent metal. As could be expected, crack formation resulted from high tensile stresses in the heat affected zone parallel to the weld direction combined with the low ductility and fracture toughness of the alloy.

10 mm

Fig. 9 Crack in as-welded titanium aluminide sheet.

Crack formation often occurred well after the coupons were removed from their fixture, even as much as 24 hours later. This delayed fracture made it near impossible to determine the effect of weld parameter changes such as pre and post heat of the coupons, weld speed, and constraint. XRD measurements were used to assess their effect on the residual stress distribution, as shown in Fig. 10. A rapid series of measurements using the position sensitive x-ray system could be completed before the next weld test was set up, and the results used to guide the parameter changes. This allowed us to bracket the minimum useful pre and post weld heat treatments (PWHT) to ensure sufficient resistance to crack formation. Subsequent microstructural analysis of the fusion zone was then used to produce an optimum weld. Measurement of the ultimate tensile strength from tensile specimens cut from butt welded sheets validated the tremendous improvement obtained by this cost effective approach (see Fig. 11).

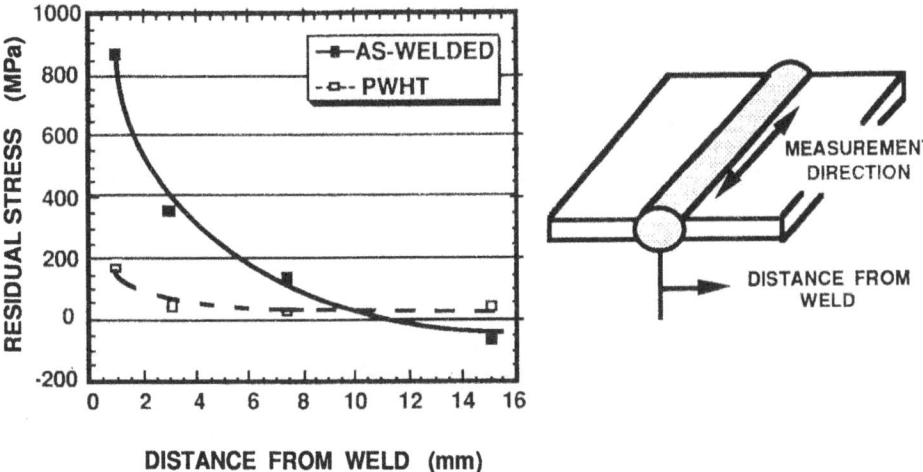

Fig. 10   Residual stress in butt welded titanium aluminide sheet.

These two examples serve to illustrate the usefulness of XRD stress measurements in solving practical service and production problems. In the case of the turbine blades, residual stress measurements were used only after all other explanations for crack formation were exhausted. Delayed cracking in the welded titanium aluminide sheets was immediately understood to result from high tensile stresses in the weld and heat affected zone. In this case, residual stress measurements were used to calibrate processing variables so as to prevent crack formation.

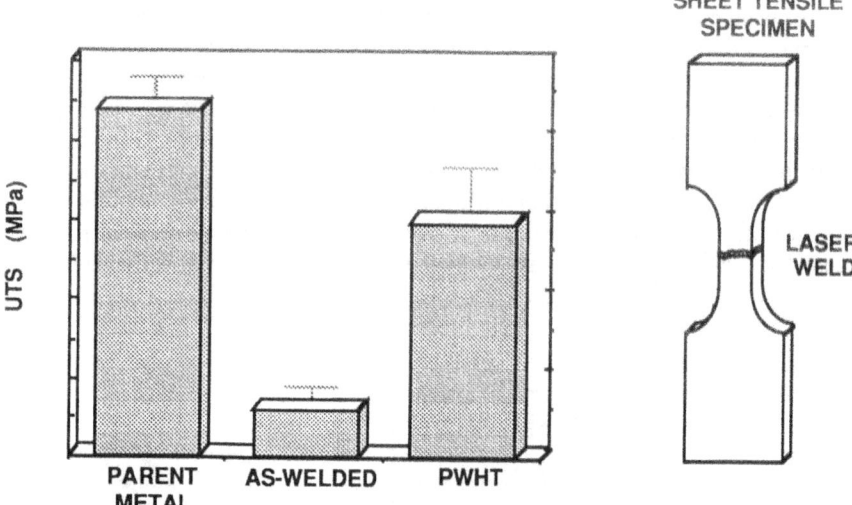

Fig. 11 Relative strength properties of welded titanium aluminide sheet.

4. CONCLUDING STATEMENTS

Practically, the use of x-ray diffraction residual stress analysis in industry is controlled not so much by capital investment, but by the difficulty in obtaining trained personnel in the interpretation of data delivered by x-ray systems. The significant theoretical improvements in understanding the diffraction technique has complicated the issue because the x-ray practitioner, using terms such as macrostress, microstress, stress gradients and non-uniform stress, and stress in individual phases, talks on a different stress scale than the engineers familiar with designing components. Thus, in my experience the dominant use of industrial x-ray residual stress measurements has been in failure analysis and verification of any remedy. Here detailed understanding is not necessary, only an acceptance by the design engineer that the data, whatever it is related to, is properly guiding the study.

Research into the arena of origin and control of residual stresses demands a more precise definition of the measured quantity. XRD stress measurements are useful in applications involving surface preparation such as machining, grinding, ion implantation, electroplating and coating. In many other important cases, such as rolling of thick plate, heat treatment of castings, and welding of thick sections, techniques measuring bulk residual stresses and the distribution through a part are more important, and lend themselves to neutron diffraction approaches.

5. ACKNOWLEDGEMENTS

The author is indebted to the support of Rockwell International through its many operating divisions for the consistent and exciting array of problems he has encountered. Support for the x-ray diffraction facility and many of the research efforts by Independent Research and Development funding is gratefully acknowledged.

6. REFERENCES

1. Shin, S.W. and Walter, G.H. (1981) 'Case Histories of Residual Stress Related Component Failures', Residual Stress For Designers and Metallurgists, L.J. Vande Walle (ed.) ASM, 1-20.

2. Chrenko, R.M. (1982) 'Thermal Modification of Welding Residual Stresses', Residual Stress and Stress Relaxation, E. Kula and V. Weiss (eds.), Plenum Press, 61-70.

3. Rice, R.C., Leis, B.N. and Tuttle M.E. (1982) 'An Examination of the Influence of Residual Stresses on the Fatigue and Fracture of Railroad Rail', Residual Stress Effects in fatigue, ASTM STP 776, American Society for Testings and Materials, 132-157.

4. Bayard, S. and Lebrun, J-L. (1987) 'Application of X-ray Fractography to a Cracked Steam Turbine Shaft', Residual Stresses in Science and Technology, E. Macherauch and V. Hauk (eds.), DGM, 935-958.

5. Weiss, W. and James, M.R. (1987) 'Residual Stresses - Fatigue and Fracture', Residual Stresses in Science and Technology, E. Macherauch and V. Hauk (eds.), DGM, 41-55.

6. James, M.R. (1989) 'Relaxation of Residual Stresses - An Overview', Advances in Surface Treatments, Vol. 4, A. Niku-Lari (ed.), Pergamon Press, 349-365.

7. James, M.R., Gnanamuthu, D.S. and Moores, R.J. (1984) 'Mechanical State of Laser Melted Surfaces', Scripta Met., 18, 357-361.

8. Green, D.J., Lange, F.F. and James, M.R. (1983) 'Factors Influencing Residual Surface Stress in a Stress-Induced Phase Transformation', J. American Ceramic Society, 66, 623-629.

9. James, M.R., Green D.J., and Lange, F.F. (1984) ' Determination of Residual Stresses in Transformation Toughened Ceramics', Advances in X-Ray Analysis-Vol. 27, J.B. Cohen et al. (eds.), Plenum Press, 221-228.

10. Green, D.J., Lange, F.F. and James, M.R. (1984) 'Residual Stresses in Al_2O_3-ZrO_2 Composites', Science and Technology of Zirconia II, N. Clausen, M. Rühle and A. Heuer (eds.), The American Ceramic Society, Columbus, 240-250.

11. Lange, F.F., James, M.R. and Green, D.J. (1983) 'Determination of Residual Stresses Caused by Grinding in Polycrystalline Al_2O_3', J. American Ceramic Society, 66, C16-C17.

12. Johnson-Walls, D., Evans, A.G., Marshall, D.B., and James, M.R. (1985) 'Residual Stresses in machined Ceramic Surfaces', J. American Ceramic Society, 69, 44-47.

13. Cohen, J.B. (1986) 'The Measurement of Residual Stresses in Composites', Powder Diffraction, 1, 15-21.

14. James, M. R. (1989) 'Residual Stresses in Metal Matrix Composites', International Conference on Residual Stresses II, G. Beck, S. Denis and A. Simon (eds.), Elsevier, 429-435.

15. Cox, B.N., James, M.R., Marshall, D.B., and Addison, R.C. (1990) 'Determination of Residual Stresses in Thin Sheet Titanium Aluminide Composites', Metal. Trans. A, 21, 2701-2707.

THE THEORY OF STRESS/STRAIN ANALYSIS WITH DIFFRACTION

I. C. NOYAN
IBM Research Division
Thomas J. Watson Research Center
Yorktown Heights, NY 10566
U. S. A.

ABSTRACT.Stress/strain analysis techniques that utilize x-ray or neutron diffraction are widely used for nondestructive testing and evaluation. In this article the basic formalism of this analysis is reviewed and its application to various classes of materials is discussed.

1. Theory

In this section the theoretical formalism utilized in diffraction based stress analysis techniques is reviewed. This formalism utilizes kinematical diffraction theory for data acquisition and linear continuum elasticity theory for data analysis. Its applicability is thus restricted to those specimens which satisfy the assumptions inherent in both of these theories.

1.1 FUNDAMENTAL EQUATIONS

All diffraction based stress analysis techniques utilize the distance between atomic planes of a crystalline specimen as an internal strain gage[1-4]. For any reflection hkl, the lattice plane spacing d_{hkl} can be determined from the angular position θ of the appropriate diffraction peak through Bragg's law; $\lambda = 2d_{hkl} \sin \theta$, where λ is the wavelength of the radiation used. This plane spacing, $d_{\phi\psi}$, is normal to the diffraction vector $L_{\phi\psi}$ which bisects the angle between the the incident and diffracted beams (Figure 1). Thus, one can define a strain, ε', along the diffraction vector $L_{\phi\psi}$;

$$\varepsilon' = \frac{d_{\phi\psi} - d_0}{d_0} . \tag{1}$$

Here d_0 is the unstressed plane spacing of the hkl planes. If one arbitrarily defines a diffraction co-ordinate system with $L_{\phi\psi}$ as one of the axes, the strains in the specimen coordinate system (S_i) can be expressed in terms of the measured strains through a second rank tensor transformation[5]:

$$\varepsilon'_{ii} = a_{ik}a_{il}\varepsilon_{kl}. \tag{2-a}$$

Here ε_{ii}' is defined along $\vec{L}_{\phi\psi}$ and the index "i" is usually chosen as "3". Substituting the appropriate direction cosines for a_{ik}, a_{il}, in terms of the angles ϕ, ψ, one obtains:

M. T. Hutchings and A. D. Krawitz (ed.),
Measurement of Residual and Applied Stress Using Neutron Diffraction, 51–65.
© 1992 All Rights Reserved. Printed in the Netherlands.

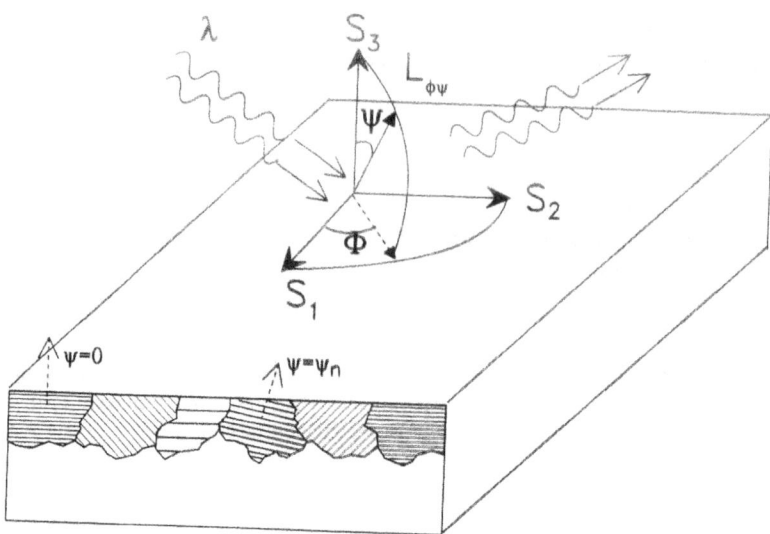

Figure 1. Definition of the sample coordinate system $\vec{S}_i$ and the measurement direction $\vec{L}_{\phi\psi}$. In polycrystalline samples, different grains diffract at different ψ tilts.

$$\frac{d_{\phi\psi} - d_0}{d_0} = \varepsilon_{11} \cos^2\phi \, \sin^2\psi + \varepsilon_{12} \sin 2\phi \, \sin^2\psi + \varepsilon_{22} \sin^2\phi \, \sin^2\psi + \qquad (2-b)$$

$$+ \, \varepsilon_{33} \sin^2\psi - \varepsilon_{33} + \varepsilon_{13} \cos\phi \, \sin 2\psi + \varepsilon_{23} \sin\phi \, \sin 2\psi.$$

Equation (2) is the fundamental equation of strain/stress analysis with diffraction. Since there are only six unknowns, one can determine d_{hkl} along six or more independent directions and solve for the strains ε_{ij} in the specimen coordinate system, S_i. The stresses σ_{ij} in S_i can be calculated through the general form of Hooke's law[5]:

$$\sigma_{ij} = C_{ijkl}\varepsilon_{kl}, \qquad (3)$$

where the C_{ijkl} are the stiffness coefficients for the material in the $\vec{S}_i$ coordinate system. It may be noted that the type of the material was not specified in this derivation. With respect to the continuum elasticity theory, equations (2)-(3) are equally applicable to random or textured polycrystalline samples, single crystals and amorphous materials. It is assumed, however, that the strains ε_{ij}', ε_{ij} in both coordinate systems are defined at the same point.

The limitations on the type of the material is imposed by diffraction through Bragg's law, which defines the directions along which one can measure d_{hkl} and obtain $\varepsilon_{ij}'[2,6,7]$. In random polycrystalline materials the spacing for the same family of planes (the same reflection) can be measured along any six independent directions, $L_{\phi\psi}$, with arbitrary ϕ and ψ. In textured polycrystalline materials, $d_{\phi\psi}$ for a given reflection can be measured only along certain directions which are defined by the orientation distribution of the crystallites for the particular texture. In the case of single crystals, the diffraction peak from each set of atomic planes (*hkl*) yields only one data point. To obtain six independent strain values, one needs to use six independent reflections[2] (here, all sets that are members of the {hkl} family that are not parallel, such as *hkl* and $\bar{h}kl$, are considered independent reflections as well). Furthermore, in the case of single crystals, the angles ϕ, ψ are determined by the symmetry of the crystal under examination. Amorphous materials cannot be analyzed since there are no diffraction peaks. However, if one could measure strains along various directions in an amorphous material by some method, one could use equations (2) to (3) to obtain the stresses in the S_i system. Consequently, for all materials classes that yield diffraction peaks, the same variation of $d_{\phi\psi}$ with ϕ and ψ is predicted by equation 2-b: If the shear terms ε_{31}, ε_{32} are zero, then the variation of $d_{\phi\psi}$ with $\sin^2\psi$ is linear (Figure 2-a). If these shear terms are finite, the $d_{\phi\psi}$ vs. $\sin^2\psi$ will exhibit curvature and "ψ-splitting" (Figure 2-b). Experimental data that exhibit such "regular" variation with $\sin^2\psi$ can, thus, be treated using equations 2-3. Oscillatory $d_{\phi\psi}$ vs. $\sin^2\psi$ behavior (Figure 2-c), frequently observed experimentally, cannot be predicted or analyzed without further extension of the basic formalism.

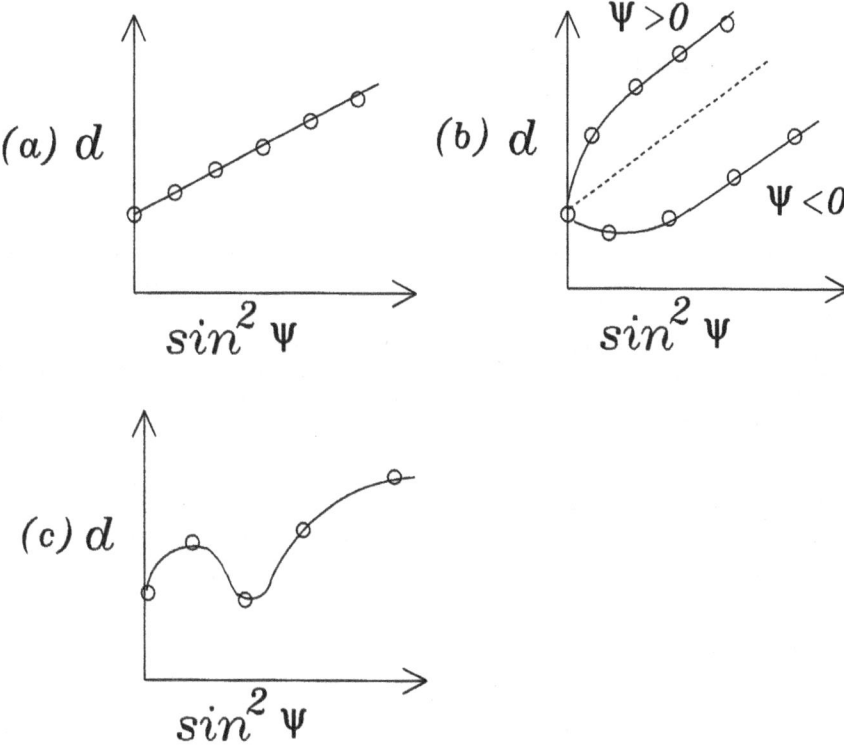

Figure 2. Types of $d_{\phi\psi}$ vs. $\sin^2\psi$ plots: Figures 2-a, 2-b depict regular behavior which can be predicted by equation 2-b. Figure 2-c depicts oscillatory behavior.

1.2 THE DIFFRACTION AVERAGE

The fundamental causes of oscillatory $d_{\phi\psi}$ vs. $sin^2\psi$ behavior can be discussed best by analyzing the way diffraction obtains strain data from polycrystalline specimens [2,7,8]. In addition to defining the direction of the vector $L_{\phi\psi}$ along which the strains ε_{ij}' are measured, diffraction conditions also define the material volume from which this information is obtained. This volume depends on the type of the material, the radiation employed and the diffraction geometry. In case of polycrystalline materials, different sets of grains diffract at the different ψ tilts (Figure 1). Thus, the plane spacing at any ψ tilt is an average value, where the average is taken over all the grains in the particular set. This average can be expressed as:

$$<d>_{\psi_n} = \frac{\sum_{k=1}^{N_j}\left\{\frac{\int_{V_k^{coh.}} d(x,y,z)_k f(z)dV}{\int_{V_k^{coh.}} f(z)dV}\right\}}{\sum_{k=1}^{N_j} V_k^{coh.}}$$

(4)

Here $d(x,y,z)$ is the plane spacing at point (x,y,z) in the kth grain diffracting at $\psi = \psi_n$. $V_k^{coh.}$ is the volume within this grain that is contributing to the position of the diffraction peak (which means that regions such as grain boundaries that have rapidly varying strains will be excluded), $f(z)$ is the (absorption) function relating the variation of diffracted intensity to depth $z(\psi)$ for the particular ψ tilt and the summation is over the grains diffracting at the tilt angle ψ_n. Since the summation is over different sets of grains for each ψ tilt, the plane spacings, $d_{\phi\psi}$, measured at the various ψ tilts originate from subsets of the total irradiated volume that may be completely or partially mutually exclusive[II]. Even though these subsets are easily separated in diffraction space, they cannot be easily identified in real space: The constituent grains of these subsets are intimately mixed, and not all regions within a correctly oriented grain contribute to the diffraction profile. Thus, for polycrystalline specimens with grain sizes much smaller than the beam size, the diffraction average at any given ψ tilt can not be mapped into a discrete continuous volume in real space.

[II] For any reflection hkl, the diffracting subsets are mutually exclusive and independent if the angles between the poles of the members of the {hkl} family do not coincide with the differences between the ψ angles used in the measurement. For example, in the cubic system, the angles between the h00 poles are 90°. Thus, for all h00 reflections, the subsets diffracting at all ψ tilts between 0 to 89 degrees are completely independent. For the general case, one can easily calculate the angles between the poles of the {hkl} family of planes, and design the experiment such that at the ψ angles used, the diffracting subsets are mutually exclusive. On the other hand, if the angles between any two poles of the {hkl} planes coincide with the difference between any two ψ tilts, strain information from some of the grains may be contained in multiple ψ tilts. Such volumes are termed "partially mutually exclusive" in this article since volumes diffracting at these ψ tilts may also have regions that are independent.

Because of these considerations, if one uses six independent plane spacings on the left side of equations (2) and solves for ε_{ij} in the S_i system, one is implicitly making two assumptions:

1. The average strains obtained by substituting the $d_{\phi\psi}$ from equation (4) into equation (1) are tensor quantities.
2. The strains in the $\tilde{S}_i$ coordinate system are homogeneous within the plane defined by $\tilde{S}_1$ and S_2.

To determine how well these assumptions correspond to reality, the local and average stress/strain states in mechanically loaded polycrystals must be considered.

1.3 THE STRAIN/STRESS STATE IN POLYCRYSTALLINE MATERIALS UNDER LOAD

It is well known that the mechanical response of polycrystalline materials is inhomogeneous on a scale comparable to the grain size [6,11-15]. Consider the case where a homogeneous traction σ^0 is applied to the specimen along S_1 (thus $\phi = 0$). At a point $A(x,y,z)$ in the material, any component of the strain tensor can be expressed as[8,9,16]:

$$\varepsilon_{ij}^t(x,y,z) \;=\; \varepsilon_{ij}^0 + \varepsilon_{ij}^{in}(x,y,z) + \varepsilon_{ij}^r(x,y,z) \tag{5}$$

where ε_{ij}^0 is the homogeneous elastic strain that would be observed if the stress σ_{ii}^0 was applied to a homogeneous isotropic material, ε_{ij}^{in} is the interaction term describing the strains arising from elastic incompatibility (i.e variation of elastic moduli with distance around point A), and ε_{ij}^r is the residual strain at A (which is dependent on the deformation history of the specimen). The strain components can be written in terms of the stress σ_{11}^0 through the isotropic Hooke's law[III]:

$$\varepsilon_{11}(x,y,z) \;=\; \sigma_{11}^0 \left[\frac{1}{E} + K_1(x,y,z) \right] + \varepsilon_{11}^r(x,y,z),$$

$$\varepsilon_{22}(x,y,z) \;=\; \sigma_{11}^0 \left[-\frac{v}{E} + K_2(x,y,z) \right] + \varepsilon_{22}^r(x,y,z), \tag{6}$$

$$\varepsilon_{33}(x,y,z) \;=\; \sigma_{11}^0 \left[-\frac{v}{E} + K_3(x,y,z) \right] + \varepsilon_{33}^r(x,y,z).$$

Here E, v, are the Young's modulus and Poisson's ratio for the (isotropic) material and $K_i(x,y,z)$ are the proportionality constants between the applied load and the resulting reaction strains at point A. These proportionality constants depend on the shape, size and orientation of the regions surrounding point A. In the case of isotropic matrix and precipitates, where the precipitates are ellipsoidal in shape, $K_i(x,y,z)$ can be written in terms of the Eshelby tensors[6]. The general case has not yet been expressed in a similar analytical expression. The local (elastic) residual strain due to previous plastic deformation, $\varepsilon_{ij}^r(x,y,z)$, forms due to the mutual elastic constraint of local regions during inhomogeneous plastic flow. For example, such partitioning may be due to the difference in the yield points of the interior of the grain and the regions bordering the grain boundaries.

Thus, in the general case, the diffraction average of the local strains $\varepsilon_{ij}(x,y,z)$ given by equation (6) should be substituted for the strains ε_{ij} on the right side of equation (2). This mean strain $< \overline{\varepsilon_{li}} >$ within the diffracting crystallites at any ψ tilt can be expressed by an equation similar to equation (4):

[III] In this equation, it is assumed that the cross terms $\varepsilon_{ij}(x,y,z) = \sigma_{11}^0 K_{ij}(x,y,z)$ (where $i \neq j$) are zero for ease of treatment. If these components are finite, the treatment will follow along the same lines but will contain additional terms.

$$
<\overline{\varepsilon_{ii}^{t}}> = \cfrac{\sum_{k=1}^{N_{j}}\left\{(\int_{V_{k}^{coh.}}(\varepsilon_{ij}^{t}(x,y,z))_{k}\,f(z)dV)\div(\int_{V_{k}^{coh.}}f(z)dV)\right\}}{\left\{\sum_{k=1}^{N_{j}}V_{k}^{coh.}\right\}},
\qquad (7-a)
$$

$$
= \sigma_{ii}^{0}[M + \overline{K}_{i}(\psi)] + <\overline{\varepsilon_{ij}^{r}}>_{\psi}
\qquad\qquad
M = \begin{cases} \dfrac{1}{E} & i = 1 \\[2mm] -\dfrac{v}{E} & i \neq 1 \end{cases}.
\qquad (7-b)
$$

In (7-b), the terms $\overline{K}_{i}(\psi)$ describe the average inhomogeneous elastic response of the subset of grains diffracting at ψ_{n} to an applied stress σ^{0}, and $<\overline{\varepsilon_{ij}^{r}}>_{\psi}$ is the average residual stress within this subset. It must be noted that equations (5-7) are also valid for those specimens which do not have an applied stress at the boundaries but do contain residual macrostresses[2]. In such a case, the term σ_{ii}^{0} is the macrostress term, applied by the macroscopic constraint of one part of the body on another (such as the constraint the bulk of the material applies on the surface layers of a shot-peened specimen).

From equations (5-7) and (1-2) one can obtain the relationship between the local strain distribution within the diffraction volume and the average strain measured by diffraction along the direction $L_{\phi\psi}$[2,8-10]:

$$
\frac{<d>_{\psi_{n}} - d_{0}}{d_{0}} = \{\sigma_{11}^{0}\{\frac{1+v}{E} + \overline{K}_{1}(\psi) + \overline{K}_{3}(\psi)\} + <\overline{\varepsilon_{11}^{r}}>_{\psi} - <\overline{\varepsilon_{33}^{r}}>_{\psi}\}\sin^{2}\psi +
$$

$$
\sigma_{11}^{0}\{-\frac{v}{E} + \overline{K}_{3}(\psi)\} + <\overline{\varepsilon_{33}^{r}}>_{\psi}.
\qquad (8)
$$

In contrast with equation (2-b), equation (8) can describe both regular and oscillatory variation of $d_{\phi\psi}$ with $\sin^{2}\psi$. Equation (8) predicts linear variation of $d_{\phi\psi}$ with $\sin^{2}\psi$, if both the average elastic interaction parameters ($\overline{K}_{i}(\psi)$) and the average residual strain terms due to previous treatment ($<\overline{\varepsilon_{ij}^{r}}>_{\psi}$) are constant with ψ. Consider first the elastic interaction terms. $\overline{K}_{i}(\psi)$ can be constant with ψ for two cases:

1. The material is truly homogeneous and isotropic ($K_{i}(x,y,z) \equiv 0$).

2. The local elastic response of the material is inhomogeneous ($K_{i}(x,y,z) \neq 0$), but the average interaction constants are equal for all ψ tilts ($\overline{K}_{i}(\psi) \neq f(\psi)$). Here, the average response coefficients, $\overline{K}_{i}(\psi)$, might still be zero, even though $K_{i}(x,y,z)$ are not. Then, the average response of the material will be equal to that of a truly isotropic material. If $\overline{K}_{i}(\psi)$ is finite and constant with ψ, the Young's modulus and Poisson's ratio measured by diffraction techniques will be different from those calculated from single crystal constants using the standard (Voigt, Reuss, Neerfeld-Hill or Kroner) approaches[2,4,6,9,10].

Similar considerations also apply to the variation of $<\overline{\varepsilon_{ij}^{r}}>_{\psi}$ and $\varepsilon_{ij}^{r}(x,y,z)$ with ψ.

On the other hand, if either $\overline{K}_{i}(\psi)$ or $<\overline{\varepsilon_{ij}^{r}}>_{\psi}$ in equation (8) are functions of ψ, $d_{\phi\psi}$ vs. $\sin^{2}\psi$ behavior will be oscillatory. The variation of these terms with ψ indicates that the average strains in the sample coordinate system are inhomogeneous (equation 7-b). Such inhomogeneity may be due to elastic interaction effects ($\overline{K}_i(\psi) = f(\psi)$), or due to the inhomogeneous distribution of plastic deformation, which then causes the formation of an inhomogeneous residual strain field in the S_i coordinate system ($< \overline{\varepsilon_{ij}} >_\psi = f(\psi)$). Because of the complexity of the deformation fields encountered in manufacturing processes, the experimental separation of elastic and plastic contributions to oscillatory behavior is not always possible. For those specimens which can be mechanically loaded in-situ on a diffractometer, one can determine the average "Poisson's ratio" for the sets of grains diffracting at various ψ tilts from which the average interaction coefficients $\overline{K}_i(\psi)$ may be calculated [16]. This method yields a measure of the inhomogeneous elastic response to a (known) applied load. There is no comparable technique for separating the effect of inhomogeneous plastic flow on oscillations in $d_{\phi\psi}$ vs. $sin^2\psi$.

2. Numerical Modelling and FEM Analysis

One approach that can be used to identify the contributions of elastic incompatibility and plastic inhomogeneity to oscillations in $d_{\phi\psi}$ vs. $sin^2\psi$, is to model the mechanical response of a hypothetical specimen with known local elastic constants and yield parameters into the plastic regime. One can then obtain the local stress/strain values from those crystallites that diffract at a given ψ, form the ψ-average, calculate the $d_{\phi\psi}$ vs. $sin^2\psi$, and correlate the oscillations in the $d_{\phi\psi}$ vs. $sin^2\psi$ to the local and average stress/strain values calculated in the model[17,18].

2.1 MATERIAL DEFINITION

Cu, which has a high relative degree of anisotropy was selected for the simulation. This ensured a high degree of elastic incompatibility across grain boundaries and maximized $\overline{K}_i(\psi)$. The (311) reflection was chosen as the diffraction peak. This peak is from a non-special low index family and can be used for stress analysis with Fe K_α radiation. For $\phi = 0$, the grains diffracting at various ψ-tilts were defined in the following manner: Since Cu has cubic symmetry, the lattice direction $R = h\vec{x}_1 + k\vec{x}_2 + l\vec{x}_3$ must be coincident with $L_{\phi\psi}$ for any grain diffracting with the hkl reflection at tilt ψ. Then, the lattice direction $D = a\vec{x}_1 + b\vec{x}_2 + c\vec{x}_3$ coincident with the load direction (in this case S_1), must satisfy the dot product:

$$\cos \{90 - \psi\} = \frac{ha + bk + cl}{\sqrt{h^2 + k^2 + l^2} \sqrt{a^2 + b^2 + c^2}} \tag{9}$$

The unknown indices (a,b,c) of the lattice vector $\vec{D}$ coincident with $\vec{S}_1$ can be determined from equation 9. Since, in the case of a polycrystal, equation (9) can be satisfied by many sets of (a,b,c) for a given ψ, it was solved numerically for the ψ-tilts 18.43°, 26.57°, 33.21°, 39.23°, 45.00°, 56.79° and 71.57° for all integers between -40 to 40. For $\psi = 0$, the range -5 to 5 was used*. Since the angles between the poles of the {311} family (ΔP) are 0°, 35.10°, 50.48°, 62.96° and 84.78°, ($\psi_k - \psi_l$) ≠ ΔP for any combination. Thus, the diffraction volumes defined are completely mutually exclusive. The resulting matrix of S_1 for each ψ-tilt under analysis is shown in Table I.

The lattice vectors along S_2 and S_3 axes are defined as follows: Since the load is assumed to be applied along S_1, $\phi = 0$ and the transverse direction S_2 is normal to the plane defined by S_1 and

---

* For $\psi \neq 0$, only a small number of (a,b,c) could satisfy *Eq (9)*, while for $\psi = 0$, there were a large number of fits.

TABLE I: Lattice directions coincident with $\vec{S}_1$ for the Cu grains diffracting at various tilt angles.

ψ	Lattice Vector along $\vec{S}_1$				
0	$(\bar{1},1,2)$	$(\bar{1},\bar{2},5)$	$(\bar{1},0,3)$	$(0,\bar{1},1)$	$(3,\bar{4},5)$
18.43	$(\bar{2},9,38)$	$(27,\overline{20},\overline{20})$	$(0,1,20)$	$(12,\overline{16},1)$	
26.57	$(1,19,27)$	$(17,\overline{21},19)$	$(25,5,\overline{21})$	$(7,\bar{8},18)$	
33.21	$(27,\overline{25},28)$	$(25,\overline{23},17)$	$(3,14,16)$	$(18,4,\overline{11})$	
39.23	$(10,7,\overline{29})$	$(7,6,\bar{5})$	$(26,\overline{17},5)$	$(5,\bar{2},9)$	
45.00	$(39,22,\overline{22})$	$(14,31,22)$	$(10,7,22)$	$(18,\overline{10},7)$	
56.79	$(32,3,37)$	$(23,21,30)$	$(31,\bar{3},30)$		
71.57	$(40,27,27)$	$(31,0,7)$	$(29,0,13)$	$(28,1,15)$	
90.00	$(3,1,1)$				

$\vec{L}_{\phi\psi}$(Figure 1). Thus, the lattice vectors along $\vec{S}_2$, $\vec{S}_3$ can be determined from the cross products:

$$\vec{S}_2 = \vec{D} \times \vec{R} \quad ; \quad \vec{S}_3 = \vec{S}_1 \times \vec{S}_2 \tag{10}$$

From equations (9) and (10) the specimen coordinate system in any grain diffracting at a particular angle ψ can be defined in terms of the lattice directions. Equation (10), when applied to Table I, yields 32 grains that are completely defined with respect to the diffraction stress analysis. For example, for a Cu grain diffracting at $\psi = 0$ with the 311 reflection, $S_1 = (\bar{1}, 1, 2)$, $S_2 = (1, \bar{7}, 4)$ and $S_3 = (\overline{18}, \bar{6}, \bar{6})$. The stress/strain response of these grains were then examined in various coupling and load configurations.

2.2 REUSS APPROXIMATION

If these grains are under an applied stress σ^l_{11} in a material that obeys the Reuss approximation (i.e. constant stress in all the crystallites with no interaction effects[2,4,5,17]), the strain along the (311) direction for each crystallite can be determined by calculating the strains in the S_i coordinate system through Hooke's law in terms of the compliances:

$$\varepsilon_{ij} = S_{ijkl}\sigma_{kl} \tag{11}$$

and using the second-rank tensor transformation given by equation (2). The plane spacing along $L_{\phi\psi}$ is then determined from equation 1. The results of this calculation for the matrix given in Table I is shown in Figure 3 for an applied stress $\sigma^l_{11} = 100MPa$. As expected, the d_ψ values from different grains diffracting at a given ψ tilt span a significant range. At $\psi = 0$, the maximum difference is 0.0005 Å, which corresponds to a strain difference of .05%. The actual $d_{\phi\psi}$ vs. $sin^2\psi$ plot obtained from such a specimen would be the mean value of the d_ψ at each ψ tilt, which is shown in as the dashed line in Figure 3. If a linear least-squares approximation is fitted to this "mean" data over the entire ψ-range; $\psi\varepsilon(0,71.53°)$, or over a narrower range; $\psi\varepsilon(0,56°)$ and the traditional analysis is applied, one obtains for σ_{11} 106 and 110 MPa, respectively, which are within the normal error limits of a typical measurement. A similar result is observed even when the oscillations are more severe. If the biaxial stress tensor

$$\sigma_{ij} = \begin{bmatrix} 100 & 0 & 0 \\ 0 & 100 & 0 \\ 0 & 0 & 0 \end{bmatrix}$$

is applied to the same material, the variation of strain from grain to grain is significantly larger as shown in Figure (4-a). The variation of the mean plane spacing with $sin^2\psi$ is also highly oscillatory. However, conventional analysis applied to the mean $d_{\phi\psi}$ vs. $sin^2\psi$ trace shown in Figure 4-a yields 88 MPa for σ_{11}, which is well within the typical error limit for the technique.

The results summarized above seem to indicate that conventional analysis applied to oscillatory $d_{\phi\psi}$ vs. $sin^2\psi$ yields acceptable results. However, texture effects are not at all taken into account in the above calculation. With texture, certain grains will be more populous at certain ψ tilts. Thus, any curve that connects the points in the the the area bounded by the dashed lines in figure 4-a can be observed, where the points to be connected are selected by the texture function. Such selectivity can yield very erroneous results. In figure 4-b, two such selected $d_{\phi\psi}$ vs. $sin^2\psi$ traces are shown. Both of these traces exhibit significantly better linearity vs. $sin^2\psi$ than the mean $d_{\phi\psi}$ vs. $sin^2\psi$ trace in Figure 4-a. However, the upper trace in Figure 4-b exhibits negative slope (which yields a *compressive* stress of -83 MPa), while the stress calculated from the slope of the lower trace is highly tensile (246 MPa). Thus, the error in both of these cases is well over 100% and the sign is wrong in the first case. In comparison, the error in the stress obtained from the mean $d_{\phi\psi}$ vs. $sin^2\psi$ shown in figure 4-a was only 12 MPa, even though the trace was highly oscillatory. The reason for such anomalous behavior lies in the number and type of grains contributing to the average values at each ψ tilt. In Figure 4-a, the four grains contributing to the average plane spacing at each ψ tilt have crystallographic orientations along S_1 that vary from those close to the 111 to those close to the 100 . Thus, a more or less random average is taken at each ψ tilt. Texture, on the other hand, results in a non-random average, with a concomitant error in the stress value if the traditional analysis with

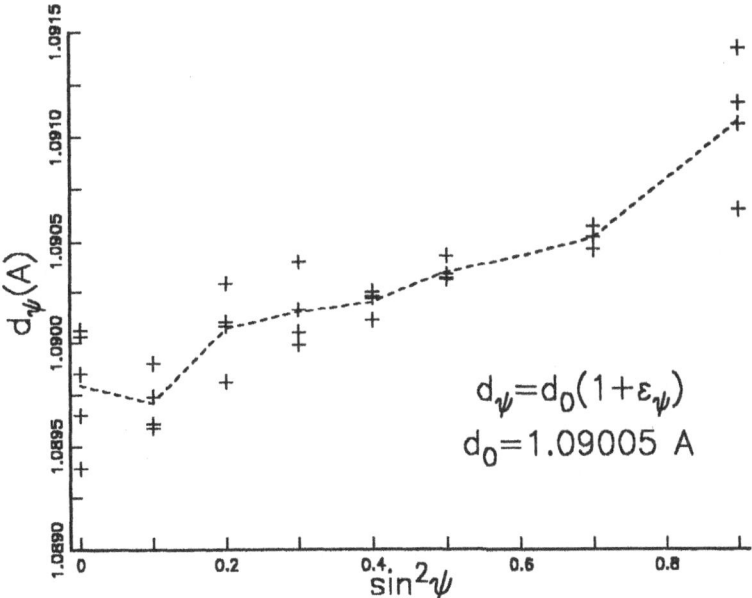

Figure 3. Variation of d_ψ with $sin^2\psi$ for the individual grains (+), and of the ψ-average (---) for the set of grains shown in Table I (Reuss limit, $\sigma^0_{11} = 100$ Mpa).

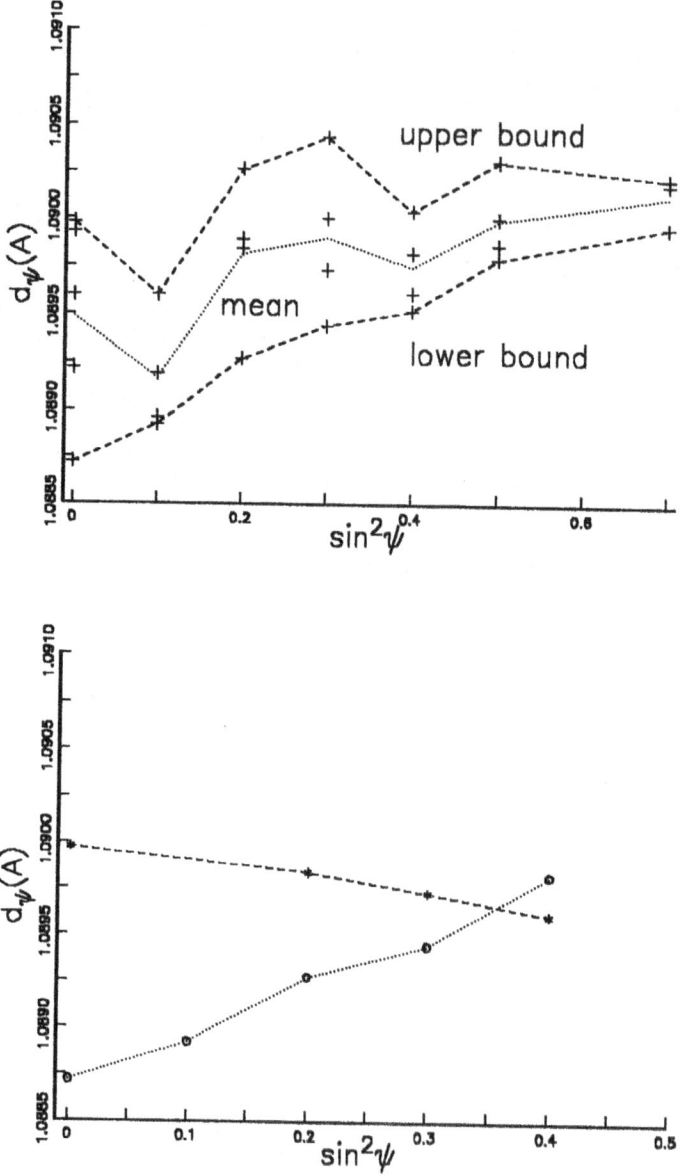

Figure 4. Variation of d_ψ with $\sin^2\psi$ for the set of grains shown in Table I (Reuss limit, bi-axial load; $\sigma^0{}_{11} = \sigma^0{}_{22} = 100$ MPa). In figure 4-a, the traces show the upper, lower and mean $d_{\phi\psi}$ vs. $\sin^2\psi$, while in figure 4-b two extreme cases that can be enhanced by texture are depicted.

isotropic elastic constants are used. Another conclusion from figure 4-b is that, one should treat even linear $d_{\phi\psi}$ vs. $sin^2\psi$ from textured specimens with caution.

2.3 FINITE ELEMENT MODELLING

The Reuss approximation, if obeyed in real materials, would cause voids between grains, since this model does not provide for constant displacements across grain boundaries [5]. This "compatibility" requirement causes the reaction strains/stresses that are represented by the terms "$\varepsilon_{ij}^{r}(x,y,z)$" and "$K_i(x,y,z)\ \sigma^0$" in equations (5) and (6). The magnitude and sign of the reaction stresses/strains and their effects on oscillations in $d_{\phi\psi}$ vs. $sin^2\psi$ was modelled by using finite element (FEM) analysis [17,18]. For these calculations, the two grains for each ψ tilt that showed the maximum difference in the effective elastic modulus along S_1 were placed at random (via a random number generator) in a 100 grain mesh. To account for some degree of inhomogeneity in plastic flow, the Schmid factors $m = cos\ \chi\ cos\ \gamma$, which depend on the orientation of the slip plane (defined by the angle χ) and the slip direction (defined by the angle γ with respect to the load direction) were also calculated and inputted into the model such that yield would start in the particular grains when the applied load exceeded the critical resolved shear stress. The remaining (non-diffracting) grains were assumed to be isotropic in elasticity and in yield. The effective elastic constants and Schmid factors for the diffracting grains and the average values for the non-diffracting grains are shown in Table II.

This mesh was stepwise loaded along $\vec{S_1}$ up to 345 MPa, which corresponded to $\simeq$ 8% total strain (elastic plus plastic) and unloaded**. The variation of the total FEM stress in the diffracting grains along S_1 is shown in figures (5 a-d) for various applied stress values. The deviation from the applied stress in the individual grains is due to the presence of interaction stresses which are calculated by the FEM code as it keeps the displacements across the element (grain) boundaries constant. The interaction stresses increase with increasing load in the elastic regime (Figures 5a,b). Plastic flow decreases the "net" deviation from the applied stress (Figure 5-c). That is, the sum of the interaction stresses (due to elastic incompatibility and plastic inhomogeneity) and the applied stress is close to the net applied stress in all grains. However, due to the inhomogeneous partitioning of yielding, there may be significant residual stresses $<\overline{\varepsilon_{ij}}>_\psi$ in the grains. These "Heyn" stresses[11,12,19] manifest themselves after the mesh is totally unloaded (figure 5-d). The Heyn stresses are biaxial and change sign from grain to grain (as is required by the equations of equilibrium[2,5]). One can see from Figure (5) that the magnitude of these stresses can be quite large (-75 to 125 MPa in this model).

The $d_{\phi\psi}$ vs. $sin^2\psi$ plots corresponding to the stress fields shown in Figures 5a-d are shown in Figures 6a-d. Again, a range of $d_{\phi\psi}$ vs. $sin^2\psi$ are possible within the area bounded by the data points and the particular $d_{\phi\psi}$ vs. $sin^2\psi$ observed from a given specimen will depend on the populations of the respective grains diffracting at each ψ tilt. If a straight line is fitted to the mean of the plane spacings at each ψ tilt, the traditional analysis (equations 2-b and 3) yields quite reasonable values (Table III). However, this mean is for a specimen in which the weakest grains (100 along the loading direction) are as probable as the strongest grains (111 along the loading direction). Again, this approximates a random specimen. Thus, the deviations in Table III cannot be extrapolated to the expected results from a textured specimen. Another conclusion that can be drawn from comparing figures 5 and 6 is that, the deviation of a given point $d_{\phi\psi}$ from linearity cannot be directly related to the average stress present in the diffracting crystallites at that ψ tilt. For example, compare the first two points in Figures 5-b and 6-b: In both cases the total spread in net stress is about 100MPa. However, the $d_{\phi\psi}$ vs. $sin^2\psi$ values from the individual crystallites at $sin^2\psi = 0$ are much further apart than those at $sin^2\psi = 0.1$.

**Further details about the modelling are given in references 17, 18.

62

TABLE II: Material constants used in the FEM model.

ψ	$\vec{S}_1$	$E_{\vec{S}_1}$(GPa)	Sch. F.	$\vec{S}_1$	$E_{\vec{S}_1}$(GPa)	Sch. F.
0	$(3,\bar{4},\bar{5})$	167.0	0.39	$(\bar{1},0,3)$	80.9	0.49
18.43	$(27,\overline{20},\overline{20})$	175.9	0.34	$(\bar{2},9,38)$	74.3	0.48
26.57	$(17,\overline{21},19)$	186.0	0.33	$(7,\bar{8},18)$	112.2	0.44
33.21	$(25,\overline{23},17)$	176.8	0.37	$(18,\bar{4},\overline{11})$	116.1	0.49
39.23	$(7,6,\bar{5})$	179.1	0.36	$(10,7,29)$	90.2	0.48
45.00	$(14,31,22)$	145.8	0.44	$(10,7,22)$	106.7	0.47
56.79	$(23,21,30)$	174.5	0.36	$(32,3,37)$	128.7	0.45

Isotropic Values:
  $\bar{E} = 129.8$(GPa) $\sigma_{yield} = 255$ (MPA)   Strain Hard. Expon. = 0.3

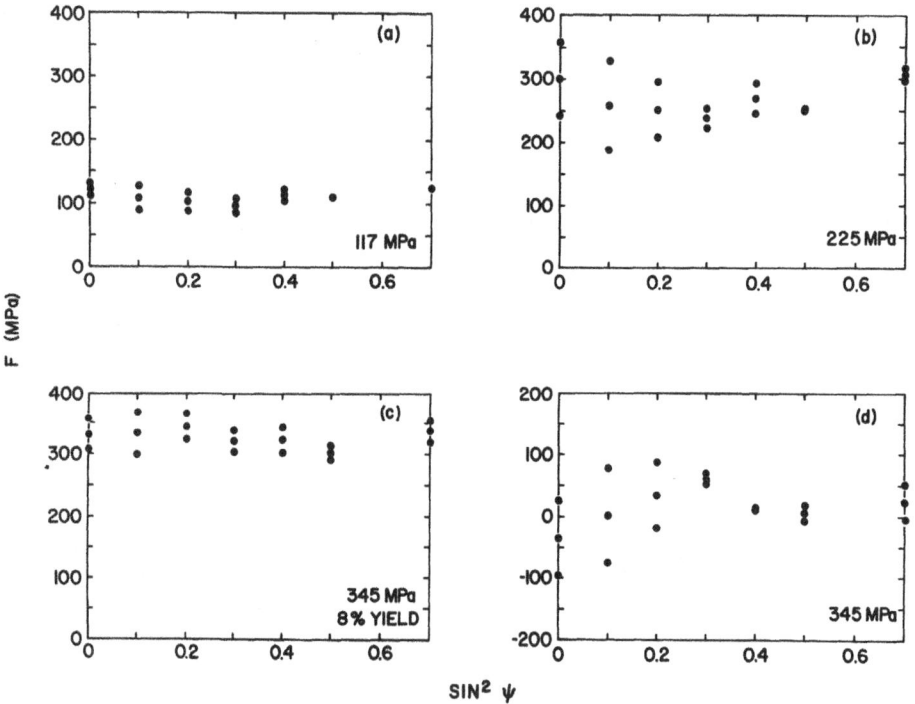

Figure 5. The variation of the total (applied plus interaction) stress in the diffracting grains of the FEM mesh as a function of the applied load (shown in the lower right corner of each plot).

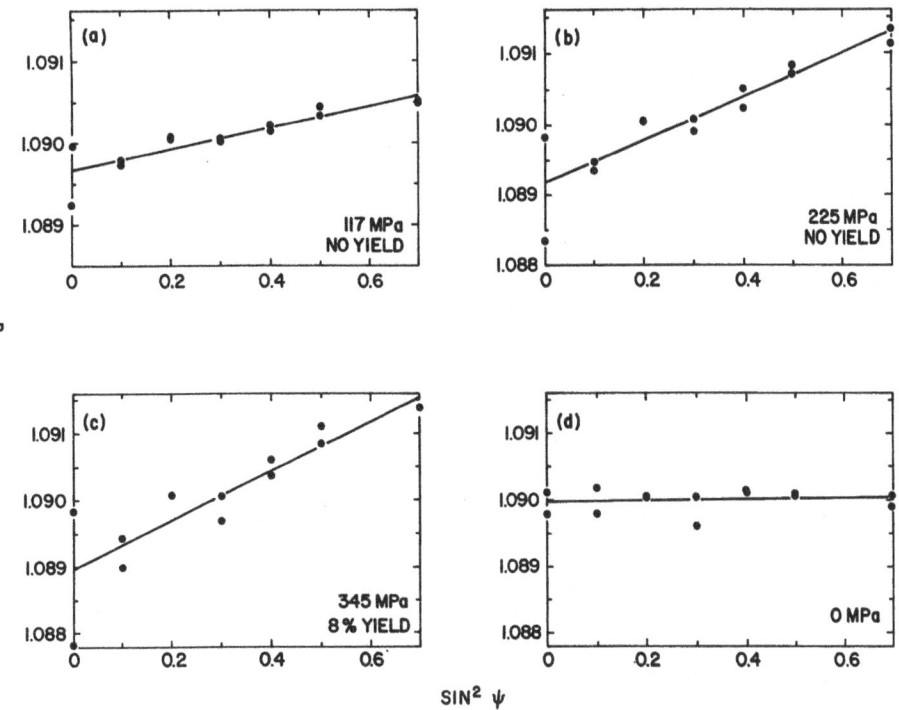

Figure 6. The $d_{\phi\psi}$ vs. $sin^2\psi$ plots corresponding to the stress profiles shown in figure 5.

TABLE III: Applied loads (σ^0_{11}) and the stresses determined from the slope of the (simulated) $d_{\phi\psi}$ vs. $sin^2\psi$ data obtained from the FEM modelling.

σ^0_{11} (MPa)	σ_{x-ray}(MPa)	Error(MPa)
117	104	13
255	243	12
345	297	48
0 (relaxed)	7	7

3. Summary and Conclusions

It should be emphasized that the models presented in the above treatment are just a first order approximation of the complex stress/strain states that exist in real polycrystalline specimens. For a better approximation, actual grain shapes rather than rectangular elements could be used, slip systems and parameters closer to, say, the Taylor slip model[20]could be defined and one could attempt to model a "real" orientation distribution for the set of grains, as opposed to a calculated one, which is the approach used here. However, the formalism described above is a self-consistent check of the diffraction stress analysis, where one obtains the local stress/strain state in a hypothetical specimen from the FEM analysis, calculates the d_ψ in various grains, forms the $d_{\phi\psi}$ vs. $sin^2\psi$ data, analyzes the $d_{\phi\psi}$ vs. $sin^2\psi$ data with the traditional stress analysis technique and compares the result to the input. As such, the results may serve as an existence proof of certain pitfalls inherent in the technique. The results may be summarized as follows:

1. Interaction stresses caused by elastic and plastic inhomogeneity and incompatibility effects can be comparable in magnitude even for simple uniaxial testing involving plastic deformation.

2. The diffraction average of such inhomogeneity effects cause oscillations in d_ψ vs. $sin^2\psi$ plots.

3. The strain distribution is inhomogeneous within the grains diffracting at a given ψ tilt as well as those diffracting at different ψ.

4. Analysis of oscillatory $d_{\phi\psi}$ vs. $sin^2\psi$ plots from textured samples assuming "regular" $d_{\phi\psi}$ vs. $sin^2\psi$ behavior can be very misleading.

5. Similarly, "linear" $d_{\phi\psi}$ vs. $sin^2\psi$ plots from some reflections of a highly textured specimen might yield erroneous results. Such analysis might even yield an incorrect sign for the stress field(Figure 4-b).

6. On the other hand, for non-textured specimens, application of the traditional $d_{\phi\psi}$ vs. $sin^2\psi$ analysis to oscillatory data seems to yield acceptable results when there is an applied traction on the specimen.

4. References

1. Lester, H. H., Aborn, R. H. (1925-1926), 'Behaviour of iron crystals under load', Army Ordnance 6, 120-364.

2. Noyan, I. C. and Cohen J. B. (1987) 'Residual Stress, Measurement by Diffraction and Interpretation' Springer, New York.

3. Cohen, J. B., Dolle, H., James, M. R. (1980), 'Determining stresses from x-ray powder patterns', NBS Special Publication 567, 453-477.

4. Dolle, H. (1979) 'Influence of multiaxial stress states, stress gradients and elastic anisotropy on the evaluation of residual stresses with x-rays', J. App. Cryst., 12, 489-501.

5. Mura, T. (1982), 'Micromechanics of Defects in Solids', Martinus-Nijhof Publishers, Hague.

6. Schwartz, L. L. and Cohen, J. B., (1987), 'Diffraction from Materials', Springer, New York.

7. Warren, B. E. (1969), 'X-ray Diffraction', Addison-Wesley, Reading, MA.

8. Noyan, I. C., Cohen, J. B., (1984), "Determining Stresses in the Presence of Oscillations in Interplanar Spacing vs. sin psi", Adv. in X-ray Analysis, Vol. 27, 129-148.

9. Noyan, I. C., (1985), "Determination of Elastic Constants of Inhomogeneous Materials with X-ray Diffraction", Mat. Sci. & Eng., 75.

10. Zhong, R. M., Noyan, I. C., Cohen, J. B. (1986), "X-ray elastic constants and their meaning for Al and Fe', Adv. in X-ray Analysis, Vol. 29, 17-20

11. Masing, G. (1923), 'Zur Heyn'ischen Theorie der Verfestigung der Metalle durch verborgene elastische Spannungen', Wiss. Veroff. Siemens Konz., 3, 231-239.

12. Masing, G. (1926), 'Berechnung von Dehnungs und Stauchungslinien auf Grund von inneren Spannungen', Wiss. Veroff. Siemens Konz., 6, 135-141.

13. Boas, W. and Hargreaves, M. E., (1948), 'On the inhomogeneity of plastic deformation in the crystals of an aggregate', Proc. Roy. Soc., A193, 89-97

14. Deve, H., Harren, S., McCullough, C. and Asaro, R. J. (1988), 'Micro and macroscopic aspects of shear band formation in internally nitrided single crystals of Fe-Ti-Mn alloys', Acta. Metall. 36, 341-365

15. Morris, W. L., Cox, B. N., James, M. R. (1987), 'Microplastic surface deformation of Al 2219-T851', Acta Metall. 35, 1055-1065

16. Schadler, L. S. and Noyan, I. C., (1991)'A study of the inhomogeneous deformation of single phase α-brass', IBM Research Report, RC 16827, T. J. Watson Research Center, NY

17. Noyan, I. C. and Nguyen, L. T. (1988), 'Oscillations in interplanar spacing vs. $\sin^2\psi$, a FEM analysis', Adv. in X-Ray Analysis, 31,91-204

18. Noyan, I. C., Nguyen, L. T., (1989), 'Effect of plastic deformation on oscillations in "d" vs. $\sin^2\psi$ plots', Adv. in X-ray Analysis, 32, 355-364

19. Berveiller, M., Krier, J., Ruppersberg, H., Wagner, C. N. J. (1991) 'Theoretical investigation of ψ-splitting after plastic deformation of two-phase materials', Proc. of ICOTOM9, in print

20. Taylor, G. I., (1938), 'Plastic strain in metals', J. Inst. Metals, 62 307-324

Acknowledgements

The author acknowledges Prof. J. B. Cohen (Northwestern University, Evanston, IL.), with whom the author developed some of the concepts reviewed in this paper. The FEM analysis data is based on two collaborative papers (references 17,18) with Dr. L. T. Nguyen (National Semiconductor, Santa Clara, CA.) Thanks are due to Dr. L. S. Schadler (IBM Research, Yorktown) for extensive discussions and comments.

THE CALCULATION OF RESIDUAL STRESS

SÖREN SJÖSTRÖM
Department of Mechanical Engineering
Linköping University
S-581 83 LINKÖPING
Sweden

ABSTRACT. The calculation of residual stress states in four different important situations will be treated, namely heat treatment of steel by quenching, welding, rolling contact loading, and thermal cycling of metal-matrix composites.

1. GENERAL COMMENTS ON RESIDUAL STRESSES.

Residual stresses are stresses that exist in a component when the component is unaffected by any external load. Normally, one considers thermal load as an external load, which, in case of a component made of inhomogeneous material, necessitates the definition of the stress-free temperature as the 'natural' temperature.

   Basically, the calculation of residual stresses does not differ from the calculation of any other stresses. One uses the analytical or numerical tools necessary to analyse the stress state resulting from the complete load history that the component is exposed to; if, after the final unloading, there is still a stress state in the component, this is by definition a residual stress state.

   However, in this description is, in fact, buried information that actually makes residual stresses a little different. The very name 'residual stress' points at the fact that the stress state is a remainder of the stress history of the component. This leads us to the conclusion that the history cannot have been elastic (since an elastic history would, by definition, leave no trace of the load history in the component). Therefore, in all cases leading to a residual stress state, the component must have been subjected to some nonelastic process. In analysing residual stress states, one must therefore deal with nonelastic constitutive descriptions, which, in turn, generally leads to more difficult analyses.

   Among the most important residual stresses are those which originate from the manufacture of the component. Rolling, forging, casting, welding, heat treatment, grinding, and shot peening are examples of processes that give residual stress states. Of these, we will concentrate on two, namely, heat treatment of steel by quenching, and welding. Fur-

M. T. Hutchings and A. D. Krawitz (ed.),
Measurement of Residual and Applied Stress Using Neutron Diffraction, 67–91.
© 1992 Kluwer Academic Publishers.

ther, we will give a few examples of the calculation of residual stresses that arise during the use of the component, namely rolling contact residual stress, and residual stress state in the matrix of a metal-matrix composite material due to thermal cycling. The latter is, in fact, a different type of residual stress by existing on a mesomechanical rather than macromechanical scale.

For the particular processes described the author has chosen to concentrate on those which he or his nearest colleagues have own experience in. The choice of processes, and the relative lengths of their respective descriptions must therefore not be taken as indications of their relative importances.

2. HEAT TREATMENT OF STEEL BY QUENCHING.

2.1. Couplings between phenomena

The quenching of steel involves a complex thermodynamical process. For the purpose of performing a mathematical treatment of various aspects of the quenching, it is instructive to think of this process as composed of three different subprocesses, namely, a heat conduction process, a phase transformation process, and a mechanical deformation process. These three subprocesses are coupled in the way shown in Fig 1 to form the complete process.

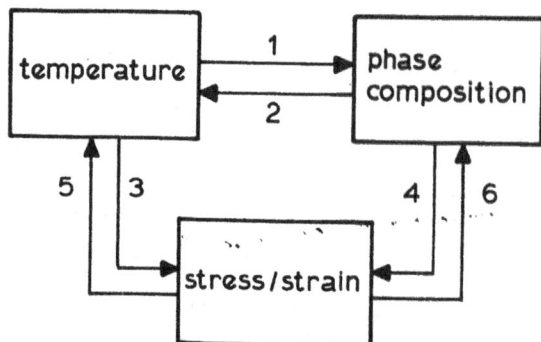

Fig 1. Diagram of couplings [Sjöström (1985)].

Coupling No 1 represents the influence of <u>temperature</u> and <u>temperature evolution</u> on the phase transformations.

Coupling No 2 represents the influence of <u>phase transformation evolution</u> on the <u>temperature evolution</u>; this influence comes from the fact that the new phase may have different thermal properties and from the fact that the phase transformations are generally accompanied by an internal heat generation.

Coupling No 3 represents the influence of <u>temperature</u> on the <u>deformations</u> through temperature expansion.

Coupling No 4 represents the influence of <u>phase transformation evolution</u> on the <u>deformations and stresses</u>; this infuence comes from phase

transformation dilatational strain, from the fact that the new phase may
have different mechanical properties and from the fact that the inner
arrangement of dislocations will be affected by the structural rearrangements involved in the transformation processes.

Coupling No 5 represents the influence of deformation rates on the
temperature evolution (through the deformation rate terms in the energy
equation).

Coupling No 6 represents the influence of deformations (and, above
all, stresses) on the phase transformation evolution, an influence that
has been studied thoroughly during the last few years.

Of these couplings, No 1, 2, 3 and 4 are usually considered as important. On the other hand, coupling No 5 can often be neglected in situations where deformation rates are not high. Coupling No 6, finally,
has not been sufficiently well understood and investigated to be taken
into full account so far.

The aim of this study has been to carry the complete modelling of
the quenching process to a higher degree of accuracy. The resulting refinement of the modelling of the phase transformations will be more
thoroughly described elsewhere; this article, on the other hand, will
concentrate on the mechanical modelling.

Analyses of the steel quenching process by modelling the temperature, phase transformation and deformation/stress parts of the problem
have been made by, for instance, Inoue and Tanaka (1975), Inoue and Raniecki (1978), Rammerstorfer et al. (1981), Inoue et al. (1981), Inoue
et al. (1985), Sjöström (1982, 1985, 1986), Denis et al. (1987a), Denis
et al. (1987b), Habraken (1989) and Järvstråt (1990). To the author's
knowledge, the only monography on the subject is that by Fletcher
(1989).

Following the ideas from Sjöström (1982, 1985, 1986), Denis et al.
(1987a) and Denis et al. (1987b), a description will now be given of the
modelling, together with a couple of representative application examples.

2.2. Temperature field calculation.

For the calculation of the temperature evolution, the energy equation,

$$\rho_0 \cdot \frac{\partial u}{\partial t} = \frac{\partial h_i}{\partial x_i} + \sigma_{ij} \frac{\partial \epsilon_{ij}}{\partial t}, \tag{1}$$

is used (for notations introduced, see the 'NOTATION' section at the end
of the article). With Fourier's law of heat conduction,

$$h_i = - \lambda \frac{\partial T}{\partial x_i}, \tag{2}$$

the following final form of the energy equation is obtained,

$$\frac{\partial}{\partial x_j}(\lambda \frac{\partial T}{\partial x_j}) = \rho_0 c_v \frac{\partial T}{\partial t} - \sigma_{ij}(\delta_{ij}\frac{\partial \epsilon^{tr}}{\partial t} + \frac{\partial \epsilon^p_{ij}}{\partial t}) + \frac{ET}{1-2\nu}\cdot\frac{\partial \epsilon^{th}}{\partial T}\cdot\frac{\partial \epsilon_{kk}}{\partial t}$$

$$- \rho_0 \cdot \frac{db}{dt} \tag{3}$$

In (3), particular attention must be paid to the functional dependence of λ, c_v, E, ν, ϵ^{th}, and ϵ^{tr} on coordinate x_i and time t via temperature T and actual volume fractions v_k of the different phases. The most obvious way of treating this is by using a linear fractioning procedure, as exemplified for λ below:

$$\lambda = \lambda(x_i,t) = \lambda[v_k(x_i,t),T(x_i,t)] = \sum_{k=1}^{6} v_k(x_i,t)\cdot\lambda_k[T(x_i,t)] \tag{4}$$

Thus, the temperature dependence $\lambda_k(T)$ of each phase k must be known.

Coupling No. 4 of Fig 1 is represented by the second, third and fourth terms of the right hand side of Eq (3), and coupling No. 5 by the dependence of material properties on v_k [Eq (4)] and by the last term of the right hand side of Eq (3). The terms corresponding to coupling No 4 can usually be neglected, since the strain rates of the quenching problem are generally very low so that the 'time rate part' of the equation is dominated by the $\partial T/\partial t$ term.

2.3. Phase transformation calculation

The phase transformations occurring during steel quenching are mainly of two kinds, namely
1) diffusional transformation from austenite to ferrite, pearlite or bainite, and
2) non-diffusional transformation from austenite to martensite.
   The mathematical modelling of these different types of transformation will be quite different, since there is a pronounced time dependence in the diffusional transformation, whereas the non-diffusional transformation is practically time-independent. The details of the phase transformation models have been published elsewhere [Sjöström (1982, 1985, 1986), Denis et al. (1987a) and Denis et al. (1987b)]; therefore, only a short description will be given here.
   A diffusional transformation can physically be thought of as a sequence of two processes: an incubation followed by a growth.
   During the incubation, a mesh of nuclei of the transformation product is formed. The main aim of a calculation model is to calculate the necessary length of the incubation period for this mesh to become sufficiently dense. If one accepts a simple additivity principle of incubation fractions, the Scheil method [Scheil (1935)] can be used; in this the incubation period of the continuous cooling transformation is considered as complete when the sum

$$S = \sum_i \frac{\Delta t_i}{\tau_{iIT}} \tag{5}$$

reaches a value of 1.0. During this part of the process there is no substantial transformation.

This, instead, occurs during the growth period, which starts at the time instant when the incubation period is completed. For the growth, the following law developed by Johnson and Mehl and by Avrami [Avrami (1939, 1940, 1941)] is often used,

$$v_k = 1 - e^{-b_k t^{n_k}} \tag{6}$$

The values of $b_k = b_k(T)$ and $n_k = n_k(T)$ can be calculated from the IT (isothermal transformation) diagram (fig 2), using the times corresponding to 10% and 90% of phase formed for the type of transformation and at the temperature in question.

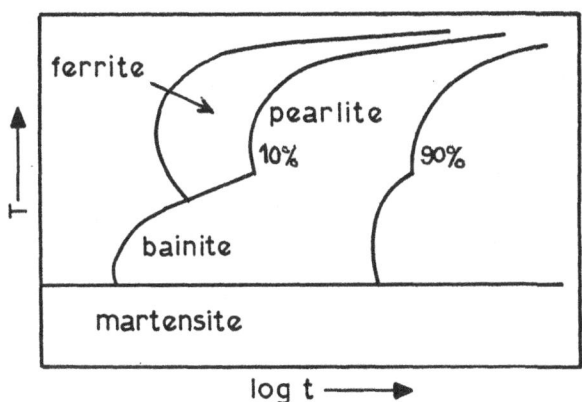

Fig 2. Typical IT-diagram [Sjöström (1985)].

The non-diffusional transformation into martensite has been treated by an equation given by Koistinen and Marburger (1959):

$$v_6 = \left(1 - \sum_{k=2}^{5} v_k\right)\left[1 - e^{-\gamma(T_{s6} - T)}\right], \tag{7}$$

where the constant γ is usually of the order of 0.010 - 0.012.

As mentioned in connection with Fig. 1, there is also an influence of the state of stresses on the transformations (coupling No 6). This influence has been investigated by Gautier (1985). It was found that for the diffusive transformation the start and end curves of the IT diagram

are moved left in the presence of any (tensile or compressive) nonhydrostatic stress state. In a martensitic transformation, the martensite
start temperature T_{S6} is lowered for $\sigma_m < 0$ and raised for $\sigma_m > 0$; there
is also reported a dependence on the deviatoric stress state making T_{S6}
increase for all $\sigma_e \neq 0$. A first attempt of modelling this stress dependence has been tried in Denis et al. (1987a) and Denis et al. (1987b),
to which the reader is referred for details.

2.4. Deformation, strain, and stress calculation.

For the mechanical modelling, it is postulated that the strain rate can
be decomposed in the following way:

$$\frac{d\epsilon_{ij}}{dt} = \frac{d\epsilon_{ij}^e}{dt} + \delta_{ij}\frac{d\epsilon^{th}}{dt} + \delta_{ij}\frac{d\epsilon^{tr}}{dt} + \frac{d\epsilon_{ij}^p}{dt}. \tag{8}$$

For the elastic strain rate, $d\epsilon_{ij}^e/dt$, the following equation has
been used,

$$\frac{d\epsilon_{ij}^e}{dt} = \frac{d}{dt}\left(\frac{1}{E}[(1+\nu)\sigma_{ij} - \delta_{ij}\nu\sigma_{mm}]\right), \tag{9}$$

where E and ν depend on x_i and t via T and v_k in the way given in Eq.
(4).

The rates of thermal and transformation strain ($d\epsilon_{ij}^{th}/dt$ and
$d\epsilon_{ij}^{th}/dt$, respectively,) are calculated by the two equations,

$$\frac{d\epsilon_{ij}^{th}}{dt} = \delta_{ij}\frac{d\epsilon^{th}}{dt} = \delta_{ij}\cdot\frac{d}{dt}\left(\sum_{k=1}^{6} v_k \epsilon_k^{th}\right), \tag{10}$$

and

$$\frac{d\epsilon_{ij}^{tr}}{dt} = \delta_{ij}\frac{d\epsilon^{tr}}{dt} = \delta_{ij}\cdot\frac{d}{dt}\left(\sum_{k=1}^{6} v_k \epsilon_k^{tr}\right). \tag{11}$$

For the time differentiations, it is important to keep in mind that
$v_k = v_k(x_i,t)$ and $\epsilon_k^{th} = \epsilon_k^{th}[T(x_i,t)]$. ϵ^{tr}, on the other hand, is a
constant, namely the transformation strain at a given temperature, for

instance, 0 °C. The combined action of ϵ_{ij}^{th} and ϵ_{ij}^{tr} is illustrated by
the typical dilatational curve of Fig 3.

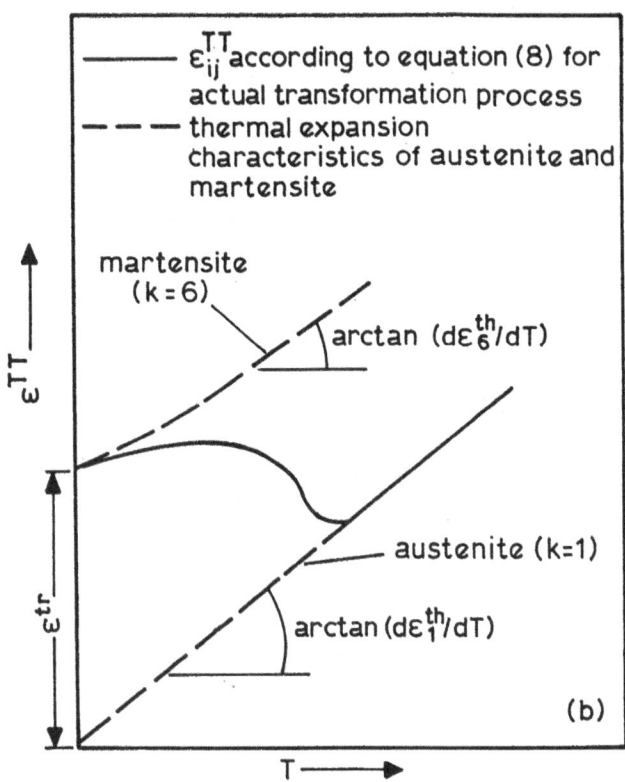

Fig 3. Typical temperature dependence of $\epsilon^{TT} = \epsilon^{tr} + \epsilon^{th}$ [Sjöström (1985)].

The description of the <u>plastic</u> strain rate $d\epsilon_{ij}^{p}/dt$ in the process
leads to some interesting developments. In addition to the conventional
plastic flow there appears a transformation plastic flow with properties
described below. In our modelling, we have therefore postulated that the
plastic strain can be decomposed into two terms,

$$\epsilon_{ij}^{p} = \epsilon_{ij}^{cp} + \epsilon_{ij}^{tp}, \qquad (12)$$

where superscript cp stands for conventional plasticity and tp for
transformation plasticity.

For the <u>conventional plasticity</u> strain rate $d\epsilon_{ij}^{cp}/dt$, the following basic assumptions are made,
(a) von Mises' yield criterion together with the associated flow rule is
   applicable,
(b) linear kinematic hardening is used,
(c) yield strength Y and hardening constant H' depend on temperature and
   phase composition; this is described in the same way as for λ in Eq.
   (4), and
(d) since the phase transformations occurring lead to significant re-
   organisation of the internal structure of the material, it is reasonable to assume that the material loses (part or all of) its
   'memory' of plastic strain accumulated during its existence as
   parent phase (i e austenite).
   These four assumptions lead to the following equations:

$$f = F(\sigma_{ij}, \alpha_{ij}) - [Y(T, v_k)]^2, \tag{13}$$

where f is the flow function. Further,

$$F(\sigma_{ij}, \alpha_{ij}) = \frac{3}{2}[\sigma_{ij} - \alpha_{ij} - \frac{1}{3}\delta_{ij}(\sigma_{mm} - \alpha_{mm})][\sigma_{ij} - \alpha_{ij} - \frac{1}{3}\delta_{ij}(\sigma_{mm} - \alpha_{mm})]. \tag{14}$$

During plastic flow,

$$\frac{df}{dt} = 0, \tag{15}$$

and

$$\frac{d\epsilon_{ij}^{cp}}{dt} = \frac{d\Lambda}{dt} \cdot \frac{\partial F}{\partial \sigma_{ij}}. \tag{16}$$

For the calculation of the kinematic hardening translation α_{ij}, the
use of Prager's basic model would give the following rate equation,

$$\frac{d\alpha_{ij}}{dt} = \frac{d}{dt}(H'\epsilon_{ij}^{cp}), \tag{17}$$

where H' is the strain hardening coefficient. The basic assumption (d)
listed above, however, calls for a modification of the hardening model.
Following, again, a line of thoughts presented in detail in Sjöström
(1982, 1985, 1986), we define a modified plastic strain μ_{ijk} to replace
the plastic strain ϵ_{ij}^{cp} as strain hardening parameter. In a case with no
plastic memory, we arrive at the following equations for $d\mu_{ijk}/dt$:

$$\frac{d\mu_{ij1}}{dt} = \frac{d\epsilon_{ij}^{cp}}{dt} \tag{18}$$

$$\frac{d\mu_{ijk}}{dt} = \frac{d\epsilon_{ij}^{cp}}{dt} - \frac{1}{v_k} \cdot \frac{dv_k}{dt} \cdot \mu_{ijk} \quad ; \quad 2 \le k \le 6 \tag{19}$$

With this hardening parameter, which must be kept track of separately for each phase, and remembering the temperature and phase composition dependence postulated for the plastic properties, $d\alpha_{ij}/dt$ can be calculated as,

$$\frac{d\alpha_{ij}}{dt} = \frac{d}{dt} \sum_{k=1}^{6} v_k H_k' \mu_{ijk}$$

$$= \sum_{k=1}^{6} v_k \left[H_k' \frac{d\epsilon_{ij}^{cp}}{dt} + \frac{dH_k'}{dt} \mu_{ijk} \right] + \frac{dv_1}{dt} H_1' \mu_{ij1}. \tag{20}$$

[Note that, according to Eq (18), μ_{ij1} can be substituted by ϵ_{ij}^{cp} in Eq. (20)].

Eqs (17) through (20) together with Eq (14) now give:

$$\frac{d\Lambda}{dt} = \frac{\dfrac{\partial F}{\partial \sigma_{ij}} \cdot \dfrac{d\sigma_{ij}}{dt} + \dfrac{dQ}{dt}}{P}, \tag{21}$$

where

$$\frac{dQ}{dt} = - \frac{\partial F}{\partial \sigma_{ij}} \left[\sum_{k=1}^{6} v_k \cdot \frac{dH_k'}{dt} \mu_{ijk} + \frac{dv_1}{dt} H_1' \mu_{ij1} \right]$$

$$- 2 \sum_{k=1}^{6} v_k \cdot Y_k \cdot \sum_{k=1}^{6} \left[\frac{dv_k}{dt} Y_k + v_k \cdot \frac{dY_k}{dt} \right] \tag{22}$$

and

$$P = \frac{\partial F}{\partial \sigma_{ij}} \cdot \frac{\partial F}{\partial \sigma_{ij}} \sum_{k=1}^{6} v_k \cdot H_k'. \tag{23}$$

Note the appearing of time derivatives of Y_k and H_k', which is a consequence of the fact that $Y_k = Y_k[T(x_i,t)]$ and $H_k' = H_k'[T(x_i,t)]$.

These equations, together with Eq (16) will give the complete flow rule for the conventional plastic flow under the assumptions listed.

Transformation plasticity is a term used for the pronounced plastic softening of the material during the phase transformation. As a result of this, considerable plastic strains arise even if the stress level is well below that needed for conventional plastic flow. The phenomenon was described quite early, e.g. by Roll (1938); the name transformation plasticity is, however, quite recent, and an exhaustive treatise of the nature and importance of the phenomenon is given by Mitter (1987).

Transformation plasticity is usually explained by two different mechanisms. The first is purely mechanical and bases on the fact that the specific volume of the transformation product is different from (generally larger than) that of the parent phase, which gives internal 'misfit' stresses. If a nonhydrostatic external stress is applied during the transformation, these local misfit stresses lead to plastic flow that can be detected even globally. This mechanism is important in diffusive as well as martensitic transformation and is usually called the Greenwood and Johnson mechanism, since it was first explained by Greenwood and Johnson (1965).

The second is a mechanism that is only important in martensitic transformation and is based on the fact that when martensitic transformation occurs under stress, the product chooses a preferential orientation. That such an orientation of the newly formed phase can lead to a plastic deformation of the transformation plasticity type was first shown by Magee (1966), and the mechanism is consequently called the Magee mechanism.

The transformation plastic strain occurs as soon as the overall stress is nonhydrostatic, while 'conventional' plastic strain can only develop as long as the stress is above the yield stress of the material. From a modelling point of view it is therefore clear that it cannot be included in the conventional plasticity model; instead the additive decomposition of the total plastic strain of Eq. (12) seems justified.

Among the models proposed for describing transformation plasticity, only those dealing with the Greenwood-Johnson mechanism can be considered as sufficiently developed to be used for the modelling of the transformation plastic behaviour of steels under triaxial stress conditions. Therefore, in the examples described below, we have chosen to include only this mechanism, using a model given by Giusti (1981).

This model was based on experiments under uniaxial constant applied stress σ_{11}, from which Giusti was able to establish the following equation for the transformation plastic strain during a martensitic transformation:

$$\epsilon_{11}^{tp} = K_6 \cdot h_6(v_6) \cdot \sigma_{11} = K_6 \cdot (2-v_6)v_6 \cdot \sigma_{11} \tag{24}$$

(K_6 in the equation is a constant). From this he postulated that the transformation-plastic strain rate in a multiaxial case would be proportional to the stress deviator, giving the following rate equation,.

$$\frac{d\epsilon_{ij}^{tp}}{dt} = 3 \cdot K_6 \cdot (1-v_6)\frac{dv_6}{dt} \cdot (\sigma_{ij} - \frac{1}{3}\delta_{ij}\sigma_{mm}) \tag{25}$$

Similar models have later been proposed by Desalos et al. (1982), Leblond et al. (1986a, 1986b, 1989) and Leblond (1989); one common property of all the models is the proportional dependence on the stress deviator, whereas the dependence on v_6 and dv_6/dt differs between the models.

Further work is being done by, for instance, Fischer (1990) on the modelling of transformation plasticity, aiming at improving the modelling for both diffusional and martensitic transformation. An important tool for this may be FEM analysis on a mesoscale, by which it is possible that even the complex interaction local stress → transformation process → transformation anisotropy of the martensitic transformation can be mastered to some extent in the future. This approach is being tested in Fischer (1990) and in recent work by Sjöström et al. (1991) and by Ganghoffer et al. (1991).

2.5. Application examples.

The above model has been used for the calculation of residual stress states in different cases. We will here show two examples:

Example 1. The quenching of a bearing roll [Järvstråt (1990)]. The roll is made of British Standard steel 52100, and the quenching is made in oil. The geometry is shown in Fig. 4. The roll was submerged into the oil vertically, which made the cooling characteristics different in its lower and upper parts. The purpose was to study the difference against a case in which the same cooling characteristics were prescribed over the whole cylinder surface. The calculations were performed with the FEM code ABAQUS, making use of the user-defined subroutine facility for describing the phase transformation model and the constitutive behaviour as described above. Results are shown in Fig. 5.

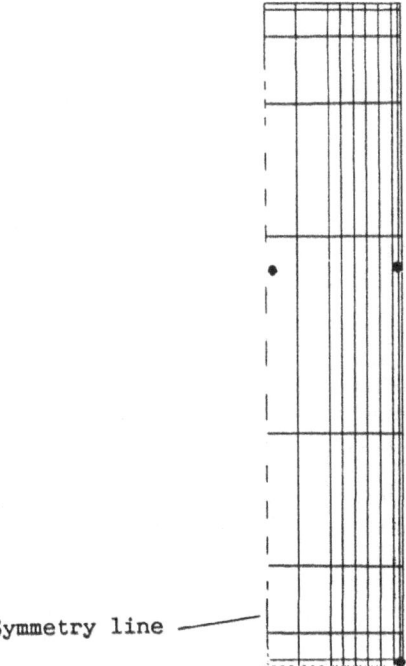

Symmetry line

Figure 4. Geometry of quenched roll [Järvstråt (1990)].

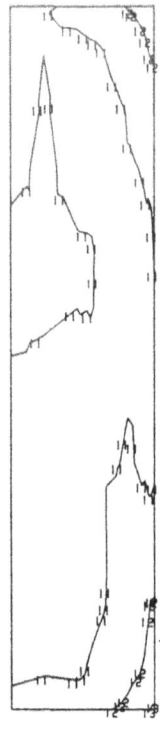

Level	MPa
1	-100
2	-90
3	-80
4	-70
5	-60
6	-50
7	-40
8	-30
9	-20
10	-10
11	0
12	10
13	20
14	30
15	40
16	50
17	60
18	70
19	80
20	90
21	100

Figure 5. Residual tangential stress in quenched roll [Järvstråt (1990)].

Example 2. Oil quenching of a long, carburised steel cylinder with 17 mm diameter, made of Swedish standard steel SS 2511 (with basic carbon content of 0.15%). The carburising has been achieved by exposing the component to a carbon-enriched atmosphere to make carbon diffuse into the component giving the carbon profile shown in Fig. 6. After quenching, the procedure leads to a hard, martensitic surface, whereas inner parts (that have the basic carbon content) are much less prone to forming martensite and usually transform into softer bainite or pearlite. This calculation was reported in Sjöström (1985), to which the reader is referred for details. Examples of the results are shown in Fig. 7. Calculations were performed for different degrees of model refinement, the corresponding results being characterised by run numbers 23→25 in Fig. 7. Of these, run No. 24 includes all the features explained above, and the calculated results come quite close to the experimental ones [by Knuuttila (1982)].

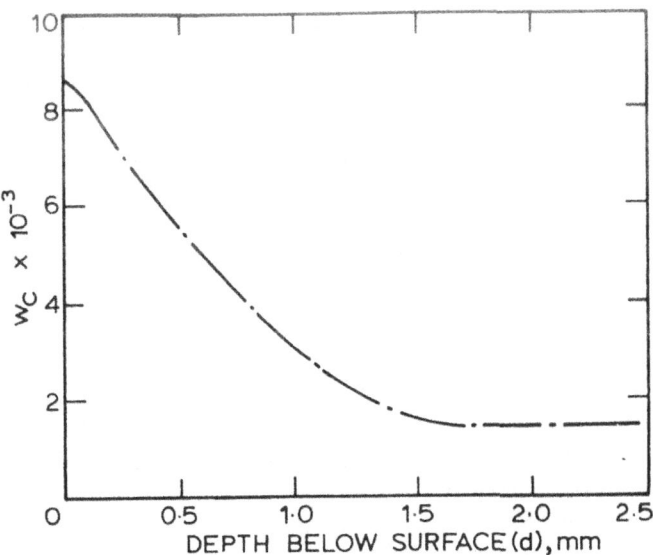

Figure 6. Carbon profile of oil-quenched long cylinder [Sjöström (1985)].

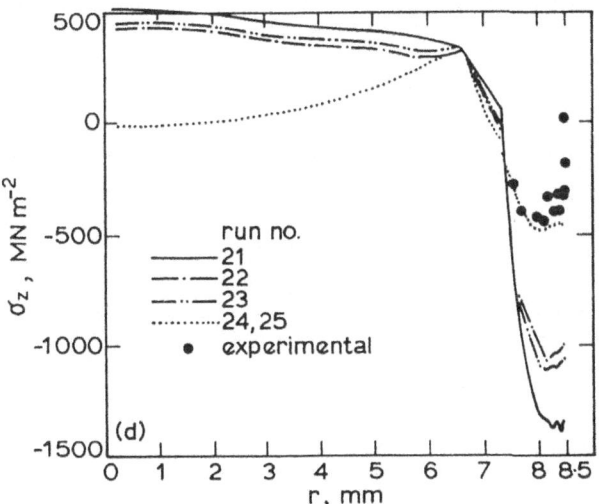

Figure 7. Residual stress state of oil-quenched carburised cylinder [Sjöström (1985)].

3. WELDING

During welding, a residual stress state is created that is usually tensile in a small region near the weld (and, to balance this, compressive in remoter regions). Basically, from a modelling point of view the welding process is quite similar to the heat treatment process; experience shows, however, that plastic flow in the region near the weld is the major origin of the residual stresses, whereas the phase transformations seem to be less important from this point of view.

In most cases it is therefore sufficient to replace the more elaborate phase transformation calculation of the previous section by a description of the thermal strain and the plastic properties of the material as a function of the temperature only, in which the changes due to the phase transformations can also be included. Further, the latent heat of the transformation can often be described by an increase of the specific heat c_v during the temperature interval during which the transformation takes place.

From this, it is obvious that the problem will still require a coupled temperature and stress calculation, performed for the complete history of the welding process (i.e. the heating → melting → solidification → cooling process). As for the quenching problem, this calls for the use of an FEM solution, using a code which can treat the nonlinear material behaviour in a satisfactory manner.

Such calculations have been performed by, for instance, Andersson (1978), Argyris et al. (1983), Josefson (1985), Josefson et al. (1988), Lindgren and Karlsson (1988), Karlsson (1989), Oddy et al. (1990), and Karlsson and Josefsson (1990).

Overviews of the status of the subject are given by Karlsson (1986), Radaj (1988), and Goldak (1989). In the recent article by Josefson (1991), the use of a simplified method is also described, by which some of the most time-consuming parts of the complete calculation model can be avoided without too serious losses of accuracy.

3.1 Application example: single-pass butt-welded pipe.

In order to give an impression of the results attainable, we will show an application from Karlsson and Josefson (1990), to which the reader is referred for closer details. The calculation is a full 3-dimensional one (i.e. the continuous welding of the circumferential weld is actually followed), and the geometry and FEM mesh are shown in Figs. 8 and 9. The material is the Swedish standard steel SS 2172 (0.18% C, 1.3% Mn, 0.3% Si, 0.3% Cr, 0.4% Cu), the welding filler material is deposited from the outside of the tube, and welding parameters (power and speed) are for a typical MIG welding process.

The thermal problem has been solved, using the FEM code ADINAT. The material properties have been given as temperature dependent, and a particular facility available in ADINAT for deactivation of elements in near-constant-temperature regions has been used.

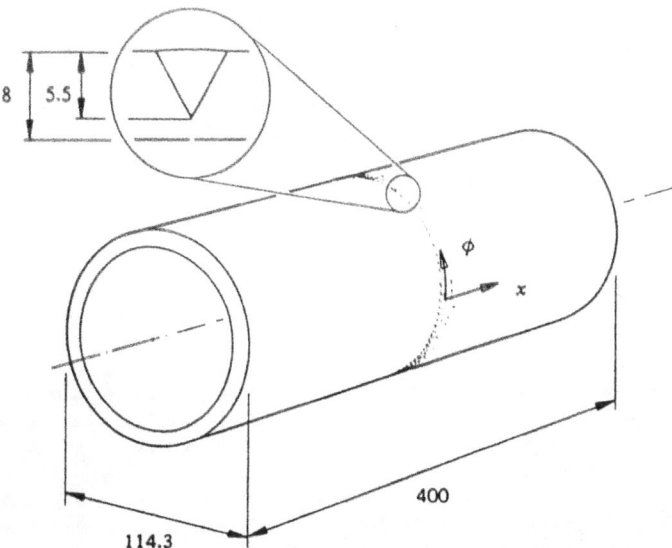

Fig 8. Geometry of butt-welded pipe [Karlsson and Josefson (1990)].

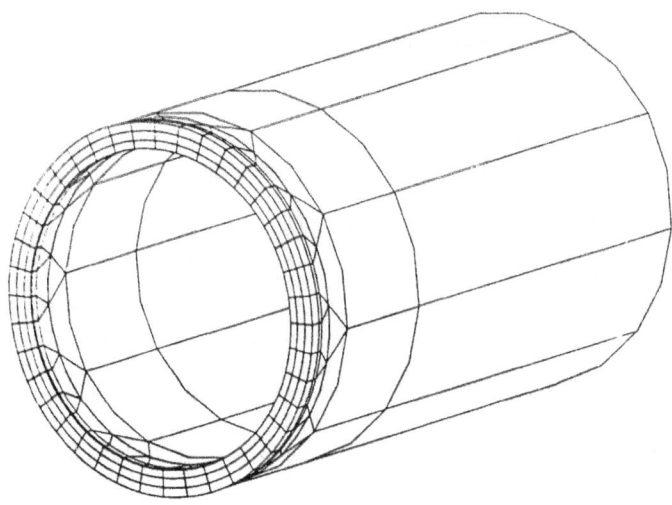

Fig 9. FEM mesh of butt-welded pipe [Karlsson and Josefson (1990)].

82

For the mechanical problem, the FEM code ADINA has been used. The mesh was the same as that for the thermal problem with the exception that a center node was added in elements likely to flow plastically (to reduce the risk of blocking in the constant-volume plastic deformation situation). Again, temperature-dependent properties were assumed.

The calculated residual stresses are exemplified by the results of Fig. 10, in which there is also a comparison with a similar but axisymmetric analysis by Karlsson (1989) and an analytical solution by Vaidyanathan et al. (1973).

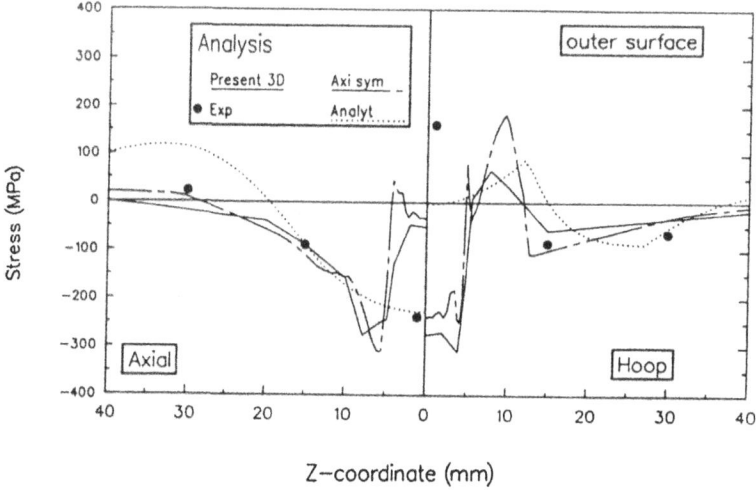

Fig 10. Calculated residual axial and hoop stresses along outer surface of butt-welded pipe [Karlsson and Josefson (1990).

The question whether transformation plasticity is important or not in welding stress calculations seems still to be open. The analysis described above has been performed without introducing any transformation plasticity model. However, in Karlsson (1989) the possible influence of transformation plasticity was tested for the corresponding axisymmetric analysis by introducing a simple yield stress lowering model. This led to some improvement of the FEM-calculated results as compared with the experimental results. However, most of the difference still remained. A final remark, valid for quenching as well as welding modelling, might therefore be that the modelling of transformation plasticity is still far from perfect.

4. ROLLING CONTACT.

During contact, local stresses are usually quite high. One particularity
of the contact stress state created in a halfspace in contact with, for
instance, a cylindrical roll is that the maximum contact stress will be
at some distance below the surface rather than at the surface itself. It
is, further, quite local, and if this maximum stress becomes higher than
the elastic limit of the material, plastic flow is introduced. This, in
turn, leaves a residual stress state after the removal of the roll.

One particularly interesting case is the rolling contact. If the
rolling process is, for instance, in a roller bearing, the over-rolling
is repeated periodically, and there will usually be plastic shakedown,
by which the deformation process is stabilised into repeating exactly
the same deformation cycle for each over-rolling cycle.

Analyses of the residual stress states of such rolling contacts
have been performed by, for instance, Rydholm (1981) and are also
treated in the monography by Johnson (1985). It is interesting to note
that, besides the obvious method of elastoplastic FEM analysis of the
over-rolling history, it is also possible to make a shortcut via Melan's
shakedown theorems, which reduces the computational effort considerably.

4.1. Application example.

An example of a rolling contact with shakedown from Rydholm (1981) will
now be shown for the rolling contact in a roller bearing. The material
is SAE 52100 modelled as linearly isotropic hardening, and the over-
rolling is simulated by the passage from left to right of the pressure
distribution shown in Fig. 11. In fig. 12 the resulting plastic zones
are shown shaded, the middle one being representative of the plastic
zone under any one of a sequence of rolls, whereas the two on the left
and right are disturbed by the artificial boundaries inserted to make
the FEM model finite. It can be seen that the central plastic zone is
practically vanishing already in the 4th cycle, indicating plastic
shakedown. Fig. 13 shows the residual stress state after shakedown.

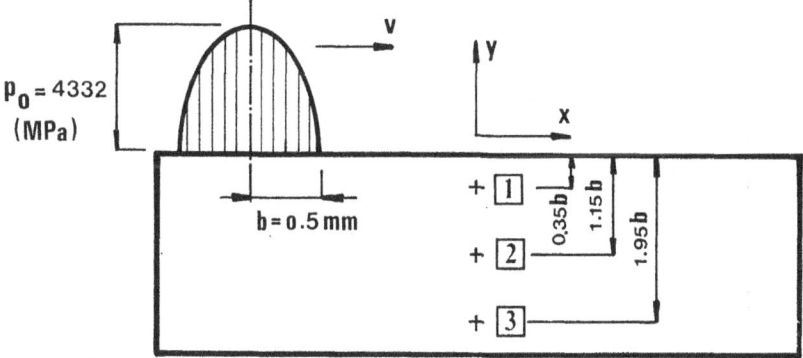

Fig 11. Geometry of the rolling contact problem [Rydholm (1981)].

84

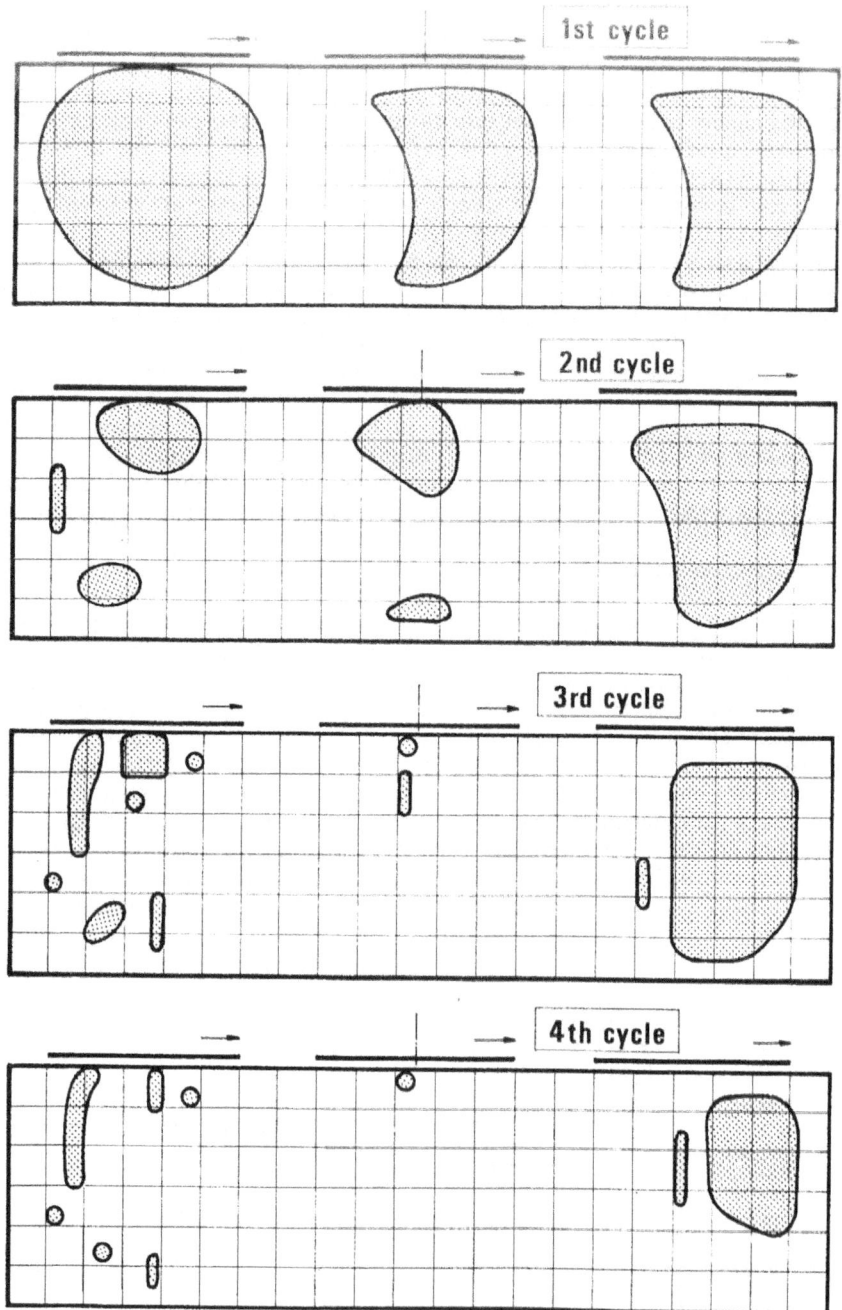

Fig 12. Shakedown 'history' of the rolling contact problem [Rydholm (1981)].

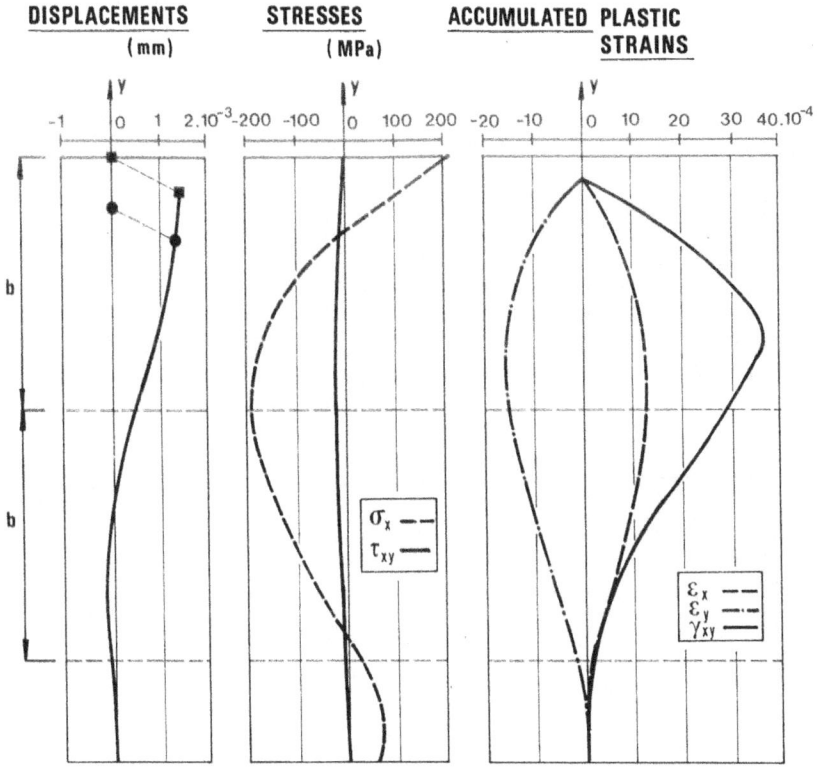

Fig 13. Residual state (displacements, stresses and plastic strains)
after shakedown in the rolling contact problem [Rydholm (1981)].

One particular feature of rolling contact is the experimental evidence of a decomposition of martensite into ferrite and carbide in the
region of maximum contact stress (a few mm below the contact surface).
This is a consequence of the transformation sequence martensite → austenite → ferrite+carbide, and the occurrence of this process is in accordance with the stress → transformation interaction described in the section on quenching.

5. THERMAL CYCLING OF A METAL-MATRIX COMPOSITE.

In a composite material, residual stresses may appear even for very low
overall loads. The reason is that the different properties of the matrix

86

and fibre give rise to local stress states that very soon reach the
level where nonelastic deformation starts.

    In this section we will briefly review an ongoing project dealing
with the thermal cycling of a metal-matrix composite (more specifically,
an aluminium matrix reinforced by silicon-carbide fibres with a length/
diameter ratio of 4). The thermal cycling scheme was 450°C → 100°C →
250°C → 100°C → 250°C → 100°C → ....The calculations have been made with
ABAQUS, and the results show that even for a quite low thermal load the
matrix will be 'overloaded' and start to yield.

    Fig. 14 shows the development of the 'pseudomacro' fibre and matrix
axial stresses and the 'overall' average axial stress during the first 4
cycles. It can be concluded that the process stabilises during these
first 4 cycles. Fig. 15 shows the distribution of residual stress in
fibre and matrix after 10 cycles .

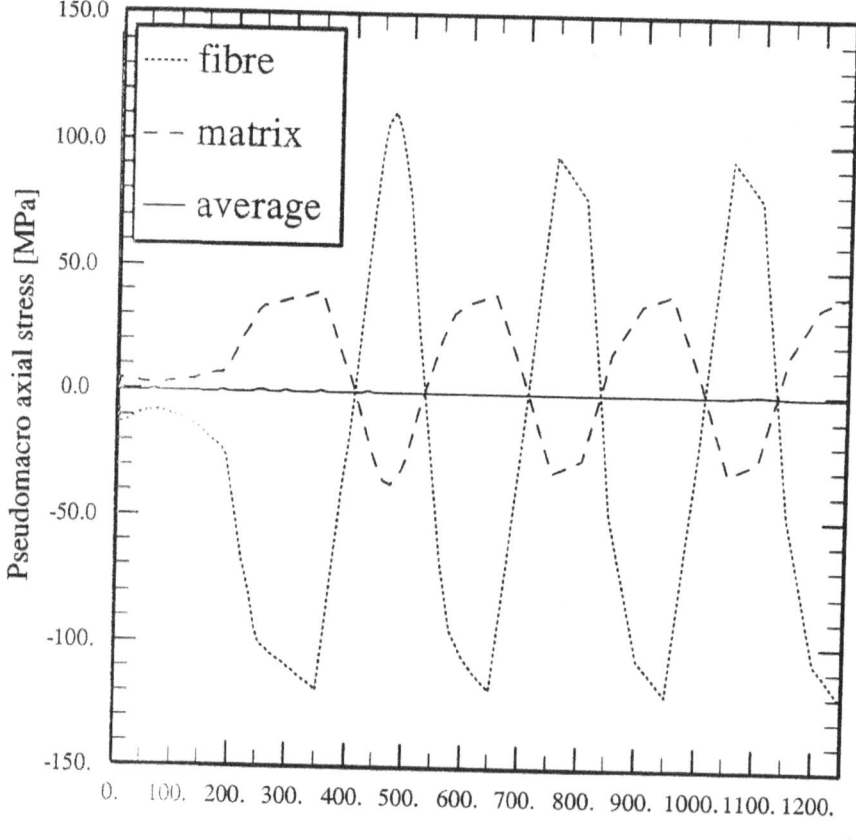

Fig 14. Development of stresses during the first 4 cycles.

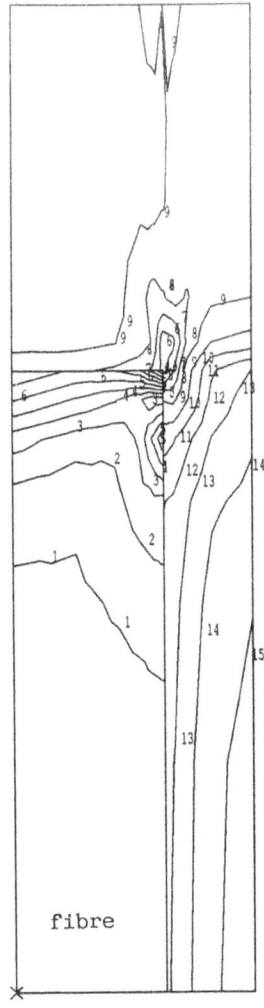

Level	MPa
1	-160
2	-140
3	-120
4	-100
5	-80
6	-60
7	-40
8	-20
9	0
10	20
11	40
12	60
13	80
14	100
15	120

fibre

Fig 15. Axial residual stress after 10 cycles.

ACKNOWLEDGMENTS

The calculations described in sections 2, 4 and 5 have been performed at Linköping University during different projects, supported by STU (the Swedish board of Technical Development), CNRS (Centre National de Recherche Scientifique, France) and the Swedish ball bearing manufacturer SKF. The calculations described in Chapter 3 have been performed by Dr. L Josefson, Chalmers University of Technology, Göteborg, Sweden, who kindly allowed me to include them in this report.

NOTATIONS

b	latent heat of the phase transformation
$b_k = b_k(T)$	parameter in the Johnson-Mehl-Avrami equation for phase No. k
c_v	specific heat at constant volume
E	modulus of elasticity
f	flow function
H'	plastic linear hardening constant
h_i	components of the heat flux vector
$n_k = n_k(T)$	parameter in the Johnson-Mehl-Avrami equation for phase No. k
S	incubation sum [Eq. (5)]
T	temperature
T_{s6}	martensite start temperature
t	time
Δt_i	length of time step i
u	density of internal energy
v_k	volume fraction of phase No. k
x_i	spatial coordinates of a Cartesian system
Y	yield strength
α_{ij}	kinematic hardening translation tensor
γ	constant used in the Koistinen-Marburger equation
δ_{ij}	Kronecker delta; $\delta_{ij} = \begin{cases} 0 \text{ for } i \neq j \\ 1 \text{ for } i = j \end{cases}$
ϵ_{ij}	infinitesimal strain tensor
ϵ_{ij}^e	elastic part ot the infinitesimal strain tensor
$\epsilon_{ij}^{th} = \delta_{ij}\epsilon^{th}$	thermal part of the infinitesimal strain tensor
$\epsilon_{ij}^{tr} = \delta_{ij}\epsilon^{tr}$	transformation part ot the infinitesimal strain tensor
ϵ_{ij}^p	plastic part of the infinitesimal strain tensor.
$\dfrac{d\Lambda}{dt}$	'rate parameter' of the flow rule [Eq. (16)]
λ	heat conductivity
μ_{ijk}	plastic accumulated strain after correction for relaxation
ν	Poisson's number
ρ_0	density
σ_{ij}	Cauchy stress tensor

σ_e von Mises' equivalent stress,

$\sigma_m = \frac{1}{3}\,\sigma_{kk}$ mean stress,

τ_{iIT} time of beginning of the isothermal growth at the temperature of time step i

Subscripts

ij   Cartesian components of second order tensors

k   Phase number. k=1 stands for austenite, k=2 for ferrite, k=3 for pearlite, k=4 for cementite, k=5 for bainite and k=6 for martensite.

Superscripts

cp   conventional plasticity
e   elastic
p   plastic
th   thermal
tp   transformation plasticity
tr   transformation

Conventions

The convention of summation (contraction) of indices appearing twice has been used

REFERENCES

Andersson B A B (1978), J. Engn. Mater. Techn. 100, 356-362.

Argyris J H , Szimmat J, William K J (1983), Proc. 3rd Int. Conf. Numerical Methods in Thermal Problems, Seattle, Washington, U.S.A., 2nd-5th August, 1988, 249-258.

Avrami M (1939), J. Chem. Phys. 7, 1103-1112.

Avrami M (1940), J. Chem. Phys. 8, 212-224.

Avrami M (1941), J. Chem. Phys. 9, 177-184.

Denis S, Gautier E, Sjöström S, Simon A (1987a), Acta Metall. 35, 1621-1632.

Denis S, Sjöström S, Simon A (1987b), Met. Trans. 18A, 1203-1212.

Desalos Y, Giusti J, Gunsberg F (1982) 'Déformations et contraintes lors du traitement thermique de pieces en acier', IRSID, RE902, Saint-Germain-en-Laie, France.

Fischer F D (1990), Acta metall. mater. 38, 1535-1546.

Fletcher J (1989), Thermal stress and strain generation in heat treatment, Elsevier, London, U.K.

Ganghoffer J-F, Denis S, Gautier E, Simon A, Simonsson K, Sjöström S (1991), Proc. 3rd Int. Conf. Residual Stresses, Tokushima, Japan, 24th-26tj July, 1991.

Gautier E (1985), These d'Etat (PhD Thesis), L'Institut National Polytechnique de Lorraine, Nancy, France, 1985.

90

Giusti J (1981), These Doctorat (PhD Thesis), Université Pierre et Marie Curie, Paris, 1981.

Goldak J (1989), Proc. Int. Conf. Trends in Welding Research, Gatlinburg, Tennessee, U.S.A., 14th-18th May, 1989.

Greenwood G, Johnson R H (1965), Proc. Roy. Soc. 238A, 403-422.

Habraken A-M (1989), These Doctorat (PhD thesis), Université de Liege, Belgium.

Inoue T, Nagaki S, Kishino T, Monkawa M (1981), Ingenieur-Archiv 50, 315-327.

Inoue T, Raniecki B (1978), J. Mech. Phys. Solids 26, 187-212.

Inoue T, Tanaka K (1975), Int. J. Mech. Sci. 17, 361-367.

Inoue T, Yamaguchi T, Wang Z (1985), Mater. Sci. Techn. 1, 872-876.

Johnson K L (1985) Contact Mechanics, Cambridge University Press, Cambridge, U.K.

Josefsson L (1985), Mater. Sci.Techn. 1, 904-908.

Josefsson L (1991), Proc. Int. Conf. Offshore Mechanics and Arctic Engineering, Stavanger, Norway, 23rd-28th June, 1991.

Josefsson L ,Jonsson M ,Karlsson L ,Karlsson R ,Karlsson T ,Lindgren L-E (1988), Proc. 2nd Int. Conf. Residual Stresses, Nancy, France, 23rd-25th November, 1988, 497-503.

Järvstråt N (1990), Thesis No. 245 (Tech. Lic. thesis), Linköping University, Linköping, Sweden.

Karlsson C T (1989), Engineering Computations 6, 133-141.

Karlsson L (1986) 'Thermal stresses in welding',in R B Hetnarski (ed.), Thermal Stresses I, Elsevier, Amsterdam, Netherlands, 300-389.

Karlsson R I,Josefsson B L (1990), J. Pressure Vessel Techn. 112, 76-84.

Koistinen D P, Marburger R E (1959), Acta Met. 7, 59-60.

Leblond J-B (1989), Int. J. Plasticity 5, 573-591.

Leblond J B, Devaux J, Devaux J C (1989), Int. J. Plasticity 5, 551-572.

Leblond J B, Mottet G, Devaux J C (1986a), J. Mech. Phys. Solids 34, 395-409.

Leblond J B, Mottet G, Devaux J C (1986b), J. Mech. Phys. Solids 34, 411-432.

Lindgren L, Karlsson L (1988), Int. J. Num. Meth. Eng. 25, 635-655.

Magee C L (1966), PhD Thesis, Carnegie-Mellon University

Mitter W (1987), Umwandlungsplastizität und ihre Berücksichtigung bei der Berechning von Eigenspannungen, Gebrüder Borntraeger, Berlin-Stuttgart, BRD.

Oddy A S, McDill J M J, Goldak J A (1990), J. Pressure Vessel Techn. 112, 309-311.

Radaj D (1988), Wärmewirkungen des Schweissens, Springer, Berlin, BRD.

Rammerstorfer F G, Fischer D F, Mitter W, Bathe K J, Snyder M D (1978), Computers & Structures 13, 771-779.

Roll F (1938), Z. Metallkunde 30, 244-245.

Rydholm G (1981), Diss. No. 61 (PhD thesis), Linköping University, Linköping, Sweden.

Scheil E (1935), Arch. Eisenhüttenwesen 8, 565.

Sjöström S (1982), Diss. No. 84 (PhD thesis), Linköping University, Linköping, Sweden.

Sjöström S (1985), Mater. Sci. Techn. 1, 823-829.

Sjöström S (1986), 'Calculation of quenching residual stresses in steel', in E. Macherauch and V. Hauk (eds.), Residual stresses, Deutsche Gesellschaft für Metallkunde, Oberursel, BRD.

Sjöström S, Ganghoffer J-F, Denis S, Gautier E, Simon A (1991), 3rd Int. Conf. Residual Stresses, Tokushima, Japan, 24th-26th July, 1991.

Vaidyanathan S, Todaro A F, Finnie I (1973), J. Engn. Mater. Techn. 95, 233-257.

COMPARISON BETWEEN FINITE ELEMENT CALCULATIONS AND NEUTRON DIFFRACTION MEASUREMENTS OF RESIDUAL STRESS IN A DIAMETRICALLY COMPRESSED RING

T.M. HOLDEN, R.R. HOSBONS, S.R. MacEWEN*
AECL Research, Chalk River, Ontario, K0J 1J0

E.C. FLOWER
Lawrence Livermore National Laboratory
Livermore, California, U.S.A.

and

M.A. BOURKE and J.A. GOLDSTONE
Los Alamos National Laboratory
Los Alamos, New Mexico, U.S.A.

ABSTRACT. A detailed comparison is made between measurements of the elastic strain components in a diametrically compressed ring by neutron diffraction and the elastic strains corresponding to the residual stresses computed by finite element methods. Semiquantitative agreement is obtained but there are discrepancies between measurement and calculation. The measurements, which were made with a conventional diffractometer as well as by time-of-flight diffraction, show up the inadequacy of the assumption of a single valued stress field.

1. Introduction

The particular advantage of residual strain measurement by neutron diffraction is that results are obtained inside engineering components and may be compared directly with the stress or strain calculated by finite element methods. Modern numerical methods are key design tools for the engineer, but many assumptions are made in the calculations such as the continuum nature of the component and the reduction of the problem into a plane stress or plane strain problem. In addition material properties such as the elasticity, yield strength and the work-hardening are needed. It is therefore vital to check the accuracy of the calculations and neutrons provide an ideal probe.

---

* Present address: Alcan International, Kingston, Ontario, Canada.

M. T. Hutchings and A. D. Krawitz (ed.),
Measurement of Residual and Applied Stress Using Neutron Diffraction, 93–112.
© 1992 *Government of Canada.*

In many cases knowing the use to which the component will be put, combined with the loads which may be applied, will convince the engineer that there is a safe margin of error in the calculation. Sometimes the calculations have further complexities, such as for welds where the yield point may be so low at high temperatures that the weld cannot sustain elastic strains and some relaxation criterion has to be applied. To get even qualitative agreement may be a vindication of a model in such a case.

A critical comparison, however, calls into question some conventional practices such as the use of isotropic average elastic and plastic material properties. Metals are polycrystalline aggregates in which every grain is constrained by its neighbours in a macroscopic stress field and different crystal orientations of grains have different elastic and plastic properties in a particular sample direction. The experimentalist attempts to discern the overall macroscopic stress field from measurements of an ensemble of correctly oriented grains each with a variety of neighbouring grains. We know that the strains in a given direction, sampled with different diffraction lines, are different. To show that there is, or is not, a unique stress field requires the measurement of many diffraction lines with high accuracy.

2.   Calculations on a Diametrically Distorted Ring

2.1 THE TEST SAMPLE

Extensive finite element calculations[1,2] were carried out first to design the workpiece. From the engineering perspective, a non-trivial sample geometry was required since the model was required for calculation of the stresses in fairly complex geometries. The boundary conditions should be readily modelled and the stress system should approximate to the case of plane stress. Since the objective was to measure a macrostress field it was desirable to have a sample with macroscopically uniform elastic and plastic properties in different directions. From the experimental viewpoint the residual stress gradients should not be so steep (compared for example with a surface-deformed metal) that extraordinarily high spatial resolution would be required to sample the gradients. On the other hand, it was required that the magnitudes of the strains be about $\pm 10 \times 10^{-4}$ so that the measurements could be made with an accuracy greater than 10%.

Geometries that were considered for the test specimen included a bend specimen, a formed cup, a tensile sheet specimen with hole, and a plane stress ring which has been subject to a diametral compression. The plane stress ring produced a good distribution of stress through the thick-section of the ring and was easily modelled in two dimensions with a minimum of uncertainty at the sample boundaries. Candidate materials for the ring which were considered included copper and stainless steel. Copper, with its low yield strength, required very large deformations before elastic strains due to residual stress were in the range $(1 - 10) \times 10^{-4}$ so that stainless steel was chosen as the material.

To avoid experimental uncertainties the sample was required to be homogeneous, fine-grained and contain minimal fabrication stresses. Single phase austenitic 21Cr-6Ni-9Mn ($Fe_{0.64}Cr_{0.21}Ni_{0.06}Mn_{0.09}$) was chosen as the material. Second phase carbo-nitrides can precipitate in certain temperature regimes, and, as in many cubic materials, deformation can induce preferred crystallite orientation or texturing. A special thermomechanical procedure was developed for fabricating plate material which resulted in mean grain size of about 25.3 μm, no precipitates of second phase particles and no preferred orientation. The mechanical properties of the ring are summarized in Table 1. Rings were machined from the plate material with outer diameters (OD) of 127 mm and inner diameters (ID) of 70 mm and thicknesses of 13 mm.

Results of the preliminary finite element analysis,[1,2] in which the ring was compressed approximately to 2.5% total strain and unloaded, were used to guide where the measurements should be made. Fig. 1 gives the sample geometry and the coordinate system used to describe the strain components. The calculations indicated that the stresses would be concentrated at the 0° and 180° positions along z, North and South, and at 90° and 270° positions along y, east and west. Measurements taken from the ID to the OD at these two locations for both the radial and tangential components were expected to be characteristic of the macroscopic residual stress field. The actual test sample was compressed 3.4 mm in a mechanical testing machine.

2.2 THE CALCULATIONS

NIKE2D[3], a finite-element non-linear stress analysis code, was used for the calculations. The finite element mesh for the full calculation allowed for east-west symmetry about the vertical axis and included machined locator flats. A total of 2010, four node, isoparametric elements were used with fifteen elements from ID to OD giving a radial spatial resolution of 3 mm in the calculation. A simple rate and temperature independent plasticity model was used derived from a linear piece-wise fit of actual uniaxial test data to a total strain of 6%.

In addition, comparison calculations were also carried out with a second finite element code FEAP[4] for both plane stress and plane strain analyses of the test sample. The results of these calculations were very similar to those of the NIKE2D calculations. Calculations were also made with NIKE3D[5] to assess the assumption of plane stress. A quarter section of the ring was modelled with 8 node brick-shaped elements through one-half of the 13 mm thickness, fourteen elements from ID to OD and twenty-eight elements from North to East. The results indicated that the through-thickness stress, σ_x, was not zero and, further, that there was a stress gradient through the thickness of the ring. However the magnitudes of these stresses were less than 20% of the in-plane components. None of the additional calculations, however, can account for the discrepancies between theory and experiment.

TABLE 1.   Elastic properties of 21-6-9 stainless steel

hkl	Diffraction Elastic Constant	Diffraction Elastic Constant	Young's Modulus	Poisson's Ratio
	$S_1 + 1/2\ S_2$	S_1	E	ν
	$10^{-3}\ (GPA)^{-1}$	$10^{-3}(GPa)^{-1}$	GPa	
002	6.56	-2.18		
113	5.41	-1.61		
420	5.39	-1.60		
531	5.02	-1.41		
220 } 224 }	4.73	-1.27		
331	4.56	-1.18		
111	4.13	-0.96		
Bulk (Kröner)			196.1	0.284
Bulk[6] (Exp)			195	0.288
Bulk[a] (Exp, tension)			194	0.27
Bulk[a] (Exp, compression)			197	0.24

a. Measurements made at AECL Research on the material of the ring. The proportional limit and yield stress under tension were found to be 315 and 340 MPa, and the corresponding values under compression were found to be 320 and 340 MPa.

a.

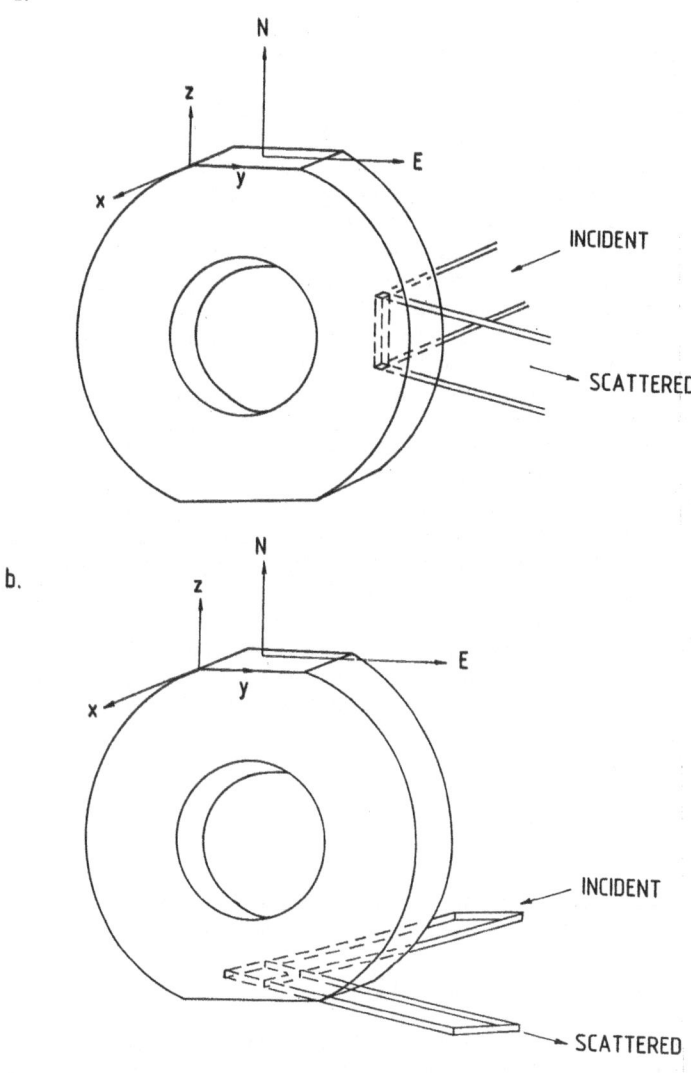

Fig. 1.   Sample and gauge volume geometry for the diametrally compressed ring.   (a) shows the arrangement for measurement of the radial strain component and (b) shows the arrangement for the tangential component.

3. Experiments

Two groups of neutron measurements were made three years apart using two very different techniques, conventional crystal diffractometry and time-of-flight measurements on a spallation neutron source. The first set of measurements was made with the L3 spectrometer at the NRU reactor, Chalk River, with a wavelength of 1.9886 Å reflected from the (331) planes of a squeezed Ge monochromator. Collimators of 0.3° angular acceptance defined the incident and scattered beam directions. The sample was mounted on a computer-controlled translation table in one of the two configurations shown in Fig. 1. For measurements of the radial strains, the neutron beam was defined by vertical slits 1 mm wide and 10 mm high in sheets of cadmium placed in the incident and scattered beams. The gauge volume, the intersection of the beams, has the form of a pillar 10 mm high with a diamond shaped section and it was positioned at the mid-thickness of the ring as shown schematically in Fig. 2(a). The configuration for measurement of tangential strains is shown in Fig. 1b, for which a horizontal slit 0.5 mm high and 5 mm wide was used. The intersection of the incident and scattered beams in the sample nearly includes the thickness of the plate as shown in Fig. 2(b). Measurements were made with the (111) and (002) reflections of the austenitic alloy since these represent the elastically stiffest and softest directions in the cubic lattice and should bracket the mechanical behaviour. The diffraction data were obtained by stepping a single detector in 0.1° steps through the appropriate angular range. The data, in the form of counts versus angular position, were fitted to a Gaussian peak on a sloping background and the spacings obtained from Bragg's law

$$d_{hkl} = \frac{\lambda}{2\sin\theta_{hkl}}. \tag{1}$$

The reference spacings d^o_{hkl} were obtained by measuring the ring prior to deformation and the load induced strain was determined from

$$\epsilon_{hkl} = (d_{hkl} - d^o_{hkl})/d^o_{hkl}. \tag{2}$$

The experimental results at the North and West locations are shown in Figs. 3 and 4. In both cases the maximum tangential strains exceed the radial strains by a factor of 4-5, and the radial and tangential strains are reversed in sign. Likewise the sign of the strain reverses between the West and North positions. Qualitatively the results are similar to a pair of bent beams which have been stressed beyond the yield point. The North case corresponds to applying the load at the top of the segment of the ring cut by the E-W equator. The West case corresponds to applying the load to the segment of the ring cut by the N-S meridian. The curves in Figs. 3 and 4 correspond to the finite element calculations and the agreement between theory and experiment with no adjustable parameters is good but not perfect as will be discussed later.

The second set of experiments was carried out on the same plate with the Neutron Powder Diffractometer at the Los Alamos Neutron

a.

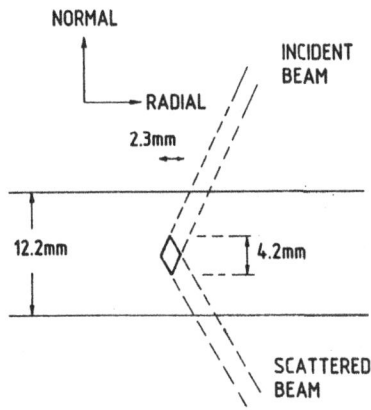

b.

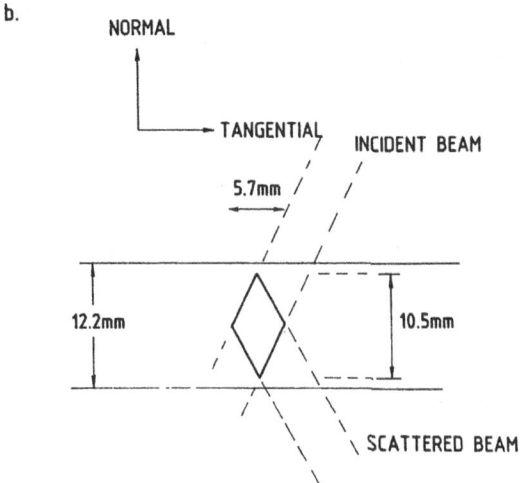

Fig. 2. Sections through the thickness of the ring showing the extent of the gauge volume for radial and tangential measurements.

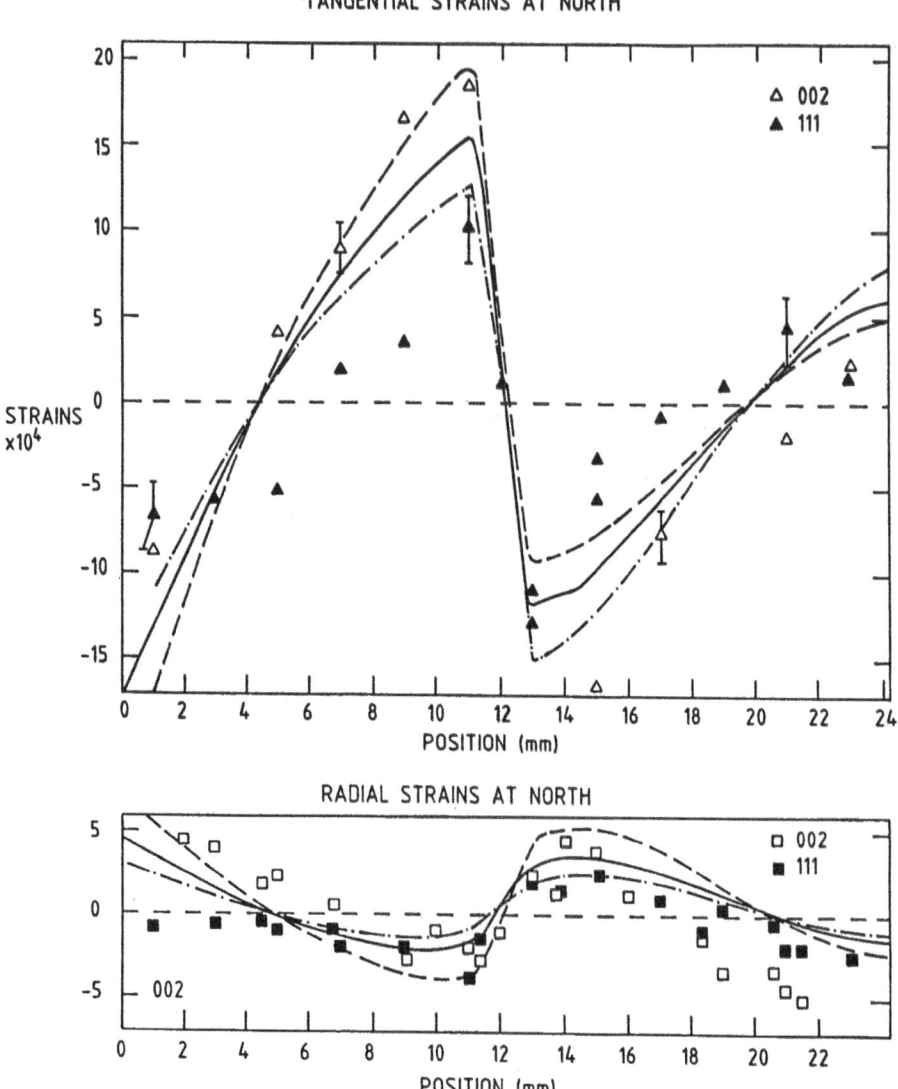

Fig. 3. The measured tangential and radial residual strains at the North location as a function of distance from the inner diameter. Closed symbols denote strain measurements deduced from (111) plane spacings and open symbols for (002) plane spacings. The solid curves are the residual strain components corresponding to the calculated residual stress computed with bulk elastic constants. Dashed and dot-dashed curves were computed with the appropriate diffraction elastic constants.

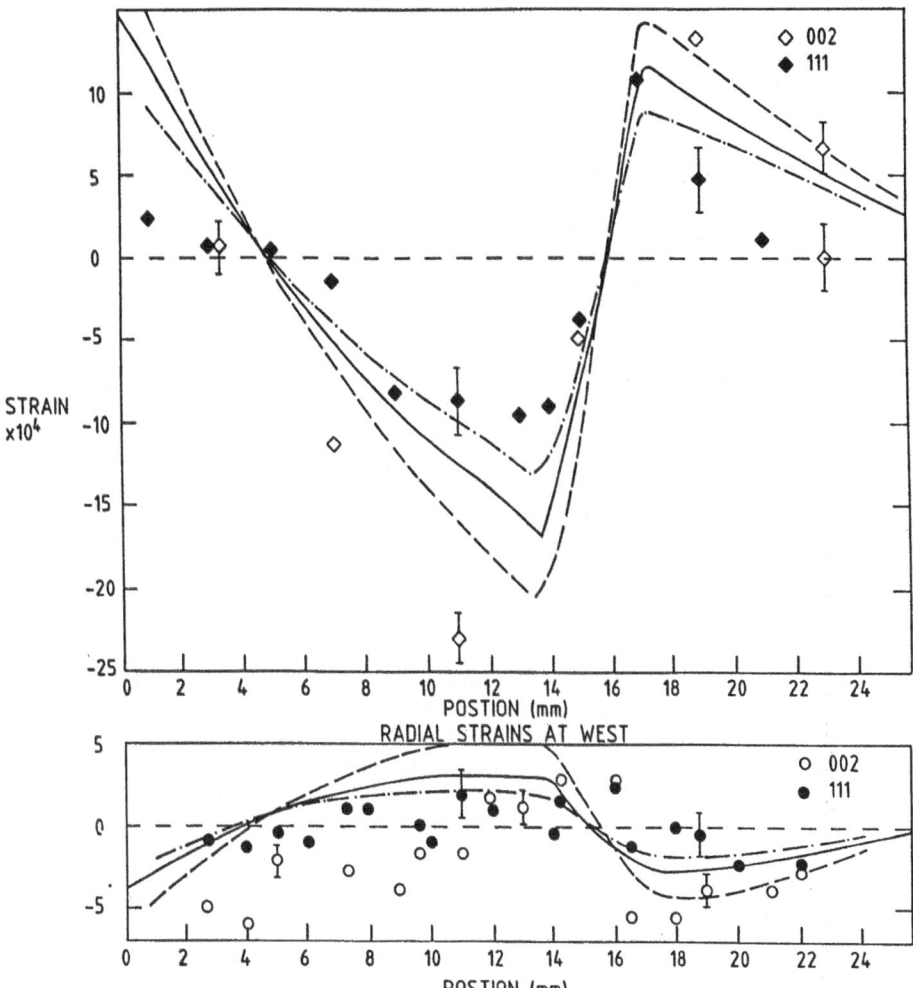

Fig. 4. The measured tangential and radial residual strains at the West location as a function of distance from the inner diameter. The symbols and curves have the same meaning as in Fig. 3.

Scattering Centre (LANSCE). The instrument makes use of the time-of-flight of a white beam travelling from the spallation source 32 m to a bank of counters at $2\theta = 90°$ to collect many diffraction peaks from the sample. In a six hour period the (002),(113),(420),(531), (220),(224),(331),(111) and (222) peaks were measured with adequate statistics. The (442) and (006), and (115) and (333) peaks superpose and are therefore ambiguous. The collection of so many peaks allows one to check consistently for the effects of elastic anisotropy. The resolution was slightly better (20-30%) than that of the first experiment. The gauge volumes for these experiments were defined by slits in 13 mm thick boron-nitride which were 4 mm wide and 2 mm high for the hoop measurements, and 2 mm wide and 10 mm high for the radial and axial measurements. This gave gauge volumes of 16 and 40 mm^3 respectively. Two further advantages of time-of-flight diffraction should be mentioned. For 90° detectors the gauge volume has a square section, and simultaneous measurement of two components at 90° is possible in the two opposing 90° banks of the NPD. The defining slit in the scattered beam has to be very close to the gauge volume to avoid parallax corrections since there is an 11° spread of the angular positions of the individual counters about the 90° setting. The detectors were calibrated by diffraction from a standard CaF_2 powder with the same configuration of slits used for the actual experiment. This is absolutely necessary for highly accurate experiments for both conventional and time-of-flight diffraction. The experimental results for all the time-of-flight diffraction peaks obtained for the tangential measurements at the West position are shown in Fig. 5. The agreement between the earlier experiments at Chalk River for the (111) and (002) reflections and the more recent experiments was within the quoted experimental errors, which is a reassuring result since the measurement techniques and possible systematic errors are completely different.

4. Diffraction Elastic Constants

The bulk values[6] of Young's modulus and Poisson's ratio for 21-6-9 stainless steel are 195 GPa and 0.288. To determine the diffraction elastic constants, single crystal elastic constants are required and the closest composition for which these were available[7] is $Fe_{0.68}Cr_{0.18}Ni_{0.14}$ for which $S_{11} = 9.9 \times 10^{-3}$ $(GPa)^{-1}$, $S_{11} - S_{12} = 13.7 \times 10^{-3}$ $(GPa)^{-1}$ and $S_{44} = 8.2 \times 10^{-3}$ $(GPa)^{-1}$. Since the Kröner model[8] matches the strains at the boundary of the crystallite within an average medium, it best approximates the case of a constrained polycrystalline assembly. The method of Dölle[9] is followed which readily permits a calculation of the diffraction constants as well as the bulk elastic constants within the Kröner model for a texture-free sample. The relationship between the strains and stresses in the hoop, radial and normal direction, is, in Dölle's notation,

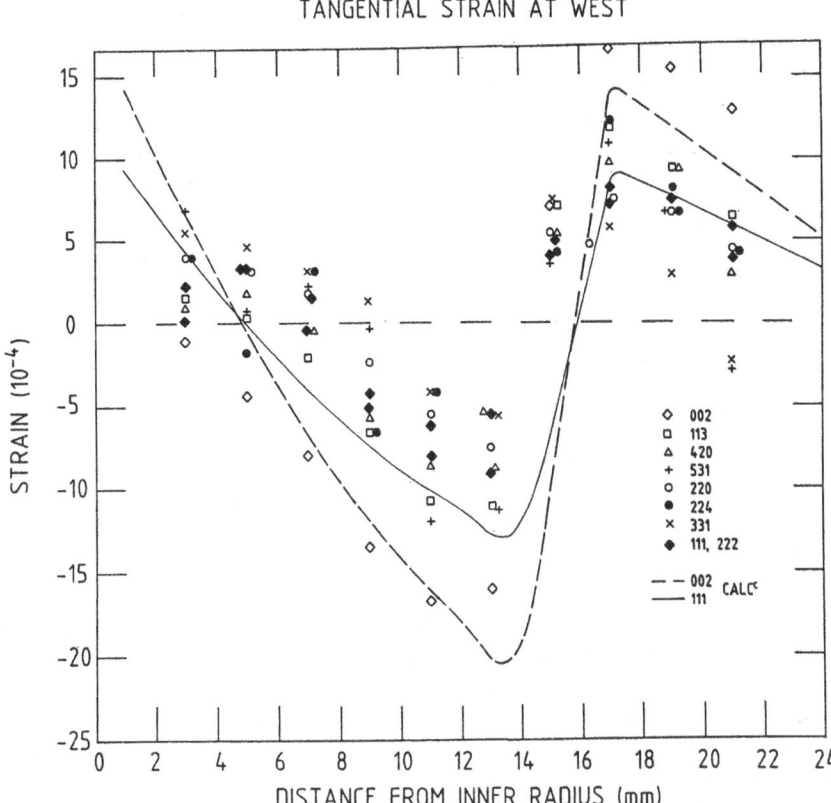

TANGENTIAL STRAIN AT WEST

Fig. 5. The measured tangential residual strains at the West location as a function of distance from the inner diameter of the ring deduced from many different plane spacings. The solid and dashed curves represent the results of finite element calculations of stress converted to strain with diffraction elastic constants for (111) and (002) planes.

$$\epsilon_H = (S_1 + 1/2\ S_2)\sigma_H + S_1\sigma_R + S_1\sigma_N.$$

$$\epsilon_R = S_1\sigma_H + (S_1 + 1/2\ S_2)\sigma_R + S_1\sigma_N. \tag{3}$$

$$\epsilon_N = S_1\sigma_H + S_1\sigma_R + (S_1 + 1/2\ S_2)\sigma_N.$$

where $\quad S_1 + 1/2\ S_2 = S_{3333} + t_{3333} - 2\ t_0\ \Gamma_{hkl}. \tag{4}$

$$S_1 = S_{3311} + t_{3311} + t_0\ \Gamma_{hkl}. \tag{5}$$

and $\quad \Gamma_{hkl} = \dfrac{h^2k^2 + k^2l^2 + l^2h^2}{(h^2 + k^2 + l^2)^2}. \tag{6}$

The S tensors are averages over the single crystal compliances and the t tensors represent the elastic interaction between the individual grain and the matrix. t_0 is a linear combination of t tensors.[9] The quantity Γ_{hkl} is the parameter that measures the anisotropy in the cubic lattice and varies from 0 for the [002] direction to 0.333 for the [111] direction. The computed diffraction elastic constants for the material are given in Table 1 together with computed values of the bulk elastic constants and are compared with experimental quantities. For the bulk elastic constants there is excellent agreement, in spite of the compositional discrepancy, giving confidence in the diffraction constants.

From the calculation of the components of the residual stress field,[3,4,5] we can compute the residual strains due to the residual stress with either the bulk elastic constants to represent the average engineering strain, or with the diffraction elastic constants to assess the size of the elastic anisotropy. The curves in Fig. 5 represent the residual strains for (002) and (111) calculated from the stresses at the West position with the diffraction elastic constants in Table 1, and the two outermost curves in Figs. 3 and 4 represent the residual strains for (111) and (002) planes while the middle curve is calculated with the bulk elastic constants.

We can also use the diffraction elastic constants to compute stresses from the measured strains. If we assume that a triaxial residual stress exists at each point in the ring, then substituting Eq. 4 and 5 into Eq. 3 we have

$$\epsilon_H = (S_{3333} + t_{3333})\sigma_H + (S_{3311} + t_{3311})\ (\sigma_N + \sigma_R)$$

$$+ t_0(\sigma_N + \sigma_R - 2\sigma_H)\Gamma_{hkl}. \tag{7}$$

Thus if we plot the hoop strains at one location, for the different (hkl) reflections versus Γ_{hkl}, then we should obtain a straight line variation whose intercept and slope gives products of linear combinations of the stresses and the elastic constants. Alternatively one may invert Eq. 3 with the appropriate diffraction elastic constants and determine the stress components from the measured strain components for each (hkl) and then compare the consistency of the deduced stresses.

5.   Comparison Between Theory and Experiment

Because of the small magnitude of the radial strains in Fig. 3 and the typical error, $\pm 1 \times 10^{-4}$, of the measurements, it is not possible to identify shortcomings in the calculation from this component. However the variation in the tangential strains is large enough to show up discrepancies.   Firstly the theory correctly predicts the position of the elastic core of the strain distribution, i.e. the rapid change-over from tensile-to-compressive strains at the West and North locations near the neutral axes.   The signs of the effects at West and North and the relative signs of tangential and radial components are correctly predicted.   The agreement between theory and experiment at the North location is good over the whole radius of the ring in that the strains calculated with bulk constants, solid curve in Fig. 3, fall between the limits provided by the (002) and (111) diffraction lines, except perhaps for a discrepancy at the ID where the compressive strains are about a factor of 2 less than the calculations.   We note however that the difference between the measured (002) and (111) strains at a given position is greater than the computed values, and this is particularly noticeable where the strains change sign away from the neutral axis near the ID and OD.

   The results at the West position differ significantly from the calculation both as regards the maximum and minimum strains near the elastic core and the tensile strain near the ID.   A possible origin of the discrepancy is an overestimate of the magnitude of the residual stress near the elastic core at the West position by about 35-50%.   This suggests that the width of the elastic region is probably greater at the West than at the North location.   This would then also help resolve the discrepancy between calculation and measurement near the ID.   When the engineering stress field passes through zero, as for example near 5 mm and 16 mm from the inner radius in Fig. 3, the corresponding strain fields for (002) and (111) must also pass through zero by Eq. 3.   We see from the data in Fig. 3 and 5, that the measured band of strains is just as wide at the crossing point as it is elsewhere.   Moreover, some strains are positive and others negative.   This suggests that there is not a unique value of stress in every grain.   It is expected that plastic flow will vary from grain to grain, characterized by different [hkl] values and will cause grain-to-grain interactions which are superposed on the macroscopic stress field.

   Fig. 5 also reveals that the spread of residual strains at any one position is always larger than that calculated from the strain field. This may be a deficiency in the Kröner elastic constants, i.e. the elastic constants should display a greater spread of values. However, it is more likely that the state of strain of each set of grains in the loaded situation is not unique and that we are seeing again the effect of grain-to-grain interactions.

6. Attempts to Deduce the Stress Field From the Measured Strain Field

The measurement of strain at a single location in many different grain orientations provides the possibility of new ways of getting the stress field. Following Eq. 4 it is logical to plot the strain components versus the anisotropy parameter Γ_{hkl}, and the results are shown in Figs. 6 and 7 for a point near the region of maximum compressive strain, 13 mm from the ID and a point where the residual strains are small but the plastic flow greatest, 3 mm from the ID. To within the experimental errors which range from $\pm 0.7 \times 10^{-4}$ to $\pm 2.3 \times 10^{-4}$, and which are much the same for a given reflection independent of which component is being examined, a straight line can be fitted to the data which shows the influence of the anisotropy of the elastic response. The slopes and intercepts derived from the data together with the errors are given in Table 2 where they are compared with quantities derived from the calculated stresses. The experimental intercepts are ill-defined because of the scatter in the data and because there are few unambiguous diffraction lines close to the cube edge direction. Allowing for the large errors the intercepts agree with theory, but the experimental slopes exceed the calculated slopes, i.e. the experimental spread of strains exceeds the calculated spread.

If we assume that the stress perpendicular to the plane of the plate is zero we may solve for the radial and tangential stresses from the intercept and slope. Unfortunately the experimental accuracy of these quantities is not sufficiently high to solve for the stresses unambiguously.

We may also take the three strain components for each (hkl) at every position and solve for the stresses, Eq. 3, with the appropriate diffraction elastic constants. If there is a unique value of stress characterizing the macroscopic behaviour at a point, then we should find that this procedure generates a single value of the stress, to within the errors, for the different (hkl)'s. Fig. 8 shows the tangential stress at distances for 3 and 13 mm from the ID. To within two standard deviations all the reflections, except (002), could correspond to a single stress value. The (002) however stands alone, either it is too compressive compared with the average as in the case of Fig. 8, or else it is too tensile (Fig. 5) in the case of the region near 17 mm from the ID. Inspection of Fig. 5 shows that the (002) results are always well separated from the others which tend to cluster together. This suggests that there is some special consideration for the [002] direction in the austenitic lattice. Similar apparent disagreement between (111) and (002) strains was noted in the study[10] of bent steam generator tubing where, well away from the neutral axis, the (111) hoop strain was tensile while the (002) hoop strain was compressive.

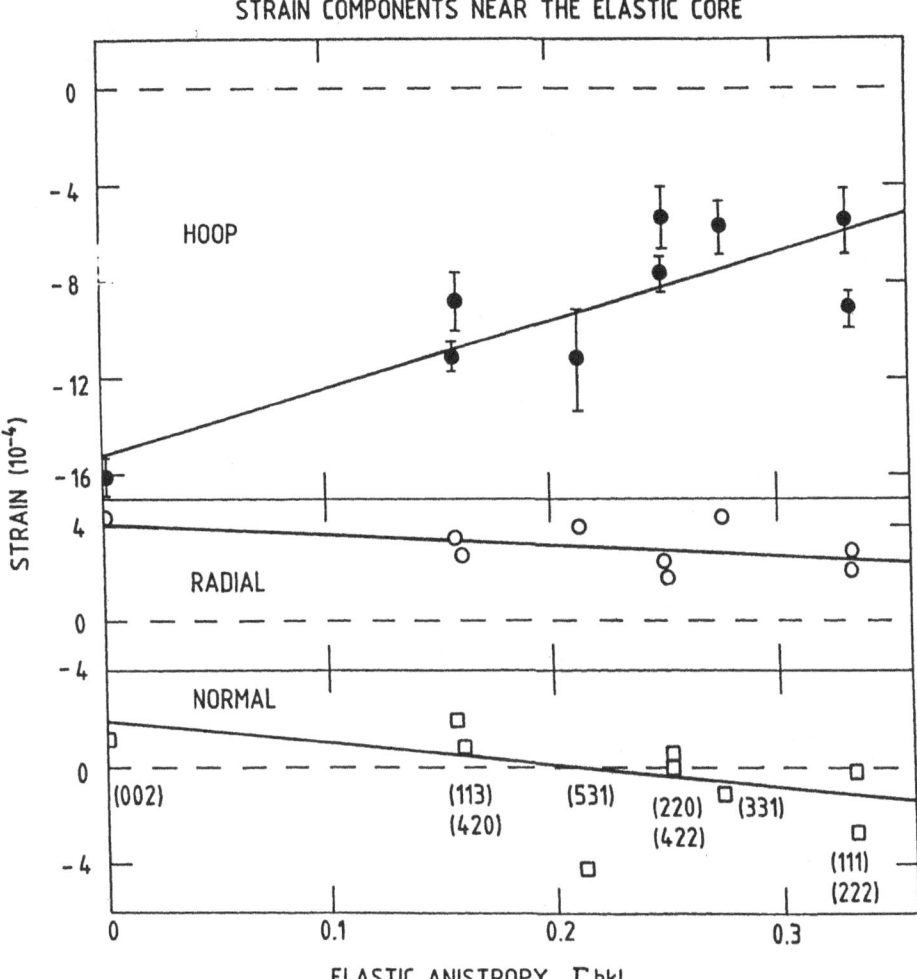

Fig. 6. Variation of the residual strains deduced from many different plane spacings plotted against the elastic anisotropy parameter for tangential (hoop), radial and normal components. The location is close to the elastic core at the West position. The results show that a straight line variation of strain with anisotropy is obtained as expected from Eqn. 7.

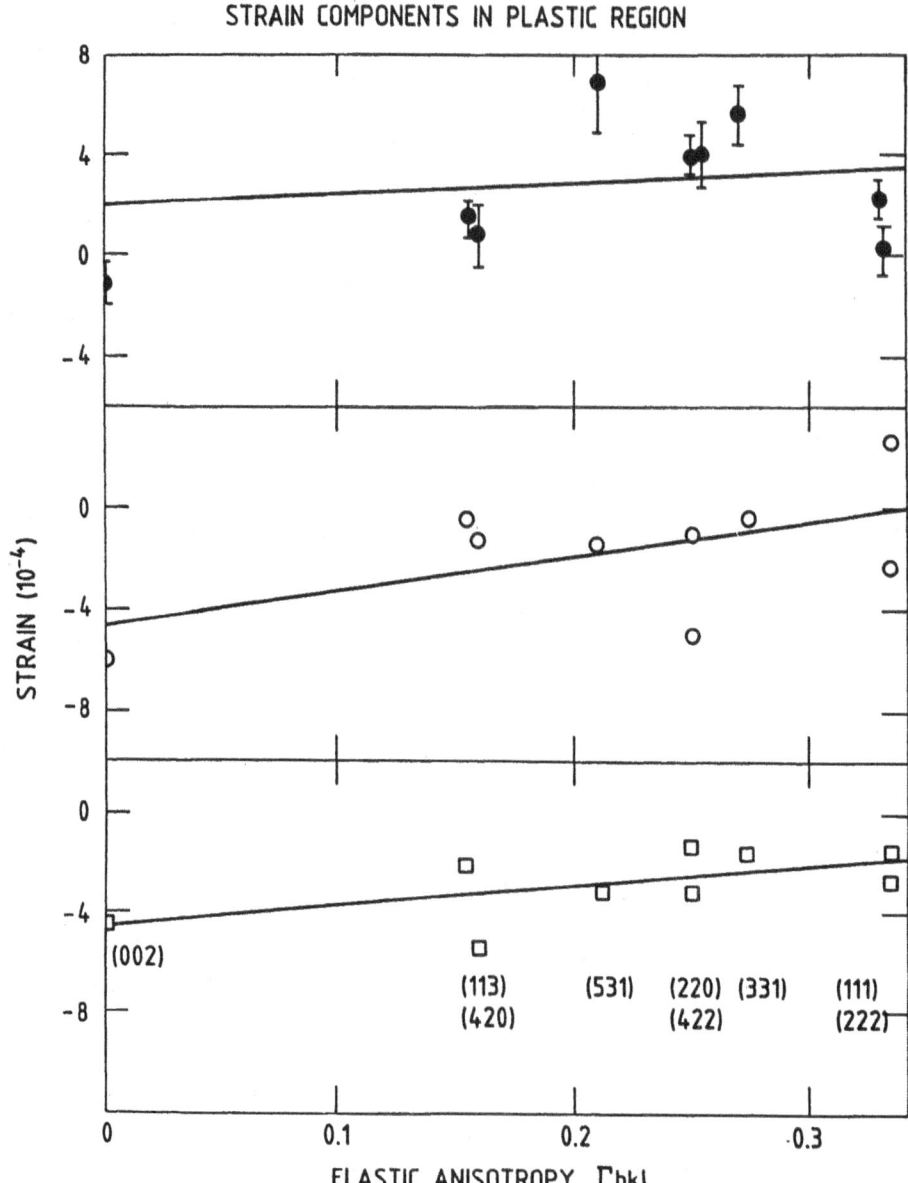

Fig. 7. Variation of the residual strains deduced from many different plane spacings plotted against the elastic anisotropy parameter for tangential, radial and normal components. The location is 3 mm for the ID where plastic flow is expected to be strongest.

TABLE 2. Analysis of strain variation with (hkl) at two locations in the diametrally compressed ring

LOCATION		HOOP		RADIAL		NORMAL	
		Exp	Calc	Exp	Calc	Exp	Calc
13 mm	SLOPE	28±2	23	-4±1	-10	-9±2	-12
	INTERCEPT	-15±7	-21	4±4	5	2±7	8
3 mm	SLOPE	4±2	-7	15±2	3	9±1	4
	INTERCEPT	2±9	7	-5±9	-1	-5±4	-3

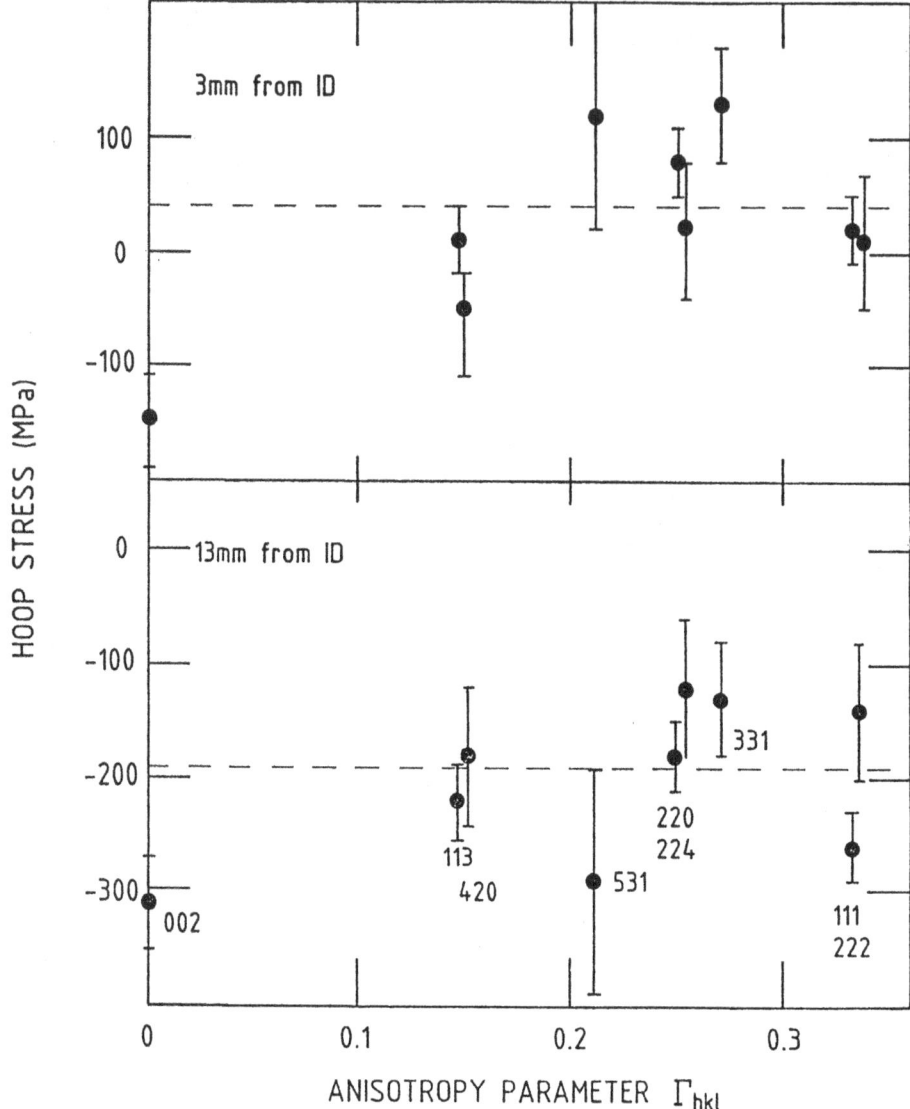

Fig. 8. Variation of the tangential residual stress deduced from the measured tangential, radial and normal residual strains with the aid of the appropriate diffraction elastic constants. The results are consistent with a single value of stress to within two standard errors except for (002) which is always several standard deviations from this value.

7. Conclusions

There appears to be semiquantitative agreement between neutron scattering measurements of residual strain and values obtained by finite element methods. A disagreement between theory and experiment shows up as an overestimate of the maximum strain and stress under load at the West position. The inadequacy of the assumption of a single-valued stress field shows up in the spread of values of strain around 5 mm at the West location where the strain goes from compressive to tensile. The spread at 5 mm is the same as the spread at neighbouring positions.

There is a greater experimental spread between the softest direction in the cubic lattice [002] and the stiffest [111] than the calculation suggests. It appears that the other reflections cluster around a single stress value, whereas the [002] direction does not conform with this behaviour. Still greater experimental precision is required in the measurements of strain to be able to test the hypothesis of a single stress field more stringently.

8. Acknowledgements

We would like to acknowledge the help of Dr. G. Dolling in the initial experiments and a valuable conversation with Dr. P. Predecki on diffraction elastic constants. Valuable experimental assistance was given by H.F. Nieman, D.C. Tennant and M.M. Potter.

9. References

1. Flower, E.C., MacEwen, S.R. and Holden, T.M., (1987) Proceedings of the Second International Conference on Advances in Numerical Methods in Engineering: Theory and Application, University of Swansea, U.K.; University of California Report UCRL-96643.

2. Flower, E.C., MacEwen, S.R., Holden, T.M. and Hosbons, R.R., (1989) Proceedings of the International Conference on Residual Stresses, ICRS2, Edited by G. Beck, S. Denis and A. Simon (Elsevier Applied Science: London), p. 209-15.

3. Hallquist, J.O., (1986) Lawrence Livermore National Laboratory Report, UCID-19677.

4. Taylor, R.L., private communication.

5. Hallquist, J.O., (1984) Lawrence Livermore National Laboratory Report, UCID-18822.

6. Ledbetter, H.M., (1977) Mat. Sci. and Engineering 29, 255.

7. Mangalick, M.C. and Fiore, N.F., (1968) Trans. Metall. Soc. Am. Mech. Eng. 242, 2363-4.

112

8. Kröner, E., (1967) J. Mech. Phys. Solids 15, 319-29.

9. Dölle, H., (1979) J. Appl. Cryst. 12, 489-501.

10. Holden, T.M., Holt, R.A., Dolling, G., Powell, B.M. and Winegar, J.E., (1988) Met. Trans. A, 19A, 2207-14.

3. ASPECTS OF FUNDAMENTAL PRINCIPLES

MACROSTRESSES, MICROSTRESSES AND STRESS TENSORS

L. PINTSCHOVIUS
Kernforschungszentrum Karlsruhe, INFP
Postfach 3640, W-7500 Karsruhe, Germany

ABSTRACT. Residual stresses are commonly classified as residual stresses of the first kind (macro residual stresses), second and third kind (micro residual stresses) according to the length scale over which they are nearly constant in magnitude and direction. Diffraction methods are sensitive to all three kinds of residual stresses, thus giving more detailed information on the stress state of a material than any other technique. On the other hand, this complicates the evaluation of diffraction data considerably. An overview is given how to detect the presence of microstresses and how to separate them from macrostresses.

At any particular point, the stress state can be characterized by a stress tensor from which the magnitude and the direction of the three principal stresses can be calculated. In case that the directions of the principal stresses cannot be inferred from the sample geometry, a complete characterization of the stress state requires the determination of lattice strains in at least six directions. The stresses can then be evaluated from the strains via a tensor calculus using tabulated or specifically determined elastic constants. Appropriate elastic constants, which describe the elastic response of the material for macroscopic dimensions or single grains, allow the determination of both macro- and microstresses.

1. Introduction

In characterizing the stress state of a material one has to bear in mind that stresses may strongly fluctuate from grain to grain or even within a grain. Often, engineers are only interested in averages over macroscopic distances, so-called macrostresses, as for these macrostresses a broad knowledge exists to relate their size and sign to the performance of the material. However, even in this case the fluctuations of the stresses on a microscopic scale, the so-called microstresses, should not be ignored, as diffraction methods like neutron stress analysis never give directly macrostresses, but only a superposition of macro- and microstresses. In such a case, microstresses are often viewed as a complication and methods are looked for which minimize the influence of microstresses. There are, of course, multi-phase materials like duplex steels or fibre-reinforced materials where it is obvious that the characterization of the stress state by macrostresses is inadequate. A closer look reveals that also in (nearly) single-phase materials the microstresses are often by no means negligible. So far, the knowledge to assess the consequences of microstresses

115

M. T. Hutchings and A. D. Krawitz (ed.),
Measurement of Residual and Applied Stress Using Neutron Diffraction, 115–130.
© 1992 *Kluwer Academic Publishers.*

for the mechanical and corrosion properties is not very broad, but as failure of technical components starts on a microscopic scale, microstresses are probably as important as macrostresses. Therefore, it might often be worth to exploit the potential of the diffraction methods to get detailed information not only on macrostresses, but also on microstresses.

In the following we will give the definitions of macro- and microstresses and explain the relation between stresses and strains. Then, we will discuss various origins of microstresses and how to separate macro- and microstresses in neutron stress analysis.

2. Definitions

In the following we will deal with residual stresses (RS), but analogous distinctions between macro- and microstresses can be made for load stresses as well. One might also think to start from strains and classify these into different categories. As the use of stresses or strains is completely equivalent, we will concentrate on stresses.

It is now generally accepted that RS can be grouped into three classes according to the length scale over which the RS are nearly homogeneous [1]:

- Residual stresses of the 1st kind (RS I) are nearly homogeneous across large areas, say several grains, of a material and are equilibrated within the whole body. RS I are also called macro RS.
- Residual stresses of the 2nd kind (RS II) are nearly homogeneous across microscopic areas, say one grain or parts of a grain, of a material and are equilibrated across a sufficient number of grains. RS II are nearly homogeneous micro RS.
- Residual stresses of the 3rd kind (RS III) are inhomogeneous across submicroscopic areas of a material, say several atomic distances within a grain, and are equilibrated across small parts of a grain. 3rd kind RS are inhomogeneous micro RS.

These definitions are sufficient to describe all RS states occurring in practical cases. Usually, a superposition of RS of the 1st, 2nd and 3rd kind determine the total RS state acting at a particular point of a material. A typical example is shown in Fig. 1. Here, the major cause of RS II is the presence of two phases A and B. As will be discussed below, RS II occur in single-phase materials as well, although they are usually smaller in proportion to the RS I than in multi-phase materials. Fig. 1 is a simplification of real conditions in that the grain boundaries are assumed to be very thin. There is evidence that the grain boundary regions cannot always be neglected for the balance of forces, but it is very difficult to obtain information on their stress state.

The stress state at any particular point of the specimen can be characterized for a given system of co-ordinates by a 3×3 tensor, the so-called stress tensor [see, e.g. 2]

$$\sigma_{ik} \qquad i,k = 1,2,3 \qquad (1)$$

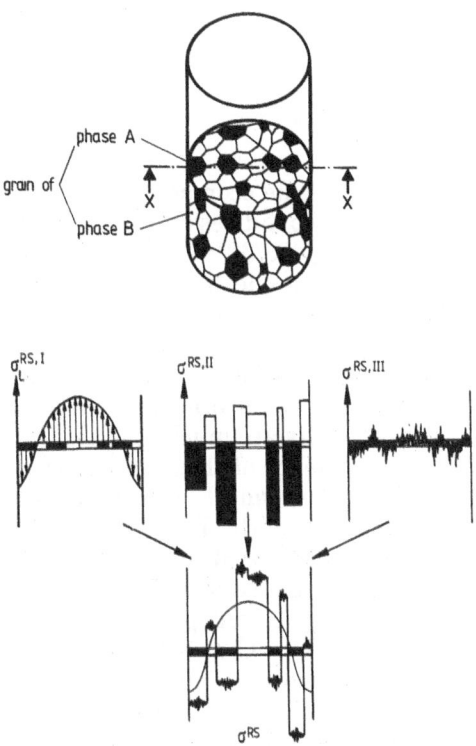

Figure 1. Illustration of a stress state composed of RS of the 1st, 2nd and 3rd kind, schematically [1].

The tensor is symmetric, i.e. $\sigma_{ik} = \sigma_{ki}$. The diagonal elements σ_{11}, σ_{22}, σ_{23} are the normal components of stress, while σ_{12}, σ_{13}, and σ_{23} are the shear components. For each tensor, there exists a system of co-ordinates in which the off-diagonal elements vanish. In this case, the diagonal elements are called the principal stresses. Each stress state can also be characterized by the 3 principal stresses and the directions of the principal stresses in respect to a reference system.

Similarly, the strain state can be characterized by a symmetric 3×3 deformation tensor

$$\varepsilon_{ik} \quad i,k = 1,2,3 \tag{2}$$

The relationship between the stress and the deformation tensor is given by Hooke's law

$$\varepsilon_{ik} = \sum_{\alpha,\beta} S_{ik,\alpha\beta} \, \sigma_{\alpha\beta} \tag{3}$$

where $S_{ik,\alpha\beta}$ are the elastic compliance constants of the crystal. Conversely, σ_{ik} are given by

$$\sigma_{ik} = \sum_{\alpha,\beta} C_{ik,\alpha\beta}\, \varepsilon_{\alpha\beta} \qquad (4)$$

where $C_{ik,\alpha\beta}$ are the elastic constants. Some economy in notation is achieved by pairing the first two and the second two indices of the $S_{ik,\alpha\beta}$ resp. $C_{ik,\alpha\beta}$ and using the equivalence $11{\to}1$, $22{\to}2$, $33{\to}3$, $23/32{\to}4$, $13/31{\to}5$, $12/21{\to}6$.

For the investigation of microstresses one has to use the single crystal values for the C_{ik} resp. S_{ik}. They are tabulated for pure compounds. There are some materials for which the preparation of single crystals has not been achieved, e.g. austenitic steel, which greatly hampers the determination of the elastic constants. There are methods to determine the elastic constants on polycrystalline aggregates, but the results are much less accurate than those obtained on single crystals. The dependence of these constants on impurities or the microstructure is usually rather small, by far smaller than the corresponding variation of the yield strength. The variation of the elastic constants with temperature is appreciable only if the temperature comes close to the melting point or a structural phase transformation.

For cubic crystals, there are only 3 independent elastic constants $C_{11}=C_{22}=C_{33}$, $C_{12}=C_{13}=C_{23}$, $C_{44}=C_{55}=C_{66}$, all other components are zero. Similarly, there are only 3 independent compliance constants S_{11}, S_{12} and S_{44}. For practical reasons $C_0=C_{11}-C_{12}-\frac{1}{2}C_{44}$ resp. $S_0=S_{11}-S_{12}-\frac{1}{2}S_{44}$ are often used, too. C_0 and S_0 are zero for elastically isotropic materials. The microstrain ε_{hkl} in the crystallographic direction (hkl) produced by a 3-axial stress state with the principal stresses σ_i ($i=1,2,3$) is given by [3]

$$\varepsilon_{hkl} = \sum_{i=1}^{3} \left(S_{12} + S_0 \sum_{i=1}^{3}{}' \alpha_i^2 \beta_i^2 + \frac{1}{2} S_{44} \left(\sum_{i=1}^{3} \alpha_i \beta_i \right)^2 \right) \sigma_i \qquad (5)$$

where α_1 are the cosines of the angles between the direction (hkl) and the crystallographic axes and β_1 are the cosines of the angles between the principal stresses and the crystallographic axes. We note that for elastically anisotropic materials the direction of the principal strains do no longer coincide with the direction of the principal stresses unless that they are parallel to the crystallographic axes.

In the case that the principal stresses point in the direction of the axes of co-ordinates and the material is elastically isotropic, the strains are given by

$$\varepsilon_{\phi,\psi} = S_{12}(\sigma_1 + \sigma_2 + \sigma_3) + \frac{1}{2} S_{44}(\sigma_3 + (\sigma_1 \cos^2\phi + \sigma_2 \sin^2\phi - \sigma_3) \sin^2\psi) \qquad (6)$$

Here ψ denotes the angle between the z-axis and the strain direction and ϕ the angle between the measuring direction and the x-direction. Hence, a plot of $\varepsilon_{\phi,\psi}$ versus $\sin^2\psi$ gives a straight line. This is illustrated for the cases $\phi=0°$ (measuring direction in the x-direction) and $\phi=90°$ (measuring direction in

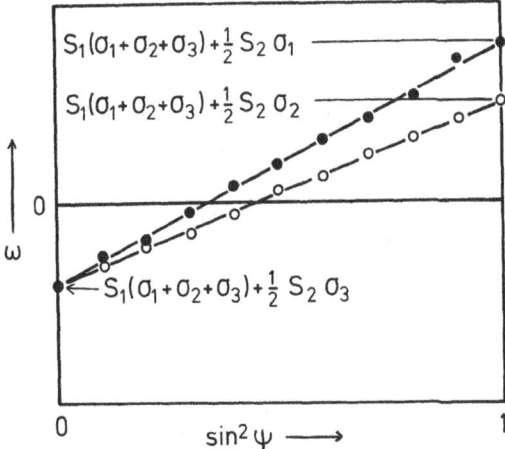

Figure 2. Schematic representation of ε vs sin²ψ-distributions for scans in the direction of the principal stress σ1 resp. σ2. ψ is the angle between the strain direction and the direction of the principal stress σ3.

the y-direction) in Fig. 2. Such plots are the basis of X-ray stress analysis, but can be used in neutron stress analysis as well. Due to the strong absorption of X-rays, only a limited ψ-range is accessible. Often, only the range $|\sin^2\psi| \le 0.5$ is investigated, but modern X-ray equipment allows to extend the ψ-range up to $|\sin^2\psi| \le 0.9$. With neutrons, the full ψ-range up to $\sin^2\psi = 1$ can be covered.

If the lattice parameter D_0 of the stress-free material is not known, $D_{\phi,\psi}$ instead of $\varepsilon_{\phi,\psi}$ is plotted and stresses can be evaluated from the slope of the D vs sin²ψ-curves. However, here only stress differences $(\sigma_1-\sigma_3)$ or $(\sigma_2-\sigma_3)$ can be determined. For X-rays this restriction is not serious, as for macrostresses the component σ_3 perpendicular to the surface is zero. For stress analysis with neutrons in the interior of the specimen there is no such boundary condition, and hence D_0 is required for a full determination of the stress state. The knowledge of D_0 poses a problem for multi-phase materials, which cannot be made stress-free by annealing, or in specimen containing gradients in chemical composition.

For macroscopically isotropic materials the notation $S_1 = S_{12}$ and $S_2 = S_{44}$ is used. S_1 and $\frac{1}{2}S_2$ are called the X-ray elastic constants (XEC's). If the material is single phase and if also the single crystal elastic response is isotropic, the XEC's can be calculated from Young's modulus E and Poisson's ratio v by

$$S_1 = -v/E \tag{7}$$

$$S_2 = 2(1+v)/E \tag{8}$$

If the single crystal elastic behavior is not isotropic, the ε vs sin²ψ-distributions can nevertheless often be approximated by eq. (6). However, in this case the XEC's depend on the Miller indices (hkl) of the reflection line used. There have been many attempts to calculate the XEC's using different assumptions on the interaction between the grains in a polycrystalline aggregate. Both Voigt's hypothesis [4] (equal strains in all grains) and Reuss' hypothesis [5] (equal stress in all grains) are still widely used [e.g. in 7] in spite of their obvious incompatibility with theory of elasticity. More realistic calculations follow the procedure of Eshelby [7] and Kröner [8], who dealt with the case of anisotropic spherical crystals embedded in an isotropic matrix [9,10]. The calculated values vary with the orientation factor Γ, which in the case of cubic materials is given by

$$\Gamma = (h^2k^2 + h^2l^2 + k^2l^2)/(h^2+k^2+l^2)^2 \tag{9}$$

For cubic systems the XEC's are expected to be given by (7) and (8) for $\Gamma = 0.2$. We note that for materials which are not strictly single-phase the XEC's depend not only on the elastic anisotropy, but also on the presence of additional phases as well as the microstructure and the texture of the material. Therefore, values reported in the literature are of limited value only, and XEC's are often determined from lattice strains observed after externally loading the specimen.

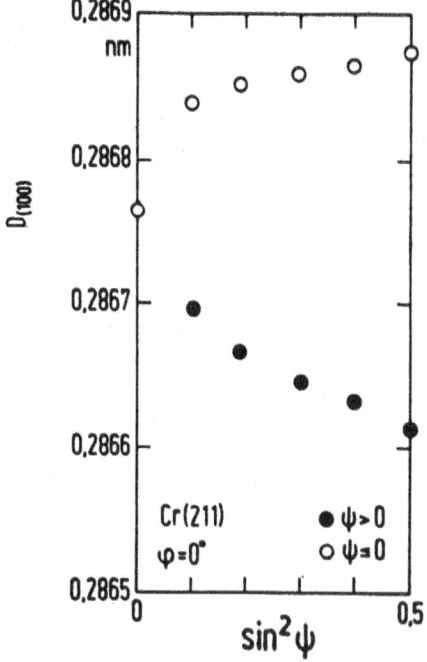

Figure 3.   D vs $\sin^2\psi$-distribution observed with X-rays at the surface of a ground C60-sample in the grinding direction [12]. The splitting of the D-values for $+\psi$ and $-\psi$ indicates, that the principal stresses are tilted away from the sample system.

In many cases of practical interest, the directions of the principal stresses can be inferred from the sample geometry, at least for lines or planes of high symmetry. In this case, the measurement of 3 strain components may be sufficient to fully determine a three-dimensional stress state. For a complicated geometry with an unknown orientation of the principal stresses, at least 6 independent strain measurements are necessary to determine the stress tensor. In practice, it is highly recommended to make more independent measurements and to extract the strain tensor from the data by a least-squares fit [11]. The stress tensor can then be calculated from the strain tensor using the above formula.

We note that the direction of the principal stresses of type II may differ from that of type I. A well-known example are the RS observed after grinding at the surface by X-rays: For measurements along the grinding direction, a so-called ψ-splitting is found, which means that the principal stresses are tilted away from the surface plane. An example is shown in Fig. 3. For RS I, however, the boundary conditions at the surface imply that two principal stresses are parallel to the surface and the third one perpendicular to the surface is zero. Therefore, RS giving rise to a ψ-splitting must contain a large component of type II with a different orientation of the principal stresses than that of the type I component.

3. Origins of microstresses

Microstresses emerge from strain incompatibilities on a microscopic scale. Such incompatibilities occur already on elastic loading if the elastic properties of the material vary from grain to grain. When an external load is applied to a polycrystalline aggregate consisting of hard and soft particles, the hard particles are loaded stronger than the soft ones. If the total stress is tensile, the RS II is tensile in the hard particles and compressive in the soft particles (Fig. 4). The most obvious reason for a variation of the elastic response is the presence of different phases, but even within one phase, elastic anisotropy can

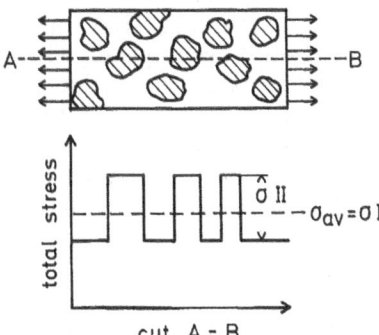

Figure 4. Illustration of a stress state in a material consisting of hard particles embedded in a soft matrix subject to a tensile load, schematically.

lead to considerable microstresses. Elastic anisotropy is small for aluminium (~10%), but rather strong in ferritic steels and even stronger in austenitic steels: Here, the ratio of Young's modulus of the hardest (111) and the softest (100) direction is a factor of 2 and 3, respectively.

After strong plastic flow, the main cause for microstresses is usually not the variation of the elastic properties from grain to grain, but the variation of the yield strength, either because the grains belong to different phases or because the yield strength depends on the crystallographic direction. Yield strength anisotropy leads to RS which may change sign from grain to grain. The development of microstresses due to yield strength anisotropy is illustrated in Fig. 5. As the yield strength depends not only on the orientation of the grain, but also on the dislocation density, microstresses may also develop because grain parts located near grain boundaries act as hard areas and those in the interior as weak areas. This has been observed in tensile deformed plain carbon steels [13].

The above mentioned microstresses are of type II. RS of type III result from all types of lattice imperfections, as interstitial atoms, vacancies, dislocations, grain boundaries, precipitations etc. (see Fig. 6). All these imperfections are rather common in most materials used in mechanical engineering, and hence very complex RS III patterns are found.

Especially after severe cold working, microstresses may easily exceed macrostresses, so that experimental data on lattice strains should be interpreted with caution. Unfortunately, the theoretical understanding of the microstresses originating from plastic flow is still in its beginnings. In particular, microstresses due to plastic anisotropy, so-called grain interaction stresses, cannot easily be predicted. Therefore, the experimentalist has to find ways to detect the presence of microstresses and to separate them from macrostresses, which is the topic of the next section.

4. Separation of macro- and microstresses

Very often, the lattice strains observed in a diffraction experiment are converted into macrostrains by tacitly assuming that microstresses are negligible. What are the criteria to justify this assumption? This is not easy to say. In fact, it is much easier to say when this assumption is not justified. Firstly, if the material under consideration contains an appreciable amount of a second phase with strongly different mechanical properties. We note that this is the case for many materials used in mechanical engineering, especially high strength materials. Minority phases are often ignored although their percentage is not very small: e.g., the cementite in a plain carbon steel with 0.5% C amounts to 7% of the volume!

The procedure to determine stresses in two- or multi-phase materials is in principle straightforward: each phase has to be investigated separately by using respective reflection lines. This sounds easy, but may be very difficult or virtually impossible, if the abundance of a phase is only ~10% or even less. In those cases where an investigation in two or more phases is feasible, the

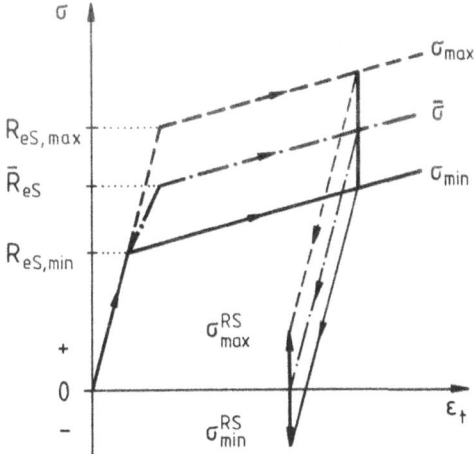

Figure 5. Development of RS II due to yield strength anisotropy [1].

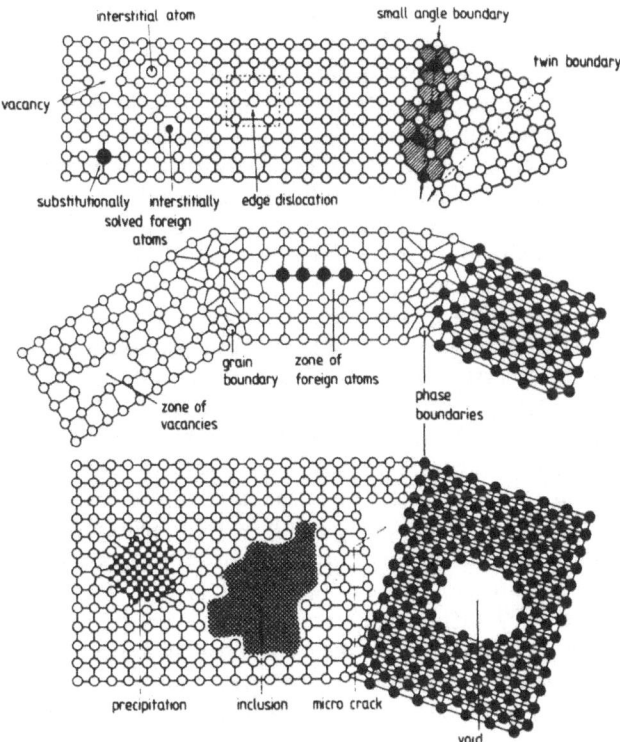

Figure 6. Different sources of RS III, schematically [1].

results are far more meaningful than macrostress values obtained by other methods.

It is often attempted to take the presence of addional phases into account by using calculated or experimentally determined XEC's. However, we note that this is justified only if the microstrains result from differences in the elastic response. If the microstrains result from differences in yield strength, the use of XEC's is completely inadequate to correct for the presence of additional phases.

In (nearly) single phase materials, a good reason to suspect the presence of strong microstresses is the observation of a marked texture. In case that the texture has not been investigated separately, one can use the intensities observed in the strain measurements to look for a texture (see e.g. Fig. 9). Experience shows that a texture produced by cold working is normally associated with considerable grain interaction stresses. In order to detect or even to characterize the grain interaction stresses, it is necessary to make measurements using many different reflection lines. Even then, the evaluation of stresses from the data is not straightforward and has to be found out from case to case. Examples will be given in a separate contribution [14,15]. In favorable cases, the stress state of the grains can be determined as a function of their orientation and macrostresses are obtained by averaging over the observed microstresses [16,17]. If only macrostresses are of interest, it is recommended to use reflection lines with high multiplicity, as these lines sample grains with very different orientations and hence are relatively little influenced by grain interaction stresses. Unfortunately, reactor based neutron diffractometers allow to use only reflection lines with low Miller indices as the diffracted intensity decreases sharply with decreasing neutron wavelength. On the other hand, diffractometers located at pulsed sources allow to use reflection lines with very high multiplicity like X-ray diffractometers (e.g. multiplicity $= 96$ for ($\{732 + 651\}$ or $\{721 + 633 + 552\}$ in case of bcc metals).

Instead of starting from the microstructure of the material under investigation one may start from the strain data to conclude the presence or absence of microstresses. In those cases, where the microstructure does not suggest the presence of microstresses, it is the only way. We note that there are also single (or nearly single) phase materials without pronounced texture which show considerable microstresses. There are several ways to estimate the influence of microstresses on the observed peak shifts:

(i) After converting strains into stresses one should check the balance of forces. Since RS are self-equilibrating stresses, the resultant force and the resultant moment produced by them must be zero. It requires the measurement of complete strain distributions across the whole specimen. If the balance of forces is fulfilled within experimental error, there is good hope, although no guarantee, that the calculated RS are of type I. An example, where such a check gave a very satisfactory result, is shown in Fig. 7 [18], whereas Fig. 8 [19] shows an example, where the observed RS cannot be interpreted as RS I. As the material was to a very good degree single phase and as special care was taken to eliminate the influence of grain interaction stresses, the discrepancy between lattice strains and macrostrains has to be ascribed to so-called deformation RS. These deformation RS are presumably

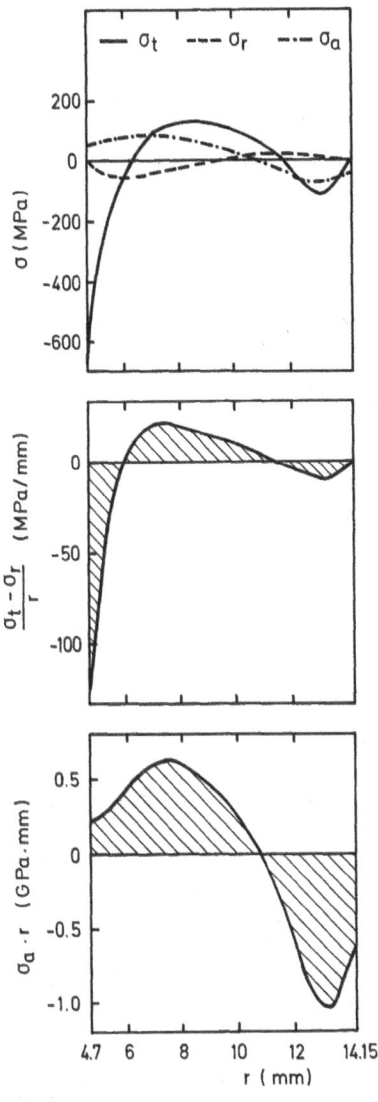

$$\sigma_t \quad --- \sigma_r \quad ---\cdot \sigma_a$$

Figure 7.
above: RS distributions in an autofrettaged steel tube calculated from neutron strain data [18].
middle: $(\sigma_t-\sigma_r)/r$ versus radius calculated from the stress values shown above. In a specimen with cylindrical symmetry, the integral of this quantity from r_i to r_a should be zero.
below: $\sigma_z \cdot r$ versus radius calculated from the stress values shown above. In a specimen with cylindrical symmetry, the integral of this quantity from r_i to r_a should be zero.

located in highly disturbed regions of the material which do not contribute to the diffraction peaks. A check of the balance of forces or by mechanical methods is the only way to detect the presence of deformation RS.

(ii) Check the linearity of the $D_{\{hkl\}}$ versus $\sin^2\psi$-distributions. Grain interaction stresses often give rise to non-linearities, in particular for reflection lines $\{hkl\}$ with low multiplicity. Examples are shown in Figs. 9 and 10. Moderate non-linearities are also expected in materials with RS I only because of elastic anisotropy [6]. Only a somewhat lengthy calculation based on the single crystal elastic properties and the observed texture can show if elastic anisotropy is sufficient to explain the non-linearities. For cubic

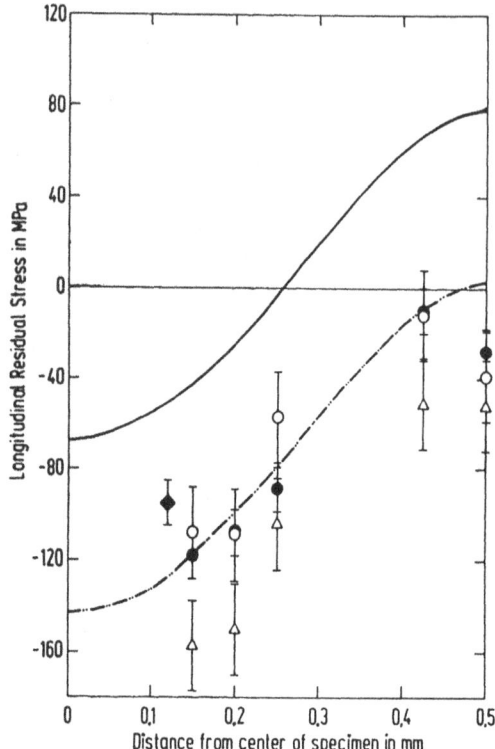

Figure 8. The full line shows the RS I in a cold rolled steel sheet vs. distance from the surface determined by a mechanical method. Data points were obtained by diffraction methods: X-rays, ●, o, Δ and neutrons ◊ using various reflection lines. X-ray values for the interior were obtained after stepwise removal of surface layers and correcting for stress relief. Scatter between the various X-ray and neutron stress values is due to grain interaction stresses. The general downward shift shows, that the RS observed by diffraction methods are a superposition of RS I and deformation RS [19].

systems, a way to avoid such a calculation is to perform measurements using the {hoo} or {hhh} reflections lines, which give linear D vs $\sin^2\psi$-distributions also in case of elastic anisotropy. Finally, we want to emphasize that the observation of a strongly non-linear D versus $\sin^2\psi$-distribution indicates the presence of large intergranular stresses, but that the observation of a nearly linear D vs $\sin^2\psi$-distribution does not necessarily imply that microstresses are negligible (see the D_{211} vs $\sin^2\psi$-distribution in Fig. 10).

In neutron stress analysis, it is very time-consuming to make measurements for many ψ. Furthermore, attenuation of the neutron beam within the sample often severely restricts the accessible ψ-range. Therefore, the following check is more useful for the neutron technique:

(iii) Perform measurements for several reflection lines {hkl} and plot the results versus the orientation factor Γ. An example is shown in Fig. 11. If the difference of the observed lattice strains can be explained by elastic anisotropy

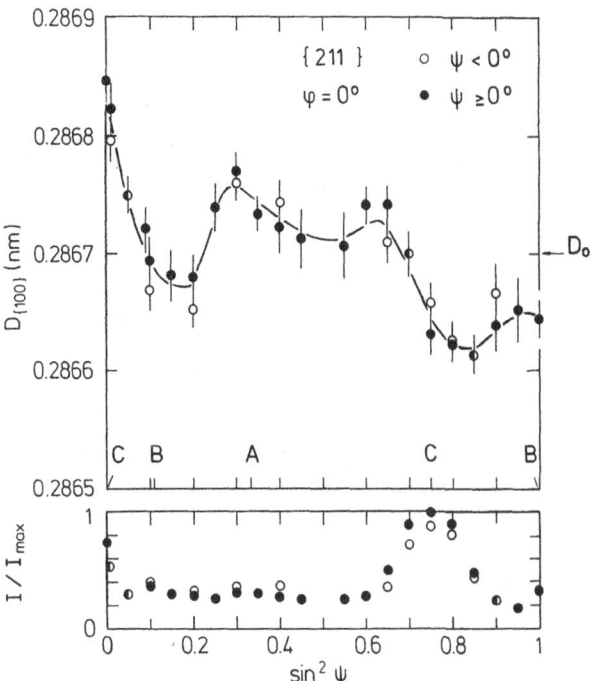

Figure 9. D vs sin²ψ and intensity vs sin²ψ observed in the interior of a cold rolled steel sheet with neutrons in the rolling direction. The oscillatory behavior of D vs sin²ψ is typical for such samples. It is partly due to elastic anisotropy, but primarily due to intergranular stresses originating from yield strength anisotropy [17].

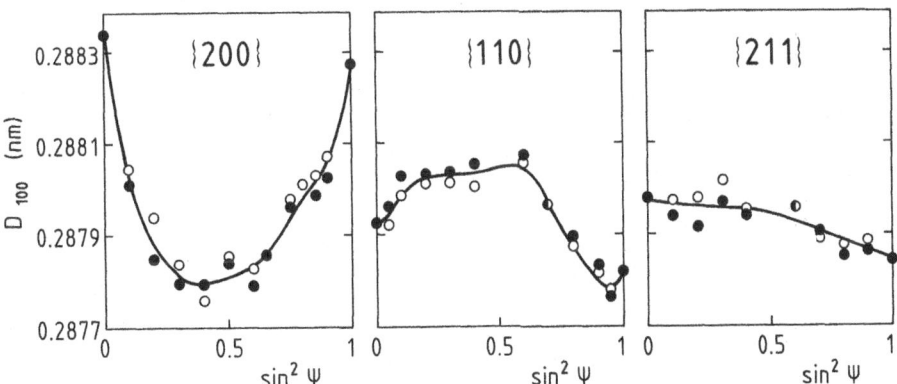

Figure 10. Examples of highly non-linear D versus sin²ψ-distributions as measured by neutron diffraction in the ferritic phase of a ferritic-austenitic steel after 12% plastic strain using different (hkl)-peaks [20].

(Fig. 11 above), there is reason to believe (but again no guarantee) that the RS are largely of type I. If not (Fig. 11, below), it indicates the presence of strong

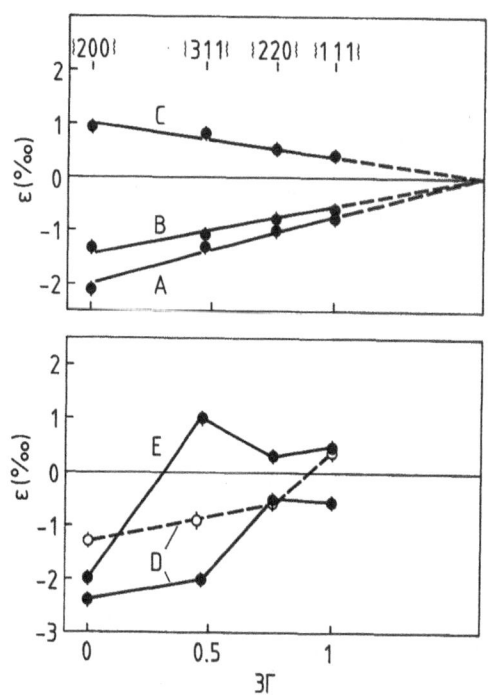

Figure 11.
Lattice strains observed by neutron diffraction in a welded plate with different (hkl) peaks versus the orientation factor Γ, above for points in the base and below for points in the weld [21]. The systematic behavior of $\varepsilon(hkl)$ vs Γ observed at points in the base indicates, that it can be explained by elastic anisotropy.

grain interaction stresses. As mentioned above, no general instructions can be given for the analysis in such a case.

In conclusion, it is already a non-trivial task to detect the presence of RS II. Looking at the microstructure and using more than a single reflection line is a big help. In this context, it is an inherent advantage of the time-of-flight technique to yield automatically data for several reflections lines. This comes all the more into play if it proves necessary to separate macro- from microstresses.

The presence of RS III is not so much a problem for the evaluation of strain data. RS III do not give peak shifts, but peak broadenings only. Peak broadenings are sometimes used to monitor changes of the density and the arrangement of dislocations after plastic deformation, annealing or recrystallization. However, there is no unique relationship between peak broadening and RS III: stress gradients in the internal probe region, grain interaction stresses and small grain size also contribute to peak broadenings.

5. Outlook

Experience shows that, there are many cases where the strains observed in a diffraction experiment do not reflect macrostresses only, but are heavily influenced by microstresses. This is sometimes seen as a drawback, as most engineers are still almost exclusively interested in macrostresses. However, it offers the prospect to learn more about the stress state of the material than can be achieved by any method which is sensitive only to macrostresses. In order to fully exploit the potential of the diffraction methods, the evaluation procedure should not aim at eliminating the influence of microstresses as far as possible, but should aim at a quantitative estimate of the microstresses. Only in this way it will be possible to create a basis for a better understanding of the origin of microstresses as well as their relevance for mechanical engineering. A better knowledge of their relevance will in turn attract more attention to them from the engineering point of view.

6. Acknowledgements

The author is indebted to Prof. Dr. E. Macherauch and Dr. B. Scholtes for helpful discussions and critical reading of the manuscript.

7. References

1. Macherauch, E., 'Origin, measurement and evaluation of residual stresses', in Residual Stresses in Science and Technology, Proc. of the Int. Conf. on Residual Stresses, 1986 Garmisch-Partenkirchen (FRG), ed. by Macherauch, E., and Hauk, V., DGM Informationsgesellschaft, Oberursel (1987), p. 3.
2. Nye, F.J., 'Physical Properties of Crystals', (Clarendon Press, Oxford 1957).
3. Glocker, R., Z. Techn. Phys. 19 (1938) 289.
4. Voigt, W., Lehrbuch der Kristallphysik, Teubner, Leipzig-Berlin (1928).
5. Reuss, A., Z. angew. Math. Mech. 9 (1929) 49.
6. Dölle, H., and Hauk, V., 'Influence of the mechanical anisotropy upon the stress evaluation by means of X-rays', Z. Metallkde. 69 (1978) 411.
7. Eshelby, J.D., Proc. Roy. Soc. A241 (1957) 376.
8. Kröner, E., Z. Physik 151 (1958) 504.
9. Sayers, C.M., 'The strain distribution in anisotropic polycrystalline aggregates subject to an external stress field', Phil. Magaz. A49 (1989) 243.
10. Windsor, C.G., 'The effects of crystalline anisotropy on the elastic response of materials', this volume.
11. Lorentzen, T., and Leffers, T., 'Strain Tensor measurements by neutron diffraction', this volume.

12. Hauk, V.M., Ouderhoven, R., and Vaessen G., 'The state of residual stress in the near-surface region of homogeneous and heterogeneous materials after grinding', Metall. Trans. A13 (1982) 1239.

13. Faninger, G., Hauk, V., 'Deformation residual stresses', Härterei-Tech. Mitt. 31 (1976) 48.

14. Pintschovius, L., 'Grain Interaction Stresses', this volume.

15. Noyan, C., 'The effect of microstresses on D(hkl) versus $\sin^2\psi$-distributions', this volume.

16. Willemse, P.F., Naughton, B.P., and Verbrauh, C.A., 'X-ray residual stress measurements on cold-drawn steel wire', Mat. Science and Engineer. 56 (1982) 25

17. Pintschovius, L., Hauk, V., and Krug, W.K., 'Neutron diffraction study of the residual stress state of a cold-roled steel strip', Mat. Science and Engineer. 92 (1987) 1.

18. Pintschovius, L., Macherauch, E., and Scholtes, B., 'Determination of residual stresses in autofrettaged steel tubes by neutron and X-ray diffraction', Mat. Science and Engineer. 84 (1986) 163.

19. Feja, K., Hauk, V., Krug, W.K., and Pintschovius, L., 'Residual stress evaluation of a cold-rolled steel strip using X-rays and a layer removal technique', Mat. Science and Engineer. 92 (1987) 13.

20. Hauk, V., Nikolin, H.J., and Pintschovius, L., 'Evaluation of deformation residual stresses caused by uniaxial plastic strain of ferritic and ferritic-austenitic steels', Z. Metallkde. 81 (1990) 556.

21. Pintschovius, L., Munz, D., and Nyilas, A., 'Neutron diffraction study of the residual stresses in austenitic steel weldments', unpublished results.

SEPARATION OF MICROSTRESSES AND MACROSTRESSES

R.A. WINHOLTZ
Department of Materials Science and Engineering
McCormick School of Engineering and Applied Science
Northwestern University, Evanston, IL 60208
USA

ABSTRACT.   In two phase or composite materials it is possible to determine both the macrostresses in the material and the average microstresses present in each phase. The stress equilibrium relations allowing this separation are outlined. It is shown that errors in the unstressed lattice parameter lead to errors in only the hydrostatic component of the stress tensor. It is thus possible to accurately determine the entire deviatoric macrostress and microstress tensors without the unstressed lattice parameters of the individual phases. Two examples of the usefulness of macrostress and microstress measurements are given. In a SiC-TiB_2 microcracking ceramic composite large hydrostatic microstresses develop during consolidation due to differential thermal contraction. Microcracks form during stressing, relaxing the thermal microstresses. Measurements of this relaxation have helped quantify the toughening in this system associated with microcracking. Measurements in both the ferrite and cementite phases of 1080 steel during low cycle fatigue show the proportion of the applied load carried by each phase. These results show that as the material yields in both tension and compression, the carbides take a higher fraction of the load and thus the stress range experienced by the carbide phase is much higher.

1. Introduction

Residual stresses can play an important role in a material's properties and performance. Therefore, it is very useful to have methods to measure the residual stresses. Neutron or x-ray diffraction offers a useful method for determining residual stresses by measuring the changes in the internal lattice spacings of a crystalline material [Noyan and Cohen (1987)]. Measurements of the lattice spacings give the strain from which the stresses can be calculated. In multiphase materials stresses will not in general be distributed equally among the phases and it is useful to define macrostresses and microstresses to describe the stress state. The macrostress is the same in each of the phases while the microstress in each phase is the deviation from their macrostress value.

Diffraction offers the unique capability to measure the both macrostresses and microstresses

131

M. T. Hutchings and A. D. Krawitz (ed.),
Measurement of Residual and Applied Stress Using Neutron Diffraction, 131–145.
© 1992 *Kluwer Academic Publishers.*

because each phase will give rise to its own set of diffraction peaks and the stresses can be determined indepenently for each phase. Stress equilibrium conditions allow these stresses to be separated into the macrostress and the microstresses. Techniques now exist for determining the entire three dimensional macrostress and microstress tensors in two phase or composite materials and have been made in a number of systems [Noyan and Cohen (1985), Winholtz and Cohen (1989), Abuhasan et al. (1990), Magley et al. (1990),].

2. Theory

2.1 Macrostresses and Microstresses

Macrostresses are the stresses traditionally associated with residual stresses and are the stresses that are measured by dissection methods. Macrostresses are the stresses that would develop in a homogeneous material. Microstresses are present because real materials are not homogeneous. The microstresses are the differences between the real stresses present and the macrostress value.

The macrostresses will vary slowly on the scale of the materials microstructure, while the microstresses vary on the microstructural scale.

Macrostresses will typically arise from different amounts of plastic deformation in different regions of a body. In bending a beam, for instance, the outer layers will plastically deform while the central portion will remain in the elastic region. This will lead to residual macrostresses in the beam after the load is removed from the beam. Microstresses result from different physical properties of the two phases such as differences in the coefficients of expansion, different plastic behavior during deformation, or differences in the elastic constants of the two phases.

The stresses at any point in a free body with no surface tractions must obey the following equations:

$$^M\sigma_{ij,j} = 0 \tag{1}$$

$$^M\sigma_{ij} \cdot n_j = 0 \quad . \tag{2}$$

Here index notation is used and summation over repeated indices is implied. An index followed by a comma and another index means differentiation with respect to the coordinate of that index. Here n_j is a unit vector in the x_j direction. Equation 1 puts limits on the gradients in the stresses and hence the possible stress fields. Any stress must obey this equilibrium equation at every point within the body. Equation 2 simply states that in the absence of surface tractions the stress components perpendicular to the surface must be zero. With diffraction however, the stress at a point cannot be measured. Instead, the average stress over some irradiated volume is measured. When we do this we see that the average microstresses which can be measured with diffraction do not necessarily follow these relations. Using these relations it can be shown that each stress component must integrate to zero over the volume of a free body without surface tractions:

$$\int_v \sigma_{ij} dV = 0 \quad . \tag{3}$$

Henceforth, it will be understood that microstress refers to an average in a phase over an irradiated region. The total stress measured in any region is by definition the sum of the macrostress in the material and the microstress in that phase. For a two phase material we can write such an equation for each phase:

$$\langle {}^t\sigma^{\alpha}_{ij} \rangle = {}^M\sigma_{ij} + \langle {}^{\mu}\sigma^{\alpha}_{ij} \rangle \qquad (4)$$

$$\langle {}^t\sigma^{\beta}_{ij} \rangle = {}^M\sigma_{ij} + \langle {}^{\mu}\sigma^{\beta}_{ij} \rangle \qquad . \qquad (5)$$

The macrostress ${}^M\sigma_{ij}$ is the same in each phase while the microstress will be the difference between the macrostress and the total stress measured for that phase. Carets on the microstresses and total stresses indicate that they are averages over the diffracting volume. It is assumed that the macrostress only has small variations within the diffracting volume. This assumption need not always be true.

Equation 3 shows that any positive residual stresses must be balanced by negative stresses in another region of the material. Macrostresses will balance over different macroscopic regions of a body, while the microstresses must balance locally. Locally here would mean a material volume necessary to carry out a diffraction experiment. Using Equation 3 and averaging over each phase in a region it can be shown [Noyan (1983)] that the microstresses must follow the relation:

$$(1-f)\langle {}^{\mu}\sigma^{\alpha}_{ij} \rangle + f\langle {}^{\mu}\sigma^{\beta}_{ij} \rangle = 0 \quad . \qquad (6)$$

Here f is the volume fraction of the β phase and (1-f) is hence the volume fraction of the α phase. This equation shows that the microstresses must be of opposite sign in the two phases and that the microstresses will be larger in the phase with the smaller volume fraction. Each component of the microstresses obeys Equation 6 independently of the others. The microstresses as measured by diffraction do not necessarily obey Equation 1 because they are averages. Equations 4, 5, and 6 provide the relations needed to separately measure the macrostresses and microstresses.

2.2 Measurement of Stresses with Diffraction

The measurement of stresses with diffraction uses the precise measurement of lattice spacings with Bragg's law

$$\lambda = 2d \sin\theta \quad . \qquad (7)$$

Knowing the wavelength of the radiation used and precisely measuring the angle of a Bragg reflection the lattice spacing d can be determined. Using the unstressed lattice parameter for a given phase, d_0^{α} for the α phase, the strain along a given direction can be determined as

$$\langle \varepsilon^{\alpha}_{\phi\psi} \rangle = \frac{\langle d^{\alpha}_{\phi\psi} \rangle - d^{\alpha}_{0}}{d^{\alpha}_{0}} \tag{8}$$

Here ϕ and ψ define the direction of strain measurement along the L_3 axis in terms of the S_i or sample coordinate system as shown in Figure 1. The strain measured perpendicular to the L_3 axis can be written in terms of the strain components referenced to the S_i coordinate system.

$$\langle \varepsilon^{\alpha}_{\phi\psi} \rangle = \langle \varepsilon^{\alpha}_{11} \rangle \cos^2\phi \sin^2\psi + \langle \varepsilon^{\alpha}_{22} \rangle \sin^2\phi \sin^2\psi$$

$$+ \langle \varepsilon^{\alpha}_{33} \rangle \cos^2\psi + \langle \varepsilon^{\alpha}_{12} \rangle \sin2\phi \sin^2\psi$$

$$+ \langle \varepsilon^{\alpha}_{13} \rangle \cos\phi \sin2\psi + \langle \varepsilon^{\alpha}_{23} \rangle \sin\phi \sin2\psi \; . \tag{9}$$

Each component of the strain tensor varies differently with the angles ϕ and ψ. By making measurements of $\langle \varepsilon^{\alpha}_{\phi\psi} \rangle$ at a number of ϕ and ψ the components of the strain tensor $\langle \varepsilon^{\alpha}_{ij} \rangle$ can be determined from their different angular dependencies by a least squares procedure (Winholtz and Cohen (1988)). After determining the strain tensor in a material for a given phase the stress tensor can be determined from the relation:

$$\langle {}^{t}\sigma^{\alpha}_{ij} \rangle = \frac{1}{S_2/2} [\langle \varepsilon^{\alpha}_{ij} \rangle - \delta_{ij} \frac{S_1}{S_2/2 + 3S_1} (\langle \varepsilon^{\alpha}_{11} \rangle + \langle \varepsilon^{\alpha}_{22} \rangle + \langle \varepsilon^{\alpha}_{33} \rangle)] \; . \tag{10}$$

Here δ_{ij} is the Kronecker's delta function and S_1 and $S_2/2$ are the diffraction elastic constants [Noyan (1985a), Rudnik et al. (1988), and Butler et al. (1989)]. Using Equations 4, 5, and 6 we can solve for the macrostress and the microstress in each phase in terms of the total stress in each phase which can be experimentally determined with diffraction measurements:

$$\langle {}^{\mu}\sigma^{\alpha}_{ij} \rangle = f[\langle {}^{t}\sigma^{\alpha}_{ij} \rangle - \langle {}^{t}\sigma^{\beta}_{ij} \rangle] \tag{11}$$

$$\langle {}^{\mu}\sigma^{\beta}_{ij} \rangle = (1-f)[\langle {}^{t}\sigma^{\beta}_{ij} \rangle - \langle {}^{t}\sigma^{\alpha}_{ij} \rangle] \tag{12}$$

$${}^{M}\sigma_{ij} = (1-f)\langle {}^{t}\sigma^{\alpha}_{ij} \rangle + f\langle {}^{t}\sigma^{\beta}_{ij} \rangle \; . \tag{13}$$

Each of these equations is valid for each of the six components of the stress tensor. While elastic strains are directly measured with diffraction it is the stress equilibrium that allows the stresses to be separated into the macrostresses and microstresses. The measured strains must be converted into stresses to make the separation. This points out the need for accurate elastic constants to determine the macrostresses and microstresses. Consequently, the values used for the elastic con-

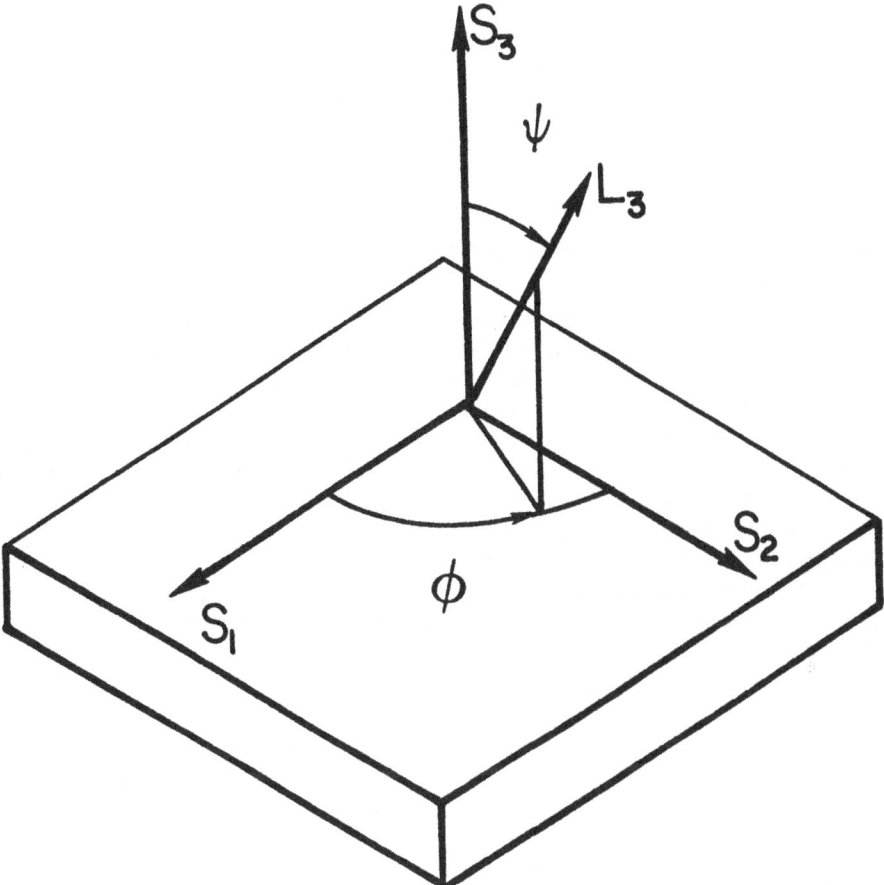

Figure 1. Coordinate systems used for diffraction stress measurements. The S_i system is the sample coordinate system to which measured stresses will be referenced. The L_3 axis is the direction along which the strains are measured with diffraction.

stants in a stress measurement should always be reported.

If one is interested only in the microstresses, the penetrability of neutrons can be used along with Equation 3 to eliminate the macrostresses from the measurement. If the entire sample can be irradiated the average macrostress must be zero and only the average microstresses will be measured. Since neutrons have a high penetrability in most materials, the entire specimen can be sampled and the average macrostress will be zero.

2.3 Uncertainties in the Unstressed Lattice Parameters

Determination of the unstressed lattice parameters for the phases can often present problems [Noyan (1985b)]. The unstressed lattice parameter is a function of the stacking fault density, the exact composition of the phases, and even the alignment of the diffractometer. In two phase or composite materials, the unstressed lattice parameters can change with diffusion or chemical reactions that take place as the material is heat treated or as the components are consolidated. Filing to obtain a stress free powder can change the stacking fault density. Noyan (1985b) reviews a number of methods for determining unstressed lattice parameters, but it is not always possible to determine them. It is thus desirable to know the errors involved in stress measurements where there are uncertainties in the unstressed lattice parameters.

Equation 10 can be substituted into Equation 9 to give the measured strains in terms of the sample stresses. It is also useful to separate the stress tensor into its hydrostatic and deviatoric components

$$\sigma_{ij} = \delta_{ij}\tau_{H} + \tau_{ij}. \tag{14}$$

The hydrostatic stress τ_{H} is given by

$$\tau_{H} = (\sigma_{11} + \sigma_{22} + \sigma_{33})/3. \tag{15}$$

The deviatoric stresses are given by

$$\tau_{ij} = \sigma_{ij} \qquad : i{\neq}j$$

$$\tau_{ij} = \sigma_{ij} - \tau_{H} : i{=}j. \tag{16}$$

From Equations 15 and 16 note that

$$\tau_{11} + \tau_{22} + \tau_{33} = 0. \tag{17}$$

Substituting Equations 10, 14 and 17 into Equation and solving for $\langle d^{\alpha}_{\phi\psi} \rangle$ gives

$$\langle d^{\alpha}_{\phi\psi}\rangle = \langle {}^{t}\tau^{\alpha}_{H}\rangle d^{\alpha}_{0}[S^{\alpha}_{2}/2 + 3S^{\alpha}_{1}] - [\langle {}^{t}\tau^{\alpha}_{11}\rangle + \langle {}^{t}\tau^{\alpha}_{22}\rangle] d^{\alpha}_{0}S^{\alpha}_{2}/2 + d^{\alpha}_{0}$$

$$+ \langle {}^{t}\tau^{\alpha}_{11}\rangle d^{\alpha}_{0}S^{\alpha}_{2}/2(1 + \cos^{2}\phi)\sin^{2}\psi$$

$$+ \langle {}^{t}\tau^{\alpha}_{22}\rangle d^{\alpha}_{0}S^{\alpha}_{2}/2(1 + \sin^{2}\phi)\sin^{2}\psi$$

$$+ \langle {}^{t}\tau^{\alpha}_{12}\rangle d^{\alpha}_{0}S^{\alpha}_{2}/2\ \sin2\phi\ \sin^{2}\psi$$

$$+ \langle {}^{t}\tau^{\alpha}_{13}\rangle d^{\alpha}_{0}S^{\alpha}_{2}/2\ \cos\phi\ \sin2\psi$$

$$+ \langle {}^{t}\tau^{\alpha}_{23}\rangle d^{\alpha}_{0}S^{\alpha}_{2}/2\ \sin\phi\ \sin2\psi\ . \tag{18}$$

Here we have $\langle d^{\alpha}_{\phi\psi}\rangle$ (the parameter actually measured) in terms of the stresses in the sample and the unstressed lattice parameter for that phase. $\langle {}^{t}\tau^{\alpha}_{33}\rangle$ has been eliminated using Equation 17. Since the deviatoric stresses vary with ϕ and ψ while the hydrostatic stress does not the deviatoric stress tensor can be determined by using the different angular dependencies using least squares techniques. Equation 17 is then used to get $\langle {}^{t}\tau^{\alpha}_{33}\rangle$. Since it is simply a multiplier, errors in d^{α}_{0} will introduce only small errors in determining the deviatoric stresses. The hydrostatic stress, however, is contained in one of three terms that do not vary with ϕ and ψ. Only the sum of these three terms can be determined from the variation of $\langle d^{\alpha}_{\phi\psi}\rangle$ with ϕ and ψ. In the first term, the hydrostatic stress is multiplied by elastic constants. Consequently, this term will be small in relation to d^{α}_{0} which is one of the other terms which does not vary with ϕ and ψ. Thus, the unstressed lattice spacing must be known very accurately in order to determine the hydrostatic component of the stress tensor while the deviatoric components of the stress tensor are relatively insensitive to it.

Equations 11, 12, and 13 are also valid separately for the hydrostatic and deviatoric stress tensors:

$$\langle {}^{\mu}\tau^{\beta}_{ij}\rangle = f[\langle {}^{t}\tau^{\alpha}_{ij}\rangle - \langle {}^{t}\tau^{\beta}_{ij}\rangle] \tag{19}$$

$$\langle {}^{\mu}\tau^{\alpha}_{ij}\rangle = (1-f)[\langle {}^{t}\tau^{\beta}_{ij}\rangle - \langle {}^{t}\tau^{\alpha}_{ij}\rangle] \tag{20}$$

$${}^{M}\tau_{ij} = (1-f)\langle {}^{t}\tau^{\alpha}_{ij}\rangle + f\langle {}^{t}\tau^{\beta}_{ij}\rangle \tag{21}$$

In these equations τ_{ij} can stand for any component of the deviatoric stress tensor or for the hydrostatic stress τ_{H}.

While it is desirable to know the entire stress tensors, the deviatoric stress tensors may be sufficient for many purposes if the unstressed lattice parameters cannot be measured. If comparison measurements are to be made, for instance before and after a treatment, the difference in stress $\Delta\sigma$ will not contain any error due to errors in d_0. Also, if measurements are to be compared to a model calculation the deviatoric stresses may be sufficient to justify the model.

3. Examples

3.1 Microcracking in an SiC-TiB$_2$ Composite

An example of the usefulness of the separation of macrostresses and microstresses is microcracking in the SiC-TiB$_2$ system [Magley et al. (1990) and Faber et al. (1991)].

Material containing 15.2 volume percent TiB$_2$ was consolidated by pressureless sintering at temperatures in excess of 2000° C and were nearly 99% dense. A specimen was cut for testing by electrical discharge machining (EDM). For comparisons a single phase SiC standard was used. The standard was cut by diamond machining. Final polishing for both specimens was done using a sequence of diamond pastes down to 0.25 μm.

The stresses were measured with x-ray diffraction using Fe Kα radiation and the 202 diffraction peak for the TiB$_2$ phase and the 121 peak for the 6H SiC phase at 144° and 148° 2θ, respectively. The SiC in the composite consisted of a number of polytypes, 6H being the predominate one. The polytypes differ from each other in the stacking sequence of the hexagonal planes and the 6H should behave the same as all the others.

Initially the composite material contained the following stresses:

$$^M\sigma_{ij} = \begin{bmatrix} -369(24) & -28(14) & -14(5) \\ - & -375(26) & 2(5) \\ - & - & 0(9) \end{bmatrix}$$

$$^\mu\sigma_{ij}^{SiC} = \begin{bmatrix} -59(9) & 2(4) & 0(1) \\ - & -72(9) & 1(1) \\ - & - & -15(5) \end{bmatrix}$$

$$^\mu\sigma_{ij}^{TiB_2} = \begin{bmatrix} 327(49) & -18(20) & 2(7) \\ - & 399(49) & -7(8) \\ - & - & 83(29) \end{bmatrix}$$

There is a large compressive stress in the surface due to the EDM process. We see large microstresses due to the thermal coefficient of expansion mismatch between the SiC and TiB$_2$ phases. Since the stresses were measured with x-rays which penetrate to only about 50 μm below the surface (where σ_{33} must be zero from Equation 2) the microstresses are largely biaxial. Deeper in the material, they would be hydrostatic in nature; σ_{33} would be close to the value of σ_{11} and σ_{22}.

Both the composite and standard specimens were then loaded in four point bending until fracture. The stress experienced by the outer fibers along the S_1 axis where the diffraction measurements were made were 249 and 266 MPa for the composite and standard respectively. After loading the specimens, the diffraction stress measurements were repeated. In the composite specimen, the macrostresses did not change significantly. The microstresses, however, show a marked decrease due to microcracking induced by the stress:

$$
{}^{\mu}\sigma_{ij}^{SiC} = \begin{bmatrix} -45(7) & -11(3) & -2(1) \\ - & -19(7) & -4(1) \\ - & - & 2(4) \end{bmatrix}
$$

$$
{}^{\mu}\sigma_{ij}^{TiB_2} = \begin{bmatrix} 248(37) & 59(20) & 10(7) \\ - & 108(37) & 22(8) \\ - & - & -12(23) \end{bmatrix}
$$

In contrast, the stresses in the single phase SiC standard had no significant changes upon loading:

$$
\left({}^{t}\sigma_{ij}^{SiC\text{-}Standard}\right)_{Prestress} = \begin{bmatrix} -34(11) & 11(7) & 5(2) \\ - & -89(12) & 4(2) \\ - & - & -1(4) \end{bmatrix}
$$

$$
\left({}^{t}\sigma_{ij}^{SiC\text{-}Standard}\right)_{Poststress} = \begin{bmatrix} -28(11) & 30(20) & 10(2) \\ - & -110(49) & 4(2) \\ - & - & -2(3) \end{bmatrix}
$$

Since the microstresses are largely due to coefficient of thermal expansion mismatch, the hydrostatic stresses are of interest. In order to account for the surface effect which relaxes stresses perpendicular to the surface, an "effective" hydrostatic stress is computed as $(\sigma_{11} + \sigma_{22})/2$. This "effective" hydrostatic stress is plotted in Figure 2 for both phases in the composite material and for the SiC standard. The microstresses in the composite material are relaxed by nearly 60% after loading while the stress in the standard does not significantly change. This observation gives direct evidence of stress-induced microstress relief in a nontransforming brittle material and is consistent with the model of stress-induced microcracking as a toughening mechanism in the SiC-TiB$_2$ system.

3.2 Stress in the Cementite Phase of Steel During Low Cycle Fatigue

Another example of the usefulness of macrostress and microstress measurements is in the deformation of a two phase material. By measuring the stresses in both the cementite and ferrite phases of steel their individual responses to low cycle fatigue were determined[Winholtz and Cohen (1991)]. Steel is an alloy of iron and carbon and its mechanical properties depend largely on the carbides present and their interactions with the ferrite matrix. Diffraction stress measurements have long been made in steel because of its widespread use, but usually only in the ferrite phase. Since the mechanical properties depend heavily on the carbides in steel, it was desirable to measure the stresses present in this phase as well.

When a two phase material is subjected to uniaxial tensile strain along the S_1 axis, it will behave

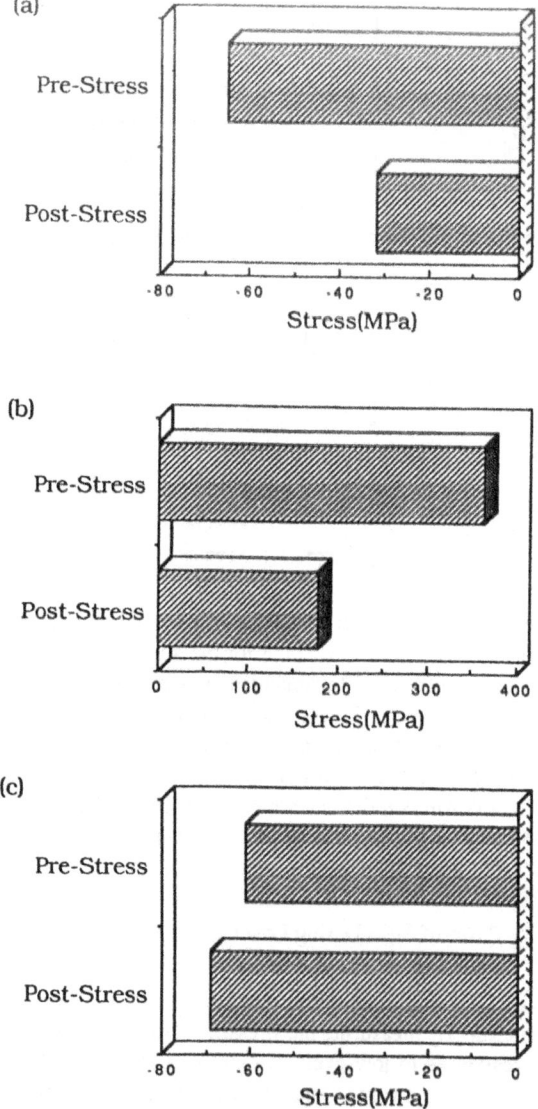

Figure 2. The "effective' hydrostatic stresses before and after loading in (a) the SiC phase in the composite material, (b) the TiB$_2$ phase in the composite, and (c) the SiC standard material.

similarly to the schematic shown in Figure 3. The bulk material will have a stress response that is elastic to the yield point where it will deform plastically with a nonlinear stress response. The stress response of the bulk material is that measured by traditional mechanical testing methods (i.e. a load cell) and is by definition a macrostress applied to the sample, τ_{11}^{Appl}. For a two phase material the average stress in the individual phases will be different than the applied stress giving rise to microstresses as shown in the figure. In this figure the elastic response of the two phases is assumed to be the same, which need not be the case.

After deformation, if the bulk stress is removed, the bulk material will relax elastically retaining a plastic offset. The individual phases will not relax to zero but the micro-stresses will be retained as shown. These micro-stresses are retained without the external load applied and can be measured via diffraction. If the bulk stress at the release of the specimen is known, the individual stress response of the individual phases can be inferred by adding back the applied stress τ_{11}^{Appl} to micro-stresses measured after the load is released. Thus, with a proper understanding of the macrostresses and microstresses the mechanical response of the individual phases can be determined even if the diffraction stress measurements cannot be made in situ on the material as it is mechanically loaded (although this could, in principle, be done).

Fatigue specimens of 1080 steel were machined and polished through 600 grit. Heat treatment was performed after polishing to remove any residual stresses due to machining and polishing. A pearlitic microstructure was obtained by austenitizing for 15 minutes at 1073 K in argon followed by cooling outside the hot zone of the furnace. Samples were examined metallographically to be sure that decarburization of the surface did not take place during heat treatment. The volume fraction of cementite was determined from the relative integrated intensity of a number ferrite and cementite diffraction peaks.

Stress measurements were made with Cr Kα x-radiation from a rotating anode source using the 211 ferrite peak and the 250 cementite peak [Winholtz and Cohen (1989)]. Fatigue loading was carried out on an MTS 100kN servo-hydraulic fatigue machine at room temperature. Specimens were fatigued in strain control with a triangular waveform and a strain amplitude of $\Delta\varepsilon/2 = 0.005$ with a strain rate of 3.3×10^{-1} s^{-1}. Load-extension hysteresis loops were recorded from the load cell and extensometer from which the applied deviatoric macrostress tensor was obtained as:

$$\tau_{ij}^{Appl} = \begin{bmatrix} 2P/3A & 0 & 0 \\ 0 & -P/3A & 0 \\ 0 & 0 & -P/3A \end{bmatrix}$$

where P is the applied force and A is the specimen cross sectional area.

The steel showed an initial distinct yield point and then quickly formed a stable hysteresis loop shown in Figure 4. Four companion specimens were run and stopped at each quarter cycle on the eleventh fatigue cycle. The specimens were then removed from the fatigue machine and the residual stresses measured. The stress measurements showed large microstresses present as well as a small macrostress because the surface of the material yielded before the interior (this effect could be eliminated by using neutron diffraction). The microstresses measured on each of the four specimens were added to the applied stress determined from the mechanical testing equipment to determine the stress in each phase at the given point on the hysteresis loop. The four measurements can then be connected to give a hysteresis loop for each phase as shown in Figure 4. We see that

142

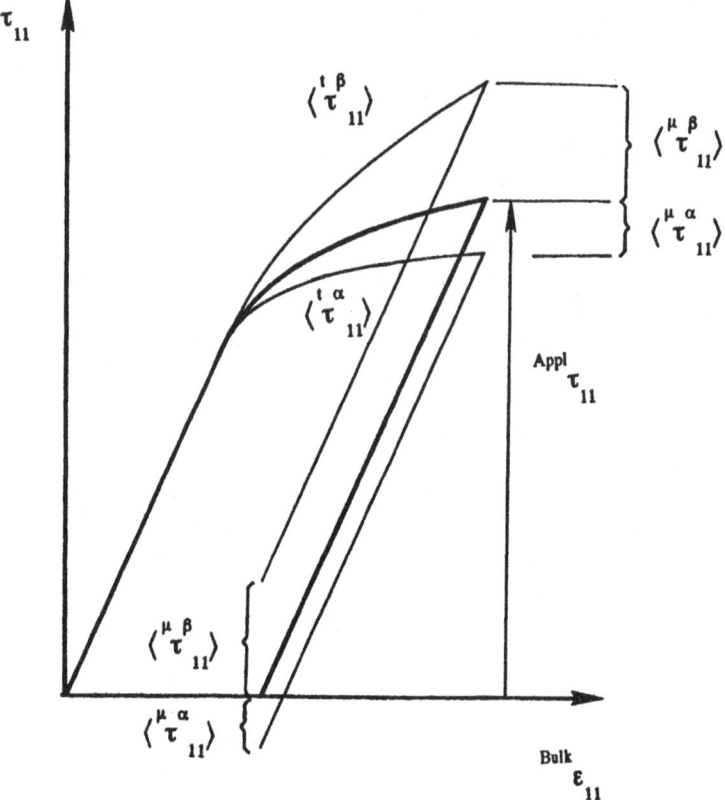

Figure 3. Schematic of the deformation of a two phase material showing the macrostresses and the microstresses in each phase.

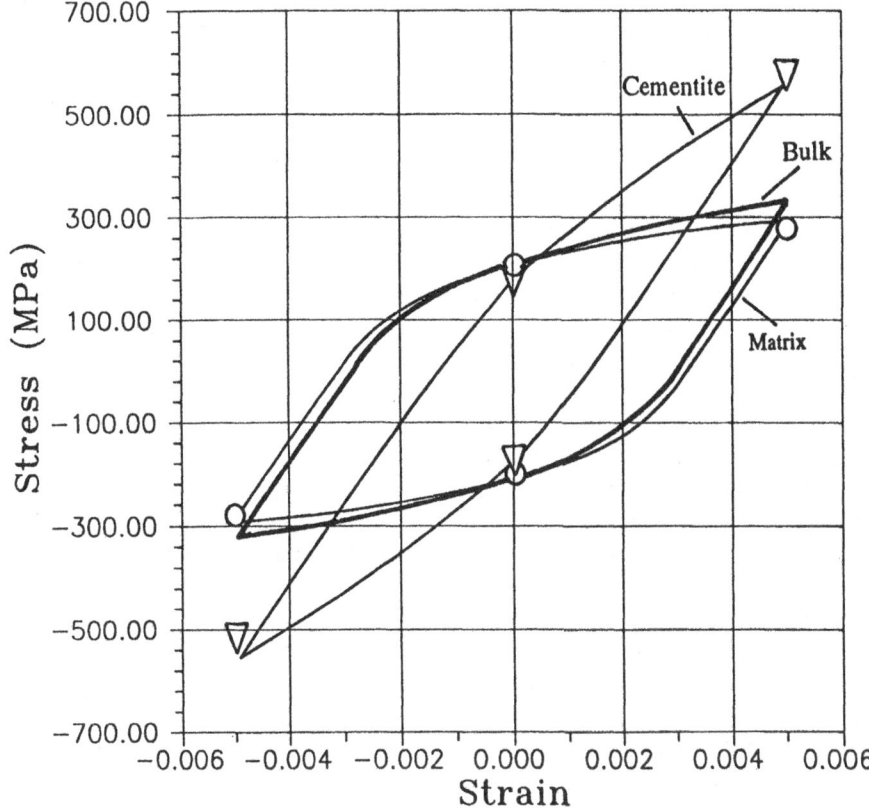

Figure 4. Stress-strain hysteresis loop for 1080 steel along with component for the ferrite and cementite phases.

144

the microstresses change continuously during the low cycle fatigue and that the stress range experienced by the cementite phase is much larger than the bulk material. Such experiments can give new insights into the mechanical behavior of two phase or composite materials.

4. Conclusions

1) Equations of stress equilibrium properly considering macrostresses and microstresses allow the macrostresses and microstresses to be separated in two phase or composite materials by using diffraction to independently measure the stresses in each phase.

2) Errors in the unstressed lattice parameter lead to an error in the hydrostatic component of the stress tensor. The deviatoric stress tensor can be accurately determined without the unstressed lattice parameter.

3) Measurements of the microstresses in a SiC-TiB$_2$ composite have proven useful in understanding stress induced microcracking as a toughening mechanism.

4) The individual phase responses can be separated in the mechanical loading of a two phase material by using diffraction to measure the microstresses present in each phase. This has been used in 1080 steel to determine individual hysteresis loops for the ferrite and cementite phases.

Acknowledgments

The author wishes to thank Professor J.B. Cohen for providing help throughout the author's work in this area. Thanks also to Professor K.T. Faber who has been very helpful in collaborations on her work in the SiC-TiB$_2$ system. The support of the Office of Naval Research under contract No. N00014-80-C-116 for this work is gratefully acknowledged.

References

Abuhasan, A., Balasingh, C, and Predecki, P. (1990) 'Residual Stresses in Alumina/Silicon Carbide (Whisker) Composites by X-ray Diffraction', J. Am. Ceram. Soc., 73, 2474-2484.

Butler, B.D., Murray, M.C., Reichel, D.G., and Krawitz, A.D. (1989) 'Elastic Constants of Alloys Measured with Neutron Diffraction', Adv. X-ray Anal., 32, 389-395.

Faber, K.T., Gu, W.-H., Cai, H., Winholtz, R.A., and Magley, D.J. (1991) 'Fracture Properties of SiC-Based Particulate Composites', to appear in Proceedings of the NATO Advanced Research Workshop on Toughening Mechanisms in Quasi-Brittle Materials.

Magley, David J., Winholtz, R.A., and Faber, K.T. (1990) 'Residual Stresses in a Two-Phase Microcracking Ceramic', J. Am. Ceram. Soc., 73, 1641-1644.

Noyan, I.C. (1983) 'Equilibrium Conditions for the Average Stresses Measured by X-rays', Met.

Trans. A, 14A, 1907-1914.

Noyan, I.C. (1985a) 'Determination of the Elastic Constants of Inhomogeneous Materials with X-ray Diffraction', Mat. Sci. Engr., 75, 95-103.

Noyan, I.C. (1985b) 'Determination of the Unstressed Lattice Parameter "a_0" for (Triaxial) Residual Stress Determination by X-rays', Adv. X-ray Anal., 28, 178-185.

Noyan, I.C. and Cohen, J.B. (1985) 'An X-ray Diffraction Study of the Residual Stress-Strain Distributions in Shot-peened Two-phase Brass', Mat. Sci. Engr., 75, 179-193.

Noyan, I.C. and Cohen, J.B. (1987) Residual Stress: Measurement by Diffraction and Interpretation, Springer-Verlag, New York.

Rudnik, P.J., Krawitz, A.D., Reichel, D.G., and Cohen, J.B. (1988) 'A Comparison of Diffraction Elastic Constants of Steel Measured with X-rays and Neutrons', Adv. X-ray Anal., 31, 245-253.

Winholtz, R.A. and Cohen, J.B. (1988) 'Generalised Least-squares Determination of Triaxial Stress States by X-ray Diffraction and the Associated Errors', Aust. J. Phys., 41, 189-199.

Winholtz, R.A. and Cohen, J.B. (1989) 'Separation of the Macro- and Micro-Stresses in Plastically Deformed 1080 Steel', Adv. X-ray Anal., 32, 341-353.

Winholtz, R.A. and Cohen, J.B. (1991) 'Load Sharing of the Phases in 1080 Steel During Low Cycle Fatigue', to be published in Met. Trans. A, .

THE EFFECTS OF CRYSTALLINE ANISOTROPY ON THE ELASTIC RESPONSE OF MATERIALS

COLIN G. WINDSOR AND TAKEO IZUYAMA*
National Non-Destructive Testing Centre
AEA Industrial Technology,
B521.2, Harwell Laboratory, OX11 ORA, UK.
*Permanent Address:University of Tokyo, Komaba, Tokyo, Japan

ABSTRACT. A brief account is given on the effects of single crystal anisotropy on the elastic response of polycrystalline materials. Measurements using neutron diffraction and a rig giving a controlled stress are described. These are able to measure the strain as a function of the stress for a number of crystallographic planes. The anisotropy effects predicted be the simple Reuss and Voigt theories will be described. Recent measurements on initially stress-free samples placed under defined stresses up to and beyond their yield point are shown These can give a valuable insight into both the elastic and plastic properties of metals.

1. Stress/strain relationships in polycrystalline materials - the effects of anisotropy

Athough the elastic properties of texture-free polycrystalline materials may appear to be isotropic, polycrystalline materials are composed of single crystals whose elastic properties are usually highly anisotropic. In particular steel has a relatively large anisotropy. The effects of crystalline anistropy become important the moment a stress is applied, for the crystallites of different orientation with respect to the stress take up differing strains. This chapter treats in more detail these effects of anisotropy. Although at first sight these appear to be a complication to the subject, in fact they are able to give important extra information on the true state of the material under load.

Single crystal anisotropy can be defined quantitatively by the matrix C_{ij} which relates the stress to the strain[1]. In an isotropic material it has a diagonal form with identical elements equal to the slope of the familiar stress versus strain curve. For a cubic crystal, there are only three independent constants C_{11}, C_{12} and C_{44}. They enable the stress to be expressed in terms of the strain through the equations

$$
\begin{aligned}
\sigma_{11} &= C_{11}\varepsilon_{11} &+& \quad C_{12}\varepsilon_{22} &+& \quad C_{12}\varepsilon_{33} \\
\sigma_{22} &= C_{12}\varepsilon_{11} &+& \quad C_{11}\varepsilon_{22} &+& \quad C_{12}\varepsilon_{33} \\
\sigma_{33} &= C_{13}\varepsilon_{11} &+& \quad C_{12}\varepsilon_{22} &+& \quad C_{11}\varepsilon_{33} \\
\sigma_{23} &= 2C_{44}\varepsilon_{23} \\
\sigma_{31} &= 2C_{44}\varepsilon_{31} \\
\sigma_{12} &= 2C_{44}\varepsilon_{12}.
\end{aligned}
\tag{1}
$$

M. T. Hutchings and A. D. Krawitz (ed.),
Measurement of Residual and Applied Stress Using Neutron Diffraction, 147–158.
© 1992 UKAEA.

Alternatively the strain may be expressed in terms of the stress through the following equations involving the single crystal compliance constants S_{ij}. Generally these constants are most often used in analytical work.

$$
\begin{aligned}
\varepsilon_{11} &= S_{11}\sigma_{11} + S_{12}\sigma_{22} + S_{12}\sigma_{33} \\
\varepsilon_{22} &= S_{12}\sigma_{11} + S_{11}\sigma_{22} + S_{12}\sigma_{33} \\
\varepsilon_{33} &= S_{13}\sigma_{11} + S_{12}\sigma_{22} + S_{11}\sigma_{33} \\
\varepsilon_{23} &= S_{44}\sigma_{23}/2 \\
\varepsilon_{31} &= S_{44}\sigma_{31}/2 \\
\varepsilon_{12} &= S_{44}\sigma_{12}/2
\end{aligned}
\tag{2}
$$

Naturally the two forms of expressing the elastic properties are equivalent and indeed the coefficients are related through the equations

$$
\begin{aligned}
(S_{11} - S_{12}) &= 1/(C_{11} - C_{12}) \\
(S_{11} + 2S_{12}) &= 1/(C_{11} + 2C_{12}) \\
S_{44} &= 1/C_{44}.
\end{aligned}
\tag{3}
$$

The elastic constants for some of the most common metals are given in table I[2]. Also given in the table is the degree of anisotropy as given by equation (4)

$$
A' = 2(S_{11} - S_{12}) / S_{44},
\tag{4}
$$

Thus iron and aluminium have anisotropy of a different sense to molybdenum. Tungsten is unusually isotropic.

The situation for a single crystal sample is very complicated, for three directions must be specified:

i)   The crystallite axes - say xyz
ii) The direction of the applied stress - given by direction cosines (uvw)
iii) The direction of the strain measurement - say hkl

In neutron diffraction the direction of the strain is specified by the reflecting planes being used, and is given by the direction cosine ratios (hkl) of these planes.

In this case the measured strain is given in terms of these direction cosine ratios and the angle Ω between the direction of strain measurement and the direction of the stress[3,4].

$$
\varepsilon = \sigma[\, S_{11}\cos^2\Omega + S_{12}\sin^2\Omega - 2(S_{11} - S_{12} - S_{44}/2)]\frac{(hkuv + klvw + klwu)}{h^2 + k^2 + l^2}.
\tag{5}
$$

Table I. Some of the elastic properties of common metals

	Modulus GPa	Poisson's ratio	S_{11}	S_{12}	S_{44}	A'
			$10^{-6}Pa^{-1}$			
Iron	211.4	0.293	7.63	-2.79	8.55	2.512
Aluminium	70.3	0.345	15.7	-5.7	35.1	1.219
Molybdenum	311.		2.8	-0.8	9.1	0.79
Copper	110.	0.33	15.0	-6.3	13.3	3.2
Tungsten	411.0	0.28	0.26	-0.07	0.66	1.0

In the case that the applied strain is parallel to the direction of measurement, then the angle $\Omega = 0$, (uvw) is the same vector as (hkl), and the equation reduces to

$$\varepsilon = \sigma[S_{11} - 2 SA_{hkl}], \tag{6}$$

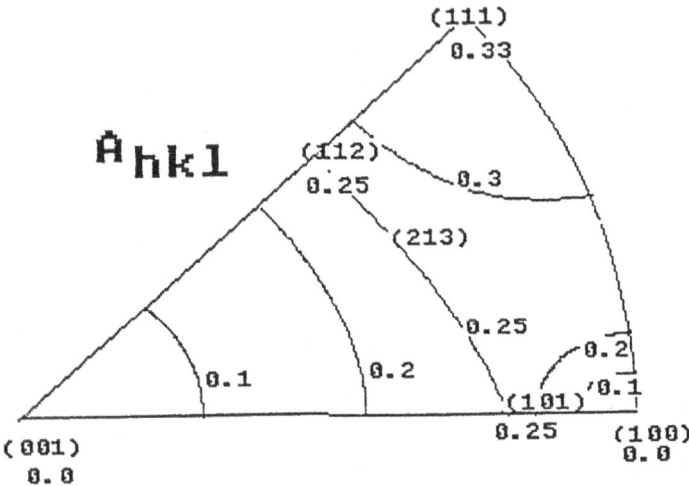

Figure 1. A stereographic projection of the anisotropy constant within the irreducible area - a single half quadrant. Some of the more important directions are labelled within brackets. The numbers show the values.

Table II. The anisotropy constants A_{hkl} for some reflections

h	k	l	ratio	A_{hkl}
1	0	0	0	0.0
1	1	0	1/4	0.25
1	1	1	1/3	0.3333
2	1	0	4/25	0.16
2	1	1	1/4	0.25
2	2	1	24/81	0.2963
3	1	0	9/100	0.09
3	1	1	19/121	0.1570
3	2	0	36/169	0.2130
3	2	1	1/4	0.25
3	2	2	88/289	0.3045
3	3	1	99/361	0.2742
3	3	2	153/484	0.3161
4	1	0	16/289	0.0554
4	1	1	33/324	0.1019
4	2	1	84/441	0.1905
4	3	0	144/625	0.2304
4	3	1	1/4	0.25
4	3	2	244/841	0.2901
4	3	3	369/1156	0.3192
4	4	1	288/1089	0.2645
4	4	3	544/1681	0.3236
5	1	0	25/676	0.0370
5	1	1	51/729	0.0700
5	2	0	100/841	0.1189
5	2	1	129/900	0.1433
5	2	2	216/1089	0.1983
5	3	0	225/1156	0.1946
5	3	1	259/1225	0.2114
5	3	2	1/4	0.25
5	3	3	531/1849	0.2872
5	4	0	400/1681	0.2379
5	4	1	1/4	0.25
5	4	2	564/2025	0.2785
5	4	3	769/2500	0.3076
5	4	4	1056/3249	0.3250
5	5	1	675/2601	0.2595
5	5	2	825/2916	0.2829

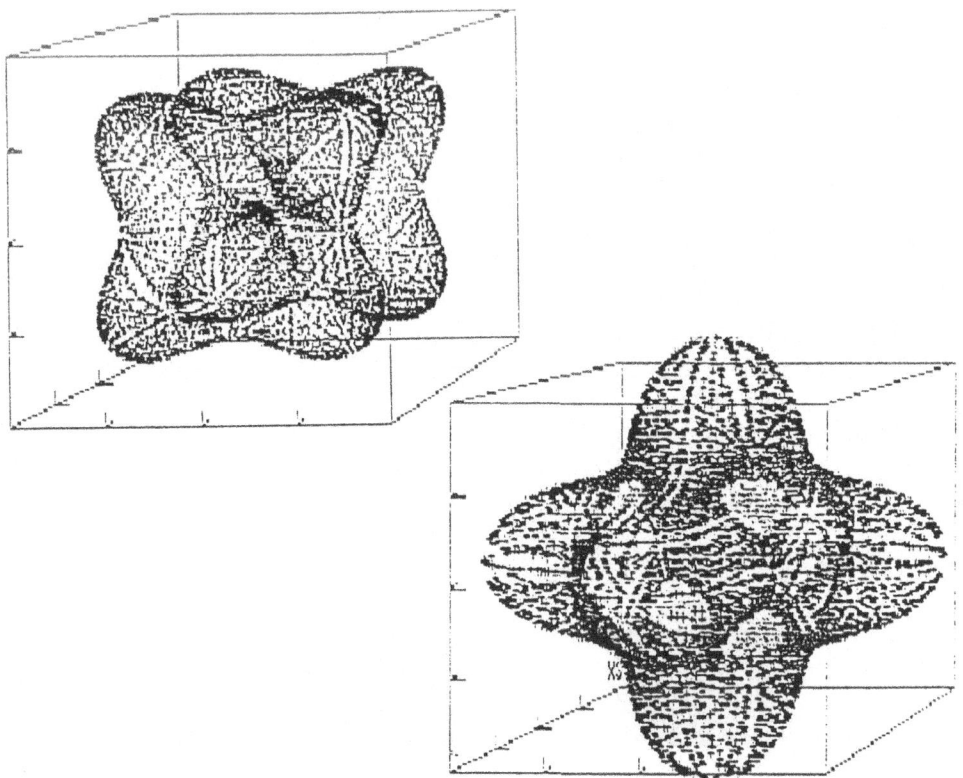

Figure 2 Three-dimensional sketches of the anisotropy factor. On the left is shown the factor A_{hkl}. On the right is shown the actual compliance constant for steel shown to scale as a function of the stress orientation.

where

$$S = S_{11} - S_{12} - S_{44}/2, \tag{7}$$

and

$$A_{hkl} = (h^2 k^2 + k^2 l^2 + l^2 h^2)/(h^2 + k^2 + l^2)^2 \tag{8}$$

The anisotropy factor A_{hkl} is very important in this subject. Some of the values for the most commonly used planes in neutron diffraction are given in table II.

The anisotropy may be plotted in various ways. A simple diagram is produced by the stereographic projection as in figure 1. With the (001) point at the origin the anisotropy factor is seen to grow from this point. The factor A_{hkl} = 1/4 given by the 110, 211, 321, 431, 532 and 541 vectors, and whenever h=k+l or h=k-l, is seen to represent an important line through the diagram. Figure 2 shows the anisotropy factor plotted out over all directions in a three-dimensional representation. On the left is shown the factor A_{hkl} alone. It has its largest lobes pointing towards the 8 <111> directions, and falls to zero along the 6 <100> directions. On the right is shown the actual anisotropy for mild steel. Because the factor S is positive, the compliance has its maximum values in the places where A_{hkl} is smallest, that is in the <100> directions. The least compliant directions are along the <111> directions where A_{hkl} is a maximum. The figure is to scale showing how large the anisotropy is. The difference between the maximum and minimum compliances for mild steel is some 49%.

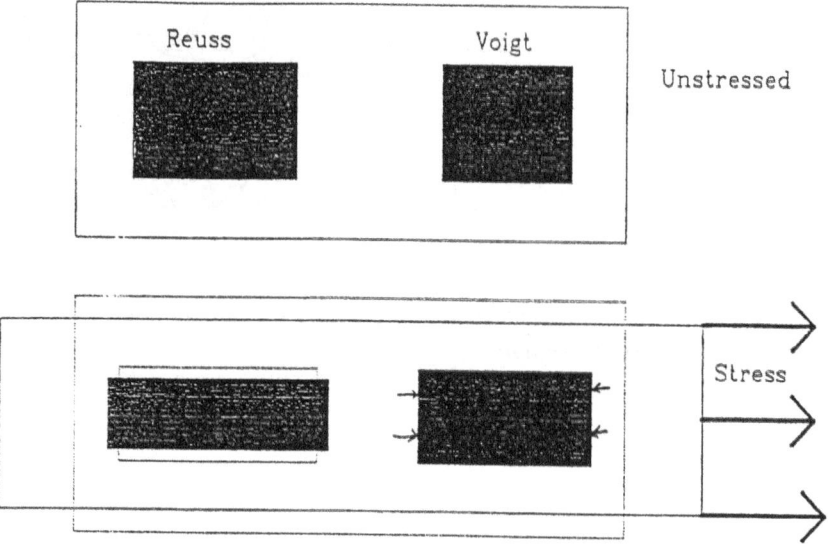

Figure 3. The distortion of crystallites having anisotropic moduli within an isotropic elastic matrix. In the Reuss limit (left), the stress is assumed constant, so that the crystallite is distorted relative to the matrix and there is a discontinuity in the strain. In the Voigt limit the strain is assumed constant, with the result that the stress in the crystallite is different from that in the matrix.

2. Strain anisotropy in polycrystalline samples.

The effects of anisotropy in a powder sample are complicated and there is no exact theory, only limiting approximations[5]. The problem is that each crystallite is imbedded in a matrix of other crystallites. Consider a stress applied to the whole assembly of crystallites, as in figure 3. One limiting case, the Reuss limit assumes that the stress distribution is uniform and that each crystallite will at first respond to this constant stress, and develop a strain proportional to the anisotropic modulus for its particular orientation. However a problem then develops at the grain boundary interface between crystallites, for the strain will change discontinously across the interface as shown in (i). The crystallites will no longer pack together, and in practice the surrounding matrix will exert forces on the crystallite leading to an extra local stresses depending on each crystallite.

The opposite limiting case is the Voigt approximation where the strains are assumed to be constant across the assembly of crystallites. The matrix strains uniformly under the applied stress, so that there is no problem at the interface. However the crystalline anisotropy means the stress developed in each crystallite under the strain will be different. The assembly will have a distribution of stresses.

The response may be calculated for both limits. In the constant stress Reuss limit, the measured strain for a given crystallographic direction [hkl] when the scattering vector Q is parallel to the stress will be equal to the single crystal strain averaged over all possible crystallite orientations whose hkl planes have normals parallel to the chosen stress direction. In fact equation 12 shows that there is no strain variation for a single crystal rotated in this way and the polycrystalline strain is given by

$$\varepsilon^{\parallel}_{poly} = \sigma[S_{11} - 2SA_{hkl}].\tag{9}$$

The isotropic average of the A_{hkl} factor is important as being representative of the average matrix material in many models. Defining p, q, r as the direction cosines corresponding h, k, l, so that $p^2 + q^2 + r^2 = 1$, then $A_{hkl} = A_{pqr}$ may be written

$$A_{pqr} = (p^2 q^2 + q^2 r^2 + r^2 p^2)$$

$$= 1/2 [1 - (p^4 + q^4 + r^4)].$$

Each of the spherical average values $<p^4>$, $<q^4>$, $<r^4>$, has the value

$$<p^4> = <q^4> = <r^4> = \int p^4 \, dp = [p^5/5]_0^1 = 1/5.$$

All three are equal so that the overall spherical average is

$$<A_{pqr}> = 1/2 [1 - 3<p^4>] = 1/2 (1-3/5) = 0.2 \tag{10}$$

154

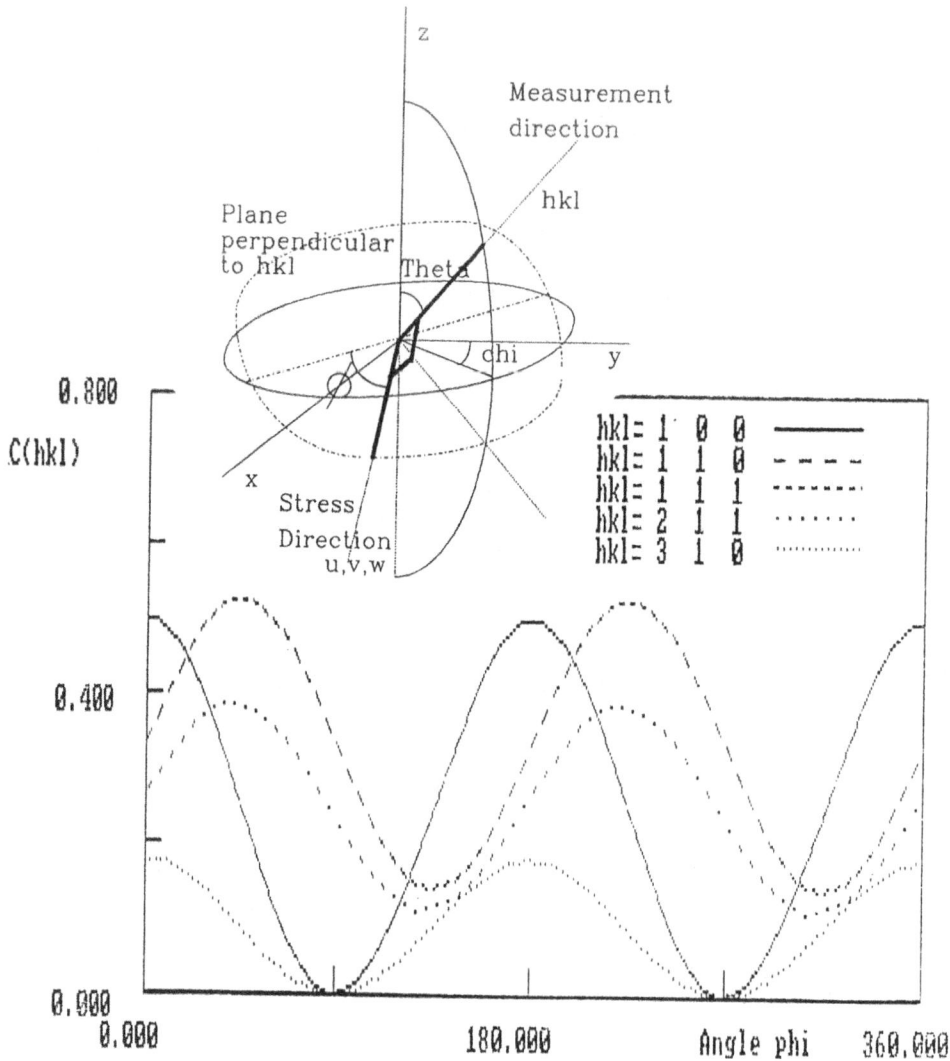

Figure 4. The anisotropy of the effective modulus C_{hkl} for crystallites perpendicular to the stress direction as a function of the angle phi of the crystallite axes in the plane where the stress is perpendicular to the measurement direction. The various axes are defined inset. The crystal axes are x,y,z. The measurement direction is along hkl. The stress direction is along u,v,w. The dashed ellipse is the plane perpendicular to the measurement direction. Within this plane crystallites of all orientations φ contribute to the measurement.

When Q is perpendicular to the stress direction, the average must be taken over all crystallite directions having hkl planes lying perpendicular to the stress direction. This time this average can be made with the result

$$\varepsilon_{poly}^{\perp} = \sigma[S_{12} + SA_{pqr}].\tag{11}$$

This average has been published many times, but its derivation is quite complicated. The inset of figure 4 shows the axes that must be considered. Besides the crystal axes x,y,z and the measuring plane direction cosines p,q,r which have already been considered, an average must be made over all stress directions u,v,w which are perpendicuar to p,q,r. Equation 5 for the strain in the perpendicular direction may then be written

$$\varepsilon = \sigma[\ S_{12} + S\ (p^2u^2 + q^2v^2 + r^2w^2)\ ].$$

The direction cosines u,v,w may be written in terms of the angle φ around the plane perpendicular to p,q,r. It may be found that

$$u^2 = 1/2\ (p^2 + r^2) + 1/2 \cos 2\varphi\ (\ q^2 + r^2p^2)/(1 - r^2) + \sin 2\varphi\ pqr/(1 - r^2),$$

$$v^2 = 1/2\ (q^2 + r^2) + 1/2 \cos 2\varphi\ (\ p^2 + r^2q^2)/(1-r^2) - \sin 2\varphi\ pqr/(1-r^2),$$

$$w^2 = 1/2\ (p^2 + q^2) - 1/2 \cos 2\varphi\ (\ p^2 + q^2).\tag{12}$$

The perpendicular anisotropy depends on the factor

$$p^2u^2 + q^2v^2 + r^2w^2 =$$
$$A_{pqr} + \cos 2\varphi\ [\ p^2q^2 - 2r^2(p^4 + q^4)/(1-r^2)] + \sin 2\varphi\ pqr.\tag{13}$$

This equation has many interesting properties. The average over orientations perpendiular to the measuring direction p,q,r is trivially made since the sine and cosine both average to zero and give the result of equation 11. The [001] direction is unique in that all terms are zero at all angles, so that the anisotropy amplitude is zero as well as its average. Other planes show an sinusoidal variation in the factor, and hence in the modulus as illustrated in figure 4.

3. The measurement of crystalline anisotropy in powder samples

The two equations (2) and (3) above can be checked by measuring the parallel and perpendicular strains for a series of different reflections hkl, whose anisotropy factor A_{hkl} spans over the range required[6]. Certainly it is desirable to include at least one of the [001] and [111] series planes which define the limits of the anisotropy variation. To perform such measurements a uniform, defined stress needs to be applied to a sample, and the strain measured as a function of stress within the elastic region. A stress rig cab deliver a predetermined stress to a machined sample of typically 60mm length and 3mm diameters. The long, thin geometry ensures that the stress is very nearly uniform over the area of the sample seen in the diffraction experiment.

The measurements in the perpendicular direction can be performed particularly easily on a pulsed neutron diffractometer such as the Rutherford Appleton HRPD instrument.

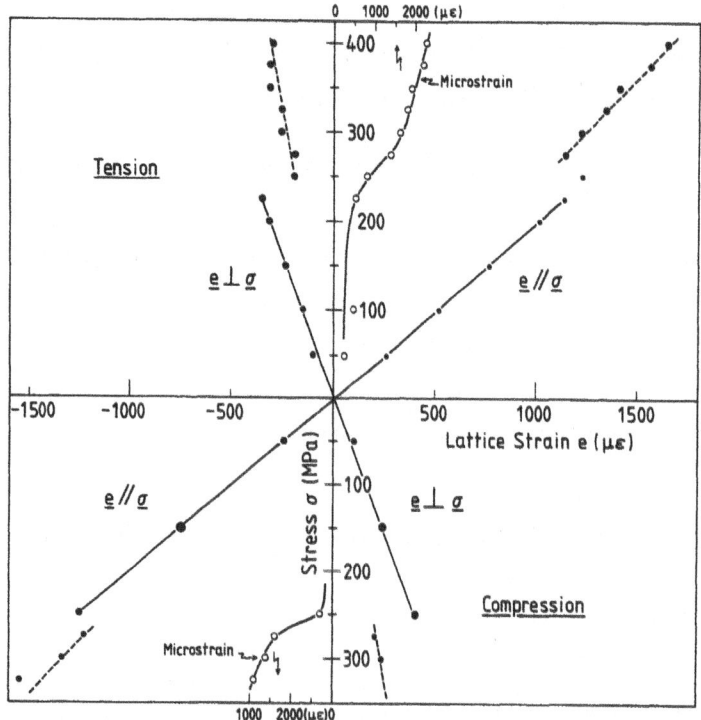

Figure 5. The stress strain curves for the 310 reflection of annealed mild steel as measured in tension (upper) and compression (lower), and with the strain both parallel and perpendicular to the stress. The measurements are made at the Institute Laue Langevin Grenoble. [7]

The source gives a pulsed neutron beam which allows the simultaneous measurement of many reflections, up to around 15. To obtain the measurements in the parallel direction, the scattering angle should be in the range 30 to 60°. Instruments giving the necessary resolution on a pulsed source are not commonplace, and most work has in practice been performed on reactor diffractometers. There is no problem in obtaining any one peak in a fixed wavelength scan in the conventional way. However at any one wavelength, there can be a problem in obtaining enough reflections at suitable angles to get an anisotropy plot containing sufficient peaks. One solution is to use an orientated monochromator plane such as the germanium (111) which simultaneously transmits neutrons from the (333) and (555) planes. The presence of neutrons of three wavelengths simultaneously in the beam causes little problem of overlapping peaks in a simple structure such as ferritic steel, and avoids the unwanted poitioning errors inevitable when resetting the monochromator planes.

Figure 5 shows the stress strain curves for the 310 reflections measured both parallel and perpendicular to the stress on the D1A diffractometer at the ILL. The upper set of curves correspond to tension and the lower set to compression. It is seen that the results for tension and compression are in very close agreement with eachother.

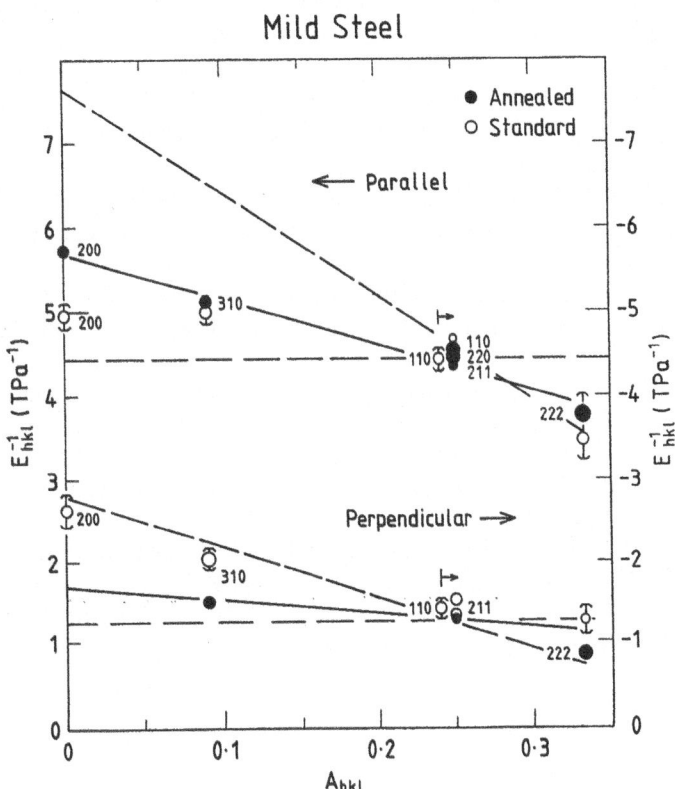

Figure 6. The inverse of the elastic moduli for a series of reflections in annealed and standard steel measured both parallel and perpendicular to the applies stress. The curves are plotted against the anisotropy factor and show a roughly linear behaviour. [7]

158

To analyse results such as these the slopes of the curves within the elastic region are obtained and plotted against the anisotropy factor as in figure 6. It can be seen that the linear relationship predicted by the Reuss and Kroner-Eshelby[8] theory is approximately satisfied, but that the slope has a characteristic value of only around 44% of the Reuss value. However the linear relationship is not fully obeyed to within the accuracy of the measurements. This is illustrated by the fact that the 110, 220, and 211 peaks which all have the same A_{hkl} do not coincide.

4. Conclusions

In contrast to the situation beyond the plastic limit, the elastic response of crystallites within metals under strain is now established. The crystallite behaviour is quite close to the predictions of the Kroner-Eshelby model, with moduli of around 44% of the Reuss value.

This research was supported by the Corporate Research Programme of the UK Atomic Energy Authority.

[1]   Kittel,C., (1986), Introduction to Solid State Physics, Wiley, New York,
[2]   Hertzberg, R. W. (1976) Deformation and Fracture Mechanics of Engineering Materials, Wiley, New York.
[3]   Dolle, H., J. (1979), 'The Influence of Multiaxial Stress States, Stress Gradients and Elastic Anisotopy on the Evaluation of Residual Stresses by X-rays' Appl. Cryst. **12,** 489-501.
[4]   Allen,A. J., Hutchings, M.T., Windsor,C. G., and Andreani,C.,(1985), 'Neutron diffraction methods for the study of residual stress fields', Adv. Phys. **34,** 445-443.
[5]   Sayers, C. M. (1984), 'The strain distribution in anisotropic polycrystalline aggregates subjected to an external stress field' Phil. Mag., **49,** 243-262.
[6]   Windsor, C. G., (1987), Chapter 3 'Experimental Techniques' in Methods of Experimental Physics, Vol **21A,** Ed Skold, K, and Price D. L., Academic Press, New York.
[7]   Hutchings, M. T.,(1990), 'Neutron Diffraction Measurements of Residual Stress Fields - the Answer to the Engineers' Prayer?', Applied Neutron Scattering Symposium December 1989, Nondestructive Testing and Evaluation **5,** 395-414.
[8]   Kroner, E., (1967), J. Mech. Phys. Solids, **15,** 319-329.

INVESTIGATION OF LARGE GRAINED SAMPLES - PRINCIPLES

W. REIMERS

Hahn-Meitner-Institut Berlin
Dept. N5
Glienicker Str. 100
W-1000 Berlin 39
Germany

ABSTRACT.   In order to analyse the residual stresses in large grained materials, the single crystal anisotropy has to be considered in calculating the stresses from the experimental strain data. Therefore an evaluation of the crystallite under study is necessary. The procedure for establishing the orientation matrix from two observed reflections is described. Using a four-circle-diffracto-meter the knowledge of the orientation matrix then allows the calculation of the diffractometer setting for every given reflection (hkl). So this method is well suited for an automized measuring routine which gives the strain tensor based on the experimentally determined diffraction angles 2θ for several reflections of an individual grain. For the refinement of the strain tensor components the least-squares method is used. The stress tensor referred to the crystal axes system is then calculated using the single crystal elastic constants. To compare the stress states in different crystallites, their stress tensors have to be transformed into a common reference system.

1. INTRODUCTION

The analysis of residual or applied stresses by means of diffraction techniques is based on the measurement of crystalline lattice spacings d_{hkl} in different sample directions. When polycrystalline specimens are exposed to a monochromatic X-ray or neutron beam and all properly oriented crystallites contribute to the diffraction intensity. Since the reflecting crystallites have different orientations, the elastic constants needed for the calculation of the stress tensor out of the strain tensor have to represent a volume weighted average over the anisotropic elastic constants of the reflecting crystallites. For the case of an isotropic orientation distribution, the relationship between stresses and strains is described by the X-ray elastic constants (XEC) s_1 and s_2 /1/.

M. T. Hutchings and A. D. Krawitz (ed.),
Measurement of Residual and Applied Stress Using Neutron Diffraction, 159–170.
© 1992 *Kluwer Academic Publishers.*

When studying textured or large grained materials, two
problems arise: first of all, the Debye-Scherrer cones
obtained as reflected intensity distribution from fine
grained materials may be split into localized Bragg-
reflections, so that reflected intensity can only be
measured at selected sample orientations. Furthermore,
for the evaluation of stresses from the experimental
strain data, the elastic anisotropy of the reflecting
volume has to be considered.

For textured materials, the complete set of crystallite
orientations for each measuring direction is obtained from
the orientation distribution function /2,3/. The anisotropic elastic constants for each diffractometer position may
then be calculated by averaging the single crystal
constants using this quantitative representation of the
texture /4/. For the case of large grained materials
spatially defined Bragg-reflections are observed which can
be attributed to individual crystallites. So for the correct analysis of the stresses present the orientation of
the crystallites under study has to be evaluated.

Using X-ray diffraction, the problem has been attacked in
the past by applying the film technique in combination
with the inclination /5/ or the rotation /6/ method. Due
to the insensitivity of the film technique, however, the
investigations had to be limited to very large grains (∅
some mm). Another method of stress analysis in large
grained materials was based on a three-circle-diffracto-
meter with a counter detector /7/. Here, the evaluation of
the orientation matrix turned out to be tedious and was
performed off-line graphically. A more systematic procedure for setting up the orientation matrix and for the
measurements is possible by using a four-circle-diffracto-
meter. The procedure /8/ consists of establishing the
orientation matrix from two observed Bragg reflections
/9/. With this information, the diffractometer setting for
any reflection (hkl) can be approximately calculated.
Measurements of their precise spatial orientations and
especially of their diffraction angles yield the data for
the strain calculation. Using single crystal elastic
constants the stresses in the crystallites can then be
calculated, and if necessary transformed to a common
reference system. Using X-ray diffraction, crystallite
sizes down to ∅ $\approx$ 40 μm could be measured. The same
algorithms are used for neutron diffraction stress
investigations. Here, crystallite sizes down to ∅ $\approx$ 1 mm
can be studied.

2. EVALUATING THE ORIENTATION MATRIX

The geometry of a four-circle-diffractometer is illustrated schematically in fig. 1. The crystallite selected for the study is located at the instrument center. Using X-rays the selection can be done either by collimating the beam or by shielding the sample surface by thin Pb-foils. The selection of the crystallites by means of neutrons is done applying the 90° degree scattering technique which allows the definition of the volume to be investigated.

The incident and diffracted beams are in the horizontal plane. The counter (here: det) is moved in this plane about the vertical instrument axis ω and makes angle 2θ with the primary beam direction (here: col). The whole Eulerian cradle may be rotated around the vertical axis by ω. The χ-axis is in the horizontal plane and makes an angle ω with the primary beam direction. The sample is rigidly attached to the φ-shaft which is supported by the χ-ring. The rotation sense for the rotation movements is defined mathematically positive.

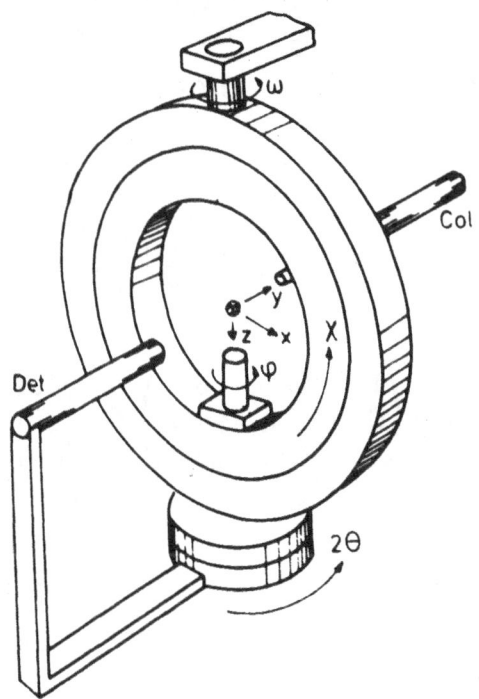

x,y,z Laboratory system 2θ, w, χ, φ Eulerian angles

Fig. 1: Four-circle-diffractometer

The orientation of the crystallite under investigation is known when it is possible to relate its axes system to the fixed laboratoy system. This coincides with the ϕ -axes system when all instrument angles are zero.

x_{Lab} parallel primary beam direction
y_{Lab} parallel to the direction of the reflected beam for $2\theta = 90°$
z_{Lab} parallel to the ω - axis

For transforming the systems it is convenient to define them in terms of cartesian axes. A crystal direction $\underline{h}$ is therefore described by a cartesian crystal direction $\underline{h}_c$:

$$\underline{h}_c = B\underline{h}$$

$$\text{for} \quad \underline{h} = \begin{pmatrix} h \\ k \\ l \end{pmatrix}$$

$$\text{and} \quad B = \begin{bmatrix} b_1 & b_2\cos\beta_3 & b_3\cos\beta_2 \\ 0 & b_2\sin\beta_3 & -b_3\sin\beta_2\cos\alpha_1 \\ 0 & 0 & 1/a_3 \end{bmatrix}$$

a_i, α_i and b_i, β_i are the direct and reciprocal lattice parameters, respectively. Since $\underline{h}$ shall be described in terms of the ϕ -axes system also, the instrumental angles ω, χ, φ , where the reflection is observed, are represented in form of cartesian axes systems Ω, X, ϕ , which coincide when the angles are all zero:

$$\Phi = \begin{pmatrix} \cos\varphi & -\sin\varphi & 0 \\ \sin\varphi & \cos\varphi & 0 \\ 0 & 0 & 1 \end{pmatrix} \quad X = \begin{pmatrix} 1 & 0 & 0 \\ 0 & \cos\chi & -\sin\chi \\ 0 & \sin\chi & \cos\chi \end{pmatrix} \quad \Omega = \begin{pmatrix} \cos\omega & -\sin\omega & 0 \\ \sin\omega & \cos\omega & 0 \\ 0 & 0 & 1 \end{pmatrix}$$

The angular set ω, χ, φ gives a diffractometer position for reflected intensity. The corresponding crystal direction can then be described in the ϕ -axes system in form of an unit vector $\underline{y}_\phi$.

$$\underline{u}_{\underline{\phi}} = \tilde{\underline{\Phi}}\ \tilde{X}\ \tilde{\Omega} \begin{bmatrix} 0 \\ 1 \\ 0 \end{bmatrix}$$

$\sim$ = transposed matrix

In the case of residual stress measurements, the material investigated and hence the cell parameters are known approximately. So the diffracting angle for a reflection $\underline{h}$ can be calculated and the detector is positioned. The spatial orientation of a reflection $\underline{h}_1$ is then searched by a systematic variation of φ and χ with fixed ω. The plane of a second reflection $\underline{h}_2$ is then defined by its angular relationship to the first reflection, so that this plane can be measured. The observations of two non-complanar reflections of known indices $\underline{h}_1$, $\underline{h}_2$ is sufficient to obtain the orientation matrix U which transforms vectors in the crystal cartesian axes system into the ϕ-axes system:

$$\underline{h}_{1\underline{\phi}} = U\ \underline{h}_{1c} \qquad\qquad \underline{h}_{2\underline{\phi}} = U\ \underline{h}_{2c}$$

with
$$\underline{h}_{1\underline{\phi}}\ \|\ \underline{u}_{1\underline{\phi}}$$
$$\underline{h}_{2\underline{\phi}}\ \|\ \underline{u}_{2\underline{\phi}}$$

Because of experimental errors and uncertainties in the knowledge of the cell parameters, the orthogonal matrix U does not, in general, satisfy both conditions. Therefore, an orthogonal unit-vector triple $\underline{t}_{1c}$, $\underline{t}_{2c}$, $\underline{t}_{3c}$ in the crystal cartesian axes system is defined where $\underline{t}_{1c}$ is parallel $\underline{h}_{1c}$, $\underline{t}_{2c}$ lies in the plane of $\underline{h}_{1c}$ and $\underline{h}_{2c}$ and $\underline{t}_{3c}$ is perpendicular to $\underline{t}_{1c}$ and $\underline{t}_{2c}$. In the ϕ-axes system another triple $\underline{t}_1$ $\underline{t}_2$, $\underline{t}_3$ is defined based in the same way on $\underline{u}_{1\varphi}$ and $\underline{u}_{2\varphi}$. Since these two unit vector triples are orthogonal by definition they can be superimposed exactly onto each other:

$$\underline{t}_{i\phi} = U\underline{t}_{ic} \qquad i = 1,2,3$$

In matrix notation:

$$T_\phi = UT_c$$

Then follows:

$$U = T_\phi \, \tilde{T}_c$$

3. ANGLE CALCULATIONS FOR ANY REFLECTION (hkl)

After having evaluated the orientation matrix by two reflections, the coordinates in the ϕ-axes system can be calculated using the orientation matrix UB for any reflection (hkl)

$$\underline{h}_\phi = UB\underline{h}$$

with

$$\underline{h}_\phi = \begin{pmatrix} h_{\phi 1} \\ h_{\phi 2} \\ h_{\phi 3} \end{pmatrix}$$

For the detector position the Bragg-equation is used. The interplanar lattice spacing d is given by:

$$d = \frac{1}{\left(h_{c1}^2 + h_{c2}^2 + h_{c3}^2\right)^{1/2}}$$

The diffractometer setting for the symmetric position ($\omega = 2\theta/2$) is then obtained by:

$$2\theta = 2 \ arc \ sin \left(\frac{\lambda}{2} \ (h_{c1}^2 + h_{c2}^2 + h_{c3}^2)^{1/2} \right)$$

$$\omega_0 = 2\theta/2$$

$$\chi_0 = arc \ tan \left(h_{\Phi_3} \ / (h_{\Phi_1}^2 + h_{\Phi_2}^2)^{1/2} \right)$$

$$\varphi_0 = arc \ tan \left(h_{\Phi_1} \ / h_{\Phi_2} \right)$$

with λ = wavelength
and index o for the symmetric position

Due to absorption effects it is sometimes useful to rotate the crystal around the scattering vector by the angle Υ until a more favorite diffractometer setting is reached. With $\Upsilon = 0°$ the rotation matrix R_o for the bisecting position is defined by:

$$R_0 = \Omega_0 \ \chi_0 \ \Phi_0$$

The new rotation matrix is then calculated

$$R_\tau = \tau \ R_0$$

with

$$\tau = \begin{pmatrix} cos\tau & 0 & sin\tau \\ 0 & 1 & 0 \\ -sin\tau & 0 & cos\tau \end{pmatrix}$$

The Υ-rotation axis is defined to be parallel to the y_{Lab}-axis for $\omega = \chi = \varphi = 0°$.

R_Υ is then generally represented by:

$$R_\tau = R = \Omega \ \chi \ \Phi$$

The solution of the system yields the new diffractometer setting for the reflection position after the rotation by the angle Υ:

$$\omega = \text{arc } \tan \left(R_{\tau 13} \; / -R_{\tau 23} \right)$$

$$X = \text{arc } \tan \left((R_{\tau 31}^2 + R_{\tau 32}^2)^{1/2} / R_{\tau 33} \right)$$

$$\varphi = \text{arc } \tan \left(R_{\tau 31} / R_{\tau 32} \right)$$

For analysing the mosaic spread of the crystallite under study rocking curves in form of ω- scans can be measured. For evaluating the anisotropy of the mosaicity, e.g. after plastic deformation, it is necessary to measure the rocking curves around different crystal axes as rotation axes. Therefore, the diffractometer has to be positioned so that, at the same time, the reflecting conditions for the reflection $\underline{h}_1$ to be measured are fulfilled and the crystal direction $\underline{h}_2$ is parallel to the ω -rotation axis. The symmetric position for the reflection $\underline{h}_2$ is then calculated. The calculated angle φ_0 is kept fixed and X is rotated corresponding to $X_2 = X_0 + 90°$. The reflection position for the reflection $\underline{h}_1$ to be investigated is then calculated by solving the following system in respect to Ω:

$$R_0 = \Omega \; X \; \Phi$$

R_{01} is the rotation matrix for the symmetric position of the reflection $\underline{h}_1$.

4. STRAIN AND STRESS TENSOR

For the determination of the precise reflection position, a centering routine is used where the calculated angles from the UB-matrix serve as starting values. With φ fixed some cycles of independent ω, X and 2θ step-scans are performed until no more systematic shifts are observed. After each cycle, the Lehmann-Larson method /10/ is used for the definition of the peak-borders. The background corrected peak intensities are then used to calculate the peak position by the center of gravity. Depending on the radiation used also the Lorentz-, polarization- and the absorption correction /11/ has to be applied. Furthermore, the calculated peak positions can be corrected for refraction.

According to Bragg's law the refined 2θ -position gives the corresponding $d(\underline{h})$-value so that the strains in $\underline{h}$ are obtained with:

$$\varepsilon(\underline{h}) = \frac{d(\underline{h}) - d_0(\underline{h})}{d_0(\underline{h})}$$

Different possibilities for evaluating the lattice spacings of the unstressed sample $d_0(\underline{h})$ are discussed below.

For expressing the measurements $\varepsilon(\underline{h})$ in terms of strain tensor components ε_{kl}^c, the cartesian crystal axes system is chosen as a fixed reference system:

$$\varepsilon(\underline{h}) = n_k \, n_l \, \varepsilon_{kl}^c \qquad k,l = 1,2,3$$

In this notation $\underline{n}$ is a unit vector parallel to the scattering vector whose components n_k are given in the cubic system by:

$$n_k = \frac{h_k}{\sqrt{h_1^2 + h_2^2 + h_3^2}}$$

For non orthogonal crystal axes systems, an orthogonalization has to be performed:

$$\varepsilon(\underline{h}) = P_k \, P_l \, \varepsilon_{kl}^p \qquad k,l = 1,2,3$$

Here, the components p_K are defined by:

$$p_k = \alpha^{-1}_{kl}\, n_l \qquad\qquad k,l = 1,2,3$$

α_{kl} are the components of the orthogonalization matrix. For the determination of the symmetric strain tensor E at least six measurements in non-complanar directions are necessary. If more information is available a least-squares refinement can be applied. The stress tensor components σ_{ij}^c are calculated by applying Hooke's law for anisotropic, quasielastic materials. Here, the single crystal elastic constants are inserted:

$$\sigma_{ij}^c = c_{ijkl}\, \varepsilon_{kl}^c \qquad\qquad i,j,k,l = 1,2,3$$

Whereas in the triclinic crystal system 21 independent elastic constants are present, their number is reduced to three ($c_{1111} \neq c_{1122} \neq c_{1212}$) in the case of the cubic system. The resulting tensor Σ^c is referred to the car-tesian crystal axes system whose orientation is dependent on the crystal orientation. In most cases, especially for comparing the values obtained in crystals which are in direct neighborhood to each other, it is preferable to transform the stress tensor into the macroscopic sample system. As the first step, the stress tensor $\underline{\Sigma^c}$ is trans-ferred into the fixed laboratory system (index L). As transformation matrix the UB-matrix can be used:

$$\underline{\Sigma}^L = \tilde{T}\,\underline{\Sigma}^c T$$

In the second step the stress tensor $\underline{\Sigma}^L$ has to be trans-formed into the common reference system. Therefore, the angles describing the orientation of the reference system relative to the laboratoy system have to be determined. The transformation matrix T is then represented by a rota-tion matrix R, which components are given by ω, χ, φ.

5. SUMMARY

The single crystal measuring technique provides reliable data about residual streses in large grained materials without using the averaging procedures needed when the orientation of the crystals is unknown. Since the measuring technique presented is based on the diffraction properties of single crystals, the stress state in the individual crystallites is registered, so giving stress values corresponding to the sum of first and second order stresses /12/. The third order stresses are then accessible either by reflection profile evaluation or by scanning the crystals, e.g. by X-ray or synchrotron radiation.

REFERENCES

Hauk, V. (1982) 'Röntgenographische Elastizitätskonstanten (REK)', in V. Hauk, E. Macherauch (eds.), Eigenspannungen und Lastspannungen, HTM-Beiheft, Carl-Hanser-Verl., München, Wien, pp. 49-57.

Bunge, H.-J. (1965) 'Zur Darstellung allgemeiner Texturen', Z. Metallkde., 56, 872-874.

Roe, R.J. (1965) 'Description of crystallite orientation in polycrystalline materials', J. Appl. Phys., 36, 2024-2031.

Barral, M., Spauel, J.M., Maeder, G. (1983) 'Stress measurements by X-ray diffraction on textured material characterized by its orientation distribution function (ODF)', in E. Macherauch, V. Hauk (eds.), Eigenspannungen, Bd. 2, Deutsche Gesellschaft für Metallkunde, Oberursel, pp. 31-47.

Bollenrath, F., Hauk, V., Müller, E.H. (1968) 'Röntgenographische Verformungsmessungen an Einzelkristalliten verschiedener Korngrößen', Metall., 22, 442-449.

Iwasaki, I. and Murakami, Y. (1970) 'X-ray study on plastic deformation of coarse grained aluminium under tensile', J. Soc. Mat. Science, 19, 58-64.

Bol'shakov, P.P., Vasil'ev, D.M., Titovets, Yu F. (1975) 'X-ray diffraction determination of the stress and strain tensor components in coarse-grained materials', Zavodskaja Laboratoriya, 41, 1099-1102.

Reimers, W. (1989) 'Entwicklung eines Einkornmeß- und Auswertungsverfahrens unter Anwendung von Beugungsmethoden zur Analyse von Deformationen und Eigenspannungen im Mikrobereich', Habilitationsschrift, Dortmund.

Busing, W.R. and Levy, H.A. (1967) 'Angle calculations for 3- and 4-circle X-ray and neutron diffractometers', Acta Cryst., 22, 457-464.

Lehmann, M.S. and Larsen, F.K. (1974) 'A method for location of the peaks in step-scan-measured Bragg reflections', Acta Cryst., A30, 580-584.

Lipton, H. (1972) International Tables, Vol. II, The Kynoch Press, Birmingham/England, pp. 265-267.

Macherauch, E., Wohlfahrt, H., Wolfstieg, U. (1973) 'Zur zweckmäßigen Definition von Eigenspannungen', HTM 28, 201-211.

THE PLASTIC REGIME, INCLUDING ANISOTROPY EFFECTS

T. LEFFERS AND T. LORENTZEN
Materials Department
Risø National Laboratory
DK - 4000 Roskilde
Denmark

ABSTRACT. The physical background for the development of internal/residual stresses in single-phase materials (the accumulation of dislocations) is outlined, and a simple model for the calculation of intergranular (type-2) stresses is described. The practical situation for the measurement of and the separation of stresses of types 1, 2 and 3 with neutron diffraction is discussed.

1. Introduction

A number of other authors in these proceedings deal with specific examples of neutron (or X-ray) diffraction measurements of internal/residual stresses in the plastic regime and the associated anisotropy effects, e.g. Noyan, Pintschovius and Reimers for single-phase materials and Withers for composite materials. Anisotropy effects caused by elastic anisotropy as dealt with by Windsor are outside the scope of the present paper.

In order to avoid undue repetition we shall concentrate on the basic principles behind the formation of internal/residual stresses (with anisotropy) by plastic deformation and the basic problems and advantages associated with measurement of these stresses by neutron diffraction.

In principle we shall only deal with single-phase materials, but occasionally we shall refer to dual/multi-phase materials for comparison.

We shall use the term "internal stresses" in the situation where it is not essential whether there is an external stress or not. The term "residual stresses" will only be used in situations where it is essential that there are no external stresses.

2. The Microstructural Origin of the "Plastic" Stresses

At any point in a sample - whether it is a small tensile specimen or a large structural component like a pressure vessel - it applies that the stress is the sum of the stresses from all the dislocations in the sample plus the stresses caused by other phenomena. In this connection "other phenomena" may be an external applied stress, or it may be an internal stress produced by thermal effects or phase transformations (including the formation of gas bubbles). The stress contribution from the accumulated dislocations is the contribution

M. T. Hutchings and A. D. Krawitz (ed.),
Measurement of Residual and Applied Stress Using Neutron Diffraction, 171–187.
© 1992 *Kluwer Academic Publishers*.

172

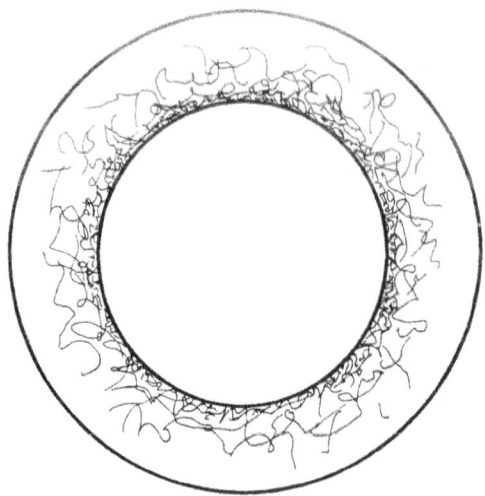

Figure 1.   Schematic illustration of the dislocation accumulation during plastic deformation in a sphere cooled from high temperature. During the initial cooling of the outer shell there is plastic deformation in such a direction as to increase the average radius of the shell (and decrease its average thickness), which leads to dislocation accumulation as sketched. The accumulated dislocastions produce mixed triaxial stresses in the shell and dilatational stresses in the centre. With further cooling the dislocations are immobilized, and they produce a residual stress when the whole sphere reaches room temperature.

associated with plastic deformation. This is illustrated schematically in fig. 1. We have a sphere of a single-phase material which is initially at a high temperature and then subsequently is cooled. First the outer shell cools and hence tries to contract, but the contraction is hindered by the central part which is still hot. This causes plastic deformation in the shell with dislocations accumulating at the boundary to the central part. With further cooling the dislocations are immobilized and hence frozen in. When the whole sphere has reached room temperature, there are dilatational stresses in the centre and mixed (triaxial) dilatational and compressive stresses in the shell. There are no "other phenomena" left, and therefore all the stresses are associated with the accumulated dislocations. The example illustrates the point that the *residual* stresses (for isothermal conditions) *caused* by plastic deformation are equal to the sum of the stresses from the accumulated dislocations - even though thermal effects have played a decisive role at an intermediate stage by introducing the plastic deformation. Obviously the subdivison of the sphere in one outer shell and one central part is artificial. In reality there is a smooth transition between the outer and the central part, but this does not change the basic conclusion.

In dual- and multi-phase materials the residual stresses caused by plastic deformation (and hence by the accumulated dislocations) are superimposed on the possible residual stresses caused by thermal effects and by phase transformations. When thermal effects or phase transformations are the cause of plastic deformation, the stresses associated with this plastic deformation are in such directions as to reduce the stresses from the thermal effects or the phase transformations. The same applies when there are thermal stresses in single-phase materials with thermal gradients. This is illustrated in fig. 1 if we take it to represent the situation when the outer shell has cooled while the centre is still hot: the dislocations introduced by plastic deformation reduce the thermal mixed stresses in the shell and the thermal compression in the centre, i.e. the dislocation configuration per se produces mixed stresses in the shell and dilatation in the centre - which is the final stress state when the whole sphere has cooled down.

Thus, the residual stresses associated with plastic deformation have a clear microstructural origin, viz. the accumulated dislocations - also when the stresses are macroscopic (type 1) as they are in the example in fig. 1 (since their range is not restricted to the scale of the microstructure). In the simple sketch there is no grain structure in the sphere, but the introduction of a grain structure would not produce any fundamental change in the overall dislocation accumulation. With a grain structure the individual dislocation is restricted to move within its parent grain, but stress transfer across the grain boundaries would still produce an effective dislocation distribution as sketched in fig. 1. However, a large fraction of the accumulated dislocations may be absorbed in the grain boundaries which would make it difficult to observe them in a microscopical investigation.

Cottrell (1964) and Ashby (1970) made a conceptual distinction between two parts of the dislocation population: the geometrically necessary dislocations and the redundant (or statistically stored) dislocations. In connection with internal stresses (at least those which lend themselves to measurements by neutron diffraction, viz. those of type 1 and type 2) only the geometrically necessary dislocations are relevant (but not all geometrically necessary dislocations are relevant). Of course there are local stresses associated with all dislocations, but the stress fields from the redundant dislocations cancel each other within rather short distances (which is not necessarily the case for the geometrically necessary dislocations). The redundant dislocations, which normally make up the largest part of the dislocation population, are the main contributors to work hardening and they produce the type-3 internal stresses.

The fact that the stress fields from the relevant geometrically necessary dislocations do *not* cancel each other within short range has two significant derived effects (apart from the long-range stresses that we deal with primarily): there is a stored energy which is larger than the sum of the dislocation energies as calculated with normal cut-off radia, and there is a Bauschinger effect. We shall deal with the stored energy and the Bauschinger effect in section 4 in connection with stresses of type 2. In principle these effects are equally relevant for stresses of type 1, but in connection with type-1 stresses they are (even more) difficult or impossible to deal with from an experimental point of view.

3. Stresses of Types 1, 2 and 3

The principle behind stress measurement by diffraction techniques is that the distortion of the lattice plane distances, the d values, are used as indicators for stresses, i.e. one measures lattice strains and then tries to convert these strains to stresses. As already implied in section 2 the internal stresses are traditionally subdivided in three groups: stresses of type 1, type 2 and type 3 (or first kind, second kind and third kind), e.g. Macherauch and Kloos (1987). The subdivision in three groups refers both to the ways the stresses (strains) are measured and to the physics behind the stresses. However, the logical grouping according to measuring technique is not fully identical to the logical grouping according to physics. In this section and in section 4 we refer to a fourth type of stresses, "type 2.5" which from the point of view of measuring technique should be counted as type 3 and from a physical point of view should be counted as type 2.

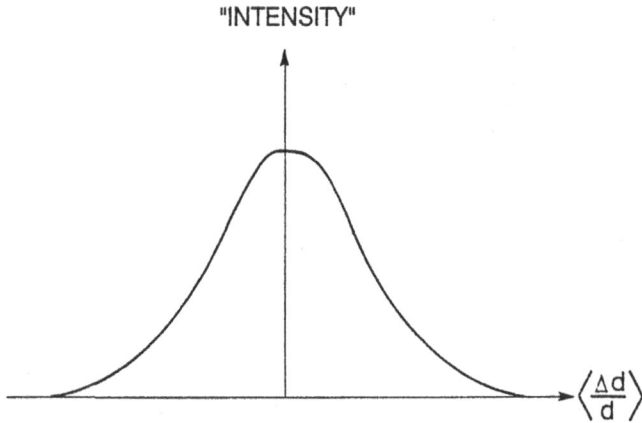

Figure 2.   Overall $\Delta d/d$ distribution in the axial direction for a residual stress state with one main axis. The y axis represents "intensity" (under quote) because it cannot in practice be measured with diffraction techniques (or any other technique).

Let us imagine a sample of any arbitrary size with an overall plastic deformation which can be referred to one axis (tensile deformation for instance). In a small volume element, a grain for instance, the lattice spacing in the axial direction is d in the unstressed state, and there is a change in lattice spacing Δd in the axial direction in the stressed state. For a start we forget the limitations inherent in our diffraction technique, and we assume that we can monitor the distribution of $\Delta d/d$ in the whole sample. In an unstressed (dislocation-free) sample the distribution would be a delta function at zero. In a sample with internal stresses the distribution would have a finite width. The centre of the distribution would be (approximately) at zero in a sample without external stresses as sketched in fig. 2, whereas the centre would be displaced from zero in a sample subjected to an external stress.

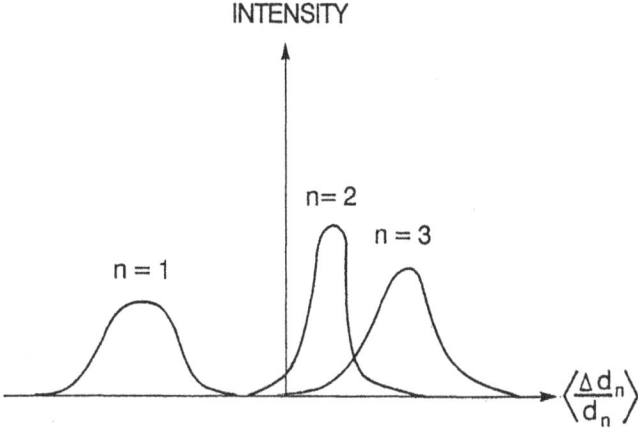

INTENSITY

n= 2

n = 3

n = 1

$$\left\langle \frac{\Delta d_n}{d_n} \right\rangle$$

Figure 3.   Partial $\Delta d/d$ distributions corresponding to the overall distribution in fig. 2. Each value of n corresponds to a specific diffraction condition (a specific d value). Now the y axis represents proper measureable intensities. The situation corresponds to a series of measurements to characterize type-2 stresses (with no macroscopic spatial variation).

Even if we could measure such an overall $\Delta d/d$ distribution for the whole sample, it would be rather uninteresting. The distributions become interesting only when we can extract partial distributions. This can be done in two different ways: (i) by the selection of special diffraction conditions and (ii) by the selection of fractions of the sample (spatial resolution). Possibly (i) is of course not a free choice; it is imposed by the diffraction technique. It leads to the selection of fractions of the sample with special orientations; from the overall $\Delta d/d$ distribution in fig. 2 we take out a number of partial distributions, $\Delta d_n/d_n$, as sketched in fig. 3. For single-phase materials each partial distribution corresponds to grains of a certain orientation. For dual/multi-phase materials each partial distribution corresponds to grains or fibres or particles of a given phase *and* a given orientation. The peak positions of the partial $\Delta d/d$ distributions according to (i) provide information about the internal stresses of type 2 which, fortunately, is a very important information in materials-science connections as described in section 4. Actually, one may take the determination of the peak positions for partial distributions according to (i) to be the experimental definition of type-2 stresses. From a physical point of view we would also be interested in the internal stresses within the grains which correspond to to subdivision of the grains in a finite number of zones with different deformation patterns (see section 4), a subdivision which is at a coarser scale than the type-3 stresses. One might use the term type 2.5. But normally the diffraction techniques do not enable us to study these stresses. With special techniques, as described by Reimers in these proceedings, we may get this information for very coarse-grained materials by adding spatial resolution according to (ii), but this is an exception. Thus, even though the peak positions for (i) distributions give us very relevant information for materials-science applications, we must accept that some relevant information is not available. In principle the width/shape of the partial distributions according to (i) reflect the internal stresses of type 3. The difficulties in the practical interpretation are discussed in 4.3.

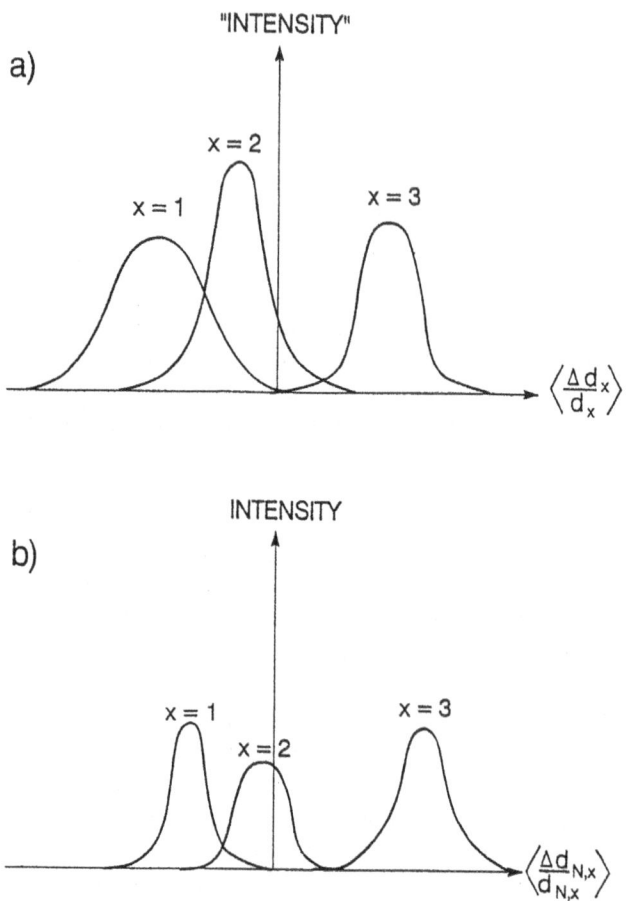

Figure 4.   The ideal (a) and the actual (b) situation for the measurement of type-1 stresses, i.e. stresses which must be described by $\Delta d/d$ distributions which are partial in space (x). Ideally one would want distributions which are only partial in x. In practice one must use distributions which are partial both in n and in x (as shown for one specific n value, N).

In engineering applications the interest is focused on internal stresses of type 1 which we monitor by the peak positions for partial $\Delta d_x/d_x$ distributions which are spatially separated according to (ii). The ideal information for this purpose is that sketched in fig. 4a, namely $\Delta d_x/d_x$ distributions which are only spatially separated according to (ii).

Unfortunately this information is not available. The diffraction techniques can only provide distributions which are partial both in n and x ($\Delta d_{n,x}/d_{n,x}$), i.e. distributions which combine (i) and (ii) as sketched in fig. 4b. Thus, for engineering applications we must accept the complication of an overlap of (i) and (ii), and hence the evaluation of type-1 stresses necessarily involves some type of deconvolution procedure as described in section 5. In the present situation where the type-2 stresses in the plastic regime are not too well understood, this deconvolution involves a significant uncertainty.

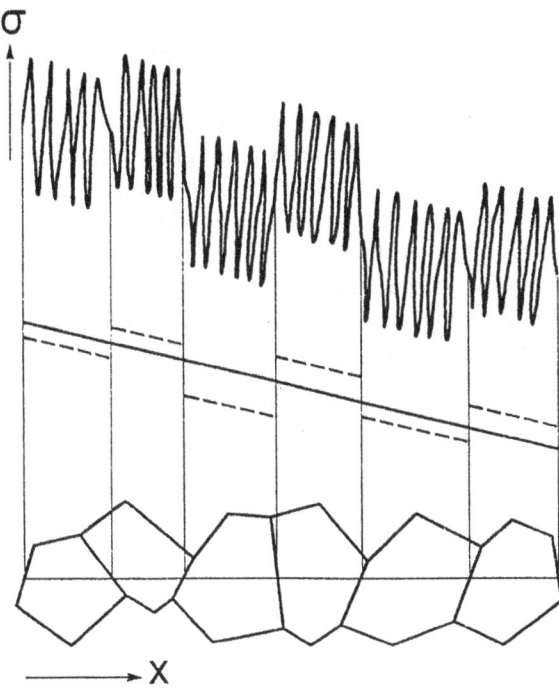

Figure 5.   Sketch of the overlap of stresses of type 1, 2 and 3 along a line through several grains in a single-phase material. The upper curve shows the stress variation as measured with a hypothetical technique partial in space only and with very good spatial resolution. The general slope (the sloping line) represents the stresses of type 1. The jumps (of the broken line) represent the stresses of type 2. The oscillations represent the stresses of type 3. The separation of the full and the broken lines from the oscillating curves is for estethical reasons; physically they should be superimposed.

Fig. 5 illustrates the superposition of internal stresses of types 1, 2 and 3 along a line in a polycrystalline sample. The overall slope refers to stresses of type 1, the jumps up and down by grain-boundary crossings refer to stresses of type 2, and the oscillations within each grain correspond to stresses of type 3. The possibility of reaching a physically relevant interpretation of type type-3 oscillations is discussed in section 4. The simple sketch in fig. 5 does not specify the exact scale of the type-3 oscillations, i.e. whether they correspond to individual dislocations or subgrain walls (Mackerauch and Kloos (1987) refer to the latter).

4. Intergranular Stresses and Other Stresses in Single-Phase Materials

In single-phase materials the only internal stresses of type 2 are the intergranular stresses. Both classical models for polycrystal deformation, the Taylor model (Taylor 1938) and the Sachs model (Sachs 1928) imply intergranular stresses: large intergranular stresses for the Taylor model (e.g. Bishop 1953) and small but finite intergranular stesses for the Sachs model (e.g. Leffers and Pedersen 1987).

The Taylor model is based on the assumption that strain continuity is maintained by a deformation pattern with identical strains in all grains. This requires one specificic stress state in each individual grain (Bishop 1953) determined by the crystallographic orientation of the grain, i.e. the model does not provide stress continuity. In the self-consistent approach first suggested by Kröner (1961) the basic logic of the Taylor-type models is expressed explicitly: one assumes that each grain interacts with a continuum matrix representiative of the average of the polycrystal and *not* with specific neighbour grains. In practice the Kröner model differs from the Taylor model by using the actual elastic properties in the interaction with the matrix (whereas the Taylor model uses infinitely large elastic constants). This makes the Kröner model an elastic-plastic model as opposed to the rigid-plastic Taylor model. The intergranular stresses calculated with the self-consistent model are not very different from those calculated with the Taylor model, but the self-consistent model provides the associated stored energy and the associated Bauschinger effect, which the Taylor model does not.

4.1. A SIMPLE SELF-CONSISTENT MODEL

A large number of models based on the self-consistent concept are described in the literature, including models in which the basic physics as suggested by Kröner have been modified (particularly in the direction of "softer interaction", e.g. Berveiller et al. (1981)). We shall quote some results from the work of Pedersen and Leffers (1987) and Leffers and Pedersen (1987). This work is based on Kröner's basic physics, but texture formation and work hardening have been included; it deals with tensile deformation (including transition to compression) of fcc materials with {111} <110> slip. The material parameters refer to copper - with an assumption of elastic isotropy and isotropic hardening according to the sum of the shears using the Voce constants suggested by Tomé et al.(1984).

Fig. 6 shows the stress-strain curve and the stored energy for a tensile experiment up to a true strain of 0.7 or 70% (the model does not consider the possibility of fracture). The computer experiment refers to a polycrystal consisting of 100 grains of initially randon orientation but with texture formation during straining. The initial and the final orientation distributions are shown as inverse pole figures in fig. 7.

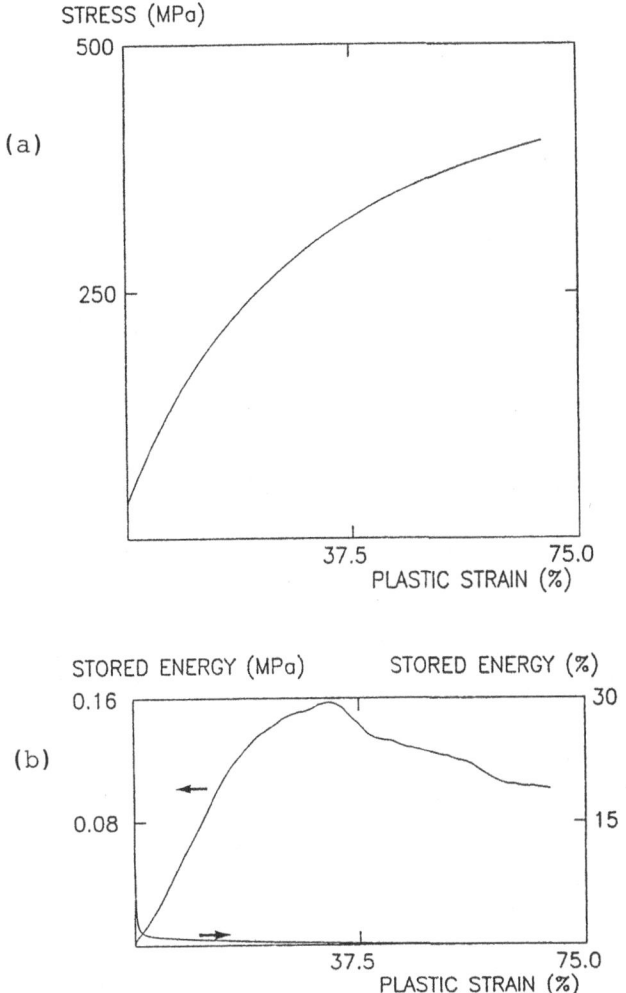

Figure 6.   Stress-strain curve (a) and stored energy versus strain (b) for a computer tensile experiment to 70% strain with a specimen consisting of 100 grains (cf. fig. 7). The stored energy (as specified in the text) is given in MPa (MJ/m³)and as a fraction of the total work done by the external forces. From Leffers and Pedersen (1987).

The stored energy in fig. 6b is the energy stored in connection with the intergranular stresses (strictly speaking the stresses between the grains and the continuum matrix), i.e. it is the energy of the geometrically necessary dislocations. The model does not enable us to derive the energy stored in the redundant dislocations, but of course the work hardening reflects this stored energy in an empirical way. The stored-energy curve in fig.

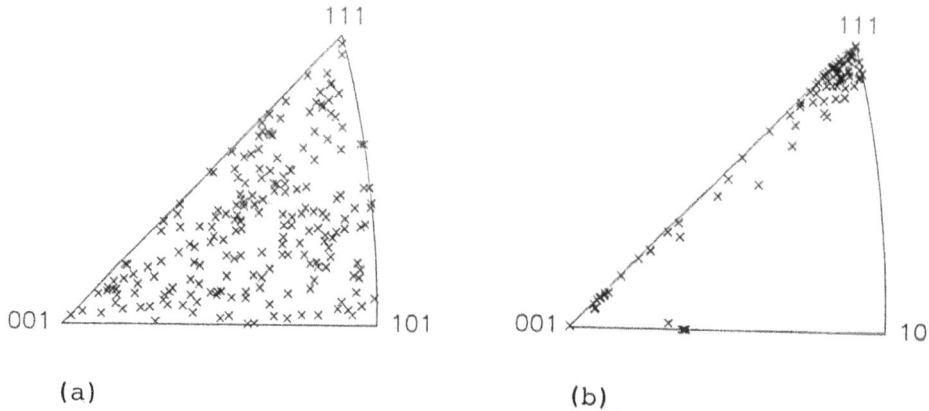

(a)   (b)

Figure 7.   The initial (a) and the final (b) orientations of the 100 grains in the computer experiment in fig. 6. From Leffers and Pedersen (1987).

6b reflects two opposite trends in the development of the intergranular stresses: they increase proportionally with work hardening, and they decrease with texture development. The reason for the decrease is that the texture development (fig. 7b) is in the direction of the two symmetrical orientations <100> and <111> which automatically fulfil the requirement of strain continuity when subjected to an external tensile stress, hence the reduction in the intergranular stresses necessary to enforce strain continuity.

Fig. 8 shows a computer Bauschinger experiment with strain reversal at a strain of 3%. The Bauschinger effect reflects the intergranular stresses. One notices that the stored energy is drastically reduced during strain reversal as an indication of a drastic reduction in the intergranular stresses (but not a reduction to a level close to zero). Fig. 9 maps the normal stresses in the tensile direction in 1000 grains in a specimen which has been strained to 3% and then relaxed to zero external stress - corresponding to the Bauschinger experiment in fig. 8 interrupted just at the transition from tension to compression. The different symbols give the stresses in the different grains. the absolutely lowest and the absolutely highest stresses are indicated: the stresses go from a compressive stress of 19.2 MPa in grains with orientations about 7° from <100> to a tensile stress of 16.1 MPa in grains very close to <111>. The symbols change 11 times between the two extremes, i.e. each symbol covers a stress range of about 3.2 MPa. Fig. 9 should not be seen as a general map of the stress level in the grains in tensile-stressed and relaxed fcc materials. It is obvious that the actual stress level depends on the material and the prestrain. But even the relative stresses depend on the texture. If a similar calculation was made for the material deformed to a strain of 70% in figs. 6 and 7, we would only find the stresses for the narrow range of orientations present (fig. 7b), and the relative stresses for the orientations actually present would not be the same as those for the corresponding orientations in fig. 9.

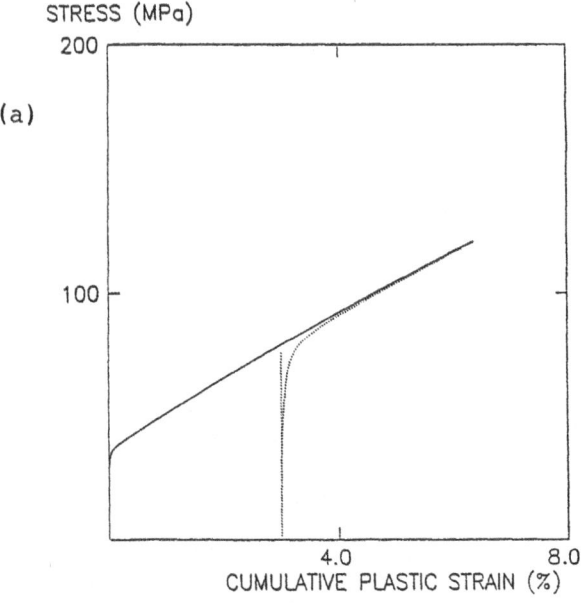

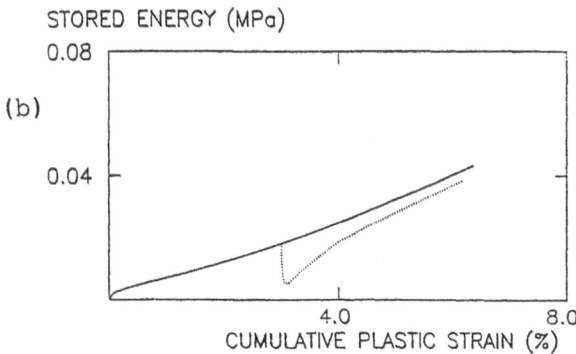

Figure 8.   Computer Bauschinger experiment with strain reversal at a strain of 3% (a) and the associated stored energy (b). A straight tensile experiment is represented by full curves; strain reversal is represented by dotted curves. In accordance with normal practice for Bauschinger experiments the negative (compressive) stresses after reversal are mirrored relative to the strain axis so that they appear positive. From Leffers and Pedersen (1987).

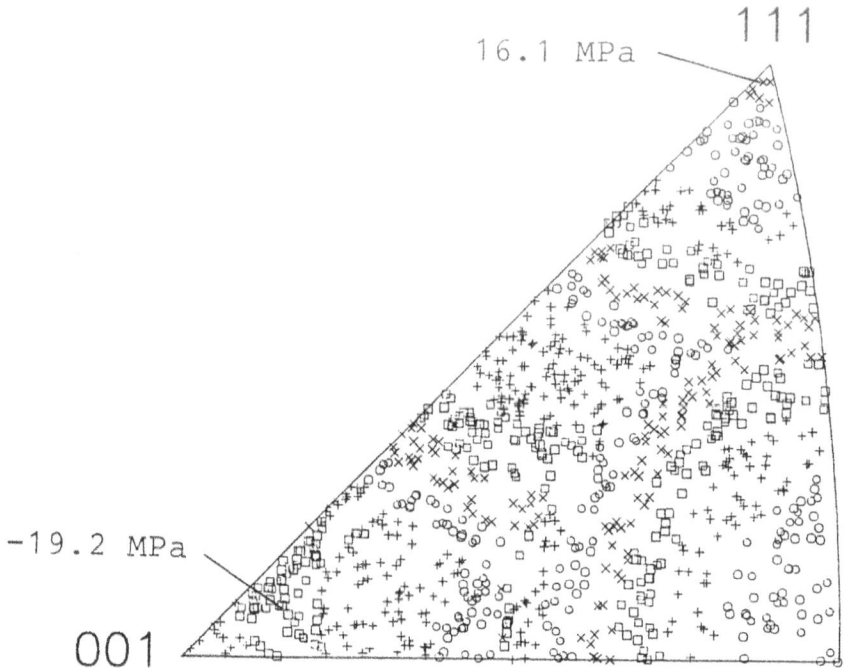

Figure 9.   The normal stresses in the tensile direction in the different (1000) grains in a computer experiment with tensile deformation to a strain of 3% followed by relaxation of the external stresses. The stresses are indicated with a cyclic code: the lowest (most negative) stresses are indicated with squares, and gradually increasing stresses are indicated with crosses, spheres, x's, squares again and so on, ending with x's for the highest (most positive) stresses. From Leffers and Pedersen (unpublished work).

As described by Windsor in the present proceedings there are also intergranular stresses in single-phase polycrystalline materials in the elastic range, caused by elastic anisotropy. In the self-consistent model (and in other models) for the plastic regime it is assumed that these "elastic" intergranular stresses are swamped by the "plastic" intergranular stresses. Of course there must be a transition range in which the two types of intergranular stresses overlap, but we have chosen to neglect this complication.

It should be noticed that the type of self-consistent model described above for the grains in single-phase materials is in principle easily modified to cover the grains in bi/multi-phase materials. For composite materials with large differences in stiffness and strength between the matrix and the reinforcing phase the stresses between the different phases are predominant, and for simplicity one normally neglects the grain structure (cf. Withers in the present proceedings).

4.2. DEVIATIONS FROM THE SIMPLE MODEL

The self-consistent model (as the Taylor model) describes the deformation of individual grains embedded in a continuum matrix without any dislocations. Thus, the model dos not allow for dislocation interaction across the grain boundaries. Neither does it allow for the fact that any given grain meets different grains at the different boundaries and hence may split up in zones with different deformation patterns, a subdivision which has been termed "grain-scale heterogeneity" by Leffers and Juul Jensen (1991). The internal stresses associated with the grain-scale heterogeneities are the stresses referred to as type 2.5 in section 3. The interaction with different conbinations of specific neighbour grains also means that the average stress level for grains of a specific orientation may vary.

Dislocation interaction across the grain boundaries is a relaxation process which tends to make the intergranular stresses smaller than those calculated with simple models as the self-consistent model used above. The intergranular stresses in the simple models are associated with geometrically necessary dislocations accumulated in or close to the grain boundaries; dislocation interactions will inevitably reduce the stresses associated with these dislocations. Pedersen and Leffers (1987) have described how dislocation interaction across the grain boundaries may change the deformation pattern in the direction of the "modified Sachs model" with a drastic reduction in stored energy and in intergranular stresses. A decrease in the level of intergranular stresses is, as a logical consequence, accompanied by an increase in the level of intergranular deformation heterogeneities: the reduced intergranular stresses are incapable of forcing the individual grains to follow the macroscopic strain.

In practice it is very difficult to assess the extent of dislocation interaction. It is obvious though that dislocation interaction across the boundaries is easiest for highly symmetrical crystal structures and less easy for less symmetrical structures. MacEwen et al. (1988) have actually observed intergranular stresses (strains) in Zircaloy-2 of a magnitude comparable to those predicted by the Taylor model (with mechanical twinning added as a supplementary deformation mode), but when it came to the details of the predicted stress (strain) evolution, they differed clearly from those observed experimentally. Intergranular stresses of Taylor/Kröner magnitude have not, to the present authors' knowledge, been observed in fcc materials. It should be mentioned that dislocation interaction poses a similar problem for the prediction of interphasial stresses in composite materials as decribed by Withers in the present proceedings.

Exactly because the intergranular stresses are difficult to prodict, and because they reflect the agreement with or the deviation from the simple Taylor/self-consistent models, measurements of intergranular stresses have a great potential in the study of the deformation mechanisms in polycrystalline materials.

4.3. MEASUREMENTS BY NEUTRON DIFFRACTION

For the measurement of intergranular (type-2) internal stresses neutron diffraction has various distinct advantages over X-ray diffraction. The most obvious advantage is that neutron diffraction samples bulk material, whereas X-ray diffraction samples a thin surface layer. X-rays will typically see the surface grains only (or the surface of the surface grains), and for the intergranular stresses the surface grains are not representative since they are not subject to the same constraint as the grains in the bulk.

In a meaningful investigation of the intergranular stresses in a plastically deformed material one must measure the stresses in the appropriate direction(s). For axisymmetric strain that means measuring the normal stresses in the tensile/compression direction, cf. fig. 9. With x-rays this can only be done with $\sin^2\psi$ extrapolation - with the complication that there is no reason to believe that the $\sin^2\psi$ plots are linear for intergranular stresses (or any other type-2 stresses), disregarding whether the material is textured or not. With neutrons the normal strains in the axial direction can be measured directly - without any extrapolation. For comparison with a calculation like that in fig. 9, the measured strains must be converted to stresses which is not quite trivial because grains with a specific crystallographic direction parallel to the tensile direction have statistically non-zero stresses perpendicular to the tensile axis. However, this is not a problem of measuring technique: the self-consistent model can, without any problems, provide the equivalent of fig. 9 with normal strains instead of normal stresses for direct comparison with neutron measurements.

When we claim that neutron-diffraction measurements do not require $\sin^2\psi$ plots, this must be taken with the appropriate qualifications. It is correct for pure type-2 stresses associated with axisymmetric strain as described above. For superimposed stresses of type 1 and type 2 we obviously need measurements in different directions; whether such measurements should be aimed at a $\sin^2\psi$ plot is an open question.

Yet another advantage of neutrons in connection with intergranular stresses as in fig. 9 is the variable wavelength: it is possible to establish Bragg conditions for a large number of (hkl) planes and hence to test many points in the unit triangle.

For internal stresses of type 3 one normally sees the routine statement that they are derived from the line broadening. However, the line broadening is produced by a combination of various physical phenomena. For ordinary single-phase polycrystals we can identify at least three different contributions: (i) variation in the level of intergranular stresses for different grains with a given orientation because the different grains interact with different neighbour grains and not with a continuum matrix (e.g. 4.2), (ii) grain-scale variations in stress within the individual grains, the stresses of type 2.5 (e.g. 4.2) and (iii) proper type-3 stresses. In practice it is impossible to make any sense of a line broadning produced by these combined effects - apart from the trivial observation that the line broadening increases with increasing strain.

Even if we consider single crystals where contributions (i) and (ii) are non-existing, it is not trivial to interpret a line broadening. Warren and Averbach (1950) considered two contributions to the line broadening: small particle size and internal stresses (of type 3), and they suggested a procedure to distinguish the two contributions. As pointed out for instance by Wilkens (1984) this distinction is artificial for deformed materials since both contributions have their origin in the same dislocation structure. As decribed by Wilkens investigation of the type-3 stresses in deformed single crystals requires line-shape analysis. For this purpose neutrons have a clear disadvantage compared to X-rays, viz. the relatively large instrumental broadening of the peaks.

We do not deny that there may be cases where line broadening (as measured with neutrons or X-rays) can provide relevant information, but the analysis of deformed single-phase materials is not among these cases.

5. Macroscopic (Type-1) Stresses

In section 3 we pointed out the obvious fact that we cannot monitor $\Delta d_x/d_x$ distributions which are partial in x only - the ideal distributions for the investigation of internal stresses of type 1. We can monitor $\Delta d_{n,x}/d_{n,x}$ distributions which are partial both in lattice orientation and in space. The measure of the type-1 stress at a point X is the peak value of $\Delta d/d$ from a $\Delta d_{N,X}/d_{N,X}$ distribution, but this includes the possible contribution from the type-2 stresses associated with the specific N value (the specific (hkl) reflection), which must be subtracted in order to get the type-1 stress.

The great majority of the stresses of type 1 which are relevant in engineering practice are in some way related to plastic deformation; this also includes the thermal stresses in welds, e.g. fig. 1. Thus, the subtraction of the "plastic" stresses of type 2 is a general problem - a problem because of our insufficient understanding of these stresses as discussed in section 4. In principle every investigation of type-1 stresses therefore ought to include an investigation of the stresses of type 2, but this is obviously not possible in practice.

As discussed in 4.2 the most important uncertainty about the type-2 stresses is the stress relaxation associated with dislocation interaction, i.e. the stresses as calculated with simple models as that in 4.1 may be reduced by dislocation interaction. In this situation one may suggest a way out: from calculations of the type presented in fig. 9 we select orientations (hkl) for which the unrelaxed stresses are close to zero, and we assume that the stresses will be low also after a general stress relaxation. Thus, we solve the problem via a clever selection of N for the $\Delta d_{N,x}/d_{N,x}$ distributions. An even better solution, which requires more measurements, would be to use a number of different n values (different (hkl) reflections) and hence to get a direct experimental indication of the importance of the choice of the reflection to be used.

But as long as the type-2 stresses are interesting from a materials science point of view, i.e. as long as they are not fully understood, we must accept some uncertainty in the measurements of internal stresses of type 1.

References

Ashby, M.F. (1970) 'The deformation of plastically non-homogeneous materials', Phil. Mag. 21, 399-424.

Berveiller, M., Hihi, A. and Zaoui, A. (1987) 'Self-consistent schemes for the plasticity of polycrystalline and multiphase materials', in N. Hansen et al. (eds.), Deformation of Polycrystals, Risø National Laboratory, Roskilde, pp. 145-156.

Bishop, J.F.W. (1953) 'A theoretical examination of the plastic deformation of crystals by glide', Phil.Mag. 44, 51-64.

Cottrell, A.H. (1964) The Mechanical Properties of Matter, Wiley, New York, pp. 277-278.

Kröner, E. (1961) 'Zur plastischen Verformung des Vielkristals', Acta metall. 9, 155-161.

Leffers, T. and Juul Jensen, D. (1991) 'The relation between texture and microstructure in rolled fcc materials', in Proceedings of ICOTOM 9, Gordon and Breach (in press).

Leffers, T. and Pedersen, O.B. (1987) 'Polycrystal calculations with a universal elastic-plastic model', in S.I. Andersen et al. (eds), Constitutive Relations and Their Physical Basis, Risø National Laboratory, Roskilde, pp. 401-408.

MacEwen, S.R., Christodoulou, N., Tomé, C., Jackman, J., Holden, T.M., Faber, J. and Hitterman, R.L. (1988) 'The evolution of Texture and residual stress in Zircalay-2', in J.S. Kallend and G. Gottstein (eds.), Proceedings ICOTOM 8, The Metallurgical Society, Warrendale, pp. 825-836.

Macherauch, E. and Kloos, K.H. (1987) ' Origen, measurement and evaluation of residual stresses', in E. Macherauch and V. Hauk (eds.), Residual Stresses in Science and Technology, DGM, Oberursel, pp. 3-26.

Pedersen, O.B. and Leffers, T. (1987) 'Modelling of plastic heterogeneity in deformation of single-phase materials', in S.I. Andersen et al. (eds.), Constitutive Relations and Their Physical Basis, Risø National Laboratory, Roskilde, pp. 147-172.

Sachs, G. (1928) 'Zur Ableitung einer Fliessbedingung', Z. verein. deut. Ing. 72, 734-736.

Taylor, G.I. (1938) 'Plastic strain in metals', J. Inst. Met. 62, 307-324.

Tomé, C., Canova, G.R., Kocks, U.F., Christodoulou, N. and Jonas, J.J. (1984) 'The relation between macroscopic and microscopic strain hardening in f.c.c. polycrystals', Acta metall. 32, 1637-1653.

Warren, B.E. and Averbach, C.L. (1950) 'The effect of cold-work distortions on X-ray patterns', J. Appl. Phys. 21, 595-599.

Wilkens, M. (1984) 'X-ray line broadening of plastically deformed crystals', in N. Hessel Andersen et al. (eds.), Microstructural Characterization of Materials by Non-Microscopical Techniques, Risø National Laboratory, Roskilde, pp. 153-168.

GRAIN INTERACTION STRESSES

L. PINTSCHOVIUS
Kernforschungszentrum Karlsruhe, INFP
Postfach 3640, W-7500 Karsruhe, Germany

ABSTRACT.   Neutron and X-ray diffraction has been used to study the residual stresses in a cold rolled steel strip, plastically strained ferritic and ferritic-austenitic steels as well as a double-V weld. It was found that the stress state of the grains varies considerably with their crystallographic orientation as a result of their anisotropic yield strength. In a heavily textured specimen, these grain interaction stresses could be characterized quantitatively, whereas in other cases only averages could be given which represent to a first approximation macrostresses or phase-specific stresses. However, a check of the balance of forces revealed that the observed residual stresses are sometimes superimposed by deformation residual stresses, which do not show up in peak shifts. They are presumably located in regions with very high dislocation density, which do not give well defined diffraction peaks.

1. Introduction

Plastic deformation of polycrystalline materials with a pronounced anisotropy of the yield strength results in the production of residual stresses (RS) which vary strongly with the orientation of the grains in respect to the strain direction. If the plastic deformation is uniform there will be no macro RS (RS I), but only micro RS arising from the interaction among grains, so-called grain interaction stresses (GIS). In practice, the plastic deformations will rarely be uniform, and hence the resulting RS will be a superposition of RS I and GIS. If the material is not single-phase, GIS will show up in each phase for which the yield strength is exceeded at least for a sub-set of grains. This is, e.g., the case for duplex steels, where the ferritic and the austenitic phase have a similar yield strength. The resultant RS state is very complex: the total RS are the sum of RS I, interphasial RS and phase specific GIS. In a diffraction experiment, all components contribute to the observed lattice strains, so that these cannot easily be converted into RS I. This may be seen as a nuisance, but conversely it offers the possibility to get more detailed information on the RS state as any other method can provide, which is sensitive to RS I only. In favorable cases, the RS state can be evaluated as a function of the grain orientation which may help to understand the deformation modes of the material. Unfortunately, the evaluation of GIS from diffraction data is still in its beginnings. In many cases only averages, which represent RS I or phase

M. T. Hutchings and A. D. Krawitz (ed.),
Measurement of Residual and Applied Stress Using Neutron Diffraction, 189–203.
© 1992 *Kluwer Academic Publishers.*

specific RS, can be given. In this contribution we report some neutron scattering studies where GIS were found to be important and describe the evaluation procedure in each case.

2. Results and Analysis

2.1. COLD ROLLED STEEL STRIP

The investigation of a cold rolled steel strip was done in collaboration with V. Hauk, W.K. Krug and K. Feja from the Institut für Werkstoffkunde, University of Aachen, who did very detailed X-ray measurements on the as-manufactured surface as well after removal of surface layers [1,2]. The starting point were the well-known oscillations observed in D_{211} vs $sin^2\psi$-distributions in the rolling direction (RD) by X-rays as well as by neutrons (Fig. 1a). Such oscillations are expected for elastic loading due to the elastic anisotropy of iron [3], but the oscillations were too strong in view of the small average slope of the D_{211} vs $sin^2\psi$-curve to be explained quantitatively by elastic anisotropy. The first neutron measurements were done with narrow slits (0.5 mm) to confine the internal probe region to the interior of the 1 mm thick specimen. For such a measurement, the peak shifts reflect both macro- and microstresses (RS I and RS II). Additional measurements with wide slits gave integrals over the whole cross-section and hence were sensitive to RS II only. The oscillations were still present, though somewhat less pronounced, showing that RS II were the major reason of the oscillations (Fig. 1b). In the following, additional measurements were carried out using the reflection lines (200), (110) und (222). The results obtained with wide slits are shown in Fig. 2. Obviously, most D_{hkl} vs $sin^2\psi$-distributions are non-linear, not only for the RD, but also for the transverse direction (TD).

The cold rolling had given rise to a pronounced texture, as can be seen from the intensity curve of Fig. 1. For the evaluation of GIS we assumed that the texture can be characterized by a few ideal orientations, i.e. {100}<011>, {311}<01-1>, {211}<01-1>, {111}<01-1> and {111}<2-1-1>. As the part of randomly oriented grains was very small, we think that the observed texture can be fairly well approximated by the ideal orientations given above. For each ideal orientation the lattice strains were taken from the D_{hkl} vs $sin^2\psi$-distributions at the intensity poles and plotted vs $sin^2\psi$ for the RD and the TD (Figs. 3 and 4). For the orientations {311}<01-1>, {211}<01-1> and {111}<01-1>, a pronounced ψ-splitting was observed for the TD. This indicates that the directions of the principal stresses do not coincide with the axes of the sample system for these grains. The lines shown in Figs. 3 and 4 were calculated from triaxial stresses fitted to the data using the single crystal elastic constants. This analysis showed that for the grains having an 110-axis parallel to the RD (the vast majority of all grains) the directions of the principal stresses perpendicular to the RD are given not by the sample system, but approximately by the crystallographic directions (1-1 0) and (001).

The RS evaluated for the interior of the specimen are the sum of RS I and RS II. RS I were calculated in two ways: (i) On the assumption, that the RS II

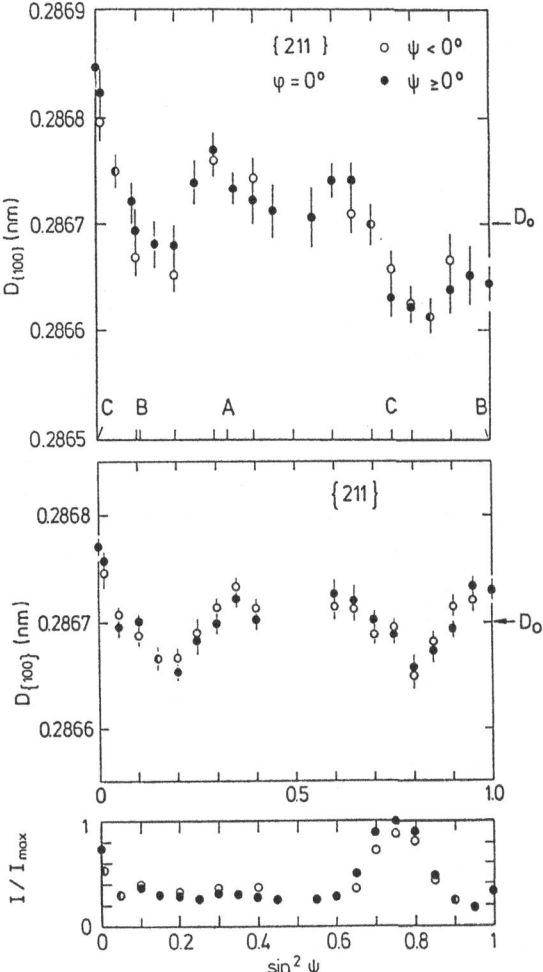

Figure 1.   a (above): Lattice spacing vs $\sin^2\psi$ measured in the interior of a cold rolled steel strip for the rolling direction using (211) peaks. A, B and C denote poles of the ideal rolling texture of iron. b (middle): the same as a), but measured with wide slits in the incident and diffracted beam to obtain averages over the whole thickness. c (below): relative intensities vs $\sin^2\psi$ of the (211) peaks [1].

are constant across the cross-section, RS I in the interior can be evaluated from the difference between the data observed with wide and narrow slits (Fig. 5). The remaining oscillations can be accounted for by elastic anisotropy and hence it seems to be justified to consider the stresses calculated from the data shown in Fig. 5 as RS I. (ii) RS I were obtained as weighted averages of the observed stresses of the different grain fractions. The values obtained by both methods were in satisfactory agreement with each other. Nevertheless, a

192

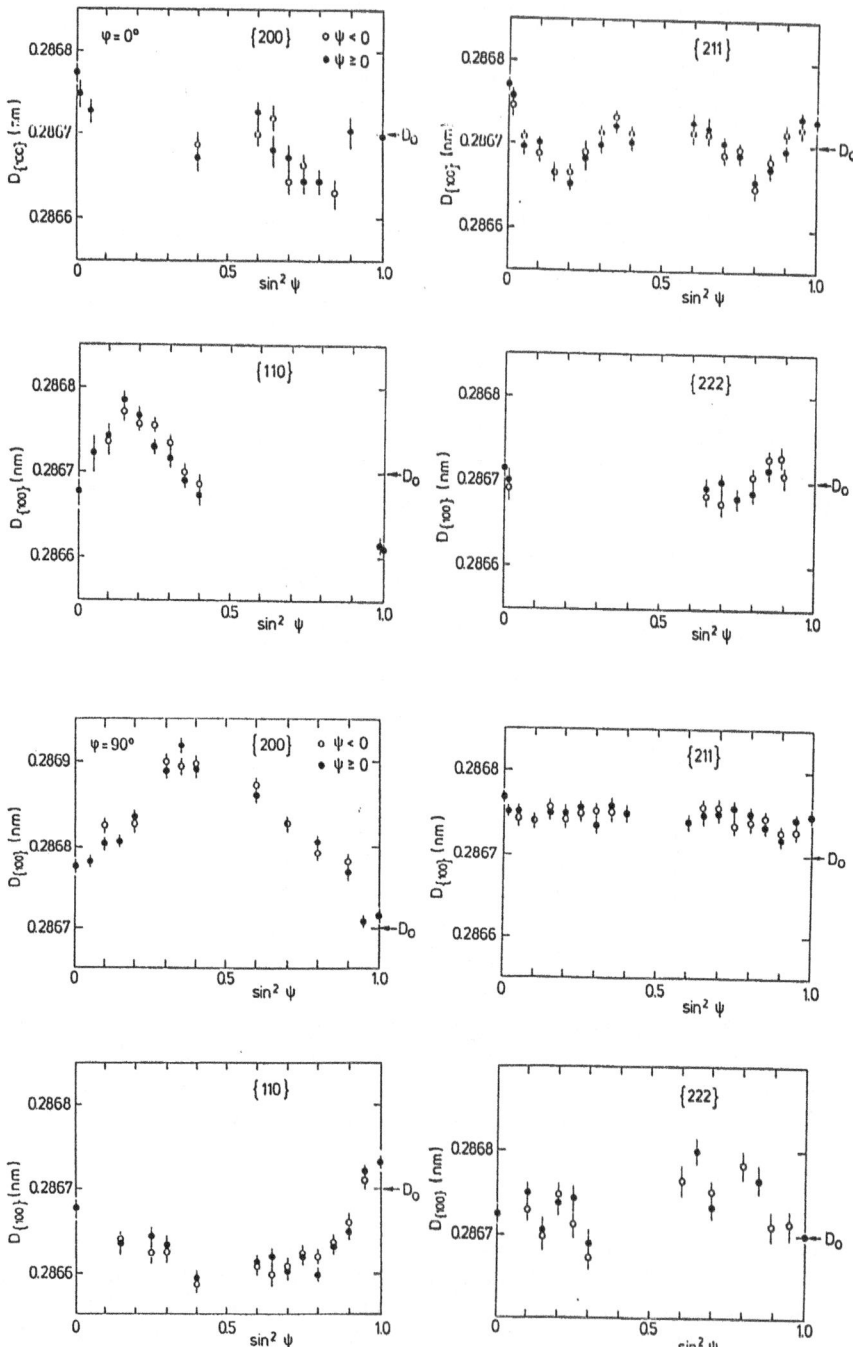

Figure 2. $D_{(hkl)}$ vs $\sin^2\psi$ in a cold rolled steel strip for the rolling direction (above) and the transverse direction (below). The D-values represent averages over the whole thickness [1].

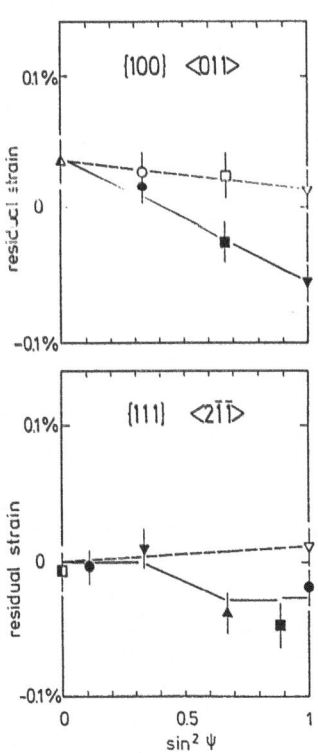

Figure 3. Residual strains vs sin²ψ in a cold rolled steel strip for grains with orientation {100}-<011> (above) and {111}-<211> (below). Different symbols refer to measurements with different (hkl)-peaks. Lines were calculated from triaxial stresses fitted to the data (— rolling direction, --- transverse direction) [1].

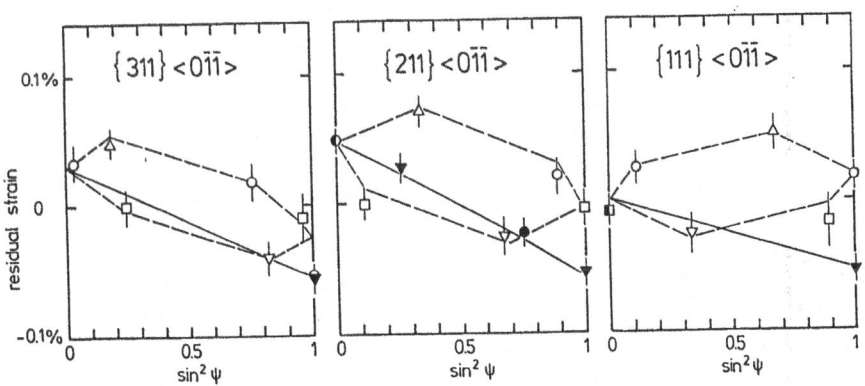

Figure 4. Residual strains vs sin²ψ in a cold rolled steel strip for grains with different orientations. Different symbolds refer to measurements with different (hkl)-peaks. Lines were calculated from triaxial stresses fitted to the data (— rolling direction, --- transverse direction) [1].

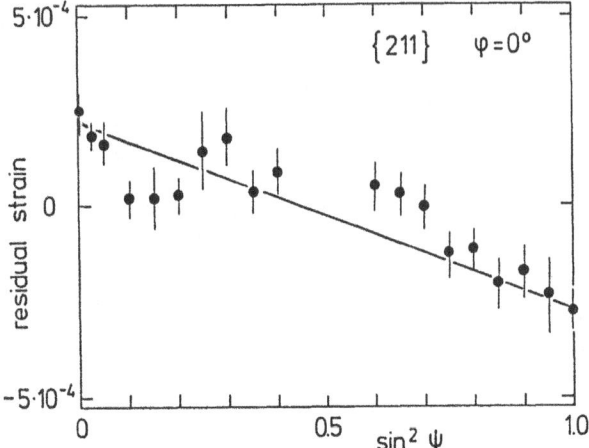

Figure 5. Residual strains vs sin$^2\psi$ deduced from the differences of the data shown in Figs. 1a and b.

further check revealed that the interpretation of the RS calculated by (i) and (ii) as RS I cannot be completely right: Starting from data which represented integrals over the whole cross-section the weighted average of the observed RS should be zero for each direction. However, a non-zero value $\sigma_1 = 30 \pm 10$ MPa was found for the RD. An even larger value was obtained from X-ray data measured after stepwise removal of surface layers [2]. As the GIS are already averaged out by the evaluation procedure and the material was single-phase to a very good degree, it is difficult to say in which regions the observed RS are balanced. We conjecture that the balance of forces is maintained in highly disturbed regions close to grain boundaries which do not contribute to the peak intensities. These regions are considered to be the location of so-called deformation RS which have been frequently found after uniform plastic deformation in homogeneous materials. An example for the observation of very strong deformation RS will be given in the next section.

2.2.   PLASTICALLY STRAINED TENSILE TEST SPECIMEN

RS were investigated in tensile test specimen made of a ferritic (25CrMo4) and of a ferritic-austenitic (X2 CrNiMoN225) steel after uniaxial plastic strain of 8% and 12%, respectively. This study was done again in collaboration with the Institut für Werkstoffkunde, University of Aachen (V. Hauk and H.J. Nikolin [4]). It is expected that uniform plastic strain will not produce RS I, which was confirmed by experiment. Consequently, the neutron measurements were performed with a wide beam giving integrals over the whole cross-section. The aim of the experiments was to separate GIS, deformation RS and in the case of the ferritic-austenitic steel, phase specific RS. The neutron results observed in the strain direction on the ferritic steel 25CrMo4 are shown in Fig. 6. The strain distributions observed in the transverse direction are all rather flat.

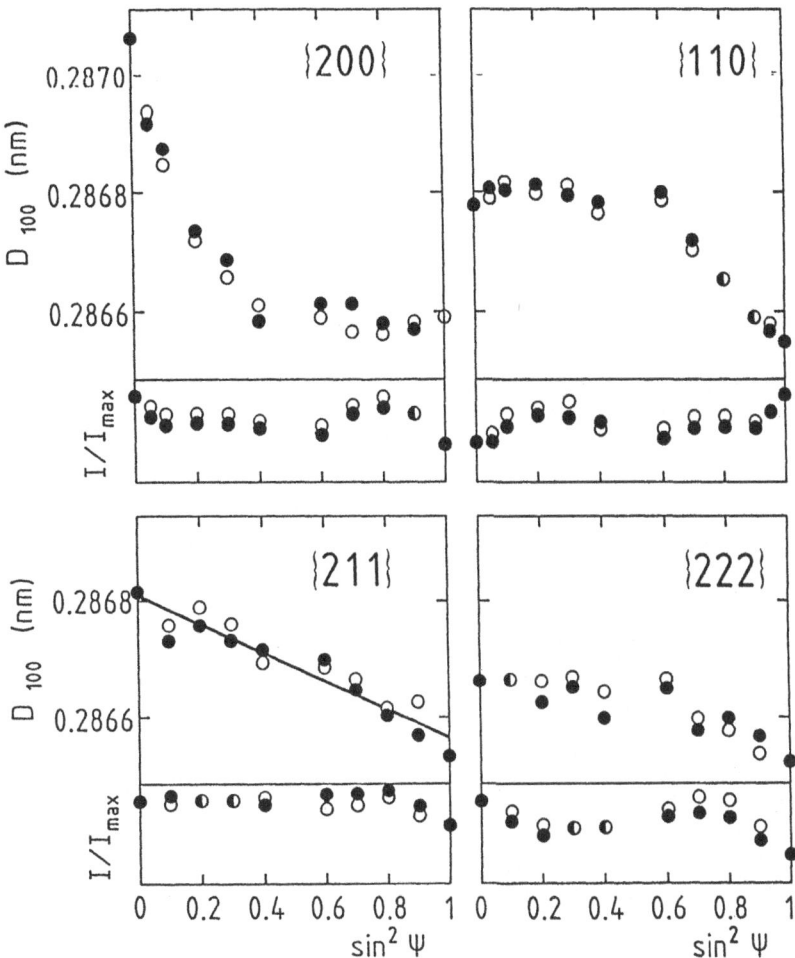

Figure 6. Lattice parameter vs. $\sin^2\psi$ as measured on the steel 25CrMo4 after 8% plastic strain in the strain direction using different (hkl)-peaks. Data were taken with a wide beam giving averages over the whole cross-section [4].

The corresponding X-ray results agree very well with the neutron results showing that the RS state is homogeneous up to the surface. The intensity curves indicate that the plastic strain produced some preferred orientations, but the texture is by far less pronounced than in the cold rolled steel strip dealt with in the preceeding section. The strong non-linearities in some of the D vs $\sin^2\psi$-curve indicate the presence of considerable GIS, but we found no way to characterize them quantitatively. Attempts to evaluate GIS from the assumption of a (110) fibre texture lead to inconsistent results, presumably because the texture was not sufficiently developed. Finally we tried to average GIS out by averaging lattice strain distributions observed for different reflection lines (Fig. 7). The apparent RS are large, i.e. 200 MPa. Obviously,

196

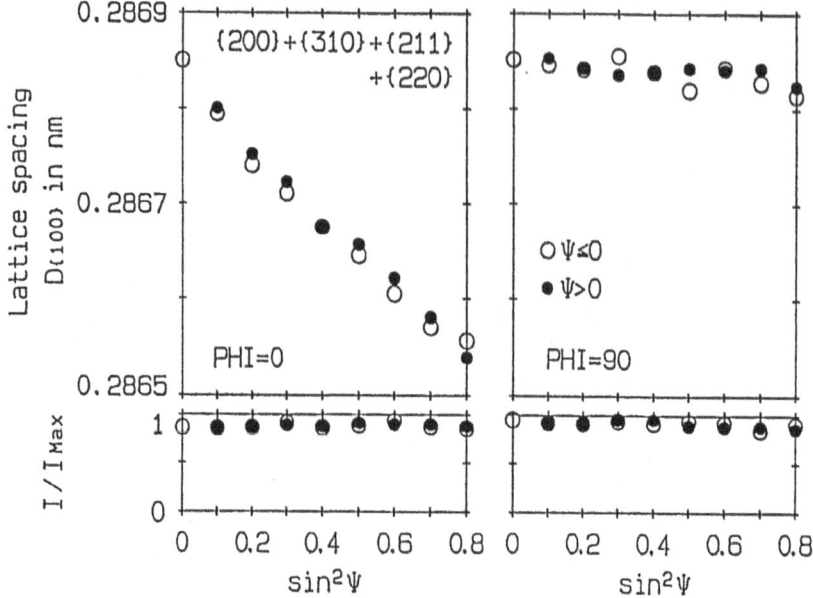

Figure 7. Lattice parameter and relative intensity vs sin²ψ as measured by X-rays on the steel 25CrMo4 after 8% plastic strain [4]. The plotted values are weighted averages of those obtained for the (200), (310), (211) and (220) peaks.

the lattice strains observed by diffraction methods are largely due to deformation RS and not to GIS.

An example of the lattice strains observed in the ferritic phase of the austenitic-ferritic steel X2 CrNiMoN225 after 12% uniaxial plastic strain is shown in Fig. 8. Examples of lattice strains observed in the austenitic phase of this material are displayed in Fig. 9. Many D_{hkl} vs sin²ψ-distributions were highly non-linear, from which the presence of large GIS can be concluded. The good agreement between X-ray and neutron results (see Fig. 9) shows again that the RS state is homogeneous up to the surface. The texture is more developed than in the ferritic steel dealt with above but still not very pronounced. In addition, considerable scatter in the intensity data indicates a somewhat irregular pattern of preferred orientations. A quantitative characterization of the GIS could not be achieved. Merely phase specific RS were evaluated from weighted averages of lattice strains observed for different reflection lines (Fig. 10). The calculated RS in the strain direction for the ferritic phase ($\sigma_x = -84 \pm 7$ MPa) and the austenitic phase ($\sigma_y = 153 \pm 5$ MPa) cancel each other largely, but not completely (the volume ratio of both phases was nearly 1:1). Deformation RS of unknown size in each phase are supposed to account for the balance of forces.

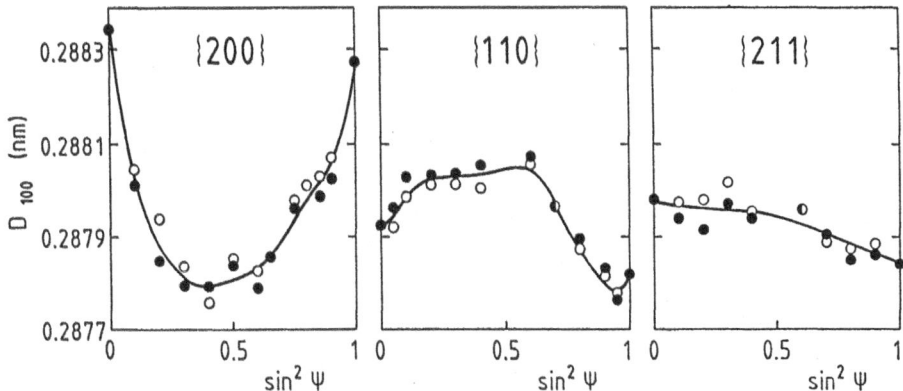

Figure 8. Lattice parameter versus $\sin^2\psi$ as measured by neutrons in the ferritic phase of a ferritic-austenitic steel after 12% plastic strain using different (hkl)-peaks [4].

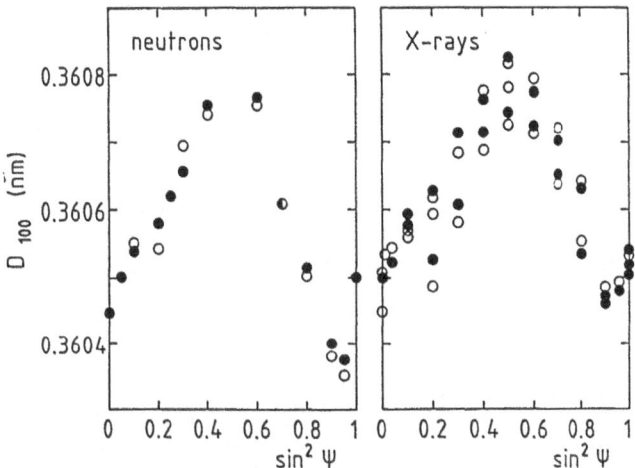

Figure 9. Lattice parameter vs $\sin^2\psi$ determined in the austenitic phase of the steel X2 CrNiMoN225 after 12% plastic strain [4]. The data on the left hand side were obtained with neutrons using the (110)-peaks. The data on the right hand side were obtained by X-rays using the (220)-peaks. Data obtained from samples cut out at different angles were converted into the system of coordinates of the original specimen.

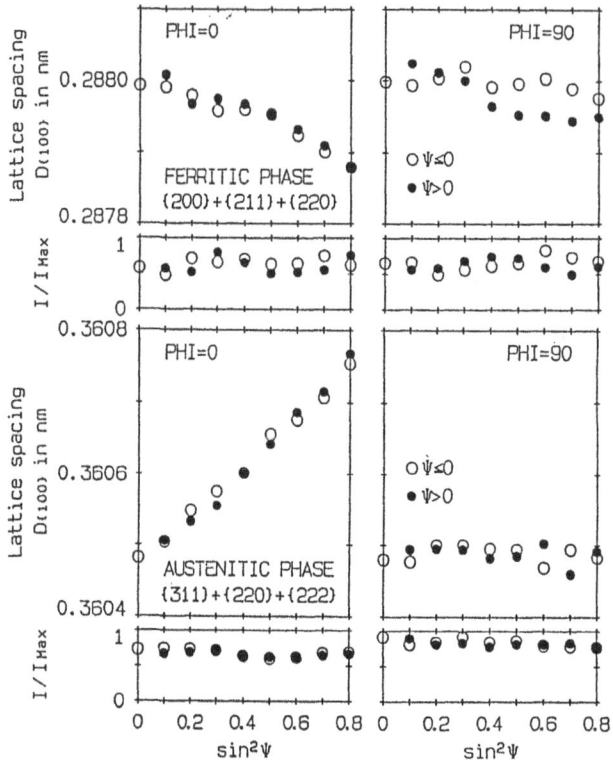

Figure 10.   Lattice strains and relative intensity vs sin²ψ measured by X-rays on the steel X2 CrNiMoN225 after 12% plastic strain. The plotted values are weighted averages of those obtained with different (hkl) peaks [4].

2.3.   A DOUBLE-V WELD

The above reported studies were based on the determination of D vs sin²ψ-distributions. This is a common technique for X-rays, but relatively little used with neutrons because of the large number of measurements needed. Furthermore, severe attenuation of the neutron beam often restricts the accessible ψ-range to such an extent that it can barely seen if D vs sin²ψ-distributions are linear or not. However, the use of different reflections will reveal important information in respect to GIS even if only the three principal strains are registered. This technique was used for the investigation of a double-V weld joining two 50 mm thick plates made of a fully austenitic CrNiMo steel [5].

50 mm are too thick even for neutrons and therefore a 5 mm thick section was cut out by spark erosion. The cross-section is shown in Fig. 11. The cutting relaxes the RS in the z-direction nearly completely, so that the specimen is

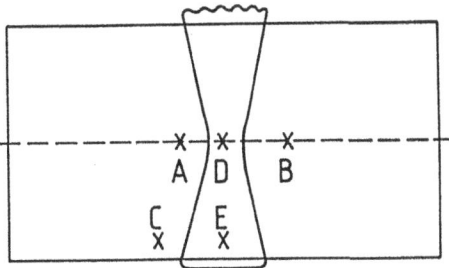

Figure 11. Diagram of the 5 mm thick section of two austenitic steel plates joined by a double-V weld. A, B etc. denote different points of measurements.

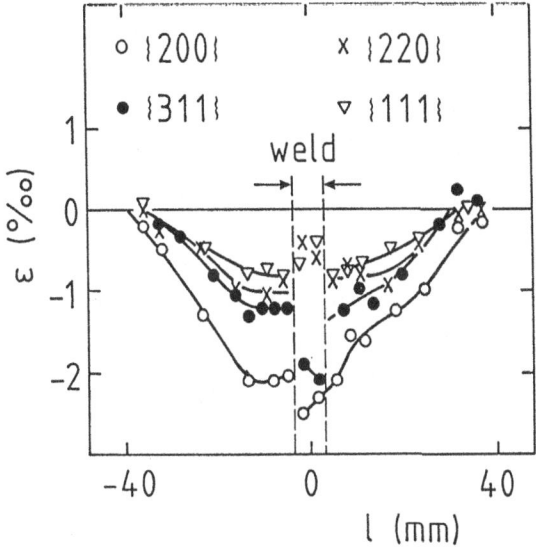

Figure 12. Lattice strain distributions observed with different (hkl) peaks along the central line across the weld of the specimen shown in Fig. 11 [5].

under plane stress. Lattice strains were measured along many lines in the weld and across the weld. The first measurements were done using the (311) reflection. However, a check of the balance of forces indicated that the observed lattice strains cannot reflect RS I only, and so additional measurements were carried out using the reflection lines (200), (111) and (220). An example of lattice strain distributions obtained with different reflection lines is shown in Fig. 12. The data observed for the base material vary in a systematic way, so that elastic anisotropy is a likely explanation for the variation.. This idea was confirmed by plotting the strains versus the

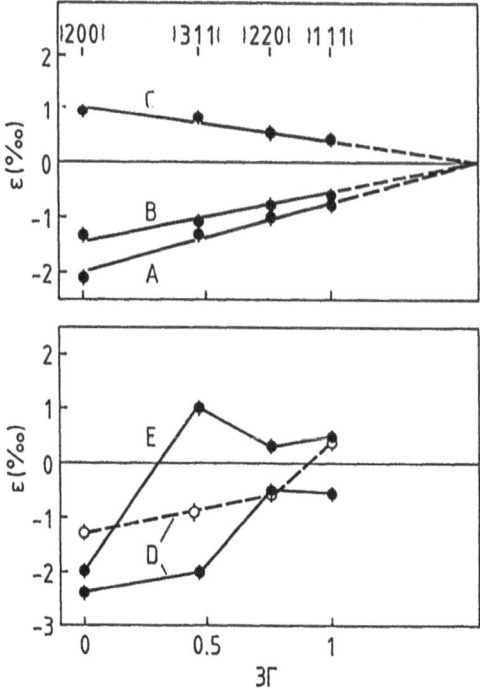

Figure 13. Lattice strains observed with different (hkl) peaks versus the orientation factor 3Γ, above for points in the base and below for points in the weld (see the corresponding labelling in Fig. 11).

orientation factor Γ (Fig. 13). However, a similar plot for strains in the weld showed little systematics with Γ. In some cases, even the sign was found to change with (hkl). Concomitantly, preferred orientations were observed in the weld, but strongly fluctuating over a range of some mm. The data gave clear evidence of large GIS, but there was no hope to characterize them because of their irregularity. In order to evaluate just RS I we used the (311) reflection line because of their high multiplicity: the higher the multiplicity the better it can be expected that GIS are averaged out. A check of the balance of forces revealed that the calculated stresses were well equilibrated for the y-component (Fig. 14). However, a similar check for the x-component showed that the weld should be under tensile stress, whereas the calculated stresses were compressive (Fig. 15). Averaging the stresses calculated from all four reflection lines used gave about zero stress, but still not the expected tensile stress of ~200 MPa.The same was true for other lines across the weld, so that the discrepancy was systematic. Possibly, deformation RS are important also here. This remained a puzzle and we suspect that the good fulfillment of the

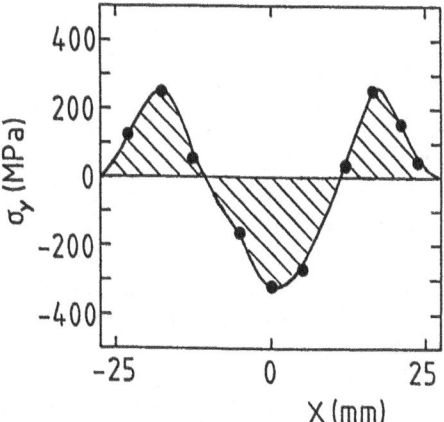

Figure 14. Residual stresses σ_y in the weld vs x calculated from (311) peaks shifts.

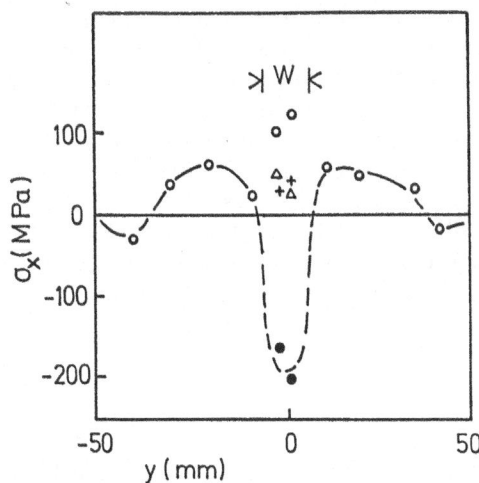

Figure 15. Residual stresses σ_x vs y along the central line across the weld calculated from different (hkl) peak shifts: circles (311), full dots (111), triangles (200), and crosses (220). The broken line is expected for the region in the weld from the data obtained for the base to get stress equilibrium.

balance of forces for the y-component may be somewhat accidental. We feel that extrapolating stress values from the base material into the weld gives more reliable estimates of the macrostresses than averaging GIS.

3. Conclusions

Large GIS are commonly encountered in ferritic as well as austenitic steels after cold working. They are accompanied by preferred orientations, but there is no clear correlation between the degree of the texture and the size of the GIS. The GIS observed in a weld were probably also linked to preferred orientations which came into being on solidification. The investigation of GIS requires the use of reflection lines with different (hkl) and preferably the determination of D vs sin²ψ-distributions. For a highly textured specimen the GIS could be characterized quantitatively. However, a general evaluation procedure for GIS which works also for weakly textured specimen is lacking. Therefore, in other cases only averages of GIS could be obtained. It was hoped that these averages represent RS I or, in the case of two-phase materials, phase specific RS but a check of the balance of forces revealed that the observed RS were not always equilibrated. Such unbalanced RS have been termed deformation RS. So far, all explanations for the observation of unbalanced RS remain speculative. It has been proposed that stress equilibrium is maintained by stresses in regions with very high dislocation density, which do not give well defined diffraction peaks, but further experimental investigations are necessary to confirm this idea.

As deformation RS can obviously not be macro RS, they may be interesting to the materials scientist, but may seem of no concern to the engineer. However, they are important at least in the sence that they complicate the interpretation of diffraction data, often more than GIS; whereas GIS can be corrected for by using many different (hkl) peaks, there is so far no way to properly evaluate macro RS from diffraction data in the presence of large deformation RS. It is recommended always to check the balance of forces, in particular if the specimen has undergone strong plastic flow.

4. References

1. Pintschovius, L., Hauk, V., and Krug, W.K., 'Neutron diffraction study of the residual stress state of a cold rolled steel strip', Mat. Science and Engineering 92 (1987) 1.
2. Feja, K., Hauk, V., Krug, W.K., and Pintschovius, L., 'Residual stress evaluation of a cold-rolled steel strip using X-rays and a layer removal technique', Mat. Science and Engineering 92 (1987) 13.
3. Hauk, V., Herlach, D., and Sesemann, H., 'Nonlinear interplanar spacing distributions in steels, their origin, calculation and regard to stress evaluation', Z. Metallkunde 66 (1975) 734.
4. Hauk, V., Nikolin, H.J., and Pintschovius, L., 'Evaluation of deformation residual stresses caused by uniaxial plastic strain of ferritic and ferritic-austenitic steels', Z. Metallkunde 81 (1990) 556.

5. Pintschovius, L., Munz, D., and Nyilas, A., 'Neutron diffraction study of the residual stresses in double-V and K-welds', unpublished results.

5. Note added in proof

Stimulated by the discussions at the meeting I looked back into the tensile test specimen made of ferritic steel (25CrMo4). Assuming that the material consists of well ordered and of highly disturbed regions, the diffraction lines will have a narrow central part, but broad wings. If the broad wings are not taken into account in determining the line position, the evaluated stresses may reflect only the stress state of the well-ordered regions, but balancing stresses in the highly disturbed regions may go unnoticed. The experiment did not really confirm this idea. It is true that the (110)-line not only broadened after 8% plastic strain, but also showed higher than Gaussian wings. However, neither was the weight in these wings very large ($\approx 5\%$) nor were they strongly asymmetric, so that the center of gravity of the total reflection line was very little affected.

RELAXATION

P. J. WITHERS
Department of Materials Science & Metallurgy
Cambridge University
Pembroke Street
Cambridge CB2 3QZ, UK.

ABSTRACT: In many situations a system can lower its energy by a redistribution of the internal stresses. In this paper I will examine such relaxation phenomena, looking specifically at the role of neutron diffraction as a means of evaluating the importance of relaxation under different loading conditions. Among the mechanisms which can be classified as relaxation processes, the following are discussed; dislocation rearrangement, interface and volume diffusion, interfacial debonding/sliding, matrix cavitation, inclusion fracture, grain boundary sliding and particle shearing. Finally, the evidence for, and the influence of, relaxation processes in Al/SiC metal matrix composites, as measured by neutron diffraction, are examined with reference to two different loading situations. Namely, the relaxation of thermal residual stresses caused by the large disparity in thermal expansion between the phases, and the relaxed flow behaviour under plastic straining.

1. Introduction

Internal stresses are commonplace in almost any material which is mechanically inhomogeneous. Typically, their magnitude varies according to the degree of inhomogeneity: for an externally loaded polycrystalline cubic close packed metal, the stress between the grains arising from the different elastic and plastic deformation characteristics of the differently oriented crystallites is relatively small, for a material consisting of two distinct phases, the stresses will often be much larger. In many situations the system can lower its energy by reducing the internal stresses or by adjusting the extent to which an applied load is carried by each of the phases. Mechanisms by which this process can occur are termed relaxational processes, and have important implications throughout many branches of materials science. In this paper I will examine relaxation phenomena, looking specifically at the role of neutron diffraction as a means of evaluating the importance of relaxation under different loading conditions. My analysis concentrates on relaxation behaviour in metal matrix composites (MMCs). Metal matrix composites provide many good examples of relaxation dominated behaviour because the inherent disparity in mechanical properties of the phases would otherwise often result in the development of very large internal stresses. The approach followed is, however, general to many different situations in which relaxation is important.

While relaxation is always driven by a tendency for the system to reduce its total energy, it is important to note that, in the absence of a suitable mechanism, energetic arguments alone cannot dictate that relaxation will occur. Among the many mechanisms which can be broadly classified as relaxation processes the following will be discussed in this paper; dislocation rearrangement,

M. T. Hutchings and A. D. Krawitz (ed.),
Measurement of Residual and Applied Stress Using Neutron Diffraction, 205–222.
© 1992 *Kluwer Academic Publishers.*

interface and volume diffusion, interfacial debonding/sliding, matrix cavitation, inclusion fracture, grain boundary sliding and particle shearing.

Finally, the evidence for, and the influence of, relaxation processes in Al/SiC metal matrix composites, as measured by neutron diffraction, will be examined through an examination of experimental results. With reference to these examples, the extent to which the exact mechanism(s) of relaxation can be identified from the neutron diffraction data will be discussed.

2. Surface Relaxation

One of the most common concerns, especially for those for those undertaking X-ray residual strain measurements is the affect on the stress state of surface relaxation. This is caused by the removal of constraint upon creating a free surface. Hanabusa, Nishioka and Fujiwara [1] looked at particulate containing systems and found that, as might be expected, a triaxial stress state could be assumed only when the analysing volume is deeper than a few 'wavelengths' (approx. the interparticle spacing) of the locally fluctuating stress (Figure 1). This condition is easily satisfied for the measurement of microstresses by neutron diffraction and thus surface relaxation need not be considered in detail here. It can, however, be a problem for the examination of fibre composites with X-rays, and demands careful specimen preparation for (fine scale) short fibre composites.

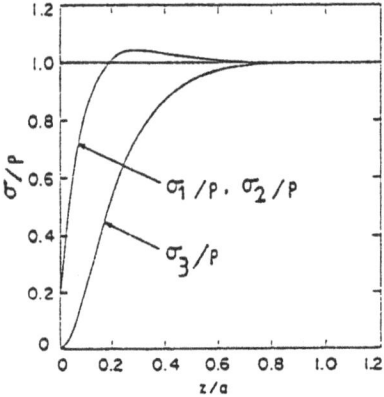

Figure 1. Surface relaxation of the thermal stress within a spherical particle containing composite as a function of the distance to the free surface [1]; a is the particle spacing, p the pressure within the particles in the bulk material.

3. Fundamentals of Relaxation

3.1 MATRIX/INCLUSION MISFITS

An insight into relaxation phenomena can be obtained from a consideration of the stress field in and around a reinforcing particle as a result of the loading environment. With regard to changes in temperature, the thermal stresses local to a reinforcing fibre are predicted to become very large indeed, even for moderate temperature changes (Figure 2). Under plastic straining similarly high stresses and stress gradients develop for quite modest plastic strains.

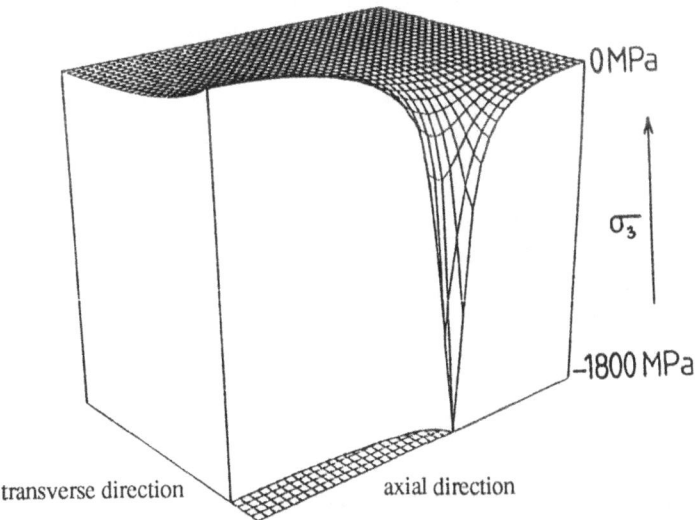

Figure 2. Plot of the spatial variation in the axial stress field (σ_3) predicted [2] to occur through the elastic accommodation of a thermally induced matrix/inclusion misfit generated by cooling (by 300K) an Al based composite containing a single aspect ratio 5 ellipsoidal particle (SiC).

In both of the above cases it is the development of a large inclusion/matrix misfit which generates the high level of internal stressing (see Figure 4 of [2]). In the thermal case, this is because the matrix and reinforcement have very different coefficients of thermal expansion (the CTE for metals is typically 2-8 times that for reinforcing ceramics). In the case of plastic deformation, it is because the plastic distortion of the matrix on straining must by matched by an elastic distortion of the reinforcement. In both instances it is clear that as the stresses build-up, the tendency for stress relaxation becomes greater.

3.2 ENERGETICS OF RELAXATION

Relaxation processes are always associated with, and driven by, an overall reduction in the energy of the system, and thus it is essential to have some idea of the thermodynamics of deformation and relaxation. Because the stress state prior to relaxation is dominated by inclusion/matrix misfits, some clarification is offered by Eshelby's theory, which allows the prediction of internal stress fields in terms of this misfit.

Consider first the stressing (by σ^A) of a single phase medium. The total energy is made up of two terms; a decrease in potential energy of the applied load as the composite extends, and an increase in the elastic energy stored in the material

$$Energy = Potential\ Energy\ Change + Elastic\ Strain\ Energy$$
$$U = -\sigma^A \varepsilon^A + \sigma^A \varepsilon^A/2$$
$$= -\sigma^A \varepsilon^A/2 \tag{1}$$

Such a system could lower its energy by creep, which would decrease the potential energy without increasing the elastic energy (under constant displacement loading this would relax the stress), or alternatively, it could lower its energy by catastrophic fracture.

In a two phase composite the stressing is nearly always uneven, typically being larger in the reinforcement, both as a result of differences in stiffness and, if plastic deformation has occurred, because of matrix plasticity. The energy of the composite is given by [3]

$$U_{Comp} = -\sigma^A \varepsilon^A/2 - f<\sigma>_I \varepsilon^{T*}/2 - f\sigma^A(\varepsilon^T + \varepsilon^{T*})/2 \tag{2}$$

where f is the volume fraction of reinforcement, ε^A is the elastic strain of the unreinforced matrix, $<\sigma>_I$ the internal stress in the inclusion, ε^{T*} the matrix/inclusion misfit strain with no applied load, and ε^T is related to the misfit strain under the applied load.

Now the potential for relaxation through an overall reduction in energy (i.e. a negative ΔU) arises from the possibility of changing the misfit between the two phases ($\Delta\varepsilon^{T*}$). It can be shown that [3]

$$\Delta U_{Comp} = -f(\sigma^A + <\sigma>_I + \Delta<\sigma>_I/2)\,\Delta\varepsilon^{T*} \tag{3}$$

Looked at simply, the first term is dominated by changes in the potential energy which occur when the misfit between the matrix and inclusion is changed through the movement of the load. The second and third terms are concerned predominantly with changes in the strain energy, and are associated with the internal stress (the microstress). In order to get a feel for how relaxation might occur, it is helpful to consider a number of examples.

3.3 RELAXATION IN THE ABSENCE OF APPLIED LOADING

In this case $\sigma^A = 0$ and the system can only decrease in energy by a reduction in internal strain energy. A good example of this situation is provided by the relaxation of thermal residual stresses. This is shown schematically in figure 3. The change in energy is given by:

$$\Delta U_{Comp} = -f(<\sigma>_I + \Delta<\sigma>_I/2)\,\Delta\varepsilon^{T*} \tag{3a}$$

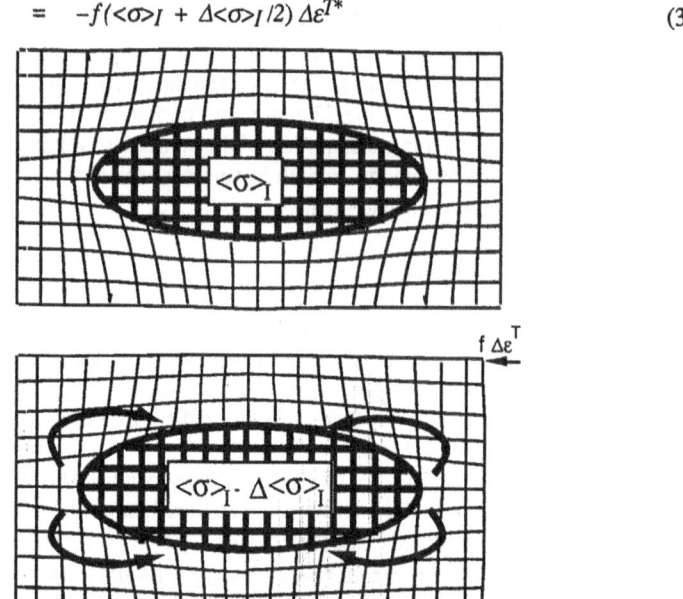

Figure 3. In the absence of an applied load, the system can relax only by a reduction in the internal strain energy; this requires the movement of matrix or inclusion material so as to decrease the shape misfit.

3.4 RELAXATION UNDER APPLIED LOADING

Whilst in theory it is possible for the composite to increase in potential energy (e.g. to contract under a tensile load) or to increase in strain energy, provided the overall change in energy is negative, the most common situation is for both the internal strain energy and the potential energy to decrease (Figure 4). It is important to note that, from a load transfer point of view, relaxation is very different from global plastic flow, which tends to *increase* the shape misfit between the phases and thus to transfer load to the reinforcement.

$$\Delta U_{Comp} \quad = \quad -f(\sigma^A + <\sigma>_I + \Delta<\sigma>_I/2)\, \Delta \varepsilon^{T*} \tag{3b}$$

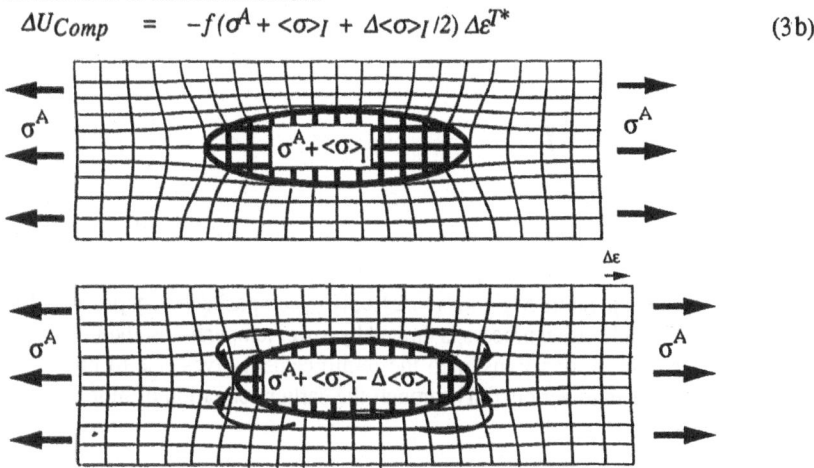

Figure 4. Under an applied load, the system can both lower the internal strain energy as well as decrease the potential energy by a reduction in the inclusion/matrix misfit. This lowers the stressing of the reinforcing phase.

4. Relaxation Mechanisms

Whilst relaxation is only thermodynamically possible if there is an overall reduction in free energy, it is important to remember that this does not mean that relaxation will necessarily take place. Relaxation also requires a suitable mechanism which can operate at a sufficient rate. Among the mechanisms which can reduce the load carried by the reinforcement, and hence can be classified as relaxation processes, are the following

- dislocation rearrangement
- diffusion
- interfacial debonding/sliding
- matrix cavitation
- inclusion fracture
- grain boundary sliding
- particle and grain boundary shearing

For any matrix/reinforcement system, the relative prominence of these processes is a complex function of variables such as temperature, inclusion aspect ratio, the applied load and the strain/heating rate. Since relaxation processes affect both the load bearing capacity at a given strain and the onset of component failure, it is important to understand the individual mechanisms so as to be able to exert some degree of control over their relative predominance. For example, diffusion-related relaxation can be limited by low temperatures and high strain rates and, as figure 5 shows, this increases the work hardening rate but decreases ductility.

210

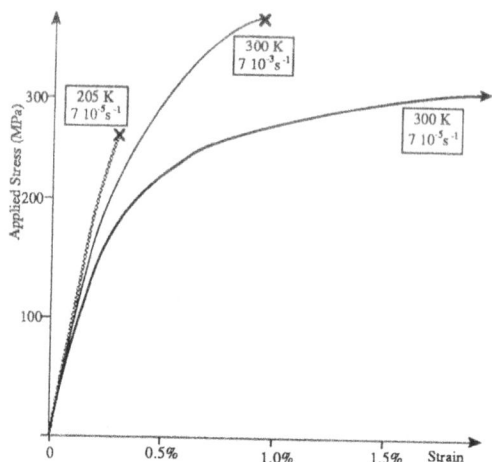

Figure 5. At room temperature Mg-11at%Li/24%SiCw composite is very prone to diffusion-aided relaxation [4]. This relaxation process is limited by lower temperatures and faster strain rates. The concomitant increase in load transfer to the reinforcement gives rise to higher rates of work hardening but results in lower ductility through the earlier activation of more detrimental stress-limiting processes.

4.1 DISLOCATION REARRANGEMENT

As shown in Figure 12 of [2], global plasticity *increases* the matrix/inclusion misfit and as such opposes the misfit-reducing tendency of relaxation. During loading, plasticity gives rise to increased straining of the reinforcement and a reduced rate of elastic straining of the matrix. This is because while the matrix deforms plastically, the reinforcement can distort only elastically. The consequent formation of a dislocation loop around a reinforcing inclusion each time the long range glide of a dislocation through the matrix is interrupted, increases the elastic distortion of the reinforcement and the stored internal strain energy of the composite. On the other hand, local *rearrangements* of these necessary dislocation structures or the stimulation of secondary dislocations can *decrease* the misfit and thus bring about relaxation. For example, consider a thermal residual stress. Typically, the inclusions are compressively stressed and the misfit can be reduced by the punching of dislocation loops [5]. These loops reduce the misfit by transporting discs of extra atoms (or vacancies) away from the inclusion (Figure 6). Such processes are energetically favoured, provided the reduction in elastic energy is greater than the effects of the increased dislocation line length.

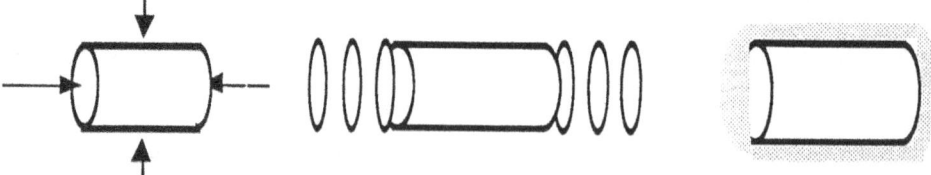

Figure 6. Plastic relaxation of a compressive thermal residual inclusion stress can take place either by the generation of prismatic dislocation loops which punch out discs of interstitials into the matrix (centre) or by the formation of a plastic zone of tangled secondary dislocations (right).

Other secondary dislocation structures are possible, such as secondary shear loops which can subsequently cross-slip. These are usually observed as dislocation tangles [6]. It may be noted that, in the context of relaxation by dislocation motion, the ability to cross-slip is often important and hence stacking fault energy is often a relevant parameter (particularly at relatively low temperatures where climb is inhibited). Aluminium has a high SFE and cross-slips easily, unlike copper for example. This may be a factor favouring the observed tendency of aluminium alloy MMCs to exhibit relatively pronounced relaxation effects, even at low temperatures.

4.2 DIFFUSION

A number of diffusion-related processes can lead to a lowering of the overall energy. Three will be discussed here:
- interfacial diffusion
- volume diffusion
- diffusion-assisted dislocation climb.

Interfacial diffusion. For most systems the metal/ceramic boundary is disordered and as such acts as a low energy channel for the movement of atoms. Gradients of hydrostatic stress provide a sustained driving force for vacancy transport (Figure 7). Because only local rearrangements can be brought about by interfacial movement, this process cannot completely relax a dilatational misfit such as that caused by cooling, but it can reduce the stress in especially high directions at the expense of others. It is quite simple to show that when interfacial diffusion is complete only hydrostatic stresses remain within the reinforcement [7].

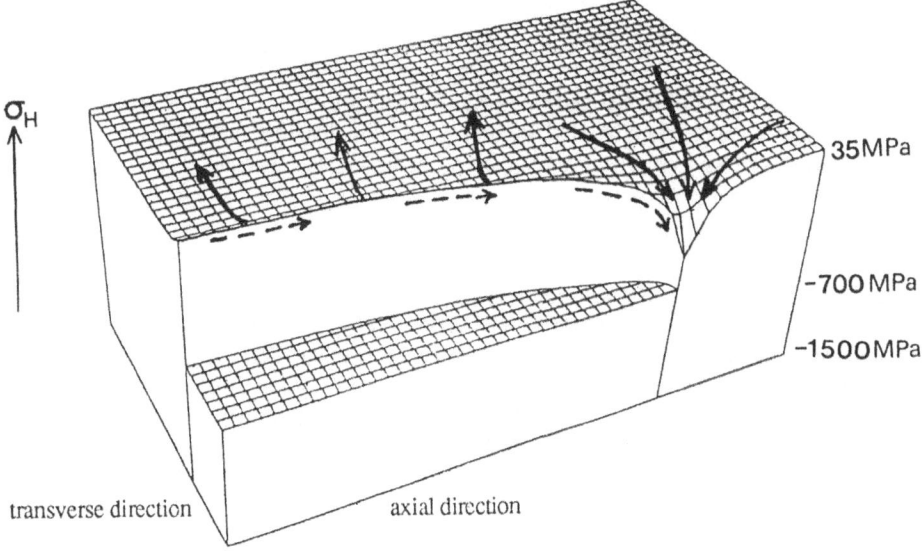

Figure 7. Though the thermal misfit for a fibre (SiC) in a matrix (Al) is isotropic, the stress state is highly anisotropic (see Figure 2). This results in gradients of hydrostatic stress (σ_H) around the inclusion/matrix interface and into the matrix. These gradients, which are shown here for a 300 K temperature drop, provide the driving force for interface (dashed lines) and volume diffusion (continuous lines) respectively.

Volume diffusion. At higher temperatures, or when interfacial diffusion is impossible or complete, volume diffusion can become important (Figure 7). For spherical particle-containing composites, it is the only diffusive mechanism available for the relaxation of thermal stresses. In addition to the larger activation energy required for volume diffusion, the greater distance required for the relaxation of hydrostatic stresses means that relaxation times are expected to be considerably longer.

Diffusion-assisted climb. Interaction between diffusive processes and dislocation motion covers a wide range of possibilities. Perhaps the most important is that of dislocation climb; this mechanism is especially important in connection with elevated temperature creep because it limits the extent of reinforcement stressing, through the removal of piled up dislocations which would otherwise elastically deform the reinforcing phase.

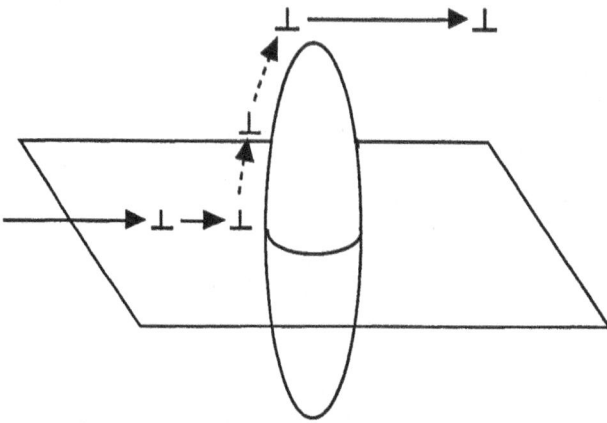

Figure 8. Load is normally transferred to the reinforcement by the accumulation of dislocation loops as a result of the interruption of dislocation glide. Provided diffusion is reasonably quick diffusive assisted climb prevents this build up and the associated load transfer.

4.3 INTERFACIAL DEBONDING AND SLIDING

It is clear that large interfacial shear stresses build up at the metal/ceramic interface and that these stresses can be reduced by some form of movement along the interface. This can occur by interfacial debonding, i.e. by the breaking of a chemical or similar bond followed by unrestricted movement, or by interfacial sliding, which provides a resistive force throughout the sliding process. Often these processes are concurrent, and the criteria governing their operation are ill-defined. Naturally, their relative importance is dependent upon the condition of the interface as well as the stress to which it is subjected [8].

4.4 MATRIX CAVITATION

Under external loading large (tensile) hydrostatic stresses can build-up in the vicinity of the fibre ends (Figure 9a). A high tensile value of σ_H will almost certainly encourage matrix cavitation. As to whether a cavity forms at the matrix interface or within the matrix itself, depends on the matrix/inclusion bond strength [10]. Recent results seem to indicate that for the Al/SiC system the interface is usually strong and that voiding generally occurs close to the reinforcement but within the matrix (Figure 9b).

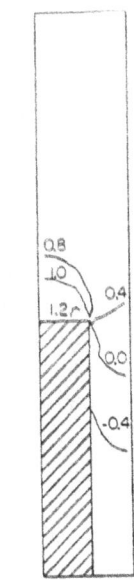 

a) b)

Figure 9. a) FEM calculation [9] for the Al/SiC$_W$ system showing how the hydrostatic component (in GPa) of the stress field can become very large near the fibre end after tensile straining (1·6%), this aids load transfer to the fibre but also encourages voiding b) bright field TEM micrograph [11] showing voids nucleated at a SiC whisker end within an Al matrix (micron marker = 0·5μm).

4.5 INCLUSION FRACTURE

As to whether the reinforcement is likely to fracture upon loading will be dependent on the magnitude of the stresses that can develop. Traditionally, for reasonably long fibres, a critical aspect ratio (s) has been specified below which the matrix shear stress (τ_{MY}) is such that the matrix is unable to stress the reinforcement to its failure strength (σ_{IU}):

$$s \quad \approx \quad \sigma_{IU} / 2\,\tau_{MY} \approx \sigma_{IU} / \sigma_{MY} \qquad\qquad (4)$$

For Al alloy matrices, this criterion leads to an critical aspect ratio (s~10) which is greater than that typically retained in real short fibre composites (~5). Unfortunately, the accuracy of this result is likely to be poor in view of the simplifications inherent in the shear lag model [2]. The Eshelby model indicates that the fibres are unlikely to fracture prior to plastic flow, but that subsequently loading of the reinforcement increases sharply. However, in many cases the activation of other relaxation mechanisms will probably limit the fibre stresses before fracture occurs.

Although ceramic whisker breakage is relatively rare, for particles greater than about 15 μm [11] particle fracture is relatively common in spite of their seemingly high UTS. Since the stresses around particles of the same shape but different sizes are not expected to be that different, it would seem likely that the origin of particle failure is not in higher stressing. Instead it seems probable that the larger particles are more flawed and thus more likely to fail at stresses below their theoretical strength than small particles or fine whiskers.

214

4.6 GRAIN BOUNDARY DIFFUSION/SLIDING

Just as the incoherent particle/matrix interface provides a low energy channel for diffusion, so do grain boundaries. This is especially so for systems for which the reinforcement is often found on matrix grain boundaries and triple points. Although apparently not the rate controlling mechanism, grain boundary sliding was observed to be an important mechanism for the attainment of the large superplastic extensions (~700%) obtained by Bieler and Mukherjee [12] in Al/SiCw.

4.7 PARTICLE AND GRAIN BOUNDARY SHEARING

Though not generally of importance for metal matrix composites, because of the high strength of the reinforcement, particle and grain boundary shearing are important mechanisms of relaxation for monolithic alloys. Particle shearing can occur for precipitate hardened systems, such as Al-Li provided the precipitates are not too well developed, whilst grain boundary cutting reduces the constraint of plastically deforming polycrystals. Essentially, dislocations become piled up at precipitates and grain boundaries causing the precipitate to be sheared or the grain boundary to be penetrated by the piled up dislocations [13], thus reducing the constraint (Figure 10).

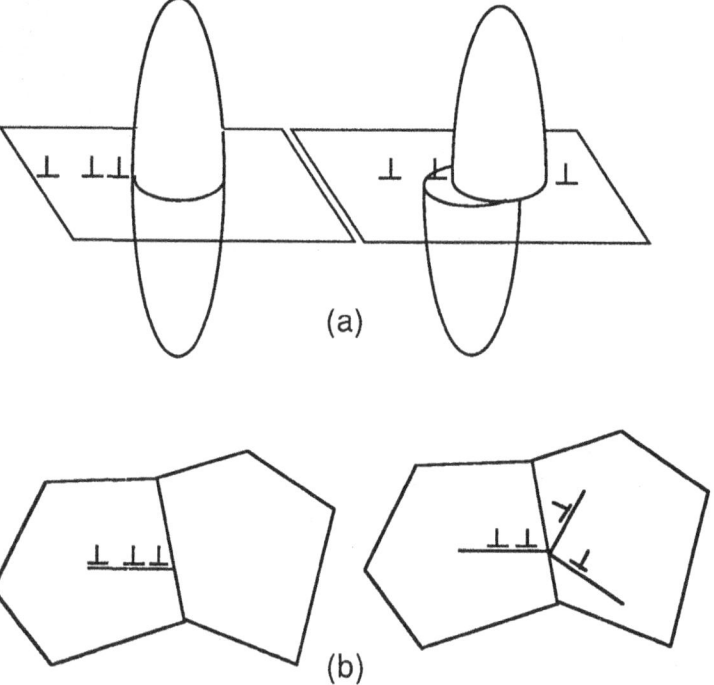

Figure 10. The relaxation of constraint through the cutting of a) shearable particles and b) grain boundaries.

5. Neutron Diffraction Studies

Three aspects of neutron diffraction make it a useful tool for the study of relaxation phenomena.
- the penetrating power of the neutron enables the construction of environmental chambers (mechanical testing rig, furnace, etc) for the monitoring of relaxation in-situ
- the technique gives the mean phase strains directly, other methods require the interpretation of overall composite performance. This is especially difficult for plastic flow when matrix work hardening complicates the stress/strain response.
- the build-up and relaxation of intergranular stresses in monolithic systems can be monitored using different reflections

One problem with the diffraction technique, however, is that it is not usually possible to unambiguously determine which mechanism or mechanisms are responsible for the observed behaviour. One simply knows the mean phase strains (microstresses), from which the extent of the net matrix/inclusion misfit arising from all the mechanisms can be calculated.

5.1 THERMAL STRESS RELAXATION

Although for the purposes of modelling thermal residual strains it is helpful to imagine that above some temperature the thermal stresses relax completely and that below that temperature they are wholly elastically accommodated, the development of thermal stress on cooling is usually much more complicated. The relationship between thermal straining and temperature for 5%SiCw/Al has been monitored by neutron diffraction [14] and is shown in Figure 11. These results show that the internal strain development is limited both by plastic deformation and time dependent relaxation.

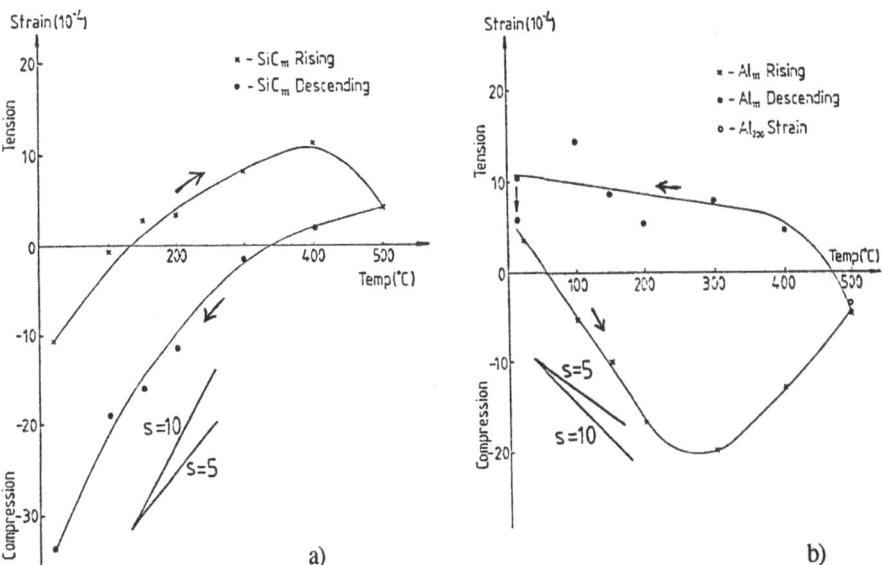

Figure 11. The axial a) fibre and b) matrix elastic lattice strains measured for an Al/5vol%SiCw composite over a thermal cycle derived from {111} lattice plane spacings [14]. The elastic strain changes predicted on the basis of the elastic accommodation of the thermal generated misfit [2] are also shown for aspect ratios of 5 and 10.

216

The extent of internal stress relaxation with time at different temperatures has been compared for SiC whisker (~0·5μm dia., 3μm long) and particulate (~15μm dia) composites over a range of temperatures (Figure 12). It is clear from the figure that in both composites below 300°C the initial strain response is largely that expected on the basis of the elastic accommodation of misfit. Above 300°C, the decrease in yield strength with increasing temperature limits the build up of internal strains. It is also clear that time dependent relaxation is marked for the whisker composite over the whole range of temperatures, while it was detected only at temperatures in excess of 500°C for the particulate composite. As to whether this is the result of the different scale of the two reinforcements or the availability of additional mechanisms of relaxation in the whisker case, is at present unconfirmed. Certainly, gradients in the hydrostatic component of the stress field around a cylindrical inclusion (Figure 7) could drive interfacial diffusion in a manner not possible around a spherical one. Two additional points might seem to corroborate the latter hypothesis. Firstly,

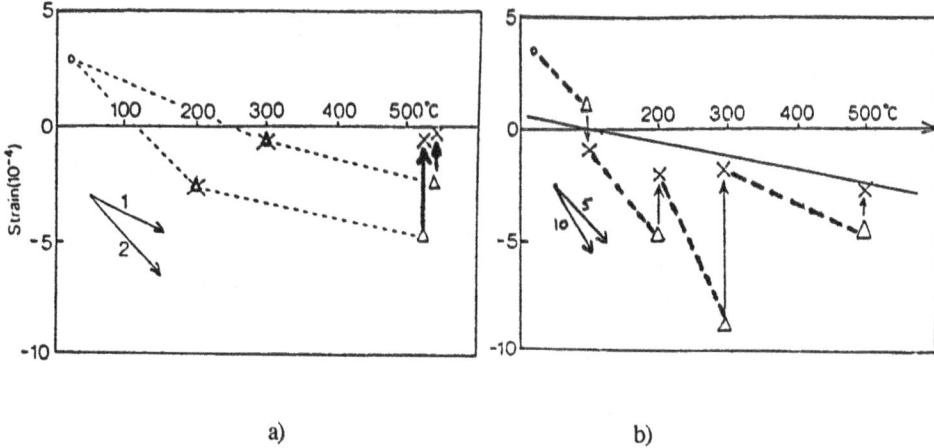

a)   b)

Figure 12. The thermally induced elastic matrix strains measured at temperature for a) 13% SiC particulate and b) 5% SiC whisker -Al composites. The triangles represent the strains first measured upon heating (within 10 mins), the crosses those made after 8-10 hrs. Consequently, the dashed lines represent the strain changes taking place on increasing the temperature, while the arrows show the extent of relaxation at a particular hold temperature. In b) the solid line represents the strain related to the hydrostatic stress component [15].

while the shape of the particulate relaxation curve (Figure 13) can be explained in terms of a single time constant for relaxation, the degree of fit for the whisker curve is unsatisfactory unless two constants are defined. Secondly, were interfacial relaxation to go to completion a hydrostatic stress in the whiskers would remain unrelaxed. This component is plotted in Figure 12b) and is in fairly good agreement with the final residual strains.

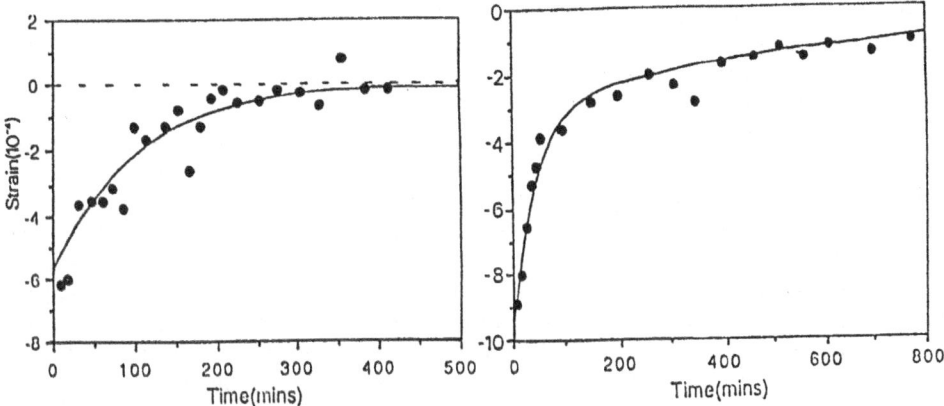

Figure 13. Strain relaxation curves for the a) particulate (537°C) and b) whisker (300°C) composites. The best fit exponential decay curves are also shown; in b) two relaxation constants are required to achieve a satisfactory fit [15].

5.2 LOAD-INDUCED RELAXATION

Under external loading very high rates of load transfer and internal stress generation are predicted. Naturally, the rate of stress build up can not be maintained indefinitely and a number of the mechanisms described above could potentially limit this.

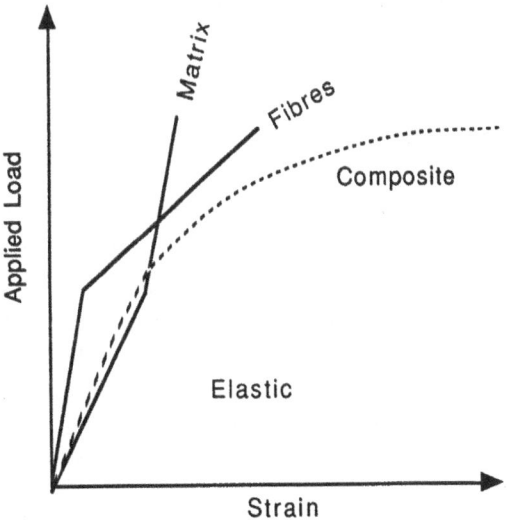

Figure 14. A schematic showing the expected lattice strain variation for the fibres and matrix. Until the onset of plastic flow the matrix strains to a greater extent than the reinforcement, but subsequent to flow the rate of load transfer would be expected to increase; this behaviour is in broad agreement with the lattice strain response measured by neutron diffraction for Al 5%SiC$_W$.

Neutron diffraction has proven to be an excellent method for the monitoring of internal strain development throughout a loading cycle. As discussed elsewhere within this symposium [2], with the onset of plastic straining there is predicted to be a marked increase in the rate at which load is transferred to the reinforcement (Figure 14). Such an increase in the loading of the reinforcement is indeed observed (Figure 15)[†]. Before examining the degree of correlation between predictive models and the experimental measurements it is useful to consider the nature of the plastic misfit. Global matrix plasticity occurs by the long range movement of dislocations: this movement is held up by the reinforcement until the stress is sufficient for the dislocation to bypass the obstacle leaving behind an Orowan loop. These Orowan loops are the simplest form of dislocation structures which can form and, provided they are sufficient in number that they can be smeared out into a continuum, create a misfit between the undistorted inclusion and the plastically deformed matrix which is equal to the plastic strain of the matrix, i.e. the matrix is assumed to deform uniformly [2]. Hence it is possible to correlate the overall plastic strain of the composite with the unrelaxed misfit caused by the impeded dislocations. This would result in a very high rate of work hardening and a very high stressing of the reinforcement, even at comparatively low plastic strains.

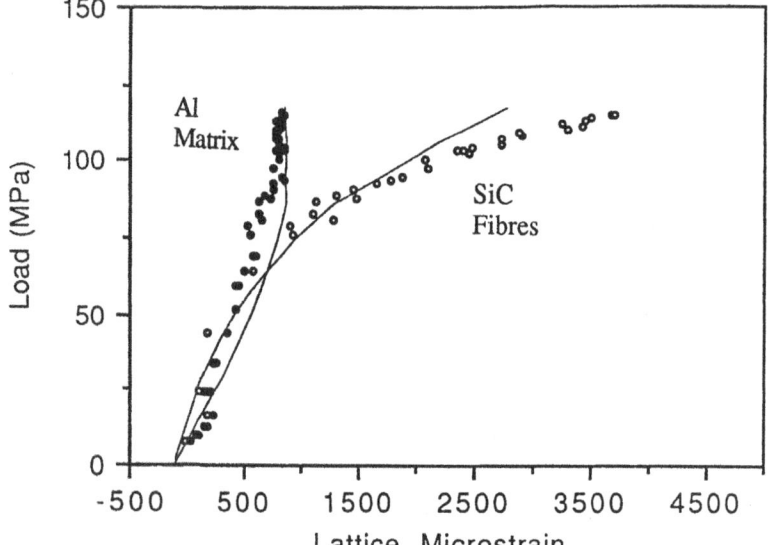

Figure 15. The observed lattice strain response modelled on the basis that the load transfer can be expressed in terms of the inclusion/matrix misfit associated with the Eshelby method [2].

Relaxation mechanisms act to relax the build up of internal stress by limiting the elastically accommodated misfit. For example, plastic relaxation could reduce the misfit through the formation of secondary dislocation structures or, more catastrophically, the elastically

---

[†] in this case it is clear that only the matrix is deforming inelastically, yet both lattice strain responses are affected from the onset of matrix plasticity; the rate of matrix elastic straining decreasing, the rate of whisker straining increasing. In a polycrystal, the reflections from plastically deforming crystallites would behave as the matrix in this example, those not yet to have reached yielding would mirror the reinforcement. Of course in a system where all the crystallites are plastically deforming equally, no load transfer would occur and so no line curving would be observed despite matrix plasticity. In most cases however, non-linearity is a sign of inelastic behaviour, though not necessarily of the crystallites contributing to the observed reflection.

accommodated misfit might be reduced by debonding of the matrix/inclusion interface. Because the net misfit is linearly related to the internal stress, it is possible to deduce the extent of relaxation from the neutron diffraction measurements. Modelling indicates that with increasing plastic strain the rate of increase in reinforcement stressing decreases or, conversely, that larger and larger plastic strains are required to generate the same increase in reinforcement stressing. This explains the decrease in composite work hardening with increased straining. Figure 15 shows the optimal lattice strain predictions from which the development of the net misfit can be deduced (Figure 16). These results show that flow is unrelaxed only for plastic strains less than 0·2%. Beyond 0·2%, the rate of increase of unrelaxed misfit falls off drastically with increased straining so that at strains greater than about 1% little load transfer occurs at all. In this region the microstress is nearly saturated and is approximately independent of further straining. Naturally, the extent to which the various mechanisms operate is dependent upon the matrix alloy, the interface structure and the reinforcement characteristics. For the commercially pure aluminium matrix studied here, the tendency for plastic relaxation is pronounced; the relaxation behaviour for engineering alloys is likely to be similar in form but different in its point of onset and magnitude.

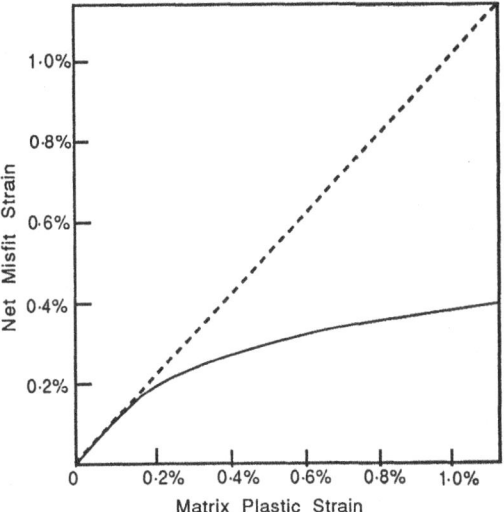

Figure 16. Beyond 0·2% plastic strain the rate of inclusion/matrix misfit generation is less than would be expected for unrelaxed plastic flow (dashed line), i.e. the rate of generation of the inclusion/matrix misfit is no longer equal to the global deformation of the matrix. It is likely that in this particular case it is plastic and diffusional relaxation which limit the rate of misfit generation and hence load transfer.

5.3 NEUTRON DIFFRACTION MEASUREMENTS AND RELAXATION MECHANISMS

As illustrated by the two previous examples, neutron diffraction measurements usually provide information about the average elastic stress within a phase in a certain direction and thus can be used to calculate the extent of the matrix/inclusion misfit and hence internal stressing. This gives one a valuable insight into the extent of relaxation, but neutron diffraction cannot give unambiguous information about the mechanisms by which this relaxation is taking place. Frequently however, a consideration of the effect of different candidate mechanisms on the

220

observed strains does clarify the matter somewhat. Consider, for example, the sense of the residual strains measured upon unloading parallel to the applied load. As seen above, *plastic load transfer* during loading tends to increase the reinforcement stress and thus leads to tensile residual inclusion stresses upon unloading (Figure 17). Since the transfer of load to the reinforcement by plastic flow is very effective, as discussed above, it is likely that the rate of inclusion stressing will be limited by *plastic relaxation*. However, plastic relaxation can only reduce the matrix/inclusion misfit caused by bulk plastic flow, and thus it can only reverse the load transfer brought about by plastic flow. Plastic relaxation thus tends to reduce the residual strains that would otherwise have been caused by plastic flow (Figure 17). *Particle breakage or debonding* would also tend to unload the reinforcement, however, they would concomitantly reduce the stiffness of the composite, and so on their own would not give rise to residual strains at all. Doubtless, *diffusional mechanisms* of relief would also occur, driven by the steep stress gradients caused by uniaxial loading. These tend to reduce the misfit generated by elastic and plastic loading and, if there is little or no plastic deformation, are able to give rise to tensile matrix residual stresses and compressive inclusion stresses (figure 17). Naturally, for diffusional relaxation to be significant, the relaxation driving stress state must be maintained for a considerable period of time.

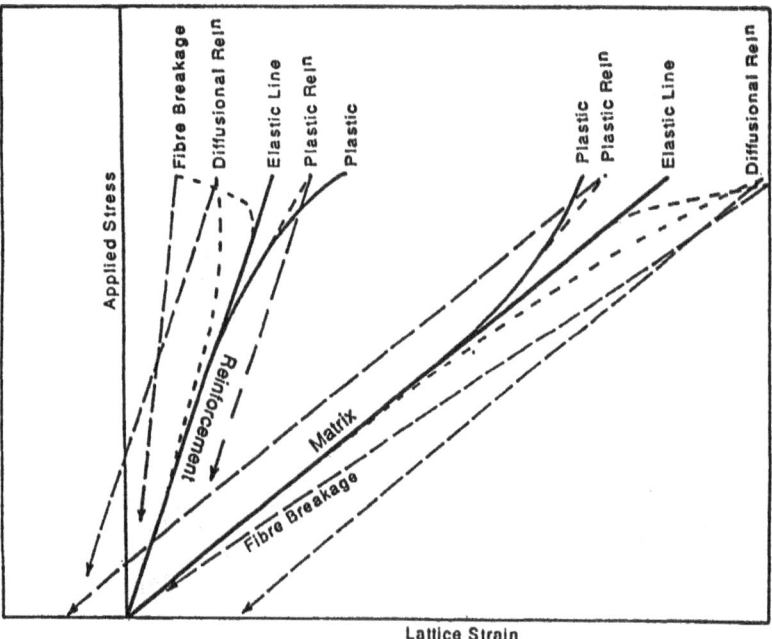

Figure 17. The elastic and inelastic strain changes expected to arise in matrix and reinforcement as a result of different load transfer mechanisms [16], the arrows indicate the final residual strains.

The variation in lattice straining over what was predominantly an elastic loading cycle was monitored by Allen et al. [16]. Whilst in good general agreement with an elastic model of load transfer, the sense of the residual strains, especially in the transverse direction, were of the opposite sense to be explained simply in terms of plastic deformation (Figure 18). Instead, arguments such as those presented above for the axial case suggest that some degree of diffusive relaxation must have taken place.

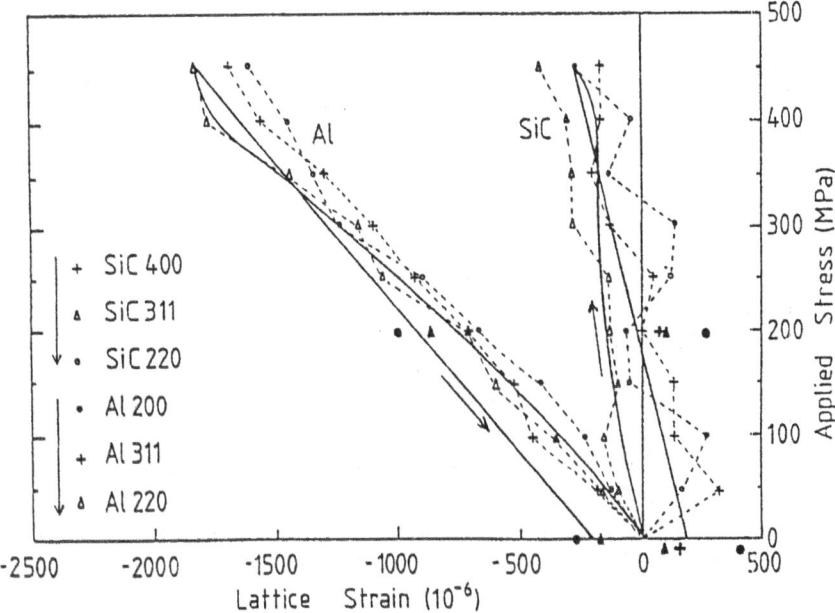

Figure 18. The sense of the residual stresses observed in Al/20% SiCp [16], can only be explained in terms of the activation of diffusional processes. These processes would increase with loading and thus cause a gentle curvature of the modelled lines away from one another as load is transferred towards the matrix. The steep curvature at high loads in the opposite direction is caused by the onset of matrix plasticity which causes a contrary transfer of load towards the reinforcement. Note that the residual strains (bold symbols) and the strains measured half way through unloading (also bold symbols) are of a sense which suggests that in this case (with limited plasticity) the internal strains are dominated by diffusional relaxation over the timescale of the experiment.

6. Conclusions

Neutron diffraction has been shown to be an effective means of monitoring the build up and relaxation of interphase and intergranular stresses. It has been demonstrated that, in the case of a metal matrix composite, the internal stresses can be very large and relaxation extensive. For a two phase composite such as this interpretation of the internal strains is fairly simple because the majority of available relaxation mechanisms are confined solely to the matrix phase. Consequently, I have discussed the principles of relaxation through an examination of the behaviour of the Al/SiC system. For a monolithic polycrystalline system, the internal stresses will develop and decay between the different crystal orientations as a complex function of strain, temperature and microstructural variables. This behaviour must be better understood if diffraction measurements on monolithic components which have undergone substantial plastic flow are to be interpreted unambiguously in terms of the macrostress variation. On the other hand, the selectivity of the diffraction technique can be viewed as a virtue, offering a special insight into the physics of the build up of intergranular stressing.

222

Acknowledgements

The financial support provided by Alcan International and The Materials Department at the Risø National Laboratory, Denmark is gratefully acknowledged. Helpful discussions with W.M. Stobbs, T.W.Clyne and N.Hansen are appreciated.

References

1. Hanabusa T, Nishioka K, Fujiwara H, (1983), 'Criterion for the triaxial X-ray residual stress analysis', Z. Metallkunde, 74, 307-313.
2. Withers P.J, (1991) 'Theory and Modelling of Composites', This Proceedings.
3. Withers P.J, Stobbs W.M, Pedersen O.B (1989) 'The Application of the Eshelby Method of Internal Stress Determination to Short Fibre Metal Matrix Composites', Acta Met., 37, 3061-3084.
4. Clyne T.W, 'Microstructural development & mechanical behaviour of whisker reinforced Mg-Li alloys', European Conf. on Comp. Mat. 3, 211-220.
5. Humphreys F.J, (1983) 'Deformation mechanisms and microstructures in particle hardened alloys', Deformation of multi-phase & particle containing materials, 4th International Risø Symposium, p 41-52.
6. Dunand D.C, Mortensen A (1990), On plastic relaxation of thermal stresses in reinforced metals, Acta metall. mater.,39,127-139.
7. Mori T, Okabe M, Mura T, (1980), 'Diffusional relaxation around a second phase particle', Acta Metall., 28,319-325.
8. De Silva R, Caldemaison D, Bretheau T, in IPCM 89, Ed. F.R. Jones, 235-241, Butterworths, (1989).
9. Christman T, Needleman A, Suresh S, (1989), 'An experimental and numerical study of deformation in MMCs', Acta Metall., 37, 3029-3050.
10. Nutt S.R, Needleman A, (1987) 'Void nucleation at fibre ends in Al-SiC composites', Scripta Met., 21, 705-710.
11. Lloyd D, (1991) 'Aspects of fracture in particulate reinforced MMCs', Acta Metall. Mater., 39,59-72.
12. Bieler T.R, Mukherjee A.K, (1988), Mechanical & Physical Behaviour in Metallic & Ceramic Composites, 9th International Risø Symposium, p 297, 5th-9th Sept.
13. Cottrell A.H. (1958) 'Trans AIME, 212, 192.
14. Withers P.J, Juul Jensen D, Lilholt H, Stobbs W.M, (1987) 'The evaluation of internal stresses in a short fibre MMC by neutron diffraction', Proc. ICCM VI/ECCM 2, Ed. Matthews F.L. et al., London, Pub. Elsevier, 2, 255-265, .
15. Withers P.J, Lilholt H, Juul Jensen D, Stobbs W.M, (1988) 'An examination of diffusional stress relief in MMCs', Mechanical & Physical Behaviour in Metallic & Ceramic Composites, 9th International Risø Symposium, p 503-510, 5th-9th Sept
16. Allen A.J, Bourke M, Dawes S, Hutchings M.T. and Withers P.J (1991), The analysis of internal strains measured by neutron diffraction in Al/SiC MMCs, sub. to Acta Metal. Mater.

INTERPRETATION OF RESIDUAL STRESS MEASUREMENTS:

SUMMARY OF DISCUSSION SESSION

T.M. HOLDEN
AECL Research, Chalk River, Ontario, Canada, K0J 1J0

and

A.V. BOWEN
RAE Farnborough, Hants, GU14 6TD, U.K.

1.   Introduction

In order to give a framework to the discussion session, we considered first what the engineer expects to obtain from diffraction measurements of residual stress and then summarized exactly what is actually measured in the experiment, namely, strains in the crystal lattice.   Any effects simulating strain caused by composition changes or temperature changes must be known by the experimentalist.   In computing the stresses from the strains we require appropriate "diffraction elastic constants" and a discussion of this topic lead to the conclusion that the purely elastic effects for a constrained polycrystalline assembly are quite well understood.   By contrast, our discussion of plasticity revealed many of the problems still left to be solved in the field.   We intended to deal with two other areas, the effects of phase transformations and thermal mismatch, which would have embraced the whole of modern work on ceramics and composite materials, but two hours was not sufficient to cover this subject.   Our last topic was to have been, "What new material science is possible", and while this was not discussed specifically the problem areas which we have identified are in fact exactly where our material science efforts have to be focussed.

Finally the authors of this discussion summary have tried to reflect the direction of the discussion accurately, but we may have missed important points or coloured the written comments in the light of our own backgrounds.

2.   The Deliverable

The engineer usually wants to know the macroscopic residual stress field, or macrostress, in his sample.   He requires the orientation and magnitude of the principal stress components as a function of position in the sample which is, therefore, an inhomogeneous distribution.[1]   Ideally, he needs to determine this non-destructively.   With this information, within a continuum mechanics approach, he can then calculate the net effect of an applied stress by adding it to the residual stress vectorially.   The

M. T. Hutchings and A. D. Krawitz (ed.),
Measurement of Residual and Applied Stress Using Neutron Diffraction, 223–231.
© 1992 *Government of Canada.*

engineer can also calculate, within the confines of fracture mechanics and knowing the relevant material parameters, the rate of propagation of cracks and the residual strength (fracture toughness) of components, i.e. the failure properties. As a rule the effects of grain-to-grain interaction stresses, sometimes known as microstresses or type II stresses (for definitions, see Ref. 1), on the behaviour are of secondary importance. However, two examples of the technological importance of these stresses can be cited. The initial transient in creep and growth behaviour of Zr alloy tubulars in a fast neutron flux is driven by the grain-to-grain stresses.[2] In corrosion studies the effect of strain on certain planes, which may be selectively attacked by corrosive agents, may be more important than the bulk strain and in general the specific (hkl) strain differs from the bulk because of grain interaction effects.

3. The Measurement

The physically relevant parameters describing the diffraction peaks are the mean position of the peak, the intrinsic width of the peak and the integrated intensity. For residual stress measurements the quantity of primary interest is the mean peak position, or equivalently the average plane spacing d_{hkl} for Miller indices (hkl). The conditions for diffraction dictate that all those grains which are correctly oriented with plane normals lying along the scattering vector, will contribute to the diffraction pattern. Within a single grain, diffraction measures the average strain. There is a distribution in strain for the set of grains oriented in the same direction, coming from the distribution of constraining environments for this set of grains. These contribute with different lattice spacings to the overall envelope of the intrinsic lineshape. This envelope may shift and change shape as a result of the applied stress. The problems of extracting the stress information lie at the heart of the deliberations of this discussion group.

The average strain measured includes the superposition of both the macroscopic residual strain field, corresponding to the stress field, as well as the grain-to-grain strains and even shorter-ranged dislocation, or type III, stresses. The superposition of these three contributions is depicted in Fig. 1. The origin of the microstresses and their experimental manifestation are discussed below. Grains with different plane normals [hkl] directed along a particular direction deform differently elastically as well as plastically under say, an applied uniaxial stress which has exceeded the yield point. When the stress is removed the plastically stiffer grains compress the other grains, and are themselves left in tension. This phenomenon occurs for any stress field so that the process of generating residual stresses produces anisotropy among the crystal orientations. The point here is that the material is not a continuum but a discrete assembly of grains each deforming differently but contributing to the net deformation. Therefore we may expect, as a rule, that different diffraction lines will yield different net stresses in a particular direction. In addition when we sample with the same diffraction line, the behaviour of grains making progressively greater and greater angles with an applied uniaxial stress exceeding the yield point, the grains will deform elastically and plastically

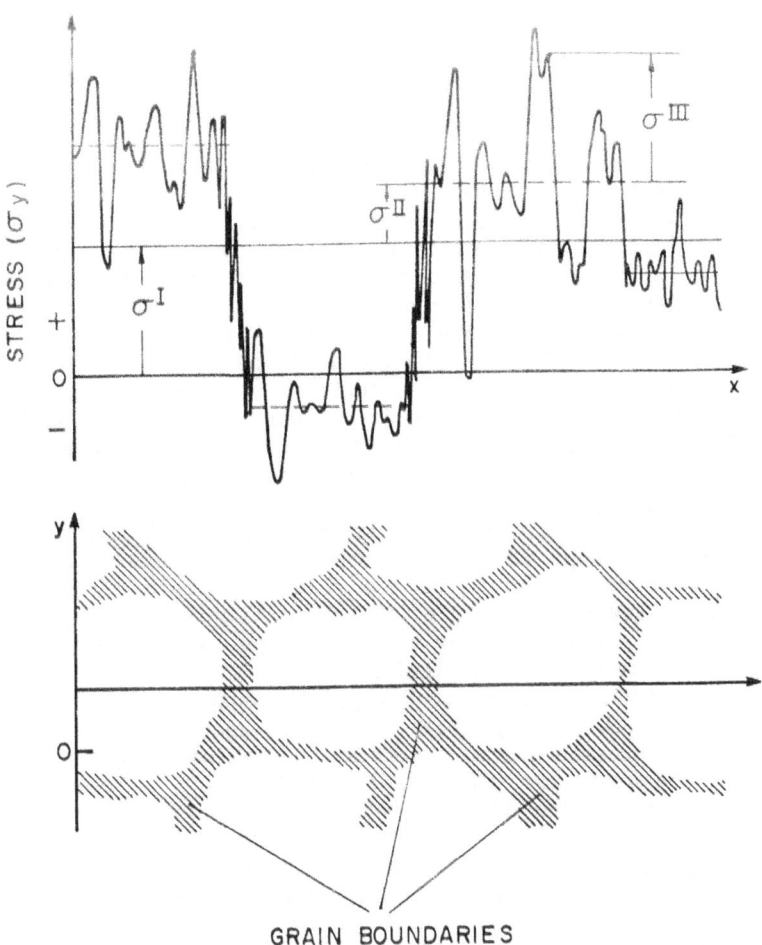

Fig. 1 Schematic representation of the three kinds of residual stresses: macrostresses (type I) extending through the material on a scale of tens of mm, grain-to-grain interaction stresses (microstresses or type II stresses) which are on a spatial scale of the grain size and short-range dislocation stresses (type III) which permit accommodation within grains.

differently as different slip systems are called into play. On unloading, the angular inhomogeneity in plastic deformation will generate grain-to-grain effects which lead in turn to the well-known[1] $\sin^2\psi$ anomalies seen in X-ray diffraction. The anomalies can also be generated by elastic inhomogeneity since if there is a macrostress in the surface layers, the variation of elastic constants in the surface grains will also cause oscillations. In general, we can get an idea of the grain-to-grain interaction stress component in the stress field by measuring several diffraction peaks.

The intrinsic peak width is also obtained from the data analysis and there may be several contributions to it. There may be a gradient of strain within the gauge volume. From an experimental viewpoint this indicates that the spatial resolution is not optimized. The variation of strain about its average value for a given orientation of grains, the microstress distribution, gives a contribution to the width as does any variation of stress within a grain (for a very elegant example of this point, see Ref. 3). Finally, there will be a contribution from dislocations (short-range, type III, dislocation stresses). Because there are several contributions to the width it is unusual to be able to extract unambiguous information about any one of them, so widths are possibly better treated at the present time as a qualitative guide rather than a definitive measurement.

The third parameter describing the diffraction peak is the integrated intensity. When the known Lorentz factors and plane multiplicities are corrected for, we have a measure of the relative number of diffracting crystallites in the gauge volume, i.e. the crystallographic texture. This information can be a guide as to how to do the experiment, i.e. if the 002 intensity is very strong in a particular direction and the 111 intensity is correspondingly weak, it makes good physical sense to characterize the residual stress with the strong peak. In some cases a family of grains can be identified corresponding to a very strong texture component and the strain tensor measured for that family of grains. Sometimes the variations in intensity from point-to-point in a sample are much greater than Poisson statistics on the number of counted neutrons and this is evidence for large grain size and hence large statistical fluctuations in the number of grains diffracting within the gauge volume.

An inherent problem in all stress measurements, either with X-rays or neutrons, is the absolute determination of lattice spacings, i.e. the establishment of the stress-free lattice spacing (d_o) of the planes being measured. There was a general feeling in the reporting back period to the Workshop that published papers should stipulate how d_o was measured.

4. Strain Response Within The Elastic Limit

It has been standard practice to derive "diffraction elastic constants" for materials and diffraction peaks of interest in order to derive stresses from strains. This is generally done by applying a uniaxial stress and measuring the parallel and perpendicular strain for a given (hkl) or making a four-point bend sample and measuring the tangential and normal strain. For a polycrystalline sample the induced strain in single grains which are constrained by their neighbours is measured. In recent years the emphasis has come to be placed on understanding[4] these results, rather than using them in a empirical way, especially with the advent of spallation sources where all reflections are collected simultaneously giving a complete picture of the crystallographic anisotropy.

If the behaviour of the response for a given (hkl) to a known stress is linear and reversible then we are below the elastic limit. If the behaviour is non-linear then we may deduce that some families of grains in the sample are yielding plastically although not necessarily the set under observation. The set under observation may be responding elastically to plastic deformation in neighbouring grains. In the linear elastic region the problem is well understood. For single crystals we need only the single crystal elastic constants, and for texture-free polycrystals the constrained response has been calculated within the Kröner model for most crystal structures.[5,6,7] The Kröner model allows one to match stress and strain at grain boundaries for spherical grain shapes in a self-consistent manner and to calculate the bulk as well as the diffraction elastic constants. The Voigt and Reuss models are incorrect oversimplifications of the texture-free polycrystal problem and there is no need to use them. The problems that remain are to calculate the Kröner model in the presence of known texture, as specified by the Orientation Distribution Function, and this can be done, in principle by straightforward numerical integration. In addition further work is required on complex grain shapes. An example of behaviour in the elastic region is shown in Fig. 2 (Fig. 2 of Ref. 4).

5. Plasticity: The Strain Response Beyond The Elastic Limit

The behaviour of single crystals beyond the yield point is different from the behaviour of single grains in a polycrystalline sample because of the constraint provided by the neighbouring grains. Because of the complexity of the yield surface in the constrained case, a satisfactory description of plasticity is still elusive; however the direction in which we should pursue the problem may be clear.

In a microscopic measurement of strain as a function of applied stress, mild steel and stainless steel are markedly different. In mild steel the elastic slope is maintained up to a yield point where there is a discontinuity when dislocations break away from solute atoms. Stainless steel shows continuous curvature and the yield point can only be defined in terms of stress to produce an arbitrary value of strain, e.g. 0.2% proof stress. An excellent discussion of this background material can be found in the book by Honeycombe[8].

In microscopic measurements of strain on a particular plane as a function of load, Fig. 3. (Fig. 6 of Ref. 4), in this case the (130) spacing in mild steel, there is a discontinuity at the yield point, but the average d-spacing continues to increase and the spread of d-spacings also increases as shown by the linewidth measurements. We could get an effective elastic constant from the slope in the plastic region, but this would be empiricism at best since we cannot yet calculate the magnitude of the increase. The concept of perfectly-elastic-perfectly-plastic behaviour is not correct at a microscopic level. More extensive measurements on the austenitic component of duplex steel (Fig. 12 of Ref.

228

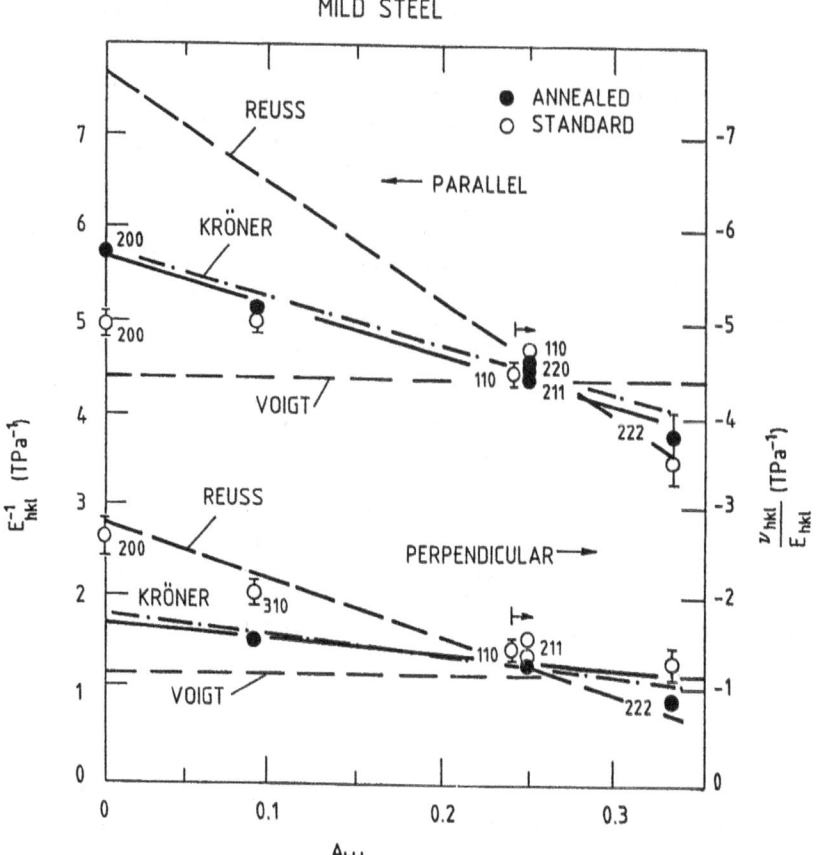

Fig. 2 Inverse of the diffraction Young's modulus for different diffraction planes plotted against the anisotropy factor[6,7] $A_{hkl} = (h^2k^2 + k^2l^2 + l^2h^2)/(h^2 + k^2 + l^2)^2$. The experimental results are compared with Kröner, Voigt and Reuss predictions for a random polycrystal with elastic constants for iron.

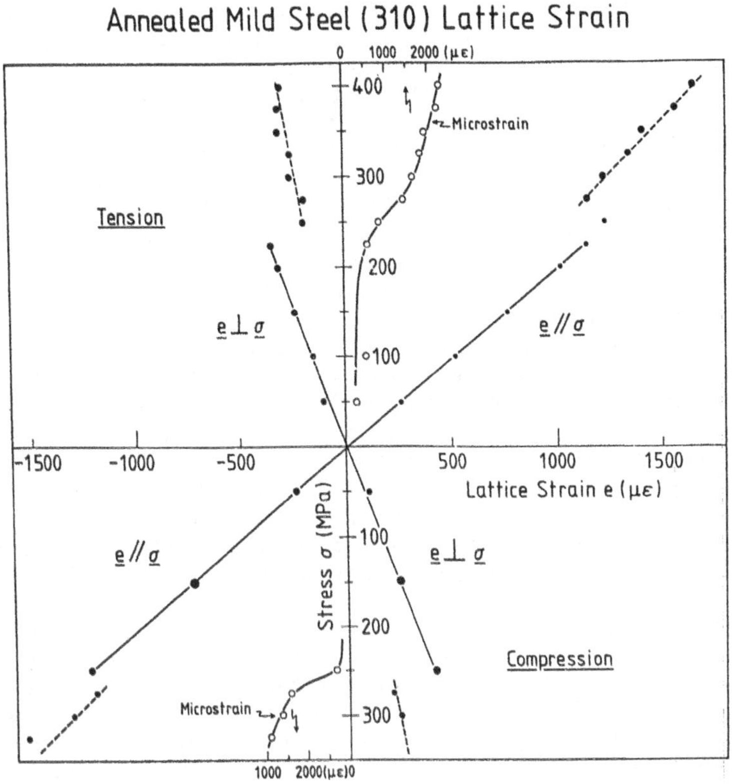

Annealed Mild Steel (310) Lattice Strain

Fig. 3 Measured lattice strain versus applied stress for mild steel in the elastic region, below the yield point, and in the plastic region above the yield point. The discontinuities at the yield point and the increase of strain with stress above the yield point are important for the understanding of the effects of plasticity. The curves labelled microstrain refer to the width of the diffraction line which signifies that a range of d-spacings associated with the (310) line exists at and above the yield point.

4) show that the [002] direction is softest in a yield situation as well as elastically, and that the [111] direction is the hardest. The apparent elastic constants here would be very different at high stresses than at low stresses and not understood at all in terms of elastic behaviour. The real problem here is to model the relative plastic response of different crystallographic directions in constrained materials, for this is what generates the grain-to-grain strains which can confuse the interpretation of strain measurements. What is the right approach for this problem? One idea is to borrow from our understanding of composite materials, such as Al:SiC, where we have two identifiable phases which are quite different as regards both elastic and plastic properties. The problem may be tackled successfully with the Kröner-Eshelby plasticity model[9,10] (see also Ref. 11), in which, schematically, we fit the inclusion into an appropriately shaped interstice in the matrix, and introduce tensorial forces to make it fit at the boundaries. We have here a phase-to-phase interaction model which is, in fact, an extreme example of a grain-to-grain interaction. We can then, conceptually at least, model the stiff [0002] direction in hexagonal close-packed materials such as Ti and Zr or the stiff [111] direction in cubic materials, in a matrix which is on average softer. Identifiable relaxation mechanisms, such as creep diffusion,[11] temperature dependence of the yield surface,[12,13] may serve to reduce the residual stress.

As a practical guide we might argue that to extract the grain-interaction stress, we should use the particular set of planes for diffraction which comes closest to the average behaviour in the Kröner plasticity model for then we are most likely to have calculated the plasticity correctly (bearing in mind the problem of matching mathematical models that are only valid at a point with results measured over a finite resolution). We include in this section a number of references[14-18] on work on modelling grain-to-grain stresses.

6. Conclusions

Residual stress measurement by diffraction is not a new field, although the neutron contribution to it is less than a decade old. By and large, X-ray stress measurements have been highly successful in specifying the nature of the surface stress state in spite of the considerations which we have dealt with in this discussion group. That is, the stress fields so obtained make good sense to the engineer in practice. One therefore is led to an intuitive position that the discrepancies caused by grain-interaction stresses are likely to be a correction term on the macroscopic stress field. However, they are important technologically and interesting from the point of view of material science investigations.

7. Acknowledgements

We would like to acknowledge very useful discussions with Drs. T. Leffers, I.C. Noyan and P.J. Withers in the preparation of this report.

REFERENCES

1. See for example, extensive discussions in "Residual Stress" by Noyan, I.C. and Cohen, J.B. (1987) (Springer-Verlag; New York).

2. Fidleris, V., Tucker, R.P. and Adamson, R.B., (1987) ASTM STP, 939, 49-85; Holt, R.A. and Causey, A.R., (1987) J. Nucl. Mat. 150, 306-18.

3. Reimers, W., Crostack, H.A., Wrobel, M. and Eckold, G., this conference.

4. Hutchings, M.T., this conference.

5. Kröner, E., Physik, Z., (1958) 151, 504-18.

6. Behnken, H., Hauk, V., Metallkde, Z., (1986) 77, 620-6 and references therein.

7. Dölle, H., (1979) J. Appl. Cryst. 12, 489-501.

8. Honeycombe, R.W.K., (1968) The Plastic Deformation of Metals", (Edward Arnold: London).

9. Eshelby, J.D., (1957) Proc. Roy. Soc. A 241, 376-96.

10. Kröner, E., (1961) Acta. Metall. 9, 155-62.

11. Withers, P.J., this conference.

12. MacEwen, S.R., Tomé, C. and Faber, J. Jr., (1989) Acta. Metall. 37, 979-89.

13. Majumdar, S., Singh, J.P., Kupperman, D. and Krawitz, A.D., J. (1990) of Eng. Mat. and Tech. 113, 51-9.

14. Berveiller, M., Hihi, A. and Zaoui, A. (1981) in "Deformation of Polycrystals" edited by N. Hansen et al. (Risø National Laboratory: Roskilde), 145-156.

15. Leffers, T., Ibid. 55-71.

16. Pedersen, O.B. and Leffers, T. (1987) in "Constitutive Relations and Their Physical Basis", edited by S.I. Anderson et al. (Risø National Laboratory, Roskilde), pp. 147-171.

17. Molinari, A., Canova, G.R. and Ahzi, S. (1987) Acta. Metall. 35, 2983-94.

18. Lepinski, P., Krier, J. and Berveiller, M. (1990) Revue Phys. Appl. 25, 361-88.

4. ASPECTS OF EXPERIMENTAL MEASUREMENT

SPATIAL RESOLUTION AND STRAIN SCANNING

P. J. WEBSTER
Department of Civil Engineering
University of Salford
Salford M5 4WT
England

ABSTRACT.   Neutron strain scanning is being increasingly used to determine internal and through surface strains in a widening range of engineering components. The locations within components that are of most concern to engineers are those at which failures are most likely to occur or to initiate. They include, in particular, regions of high stress and steep stress gradients. Extended regions of high stress often arise inside thick components where neutron beam attenuation may be a principal measuring constraint. High stresses and the steepest gradients usually occur in combination at surfaces, interfaces and sharp discontinuities where precise positioning and edge effects pose problems. In this paper the importance and practicalities of defining the size, shape and location of the "gauge volume" are discussed. Examples are presented of the methods and equipment employed to optimise measurements made at surfaces and internally in a range of components of different shapes, sizes and residual stress distributions.

1. Introduction

Stresses, and the corresponding strains, in materials are often classified as being of types 1, 2 or 3. Type 1 stresses, also known as macrostresses, vary over distances comparable and related to the dimensions of engineering components. Types 2 and 3 stresses are both classed as microstresses. Type 2 stresses have ranges comparable to the dimensions of grain sizes and the microstructure and are often a result of anisotropy, or are related to inhomogeneities and secondary phases. Type 3 stresses are intra-granular shorter range stresses which occur around dislocations and point defects [1,2].

Type 3 microstrains have such a short range that their variation cannot be spatially resolved by neutron scanning. They are a function of the material, rather than the shape, of a component. As they represent the sum of a range of values they have the effect of increasing the widths of the diffraction peaks and their spread is evaluated from measurements of peak broadenings.

Type 2 microstrains are often of concern in multiphase and composite materials in which stresses arise due to differential thermal expansion. As with type 3 strains their range is usually such that it is not possible to resolve their spatial variation by neutron scanning. Most strain measurements determine the average over the entire, or selected, volumes of each phase that comprises the material. In anisotropic materials, such as composites containing whiskers or fibres, the stresses usually exhibit substantial directional variation and measurements on these materials are often concerned with determining the directional rather than the spatial stress variation.

M. T. Hutchings and A. D. Krawitz (ed.),
Measurement of Residual and Applied Stress Using Neutron Diffraction, 235–251.
© 1992 *Kluwer Academic Publishers.*

236

Type 1 macrostrains are the type of strains of particular interest to the design engineer whose concern is the shape and fabrication of components or structures from well characterised materials. They vary throughout a component and their spatial variation can often best be determined from measurements of diffraction peak shifts by neutron strain scanning. This paper is concerned with the importance of spatial resolution in macrostress measurements and with the practicalities of neutron strain scanning, equipment design and measuring techniques.

2. Engineering considerations

Most of the earliest neutron strain measurements were made by scientists with long experience of utilising neutron scattering in condensed matter research. They established that the neutron method for measuring strain was viable and developed the theory and techniques using adapted high resolution neutron diffractometers at nuclear reactor and spallation neutron sources. Measurements were made on selected test samples upon which comparative traditional strain measurements were also made [2-8]. The science base requires continuing development but has advanced sufficiently for the technique now to be useful to a wider range of users. For some years neutron scientists have been collaborating with a few research engineers or have provided an industrial strain measuring service. A minority of engineers is now aware of the technique and a small number of the minority has an appreciation of its potential and constraints. It is evident that this number will increase together with a demand for facilities optimised for engineering measurements.

A principal concern of such engineers is, and will continue to be, the determination of stresses in critical components which have to sustain applied loads in service. The locations within components that are of most concern to engineers are those at which failures are most likely to occur or to initiate. They include, in particular, regions of high stress and steep stress gradients. Extended regions of high stress often exist inside thick components where neutron beam attenuation, positioning and access may be principal measuring constraints. High stresses and the steepest gradients usually occur in combination at surfaces, interfaces and at sharp discontinuities where precise positioning and edge effects pose problems.

Structural engineering materials are chosen for a combination of physical properties such as hardness, toughness, ductility and resistance to fatigue failure, corrosion and oxidation, which enable a component first to be manufactured and then to perform satisfactorily in use at an acceptable cost. Generally, they are complex multiphase or composite materials which respond to heat-treatments. Their neutron diffraction patterns are likewise usually complex although many steels and alloys often contain a dominant phase, with a crystal structure simply related to that of the principal component element, which can be used for strain measurements.

The shapes and sizes of structural engineering components are determined by the function that they have to perform and the stresses that they have to withstand. Their shapes are often complicated although many have axial or other symmetry elements, or some plane or straight surfaces, which can substantially reduce the problems associated with positioning and measurement.

A typical engineering sample is often one that, from a strain measuring viewpoint, is either too large or too small, is an awkward shape and has a complex multiphase structure. It is only on rare occasions that the shape of a sample and the material can be chosen for the convenience of strain measurement. In general the design of scanners must be such as to be adaptable for a wide range of components.

3. Principles of neutron strain scanning

The angles of coherent diffraction of neutrons from a polycrystalline material are determined by the Bragg equation:

$$2d\sin\theta = \lambda \tag{1}$$

where d is the interplanar spacing, 2θ is angle between the incoming and scattered neutrons and λ is the neutron wavelength. Differentiation of the Bragg equation gives expressions for the lattice strain:

$$\varepsilon = \delta d/d \tag{2}$$

$$= -\cot\theta.\delta\theta \qquad \text{for } \lambda \text{ constant,}$$

where $\delta\theta$ is the shift in Bragg angle corresponding to the lattice strain ε. The measured quantity $\delta\theta$ may be written expressly as:

$$\delta\theta = -\varepsilon.\tan\theta \tag{3}$$

or in terms of the the strain sensitivity

$$\delta\theta/\varepsilon = -\tan\theta \tag{4}$$

It is evident from equation 4 that near back-scattering geometry, as employed in the X-ray method because of strong absorption, gives the highest strain sensitivity but for neutron scanning this is at the expense of spatial resolution which is best when the reflection is measured at 90°. When neutron measurements are made at constant wavelength the optimum angle for measurement is usually in the range 90° to 120° to ensure that both spatial resolution and strain sensitivity are adequate. In time of flight measurements the ±90° angles are usually preferred so that two orthogonal measurements can be made simultaneously with the same spatial resolution and strain sensitivity.

4. The neutron strain scanner

Many neutron strain scanners have been developed from neutron powder diffractometers. One that has been adapted and extensively used for strain scanning is the high resolution diffractometer D1A at the Institut Laue Langevin that is shown in outline plan view in figure 1. D1A has a vertically focussing monochromator situated on a thermal neutron guide with a high take-off angle of 123° giving high resolution at large scattering angles and a near perfect Gaussian peak shape. Many strain measurements are made on ferritic steel components and for one of the available wavelengths, 0.19 nm, the (211) reflection which is often measured occurs at a near optimum 109°. This combination of features has made D1A very suitable for adaptation as a strain scanner.

When operated as a scanner a three axis orthogonal automated translator is mounted on the omega table to enable samples to be translated through the "gauge volume" which is positioned at the reference point at the centre of the diffractometer.

In operation the scanner measures peak position as a function of location and direction within a sample. Peak position measurement is a standard diffractometer function which is enhanced when scanning by using a Gaussian peak profile fitting routine. To measure

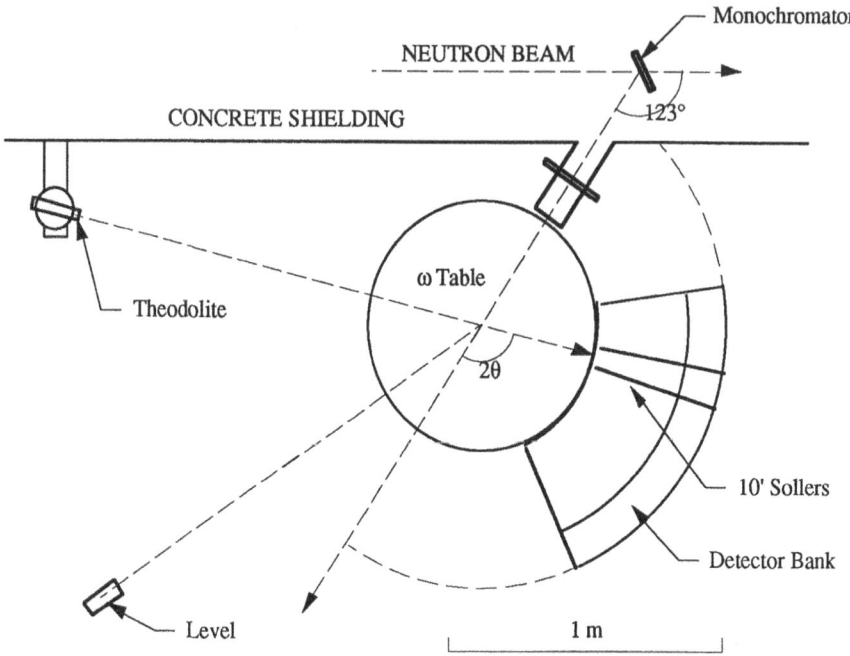

Figure 1. Plan view of the high resolution strain scanner D1A at the ILL.

the relative positions of the reference point, the gauge volume and the sample it is necessary to provide fixed reference points and to define a co-ordinate system. The orientation of the sample, in the horizontal plane, is determined from the angular position of the omega table.

4.1 THE GEOMETRICAL REFERENCE POINT

The central axis of most powder diffractometers is usually the vertical axis of the omega table upon which samples are generally mounted. The detector bank is rotated about the same axis which is also used as the vertical reference axis for scanning. A position on it usually at the height of the centre of the detectors, on D1A 500 mm above the omega table, is defined as the geometrical reference point of the scanner.

To determine the stress at points within a sample it is necessary to measure the strain in at least three orientations, and sometimes six, for each location and then to combine the data to derive the stress. In order to avoid the accumulation of errors, in particular in regions of steep stress gradients, it is essential that the reference point is very precisely defined, preferably to better than 0.1 mm. This degree of accuracy can be conveniently achieved using theodolites mounted at fixed locations. The arrangement on D1A is shown in figure 1. A theodolite rigidly fixed to a wall bracket provides one defining line descending at about 10° towards the reference point and a level defines a horizontal line at the height of the reference point. The reference point is indicated by means of a fine spike mounted at the reference height on the omega table and then positioned horizontally with a sequence of orthogonal horizontal adjustments until no translational movement of the spike is observed in the theodolite as the omega table is rotated. The theodolite and level are then focussed on the spike to define the reference point.

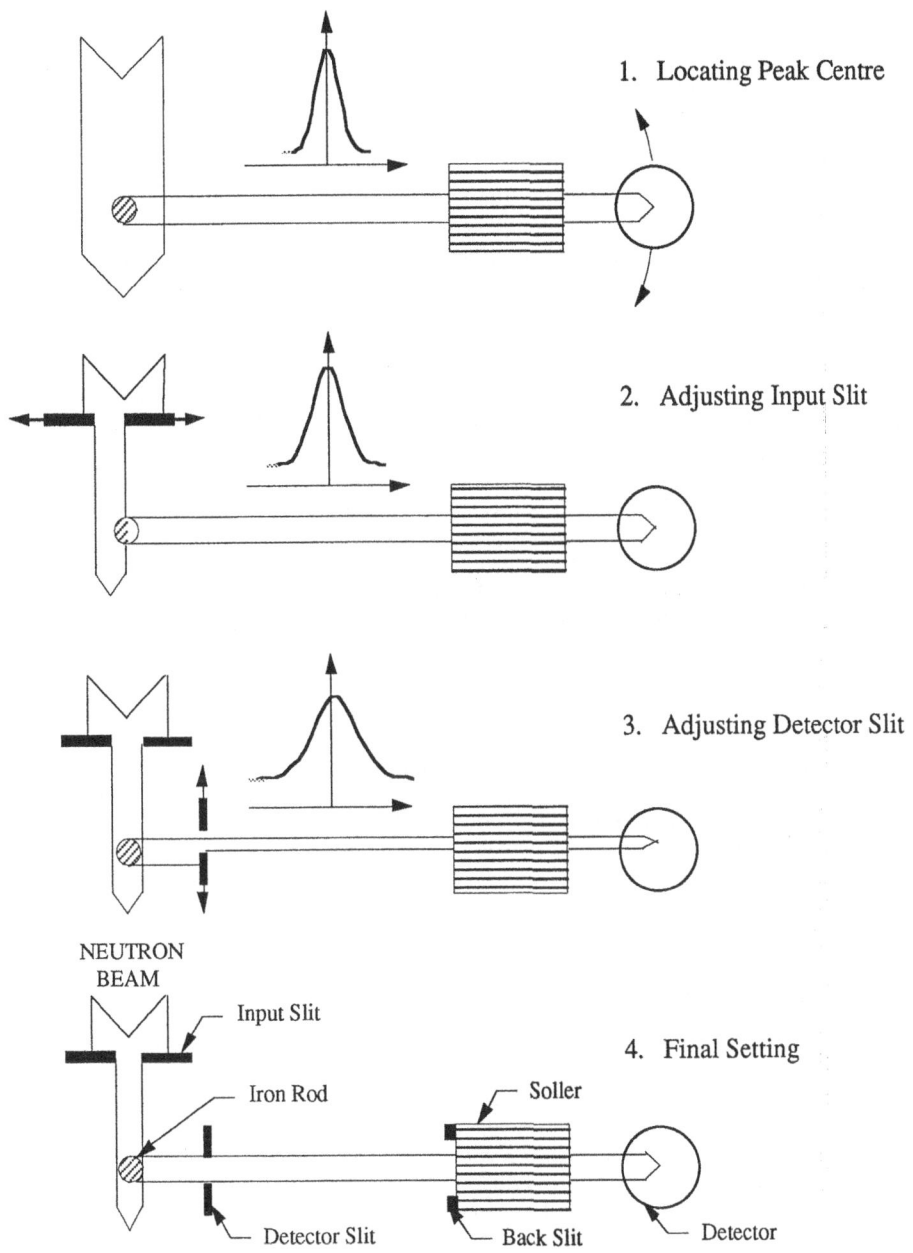

Figure 2.   Sequence of operations for centring the neutron beam and "gauge volume" at the geometrical reference point on D1A.

4.2 POSITIONING THE "GAUGE VOLUME"

The shape of the gauge volume is determined by the shapes of the input and output masks and the angle of the Bragg reflection. The strain sensitivity is proportional to $\tan\theta$ and thus there are advantages to be gained by measuring at higher 2θ angles. This is the method that is, and must be, due to the very strong absorption, adopted when X-rays are employed. However, the geometrical gauge volume at high, or low, angles is sharply elongated with a consequent loss in spatial resolution. In practice, the effective gauge volume for X-rays is not as sharply elongated as would appear from simple geometric considerations but is rapidly truncated due to the strong absorption. For neutrons, provided that the dimensions of the gauge volume are sufficiently small so that differential absorption is not a significant factor, the geometrical gauge volume approximates to the effective gauge volume. Whatever the shape it should be centred at the geometrical reference point.

Figure 2 illustrates the sequence of operations for positioning the neutron beam on D1A. A 1 mm diameter iron rod is mounted vertically in the position formerly occupied by the locating spike at the reference point and a diffraction pattern is recorded using a wide neutron beam. The centre of a suitable reflection, usually the (211), is determined and a detector is set at the centre angle. A 1 mm wide vertical slit is inserted in the input beam and the neutron count is recorded as the slit is traversed horizontally. The slit is then adjusted to the position at which the maximum neutron count was recorded which corresponds to being centred on the reference line with the neutron beam aligned along a radius. A 1 mm vertical slit fixed to the detector bank is then inserted in the output beam and similarly traversed until it is centred. Finally a wider back detector slit is fitted to eliminate stray neutrons from other locations during measurements on extended samples.

The vertical positioning of the beam is effected by inserting a square mask in the input beam and vertically traversing a horizontal slit through the beam followed by re-positioning to the reference height and setting the level.

For optimum accuracy it is preferable for the positions of the slits to be kept fixed for the duration of the experiments but this is rarely possible as optimising the gauge volume size and shape and the positioning and re-orientation of large samples often necessitates the temporary removal or radial displacements of the beam masking system. It is an important feature of scanner design that slit systems should be able to be precisely re-positioned after adjustment and that radial slit displacement does not result in horizontal or vertical beam displacement.

4.3 BEAM DIVERGENCE

Other factors affecting the size and shape of the gauge volume and the intensity distribution within it are the radial distances of the input and output masks from the reference point and the divergences of the neutron beams. Most powder diffractometers are designed to have small divergence in the horizontal plane to ensure adequate angular resolution but often have relaxed vertical divergence, or sometimes vertically focussing monochromators to increase the neutron count. The high resolution diffractometer D1A, for example, has an effective horizontal input divergence of 15' and 10' Sollers in front of the detectors and a 6° vertically focussing monochromator. The effects of the input divergences on the gauge volume are shown in figures 3 and 4.

Figure 3 illustrates the theoretical relative widths and proportions of the main beam and the penumbra for D1A for a series of commonly used slit widths, 0.5, 1.0 and 2.0 mm as a function of radial distance from the reference point. The size of the penumbra, the region of diminishing neutron flux outside the geometric slit dimensions caused by the divergence, increases with the radial distance of the mask from the reference point. The importance

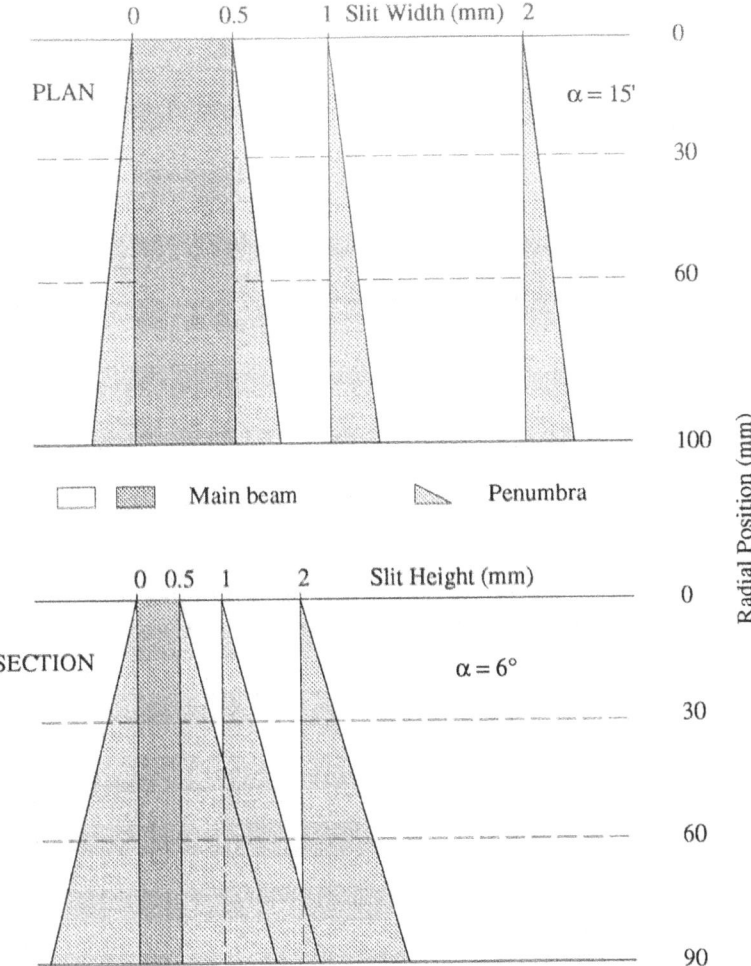

Figure 3.   The effects of horizontal and vertical beam divergence, and radial position of the input slit, on the gauge volume definition of D1A.

of the penumbra relative to the main beam increases inversely as the slit dimensions and linearly with the radial distance. Figure 4 shows the corresponding measured intensity profiles for a 2 × 2 mm mask at a radial distance of 20 mm. In the horizontal plane the divergence is small and the penumbra is significant only when precise measurements with narrow slits are being made.   The vertical divergence resulting from the focussing monochromator is usually of consequence and on D1A generally must be allowed for or reduced by additional vertical collimation. The effects of the horizontal divergence of the 10' detector Sollers are usually small relative to the input divergences and the radial position of the detector slit is not normally critical.

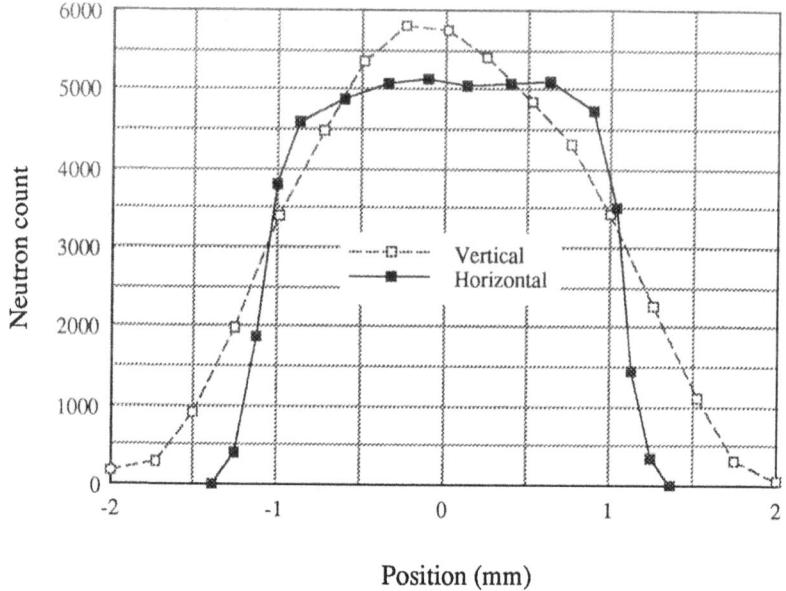

Figure 4. Intensity profiles measured horizontally and vertically across the D1A gauge volume with a 2 × 2 mm mask situated 20 mm from the reference point.

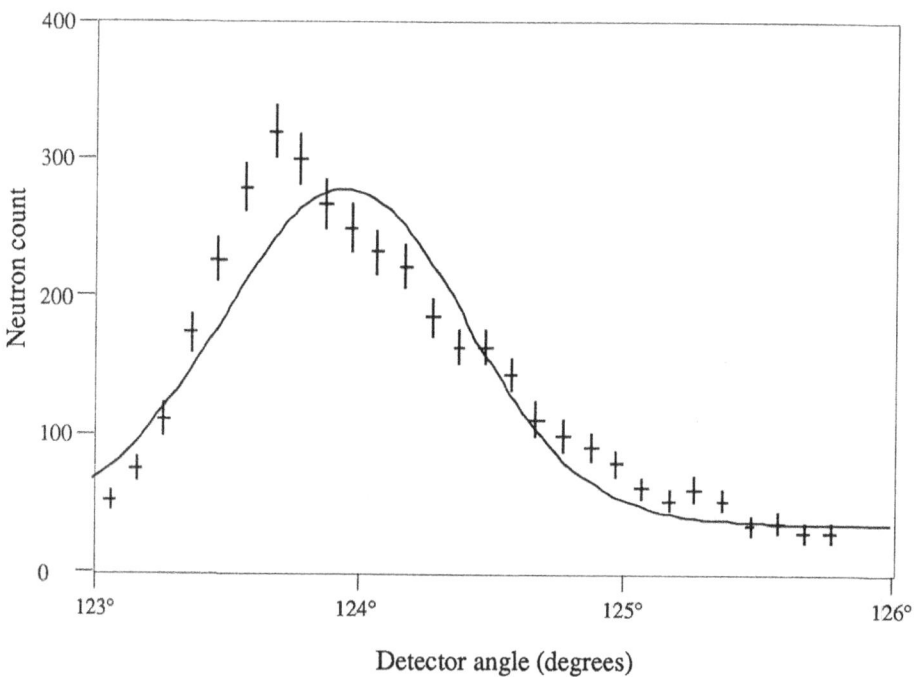

Figure 5. Asymmetric peak profile measured adjacent to a peened surface using D1A.

5. Internal strain measurements

When strain scanning measurements are made data are collected of neutrons scattered from the gauge volume. This volume should be large enough to diffract sufficient neutrons to provide a good statistical average in a reasonable time but none of its dimensions should be so long as to extend across regions with significantly different macrostresses or detail will be lost. If a gauge dimension is excessive, in regions of constant stress gradient symmetrical peak broadening will occur but if there are changes in gradient there will be asymmetrical broadening and peak distortions which are difficult to interpret even with good statistics. Generally if asymmetry is observed in a peak profile it is better to reduce the gauge size, if necessary to the minimum set by beam divergence, and to work with poorer statistics over a narrower strain range than with better statistics integrated over a wider range of strains. An example of a distorted peak from a near surface region which includes an exceptionally steep and then a reversed stress gradient is shown in figure 5 together with the best, but clearly inappropriate, single Gaussian fit.

When measurements are made in more than one orientation the shape of the gauge volume is of consequence. What is optimum for one orientation is unlikely to be the best for all. Figure 6 illustrates in plan and section the different and common areas occupied by the gauge volume for orthogonal measurements at three length to width ratios. The greatest commonality is achieved with a cubic sampling volume at a detector angle of 90°, but in components such as plates or rods, or samples with obvious symmetry, gauge volumes elongated along a direction with a low stress gradient, usually parallel or normal to a symmetry axis, may be employed to reduce counting times.

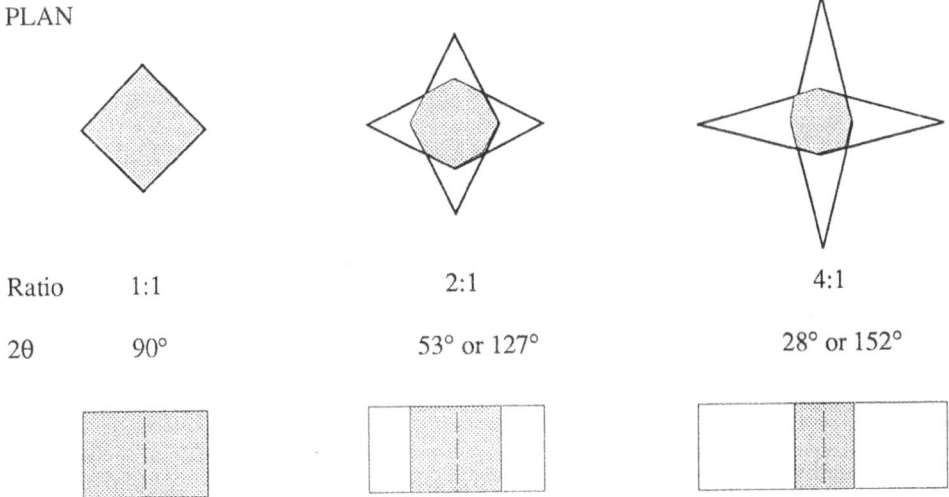

PLAN

| Ratio | 1:1 | 2:1 | 4:1 |

| 2θ | 90° | 53° or 127° | 28° or 152° |

SECTION

Figure 6. Gauge areas, shown in plan and section for aspect ratios 1, 2 and 4 showing areas of overlap for two orthogonal orientations.

244

6. Through surface strain measurements

Through surface strain measurements pose particular problems when strain scanning. The principal problems are those associated with the vanishingly small gauge volumes that are inherent when measuring just at the surface and the change in shape of the effective gauge volume that occurs when the geometrical gauge volume is only partly within the sample. For samples with flat or cylindrical surfaces the dimensions of the gauge volume can often be extended in one dimension to increase the neutron count rate without loss of spatial resolution but even so it is often necessary to extend the counting time to obtain adequate counts just at the surface where the effective gauge cross-sectional area approaches zero. The effects of changing gauge shape are illustrated in figures 7-11.

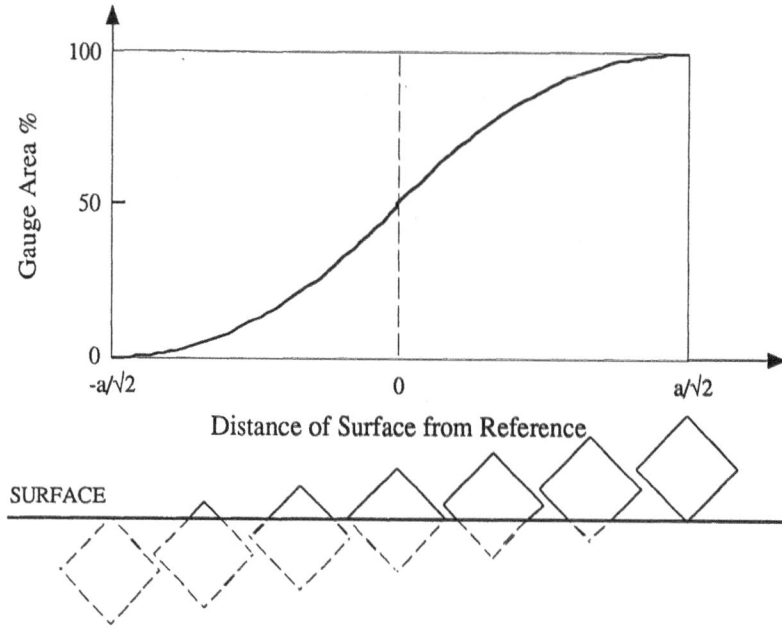

Figure 7. Effective gauge area and shape as a function of distance from the reference point for a surface passing through a square cross-section gauge volume.

6.1 EFFECTIVE GAUGE POSITION AND AREA

Figure 7 shows in plan view the size and shape of the effective gauge area, disregarding any effects due to divergence, as a surface passes through a gauge volume of square cross-section, as would occur at a measuring angle of 90° with input and output slits of equal width, a. The surface will just touch the edge of the gauge area when it is at a distance of $a/\sqrt{2}$ from the geometrical gauge centre which coincides with the scanner reference point. As the surface is moved towards the reference point the effective gauge area will be triangular in shape and will increase parabolically in size. After the reference position has been passed the effective area becomes pentagonal increasing in size parabolically inverse until at a

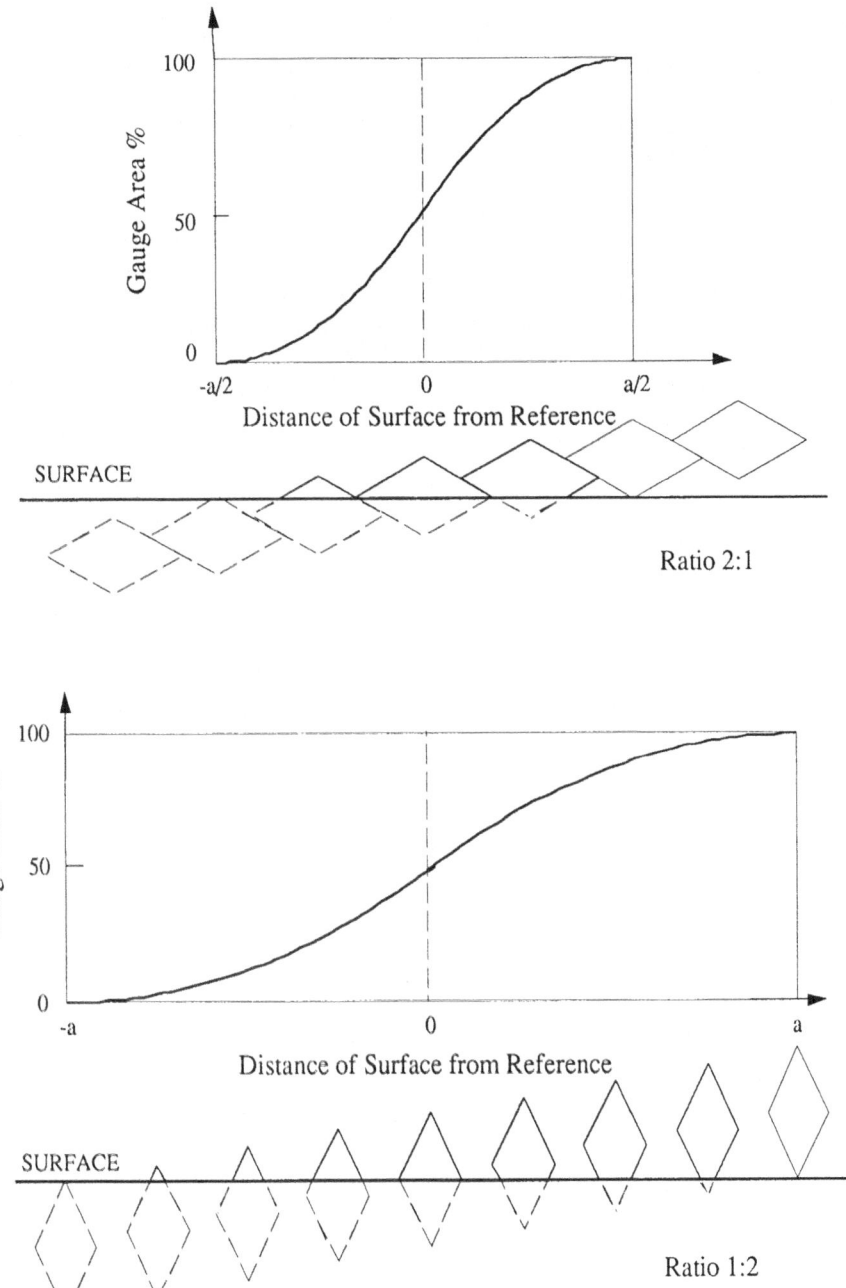

Figure 8.   Effective gauge areas and shapes as functions of distance from the reference point for a surface passing through gauge volumes angled at 53° or 127°.

depth of a/√2 it attains its full size and becomes square. When fully in the sample the centre of the gauge area is at the reference point and any displacement of the sample corresponds to the same sized, but opposite sign, displacement of the gauge area. However, when passing through the surface the centroid of the effective gauge area is displaced from the reference point and moves at a different rate to that of the the the sample translation. It is necessary at surfaces to make a correction for the displacement of the centroid of the effective gauge volume from the reference point.

Figure 8 illustrates the effect of measuring at angles other than 90°, in this case for angles of 53° or 127° at which the gauge length to width ratio is 2:1 or 1:2 depending upon whether the measurements are being made normal or parallel to the surface. When a set of orthogonal measurements is being made through a surface it is in general necessary to correct separately for the centroid displacement for each orientation. In the case illustrated, where the aspect ratio is 2:1, when the long axis is parallel to the surface the centroid correction for scanning in the normal direction is half that required for when the short axis is parallel to the surface.

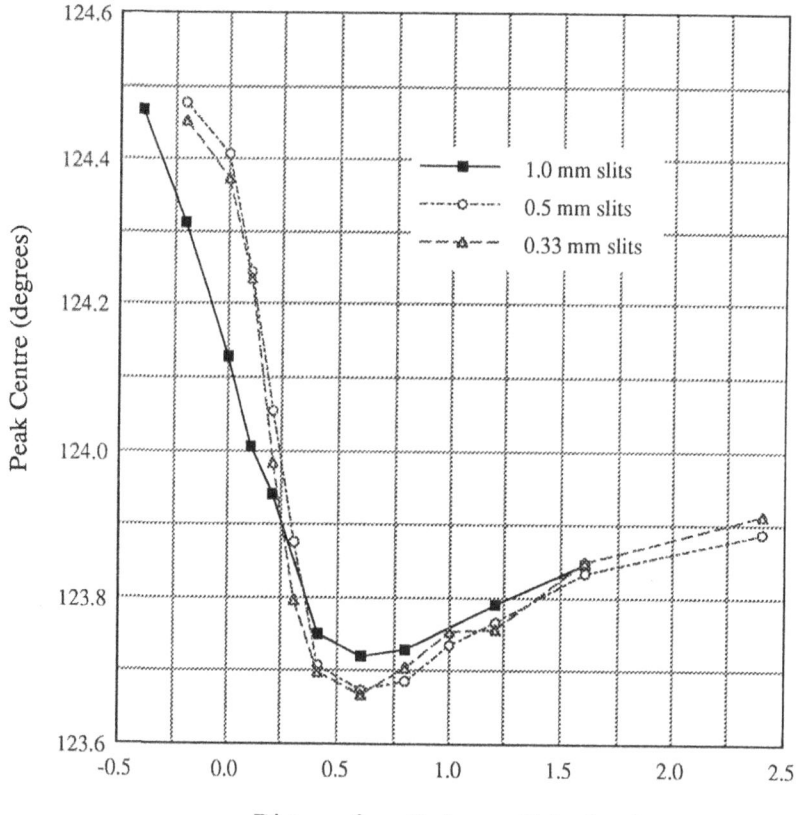

Distance from Reference Point (mm)

Figure 9.   Peak centres versus position of a peened surface relative to the reference point for three different slit widths, showing the need for gauge volume positioning corrections, divergence and averaging effects.

Figure 9 shows the effects of gauge volumes with similar shaped but different sized cross-sectional areas. In this case the gauge volumes were defined by three matching sets of long input and exit slits of widths 0.33, 0.5 and 1.0 mm respectively, with the measurements being made around 124°. The data were collected through the surface of a heavily peened plate to measure the strains in a direction parallel to the surface in a region where the residual stresses change rapidly from high surface compression to sub-surface tension before decaying to near zero in the interior. The figure shows the peak centre shift as a function of the location of the surface relative to the reference point. The 0.33 and 0.5 mm slit data are almost indistinguishable due to the small difference in their widths which is less than the blurring effects of the penumbra. Each requiries a surface centroid correction of up to about 0.2 mm. The 1 mm data on the other hand is displaced and smoothed relative to the data from the narrower slits. For the 1 mm slit the centroid correction through the surface is up to about 0.5 mm and the apparent sharpness of the minimum, corresponding to a sub-surface tensile peak, has been reduced because the averaging is over an extended sampling window.

6.2 GAUGE SHAPE EFFECTS ON DIFFRACTION PEAK PROFILES

D1A was designed as a high resolution diffractometer with 10' Soller collimation in front of each detector, making the detectors angular sensitive rather than position sensitive thus enabling diffraction patterns to be recorded from different sized powder samples not accurately positioned at the centre of the diffractometer. This configuration when adapted as a strain scanner, with a centred input slit and a detector slit attached to and moving with the detector as it scans through 2θ, defines the gauge volume that makes it is possible successfully to measure strain changes through surfaces. In some other configurations severe peak shift corrections are required to compensate for apparent shifts due to the changing shape and lateral displacement of the effective gauge volume at surfaces.

Figure 10 illustrates, for measurements made at 90°, how the effective gauge area changes shape, and position relative to the detector system, as the flat surface of a sample is scanned in the normal direction. At scan position 1 when the gauge volume is totally immersed in the sample the gauge area is square and centred relative to the detector. As the sample is withdrawn progressively through to position 8 the effective gauge area, as seen from the detector, decreases from one side and there is a lateral displacement of the centroid of the gauge area. The proportion of asymmetrical to symmetrical area subtended at the detector, which increases as the sample surface is withdrawn from the gauge volume, is illustrated in the lower part of the diagram. From scan position 7 onwards the entire effective gauge area is on one side only of the line from the reference point through the centre of the slit to the detector. This asymmetry and displacement has severe consequences for scanners that use a fixed detector slit and a static position sensitive detector.

Position sensitive detectors, with an angular resolution of about 0.1°, are often used in strain scanners, in preference to single detectors, to improve neutron counting rates when internal measurements are being made. The gain in counting rate, relative to that of a single detector scanning a Bragg peak, is often about a factor of 20 with usually only a minor loss in spatial resolution. Consequently, for internal strain measurements, the case for their use is usually overwhelming.

If, however, position sensitive detectors are used for surface measurements large, rapidly changing, spatially sensitive corrections must be made. The surface effect on the shape and position of a Bragg peak measured by a PSD varies as a function of the detector slit width, its radial position and the position and orientation of the sample surface relative to the reference point. Figure 11 shows the calculated peak shapes as a function of the

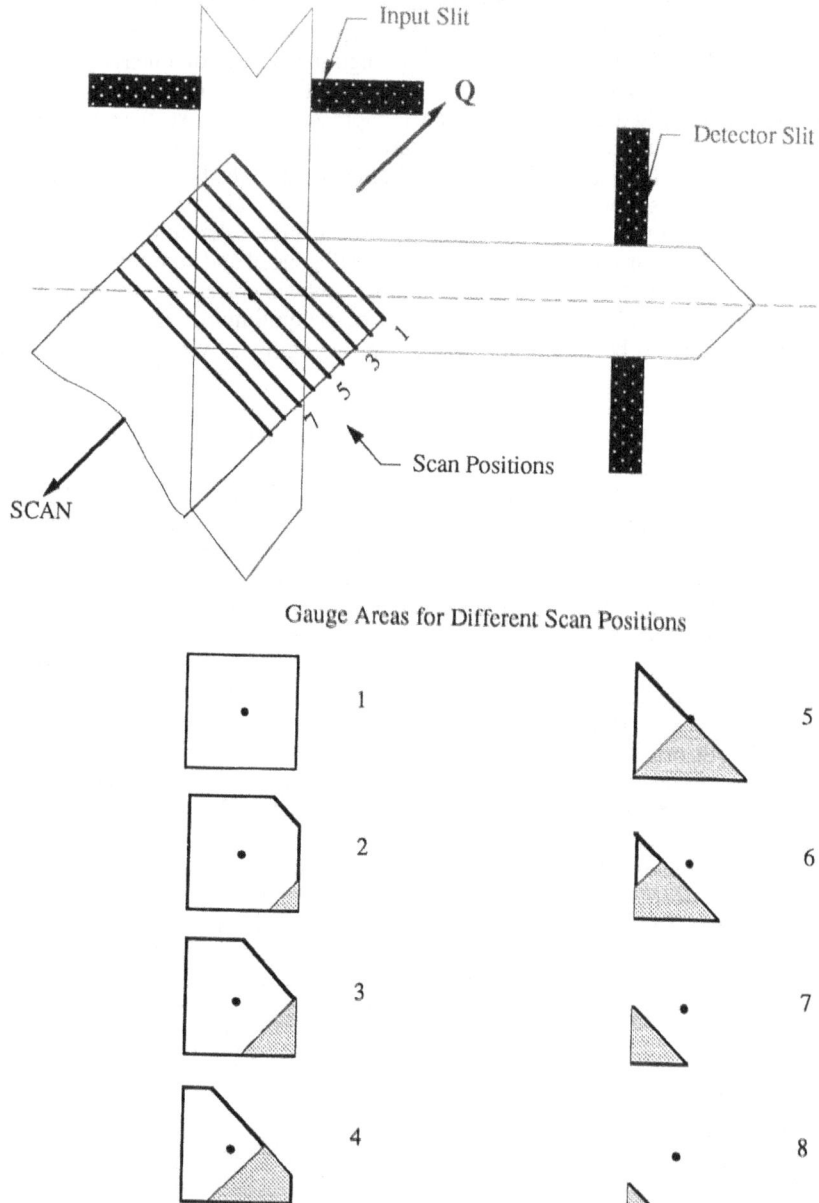

Figure 10. Change in gauge area for measurements at 90° as a plane sample surface is scanned through the reference point, showing the ratio of asymmetric (shaded) to symmetric area subtended at the detector.

2 mm slit 30 mm from the reference point

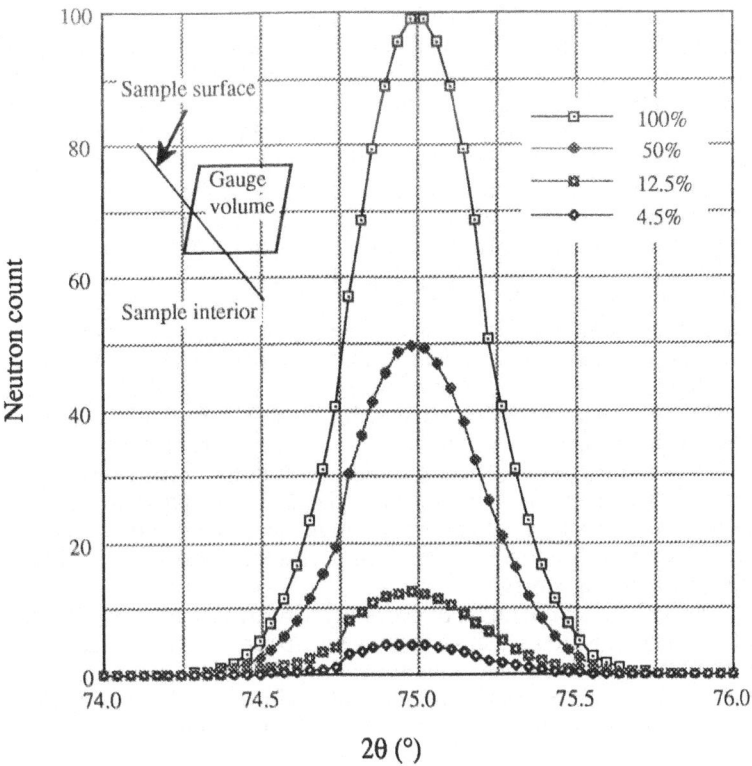

Figure 11. Calculated Bragg peak profiles at 75° for D20 at the ILL when used as a strain scanner for different proportions of the gauge volume within the sample [9].

proportion of the gauge within the sample for a not untypical strain scanning configuration for the ILL diffractometer D20 which is fitted with a PSD. The Bragg peak is at 75° and the 2 mm wide detector slit is situated 30 mm from the reference point. When the gauge volume is fully in the sample the peak is symmetric and the instrument functions as an efficient strain scanner. Through the surface, however, the peak is noticeably asymmetric and displaced. The value of the apparent shift in the peak that is actually recorded will depend upon the fitting routine used for analysis. If a symmetrical Gaussian, as routinely used for internal measurements, is employed the fit will be poor and the apparent peak shift due to the changing geometry may be greater than that observed for a large strain change. The results given by other fitting routines will vary depending upon the shapes of the functions used.

7. Summary

Neutron strain scanning is now a mature application of neutron science and is rapidly becoming an established engineering technique [10,11]. It is being applied to a steadily widening range of investigations of real engineering components of different shapes and sizes. Many of the investigations are of critical components in which large residual stresses and steep stress gradients occur. In some investigations stress gradients as steep as 4000 MPa/mm have been measured. To determine these stresses it is necessary to measure the corresponding strains in several orientations of the component, mostly in three orthogonal directions, at the same location. For a constant wavelength instrument this involves in essence two groups of measurements, firstly that of Bragg peak position, or peak shift, and secondly the position and dimensions of the gauge volume in the sample.

The accurate determination of peak position using Gaussian or other fitting routines is a standard neutron diffraction measurement limited only by the statistical quality of the data and the symmetry or otherwise of the peak. Strain induced peak asymmetry results when there is an asymmetric strain distribution within the gauge volume. If this is observed it is generally best if practicable to reduce appropriate dimensions of the gauge volume until the asymmetry is acceptably small, otherwise details of the strain pattern will be lost.

The accurate positioning of the gauge volume relative to the scanner reference position and its shape and location in the sample are at least as important as the measurement of peak position particularly in regions of large stress gradient and at surfaces, interfaces and other discontinuities such as at crack tips. In practice the reference point should be defined, by theodolites or equivalent devices, to within 0.1 mm; the centre of the incident neutron beam should pass through it and the detector should be aligned on it. Corrections for relative displacement of the centroid of the effective gauge volume are essential at surfaces and interfaces when the sample is only partially within the gauge volume. Once the positional corrections for each orientation have been made it is then possible to combine the data to derive the corresponding stresses.

If the scanner has a PSD it might be necessary to restrict its use to internal measurements because of the large position sensitive instrumental peak shifts that arise due to the change in the shape of the effective gauge volume near surfaces. Alternatively it may be possible to make an angular scan with the detector bank and process the data from each element separately to combine to some extent the speed of a PSD with the resolution of a single detector.

8. Acknowledgements

The author wishes to thank the Directors and staff of the ILL and ISIS for the provision of facilities and support for the development of strain scanners, and the UK Science and Engineering Research Council, The Welding Institute, British Rail and Rolls Royce for the resources provided by grants GR/D63196, GR/E55556, GR/F02427 and GR/F52873.

9. References

1. Eigenmann, B., Scholtes, B. and Macherauch, E. (1989) 'X-ray stress determination in ceramics and ceramic-metal-composites' in G. Beck, S. Denis and S. Simon (eds.), Proc. Int. Conf. on Residual Stresses ICRS2, Nancy, November 1988, Elsevier Applied Science, London, pp. 27-38.
2. Lorentzen, T. (1988) 'Non-destructive evaluation of residual stresses by neutron diffraction', NDT International 2, 385-388.
3. Allen, A., Andreani, C., Hutchings, M. T. and Windsor, C.G. (1981) 'Measurement of internal stress within bulk materials using neutron diffraction', NDT International 14, 249-254.

4. Krawitz, A. D., Brune, J. E. and Schmank, M. J. (1982) 'Measurements of stress in the interior of solids with neutrons' in E. Kula and V. Weiss (eds.), Residual stress and stress relaxation, Plenum Press, pp. 139-155.
5. Pintschovius, L., Jung, V., Macherauch, E. and Vohringer, O. (1983) 'Residual stress measurements by means of neutron diffraction', Mater. Sci. Eng. 61, 43-50.
6. Holden, T. M., Dolling, G., MacEwan, S. R., Winegar, J. E., Powell, B. M. and Holt, R. A. (1984) 'Applications of neutron diffraction to engineering problems', Proc. 5th Canadian Conf. on NDT, Toronto, October 1984.
7. Allen, A., Hutchings, M. T. and Windsor, C. G. (1985) 'Neutron diffraction methods for the study of residual stress fields', Adv. Phys. 34, 445-473.
8. Stacey, A., MacGillivray, H. J., Webster, G. A., Webster, P. J., Ziebeck, K. R. A. (1985) 'Measurement of residual stresses by neutron diffraction', J. Strain Analysis 20, 93-100.
9. Wang, X. (1991) Private communication.
10. Krawitz, A. D. and Holden, T. M. (1990) 'The measurement of residual stresses using neutron diffraction', MRS Bull. XV, 57-64.
11. Webster, P. J. (1990) 'The neutron strain scanner: a new analytical tool for engineers', Steel Times 218 No. 6, 321-323.

STRAIN TENSOR MEASUREMENTS BY NEUTRON DIFFRACTION

T. LORENTZEN AND T. LEFFERS
Materials Department
Risø National Laboratory
DK-4000 Roskilde
Denmark

ABSTRACT. The complete strain tensor can be determined by neutron diffraction. In this paper the required number of measurements as well as the required measurement accuracy are discussed. Calculations based on measurements on a powder sample are reported, and the results are discussed. Further a simulation is done on measurements performed along three orthogonal directions combined with the three directions in between these.

1. Introduction

The neutron diffraction technique for internal stress determination allows normal strain components to be measured in arbitrarily chosen directions. It is hence possible to establish the complete description of the strain state from any six measurements of independent strain components, without any assumptions of principal strain directions. Many publications mention this potential of the neutron diffraction technique e.g. [1], though the topic has only been delt with in more details by Priesmeyer [2]. However, while individual strain components can be measured with an accuracy often better than $\pm 100 \times 10^{-6}$, this accuracy is not sufficient to render an acceptably low error on the final strain tensor, when determined from only six different strain measurements. To establish an error estimate, an excess of six strain measurements is required, though it is not well established how many additional strain measurements are required to give acceptably low errors. It is the aim of this paper to shed some light on the above presented arguments.

In order to measure different strain components, the sample must be reoriented in the neutron beam, so the instrumentation must provide an accurate means of moving the sample around on the spectrometer sample table. In order to provide the possibility to measure the strain in six different directions, it is often necessary to dismount the sample, and manually reorient it. Any such manual sample handeling will inevitably be a source of error in the gauge volume definition, and hence the different strain components are not measured in exactly the same gauge volume. This will be a source of error in the strain tensor determination. Further, it is often of interest to determine a strain profile along a line through the sample, and it can be extremely hard to establish a configuration that allows a multiplum of strain components to be measured in conjunction with a determination of

253

M. T. Hutchings and A. D. Krawitz (ed.),
Measurement of Residual and Applied Stress Using Neutron Diffraction, 253–261.
© 1992 *Kluwer Academic Publishers.*

the strain profile. It has been one of the main objectives of the PSD based experimental set-up at Risø to combine these options.

The PSD based set-up described in [3] is equipped with a robot allowing samples to be rotated around the incident beam, hence the direction in which the strain is measured, can be altered simply by rotating the sample around the beam. This simple way of re-orienting the sample, whereby the strain components will be lying on a cone, is believed to limit the experimental scatter introduced, when unavoidably changing the gauge volume shape. However, it has been pointed out [4] that strains deduced from measurements on a single cone does not provide sufficient information to establish the complete strain tensor. It is hence required to shift to a second diffraction cone for at least one of the six required strain measurements.

The following discussion on strain tensor determination by neutron diffraction relates to the PSD based set-up at Risø , and hence measurements and calculations are based on the way the different strain componets are aquired using this instrumental configuration. However, the conclusions regarding asscuracy and necessary level of overdeterminancy is general and will be relevant for strain tensor determination using other experimental configurations.

2. Powder measurements

In order to test the accuracy in determining complete strain tensors by utilizing the PSD based experimental set-up at Risø, a series of measurements were performed on a standard compacted Al_2O_3 powder sample. The sample is cylindrical, and it is aligned on the spectrometer sample robot with the cylinder axis coinciding with the incident beam (see Fig. 1). This sample geometry will provide identical scattering conditions for all rotations around the incident beam, without any excess of material obscuring the view from the detector.

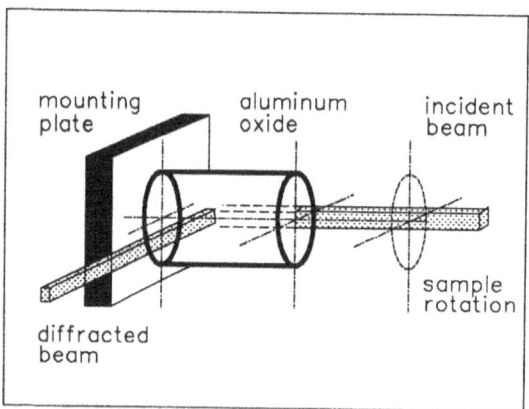

Figure 1: Experimental configuration for strain tensor determination in a standard Al_2O_3 powder sample. The cylindrical specimen is mounted with the cylinder axis coincidal with the incident beam.

In order to combine the above described means of altering the strain direction, with the aim of recording residual strains along a line through the sample, the recording of Bragg peaks is done by scanning the wavelenght instead of the diffraction angle. A total of 6 "strain components" were measured using two different diffraction cones, and table 1 shows the result of these measurements. It is noticed that all reported strains are in increments of 48×10^{-6}. This specific increment is due to the fact that the measured d-spacings normally are reported with an accuracy of 4 digits, and the differences in the measured d-spacings are on this fourth digit.

Table 1: The six strain measurements on the standard compacted Al_2O_3 powder sample. The strain components are located on two different cones corresponding to $2\theta_S = 90^o$ and 100.32^o, and spatially separated on each cone by 35° (Ω_S).

measurement no.	$2\theta_S[^o]$	$\Omega_S[^o]$	$\epsilon[\times 10^{-6}]$
1	100.32	210	0
2	90.00	0	+48
3	90.00	35	+48
4	90.00	70	-48
5	90.00	105	+144
6	90.00	140	+48

The set of equations to be solved in order to establish the strain tensor from these measurements is given by (1) where l, m and n are the direction cosines for the different strain componets:

$$
\begin{pmatrix}
e_{l_1,m_1,n_1} \\
e_{l_2,m_2,n_2} \\
e_{l_3,m_3,n_3} \\
e_{l_4,m_4,n_4} \\
e_{l_5,m_5,n_5} \\
e_{l_6,m_6,n_6}
\end{pmatrix}
=
\begin{pmatrix}
l_1^2 & m_1^2 & n_1^2 & l_1 m_1 & m_1 n_1 & n_1 l_1 \\
l_2^2 & m_2^2 & n_2^2 & l_2 m_2 & m_2 n_2 & n_2 l_2 \\
l_3^2 & m_3^2 & n_3^2 & l_3 m_3 & m_3 n_3 & n_3 l_3 \\
l_4^2 & m_4^2 & n_4^2 & l_4 m_4 & m_4 n_4 & n_4 l_4 \\
l_5^2 & m_5^2 & n_5^2 & l_5 m_5 & m_5 n_5 & n_5 l_5 \\
l_6^2 & m_6^2 & n_6^2 & l_6 m_6 & m_6 n_6 & n_6 l_6
\end{pmatrix}
\begin{pmatrix}
\epsilon_{xx} \\
\epsilon_{yy} \\
\epsilon_{zz} \\
\gamma_{xy} \\
\gamma_{yz} \\
\gamma_{zx}
\end{pmatrix}
\quad (1)
$$

Using the data given in table 1 and taking the first measurement (no 1 of table 1) as the reference measurement, this system of equations is now established as:

$$
\begin{bmatrix}
0 \times 10^{-6} \\
48 \times 10^{-6} \\
48 \times 10^{-6} \\
-48 \times 10^{-6} \\
144 \times 10^{-6} \\
48 \times 10^{-6}
\end{bmatrix}
=
\begin{bmatrix}
0.3078 & 0.1026 & 0.5896 & 0.1777 & -0.2460 & -0.4260 \\
0.5000 & 0.0000 & 0.5000 & 0.0000 & 0.0000 & 0.5000 \\
0.3355 & 0.1645 & 0.5000 & 0.2349 & 0.2868 & 0.4096 \\
0.0585 & 0.4415 & 0.5000 & 0.1607 & 0.4698 & 0.1710 \\
0.0335 & 0.4665 & 0.5000 & -0.1250 & 0.4830 & -0.1294 \\
0.2934 & 0.2066 & 0.5000 & -0.2462 & 0.3214 & -0.3830
\end{bmatrix}
\times [\epsilon]
$$
$$(2)$$

The calculation renders a strain tensor for the compacted powder sample as given in (3):

$$[\epsilon] = \begin{bmatrix} -14210 \times 10^{-6} \\ -16073 \times 10^{-6} \\ 13042 \times 10^{-6} \\ -2021 \times 10^{-6} \\ 3123 \times 10^{-6} \\ 1264 \times 10^{-6} \end{bmatrix} \tag{3}$$

Obviously the expected result is a zero strain tensor from the powder sample, and as seen from the given results, the calculated strain tensor is in great error. The errors are due to the inherent experimental scatter as well as due to the sensitivity of the system of equations. The set of linear equations is solved by the so called "Singular Value Decomposition" technique [5], whereby the coefficient matrix is decomposed into the product of three matrices; one of which (normally named W) contains parameters allowing a calculation of the "condition number" for the system. This serves to identify wether the system of equations is approaching an ill conditioned state. The condition number is calculated as the ratio of the largest and the smallest of the W-componets, and the system is said to be ill conditioned if this number is too large, and the coefficient matrix is singular if this ratio is infinite. Another way of evaluating the condition number is that the system is said to be ill conditioned if the reciprocal of the condition number approaches the computers floating point precision, e.g. 1×10^{-6} for single precision calculations. For the given set of equations, the reciprocal condition number was found to be 0.002721. Even though this level is not close to the floating point precision, it is relatively low, and indicating a system approaching an ill conditioned state.

3. Experimental scatter

As seen in the previous paragraph the strain tensor was in great error when calculations were based on the specific set of measurements, and the most desirable way of improving the accuracy in strain tensor determination, is to improve the accuracy on the individual strain measurements. For a specific strain state that for instance corresponds to a hydrostatic strain of 50×10^{-6}, then using a series of six exact strain measurements the above described calculations leads to the correct strain tensor:

measurements   calculated $[\epsilon]$

$$\begin{bmatrix} 50 \times 10^{-6} \\ 50 \times 10^{-6} \\ 50 \times 10^{-6} \\ 50 \times 10^{-6} \\ 50 \times 10^{-6} \\ 50 \times 10^{-6} \end{bmatrix} \Rightarrow [\epsilon] = \begin{bmatrix} 50 \times 10^{-6} \\ 50 \times 10^{-6} \\ 50 \times 10^{-6} \\ 0 \times 10^{-6} \\ 0 \times 10^{-6} \\ 0 \times 10^{-6} \end{bmatrix} \tag{4}$$

However, the above level of 50×10^{-6} closely resembles the achievable accuracy in strain measurements by neutron diffraction. Hence this is a common level of experimental scatter. In order to test a set of measurements which are all $\pm 50 \times 10^{-6}$ off the true strain value, an other calculation was performed simulating that the six measurements on the powder sample were all in $\pm 50 \times 10^{-6}$ error, and measured in the same directions as given for the actual powder measurements.

measurements   calculated $[\epsilon]$

$$
\begin{bmatrix}
50 \times 10^{-6} \\
-50 \times 10^{-6} \\
50 \times 10^{-6} \\
-50 \times 10^{-6} \\
50 \times 10^{-6} \\
-50 \times 10^{-6}
\end{bmatrix}
\Rightarrow [\epsilon] =
\begin{bmatrix}
-13147 \times 10^{-6} \\
-14923 \times 10^{-6} \\
11984 \times 10^{-6} \\
-1490 \times 10^{-6} \\
2923 \times 10^{-6} \\
1064 \times 10^{-6}
\end{bmatrix}
\tag{5}
$$

Again the correct result is that of a zero strain state, and the calculated tensor components are in great error. In order to arrive at a strain tensor with the error on individual strain components not exceeding $\pm 100 \times 10^{-6}$, the above error level of $\pm 50 \times 10^{-6}$ must be reduced to $\approx \pm 5 \times 10^{-6}$. However, such a low error level is currently not achievable in neutron diffraction strain measurements.

Obviously, a set of equations established from six directions which are spatially very close, provides a rather ill conditioned system of equations, which will be extremly sensitive to measurement errors. Hence, it has been suggested that an appropriate choice of strain components is that of three orthogonal directiones combined with three directions which bisects the angle between pairs of the three orthogonal axes. This approach was investigated by simulating that the initial six strain measurements given in table 1 for the powder sample, were taken in the described six directions. The coefficient matrix which has a reciprocal condition number 0.2 compared to 0.002721 given previously, is found as::

$$
\begin{bmatrix}
1.0000 & 0.0000 & 0.0000 & 0.0000 & 0.0000 & 0.0000 \\
0.0000 & 1.0000 & 0.0000 & 0.0000 & 0.0000 & 0.0000 \\
0.0000 & 0.0000 & 1.0000 & 0.0000 & 0.0000 & 0.0000 \\
0.5000 & 0.5000 & 0.0000 & 0.5000 & 0.0000 & 0.0000 \\
0.0000 & 0.5000 & 0.5000 & 0.0000 & 0.5000 & 0.0000 \\
0.5000 & 0.0000 & 0.5000 & 0.0000 & 0.0000 & 0.5000
\end{bmatrix}
\tag{6}
$$

This system is hence much more well conditioned, and the solution is a strain tensor as given by:

$$
\epsilon = \begin{bmatrix}
1 \times 10^{-6} \\
48 \times 10^{-6} \\
48 \times 10^{-6} \\
-144 \times 10^{-6} \\
192 \times 10^{-6} \\
48 \times 10^{-6}
\end{bmatrix}
\tag{7}
$$

The average error on the strain tensor components for this solution is $\approx \pm 80 \times 10^{-6}$ based on only six measurements as compared to $\approx \pm 8000 \times 10^{-6}$ for the original set of six equations, and the maximum error on any strain tensor component is 192×10^{-6} as compared to $\approx 16073 \times 10^{-6}$ for the original set of six equations. It is quite obvious that much accuracy is gained by chosing a more well conditioned set of equations, though an error of 192×10^{-6} is still unaceptably large. For this set of equations the error level for individual measurements must be reduced to $\approx \pm 25 \times 10^{-6}$ in order to assure that errors on individual tensor components does not exceed $\approx \pm 100 \times 10^{-6}$.

The conclusion on the above described calculations is hence that even with a well conditioned set of equations, a strain tensor calculation based on only six measurements, does not give an acceptably low error level on the individual strain tensor components. The required level of accuracy on individual strain measurements for six measurements to be enough, is currently not achievable by neutron diffraction strain measurements.

4. Overdetermined systems

Accepting the inherent experimental scatter, one way of improving the accuracy of the solution, and a necessary condition for establishing an error estimate on the strain tensor components, is to overdetermine the system of equations by additional measurements. In order to evaluate this option a set of simulated measurements were tested. All simulated measurements were assumed to be in $\pm 50 \times 10^{-6}$ error, and distributed on the two cones given by $2\theta_S = 70°$ and $110°$. Compared to the initial set of measurements on two cones, the angle between the cones is now increased to $40°$ compared to $10°$ previously. This was done in order to initiate calculations with a more well conditioned system with a reciprocal condition number of 0.009912 compared to 0.002721 previously. This set of six measurements was increased with another 14 simulated measurements, which were spatially distributed inbetween the initial six, and located on the same two cones. The additional measurements were all in pairs of $+$ and -50×10^{-6} corresponding to sets of measurements which continues to be exposed to an experimental scatter of $\pm 50 \times 10^{-6}$.

The result of adding these "measurements" to the system of equations is here evaluated by the average error found as the average deviation from zero of all strain tensor components. As such this error is a rough measure of the general error on the calculated strain values. It was found that the solution begins to stabilize at approximately 10-12 equations, where the average error is reduced to $\approx 70 \times 10^{-6}$, and the maximum error on any strain tensor component is $\approx 130 \times 10^{-6}$. Additional measurements up to a total of 20 did not provide dramatic improvement of the solution. It should be noticed that all these additional measurements deliberately continued to be in error. The reduction of the average error is not expected to follow a monotonic decreasing path as the fluctuations in the individual measurements are expected to be random, and hence there will be a range of possible solutions for a given number of equations.

As described previously much accuracy is gained by chosing a more well conditioned system of equations initially. However, this did not prove to be sufficient to establish the complete strain tensor from only six measurements. Also the test starting with measuring three orthogonal strain components together with three components located in between these three, was continued simulating an overdetermined system of equations by adding "measurements" in directions spatially distributed in between the initial six. This test was done with the same series of simulated measurements all beeing in error of $\approx \pm 50 \times 10^{-6}$. The results are given in figur 2 in terms the average error as a function of the number of measurements, and as seen then the initial error from only six measurements is relatively low, and not much accuracy is gained by adding more inaccurate measurements. Compared to the simulated set of measurements on two different cones as described above, the average error from only six measurements is reduced from 460×10^{-6} to only 79×10^{-6}. It is hence verified that the system when starting with a more well conditioned set of equations, errors more rapidly approaches an acceptable level. However, if the additional measurements continue to be inaccurate, it is still found that up to 12 measurements might be required for the errors on individual tensor components to reach an acceptably low level.

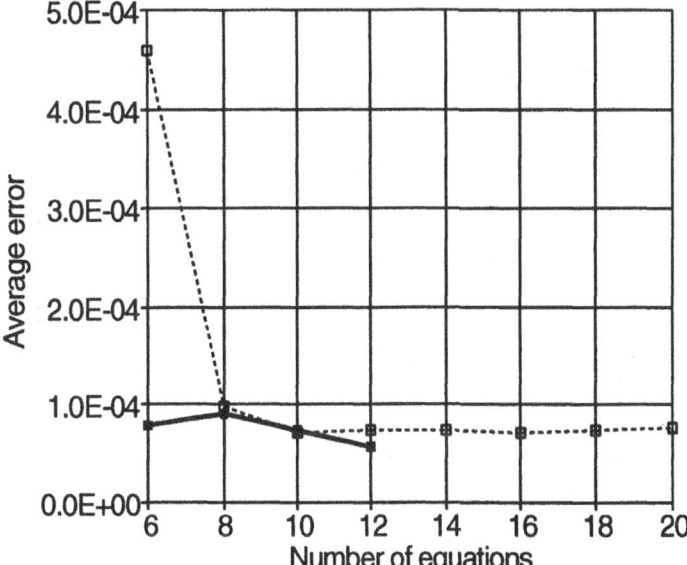

Figure 2: The average error on all strain tensor components in the Al_2O_3 powder sample as a function of the number of equations in the linear set. The results are based on ($\square$) measurements on two cones, and ($\blacksquare$) simulated measurements starting with directions along three orthogonal axises, and three directions in between these followed by directions in between these six directions.

260

5. Summary

Neutron diffraction strain measurements can be utilized to determine complete stress and strain tensors in structural components. In principle this can be done from only six different strain measurements, however, even though individual strain measurements can be done with an accuracy of $\approx \pm 50 \times 10^{-6}$, this is not sufficient for an accurate strain tensor determination. The accuracy required depends strongly on the choice of directions in which the strains are measured. Test measurements based on only two diffraction cones only separated by $\approx 10°$, rendered a strain tensor with an average error on the tensor component of $\approx \pm 8000 \times 10^{-6}$. By increasing the angle between the two cones then the calculations based on simulated experimental scatter gave an average error reduced to $\approx \pm 460 \times 10^{-6}$, which is still not an acceptably low error level, and calculations showed that the experimental scatter on individual measurements should be reduced to $\approx \pm 5 \times 10^{-6}$ in order ot give an acceptably low error level on the tensor components. It has been suggested that a reasonable choice of directions in which to measure the strains, would be three orthogonal directions and the three directions bisecting the angle between pairs of these orthogonal axises. This proved to be a much better choice of directions than the described set of directions on two cones. The average error on the tensor components was hereby reduced to $\approx \pm 80 \times 10^{-6}$, however, even though this average error level is acceptable, individual tensor components could still possess unaceptably large errors. For this choice of directions, the experimental scatter on individual measurements must be reduced to $\approx \pm 25 \times 10^{-6}$ in order to get acceptably low errors on the individual tensor components.

The described measurements and calculations have verified that in most cases it is not sufficient to measure only six strain components for a complete strain tensor determination. The experimental scatter must be reduced beyond the presently feasible level. It is hence necessary to overdetermine the mathematical problem by adding more measurements, and it has been shown that depending on the choice of directions, between 8 and 12 measurements will be sufficient in order to arrive at a strain tensor with acceptably low errors on the individual strain components.

Accepting that up to about 12 measurements are required, then strain tensor determination becomes extremly tedious, and whenever possible the problem should be reduced to three measures of strain by realistic assumptions of principal stress and strain directions.

References

[1] Allen, A.J., Hutchings, M.T., Windsor, C.G. (1985). 'Neutron diffraction methods for the study of residual stress fields'. Advances in Physics 34,4. 445-473.

[2] Priesmeyer, H.G. and Schröder, J. (1988). 'Strain tensor determination using neutron diffraction'. Presented at the Conference on Fatigue and Stress, London, U.K.

[3] Lorentzen, T., Leffers, T. and Juul Jensen, D. (1991), 'Implementation and application of a PSD set-up for neutron diffraction strain measurements'. In these proceedings.

[4] Lorentzen, T. and Christoffersen, J. (1990), 'Limitations on the strain tensor determination by neutron diffraction using a position sensitive detector', NDT International, Vol 23, No.4, p107-109.

[5] Press, W.H., Flannery, B.P., Teukolsky, S.A. and Vetterling, W.T. (1989), 'Numerical Recipes, The Art of Scientific Computing'. Cambridge University Press, Cambridge, U.K.

[6] Popov, E.G. (1978) 'Mechanics of materials', second ed., Prentice/Hall International editors, London, U.K.

INVESTIGATIONS OF LARGE GRAINED SAMPLES - EXAMPLES

W. REIMERS [a], H.-A. CROSTACK [b], M. WROBEL [b], G. ECKOLD [c]

(a) Hahn-Meitner-Institut Berlin GmbH
Dept. N5
Glienicker Str. 100
W-1000 Berlin 39

(b) Fachgebiet Qualitätskontrolle
Universität Dortmund
Postfach 500 500
W-4600 Dortmund 50
Germany

(c) RWTH Aachen und Institut für Festkörperforschung
KFA Jülich GmbH
Postfach 1913
W-5170 Jülich
Germany

ABSTRACT. The single grain measuring and evaluation technique allows the investigation of individual grains embedded in their polycrystalline matrix. Using X-ray diffraction grains at the surface of the material were analysed with a local resolution of $\varnothing \approx 50$ μm. This way the stress distribution over the coarse grain zone and the transition zone of a welding was investigated. Applying neutron diffraction grain sizes of $\varnothing > 1$ mm could be studied. For investigating the inhomogeneous elastic and plastic deformation behaviour of crystallites in the polycrystalline matrix grain clusters were analysed at different applied stress levels. Significant variations from grain to grain are found. The results concerning the elastic deformation are compared to the macroscopic deformations and to the deformation states calculated for free single crystals so indicating the influencing factors on the crystallite-crystallite interaction in the polycrystal. The plastic deformation state was analysed by rocking curves in different crystal directions which allowed the calculation of the anisotropy of the plastic deformation. Moreover, the inhomogeneity of the onset and the development of the plastic deformation as function of the external load was followed.

1. INTRODUCTION

By means of analysing the precise reflection position in the diffraction angle 2θ of selected grains, their elastic deformation and stress state can be studied. The stress value so obtained represents the sum of the first and second order stresses. So, by studying several neighboured grains, the stress variations as well as the averaged stress state (macroscopic stress) are accessible. Stresses of third order can be obtained either by analysis of the 2θ-profile or by measuring the 2θ-profile as a function of the sample translation in respect to the measuring spot. So the single grain measuring and evaluation technique allows the analysis of the residual stress state in coarse grained materials but it can also be used for studying the factors determining the deformation behaviour on the microstructural scale of polycrystals. Since the elastic properties of polycrystalline materials are basically characterized by the combination of the single crystal elasticity and the additional interactions between crystallites, the comparison of

M. T. Hutchings and A. D. Krawitz (ed.),
Measurement of Residual and Applied Stress Using Neutron Diffraction, 263–276.
© 1992 *Kluwer Academic Publishers.*

deformation distributions in grain clusters with calculated free single crystal deformations reveals the extent of the crystallite-crystallite interactions.

Plastic deformations of crystallites affect the shape of their rocking curves. Hence, the measurement of the rocking curves with different crystallographic directions as rotation axes allows the detection of plastic deformation and the investigation of its anisotropy. The so obtained experimental data could be interpreted by using the principle of minimum virtual energy which allows the calculation of the activated gliding systems and their gliding sums. Moreover, the inhomogeneity of the onset and the development of the plastic deformation is measured for investigating the crystallite-crystallite interactions also in the plastic deformation region.

2. EXPERIMENTS WITH X-RAY DIFFRACTION

X-ray diffraction experiments are especially suited for near surface investigations but also for investigations with high local resolution. Since strong stress gradients can be expected in the vicinity of weldings, often X-ray diffraction is used for their analysis. The heat impact of the welding process, however, may lead to a coarsening of the grains, so that the conventional measuring technique can no longer fulfill the demand for small measuring spot sizes. For the investigation of the stress distribution in a plasma welded steel (German grade X8 Cr17) therefore the single grain measuring technique was applied. The investigations were performed in ferrite grains as well in the transition zone with grain sizes of $\varnothing \approx 40$ μm as in the coarse grain zone. The rough positions of the crystals selected in the sample surface was defined by applying a lead diaphragm. The correct positioning of the diaphragm in the X, Y directions on the sample surface was checked by intensity measurements. The correlation of the grain investigated to the micrograph is given in fig. 1. Here, the grain boundaries are shown and the grains are numbered.

The orientations of the crystals were determined by means of analysing the angular orientations of the {200} and {110} reflections. The measurements of the interplanar lattice spacings were then performed on the {211} reflections. For the calculation of the stress tensor components the single crystal elastic constants for Fe were used. For the evaluation it was introduced $\sigma_{33} = \sigma_{13} = \sigma_{23} = 0$. From these conditions the value d_0 was refined individually for each grain. With $d_0 = 2,87264$ (25) a good agreement was found for the 9 grains investigated. This result gives evidence that the assumption presented above is fulfilled in this sample due to the small penetration depth of the Cr-Kα-radiation. Furthermore, there is no evidence for significant variations in the chemical composition.

265

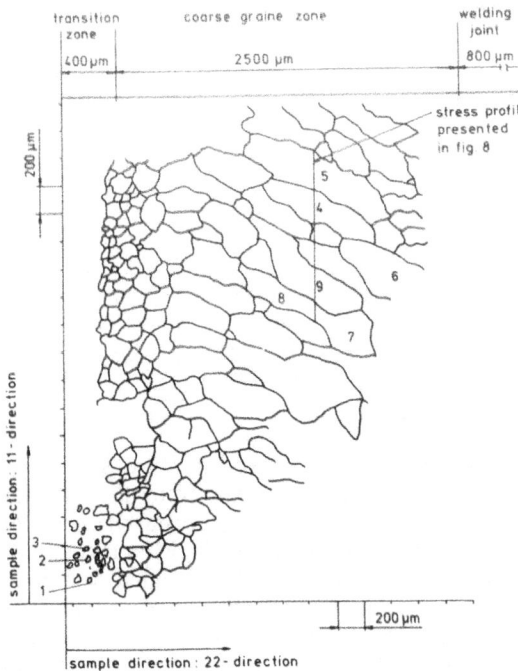

transition zone | coarse graine zone | welding joint

400 μm | 2500 μm | 800 μm

200 μm

stress profil presented in fig 8

sample direction: 11- direction

200 μm

sample direction: 22- direction

Figure 1. Grain boundaries corresponding to the micrograph and numbering of the grains investigated

The results for the three grains measured in the transition zone are given in table 1. The stress tensor components σ_{11} and σ_{22} are in the surface of the sample where σ_{11} is parallel to the undirectional welding joint and σ_{22} is perpendicular to it.

TABLE 1. Stress tensor components in individual grains situated in the transition zone of a plasma welding joint - (values in MPa, standard deviations in parenthesis)

Stress tensor components referred to the welding joint				Main stresses	
grain no.	σ_{11}	σ_{22}	σ_{12}	σ_{11}^{M}	σ_{22}^{M}
1	- 49 (48)	- 114 (46)	- 101 (46)	24 (48)	- 187 (47)
2	- 103 (39)	- 191 (38)	- 126 (38)	- 13 (38)	- 280 (39)
3	- 114 (38)	- 97 (32)	- 92 (33)	- 13 (35)	- 197 (38)
Ø value	- 89 (24)	- 134 (23)	- 106 (23)	- 1 (24)	- 221 (24)

In the transition zone comparative $\sin^2\psi$- measurements were possible due to a large vertical aperture. The mean value for σ_{22} obtained from the individual grains is quantitatively in agreement with the results of the polycrystal measurement of σ_{22} = - 158 (9) MPa. For σ_{11}, however, a mean value of - 89 (4) MPa is found which is lower in its absolute value than the result of - 140 (6) MPa obtained from the conventional measurement. This difference can be due to the comparatively small number of grains measured so that the mean value is not yet representative. This aspect is demonstrated later on by the comparatively important stress variations from grain to grain obtained in the coarse grain zone (tab. 2).

The evaluation of the deformation data obtained from the individual grains also yields the stress tensor component σ_{12}. Tab. 1 gives evidence that this component is significantly different from zero. This finding is interpreted by the heat distribution during the plasma welding process. At any time, the heat is very localized so that a radial stress component arises due to the temporary displacement of the heat impact during the welding process. The diagonalization of the stress tensor for the individual grains shows that in fact nearly an uniaxial stress is present (Tab. 1).

The results of the measurements in the coarse grain zone are summarized in Tab. 2.

TABLE 2. Stress tensor components in individual grains situated in the coarse grain zone of a plasma welding joint - (value in MPa, standard deviations in parenthesis)

Stress tensor components referred to the welding joint				Main stresses	
grain no.	σ_{11}	σ_{22}	σ_{12}	σ_{11}^{M}	σ_{22}^{M}
4	− 29 (41)	− 156 (28)	− 156 (32)	75 (38)	− 261 (32)
5	− 141 (31)	− 198 (26)	− 76 (27)	− 88 (25)	− 250 (32)
6	− 82 (17)	− 73 (17)	− 112 (16)	34 (10)	− 190 (29)
7	− 108 (36)	− 80 (28)	− 169 (31)	75 (19)	− 263 (28)
8	− 163 (28)	− 145 (24)	− 143 (26)	− 10 (5)	− 297 (29)
9	− 38 (35)	− 22 (19)	− 52 (31)	22 (28)	− 83 (35)
∅ value	− 94 (13)	− 112 (10)	− 118 (11)	18 (10)	− 224 (13)

Tab. 2 gives evidence for the extent of stress variations from grain to grain.

There is no significant change in the stress states between the transition zone and the coarse grain zone. The mean value for σ_{11} and σ_{22} of the 6 grains investigated in the coarse grain zone is lowered only by approximately 40 MPa compared to the results of the polycrystal measurement in the transition zone. So this difference is less important than the variation from grain to grain. Also, for the σ_{12} stress component no significant changes are observed between the transition zone and the coarse grain zone. The calculation of the main stresses shows again the anisotropy of the stress state.

A further application field for the single grain measuring technique using X-ray diffraction is the characterization of single crystal components, e.g. turbine blades. By means of rocking curves the homogeneity of the crystallization can be checked. The measurement of X-ray intensities allows the determination of the impact depths of surface treatments, since e.g. intensive grinding destroys the single crystal lattice and, hence, reduces the intensity concentrated in Bragg-reflections.

3. EXPERIMENTS WITH NEUTRON DIFFRACTION

3.1 Elastic Deformation

Neutron diffraction in combination with the single-grain measuring and evaluation technique is suited for analysing residual volume stresses in large grained materials or also for the investigation of crystallite-crystallite interactions. Since the deformation of the whole grain is measured, the effect of the free surface is of minor importance.

As sample material for the experiments under external load the Ni-base alloy IN 939 was chosen. The grain sizes range from ∅ ≈ 1 mm up to ∅ ≈ 5 mm. Flat tensile specimen were prepared and installed in a tensile apparatus which allows the application of defined uniaxial stresses. The applied stress is examined by calibrated strain gauges on the sample holder.

The neutron diffraction experiments were performed on the triple-axis spectrometer UNIDAS. The instrument was equipped with a full circle Eulerian cradle which allows the realization of the diffraction conditions for the individual Bragg-reflections. Single crystallites could be selected by using fine collimators and diaphragms: Soller collimators were inserted in the primary and secondary beam so that the horizontal divergency was cut down to 15'. Vertical slits were set yielding a vertical divergency of 20'. Furthermore, the horizontal beam diameter was reduced to 1.5 mm by primary and secondary slits. By applying the 90° scattering technique which implies an adaptation of the incident neutron wavelength to the interplanar lattice spacings of the orienting (hkl) reflections chosen, the location of the grain selected was well defined within the sample. More precisely, the location of the grain under study was subsequently determined by Bragg intensity measurements as a function of the position of Cd-diaphragms on the sample. These measurements were repeated for different [hkl]-directions so that the grain position was checked in X, Y, Z and the grain could be adjusted in the center of the spectrometer by means of sample translations.

In each flat tensile specimen several grains were investigated. The Bragg-reflection measurements were performed preferably for the reflection types {200} and {220}. Each reflection position was centered by step scan measurements in ω and χ. The interplanar lattice spacing was analysed at fixed 2θ-angles by $\Delta\lambda$-scans which are enabled by the combined movement of the double monochromator system (graphite (002)) and the analyser (graphite (002)). This measuring procedure allows the measurement of the stress induced shift of different Bragg types (hkl) with almost the same experimental accuracy. The grains were investigated at three stress levels within the elastic deformation region ($\sigma_{ext} = 0$; 340 MPa, 510 MPa). The comparison of the interplanar lattice spacings of Friedel-equivalents (d_{hkl} and $d_{\overline{hkl}}$) indicates that $\Delta d_{hkl}/d_{hkl}$ could be determined for most of the reflections with an accuracy of ± 0.1 /°°.

The strain values ε'_{33} (hkl) are calculated from the d_{hkl} values measured at the different stress levels σ_{ext} by means of a least-squares fit of the regression coefficient from d_{hkl} vs. σ_{ext}. The strain tensor ε_{exp} is refined including the whole set of reflections for the applied stress of $\sigma_{ext} = 510$ MPa. Using the least-squares algorithm, a weighting factor of $1/\sigma^2$ was applied, where σ is the standard deviation of each measurement as obtained from the regression coefficient. The typical agreement between the experimental and calculated values of ε'_{33} (hkl) is presented for one grain in tab. 3.

TABLE 3. Comparison between the experimental and calculated strains ε'_{33}(hkl) (in 10^{-5})

h	k	l	ε(hkl)$_{cal}$	ε(hkl)$_{exp}$	$\sigma(\varepsilon$(hkl)$_{exp})$
2	0	0	314.90	327.00	14.00
0	2	0	- 133.82	- 130.00	3.00
0	0	2	- 102.96	- 107.00	3.00
2	2	0	95.21	91.00	3.00
0	2	2	- 113.69	- 113.00	4.00
2	0	2	100.35	104.00	3.00
2	-2	0	85.00	84.00	2.00
-2	0	2	111.59	112.00	1.00
0	2	-2	- 123.08	- 122.00	5.00

The relative position of the grains in specimen 1 and the tensor component ε_{11} in the direction of the tensile axis are shown in fig. 2. For comparison, the mean macroscopical strain tensor component ε_{11} ($\varepsilon_{11} = 240 \times 10^{-5}$) is shown as a reference plane.

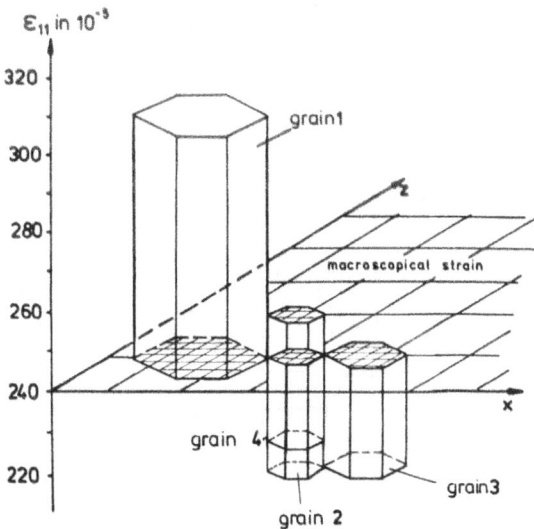

Figure 2. Comparison of the macroscopical and microscopical strain tensor component ε_{11}

The observed strain variations in the sample amount up to 30 %. The strong deformation of grain 1 is due to its special orientation where [100] the crystal direction is almost parallel to the applied stress direction. Thus grain 1 exhibits a larger longitudinal deformation than the mean value ε_{11}. The smaller grains are less deformed so that the deformations of the individual grains are almost averaged out over the four grains.

Considering the macroscopical strain value in tensile direction and the corresponding value for the free single crystal for grain 1 it is evident that in the matrix an intermediate value is realized (fig. 3).

270

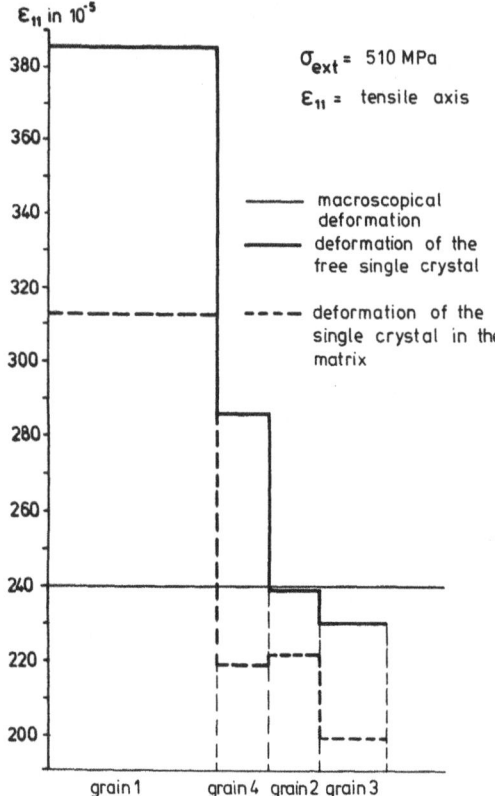

Figure 3. Strain tensor component ε_{11}

Due to the volume relationship and to the favourite orientation of grain 1, the neighbouring grains are compressed so that their elongations along the tensile axis are smaller than the free single crystal elongation and the mean value, respectively. The deformations of another set of grains (specimen 2) is shown in fig. 4.

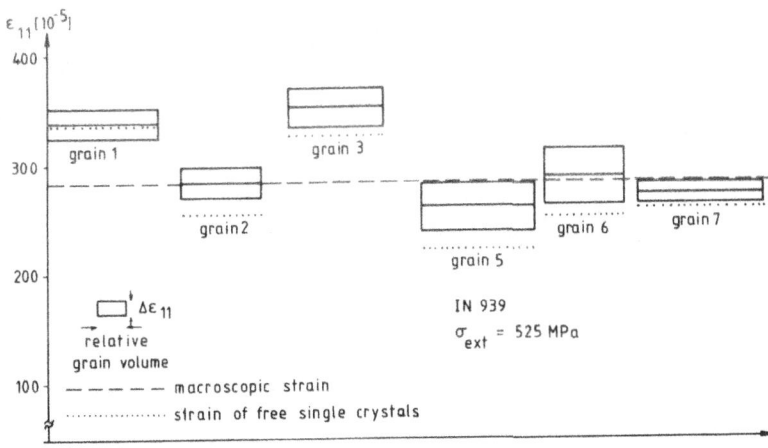

Figure 4. Strain tensor component ε_{11} for 6 grains

Here again it can be observed that the deformation in the two groups of investigated crystallites exhibit the tendency to homogeneous strains. In terms of calculated stresses (fig. 5) it turns out that the grains whose orientation lead to large deformation (grain 1 and grain 3) put an additional tensile stress on the grain 2 in between them.

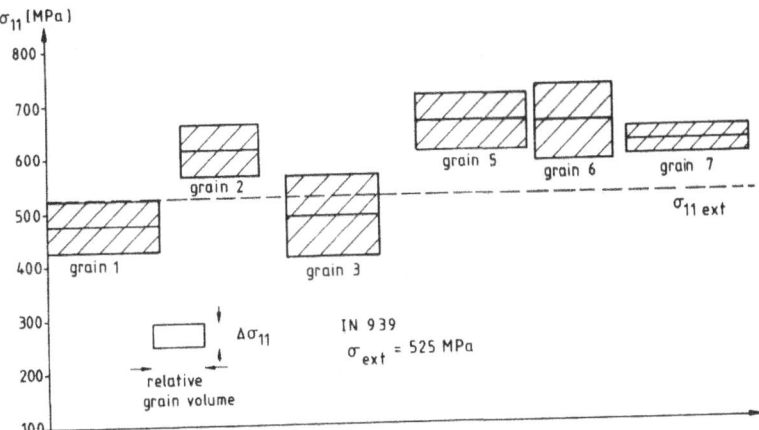

Figure 5. Stress tensor component σ_{11} in 6 grains

272

Whereas neutron diffraction gives integral deformation and stress values for the
crystallites under investigation, the strain-stress distribution over the grain diameter can
be studied by synchrotron radiation, whose parallelity allows a quasi pointwise
registration of the deformation state. First experiments performed at HASYLAB,
DESY, also on the Ni-base alloy IN 939 gave evidence for inhomogeneities over the
grain diameter (fig. 6), which are in the same order of magnitude as the variations
found from grain to grain.

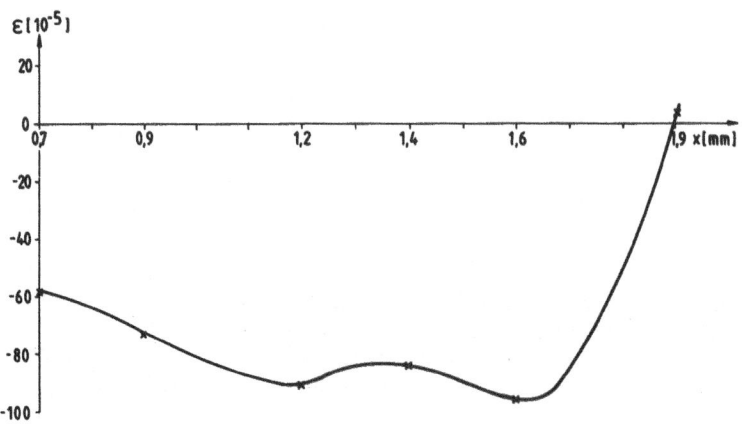

Figure 6. Strain distribution along the diameter of one grain. Synchrotron measurement
at the surface.

Further experiments are planned which are dedicated to the investigation of the strain
stress course near the grain boundaries.

3.2 Plastic deformation

The investigations performed with neutron diffraction in the elastic deformation region
were extended into the plastic deformation region measuring the same grains. The
plastic deformation of the sample material leads to significant changes in the rocking
curve (fig. 7). The intensity as well as the full width half maximum of the reflections
are affected.

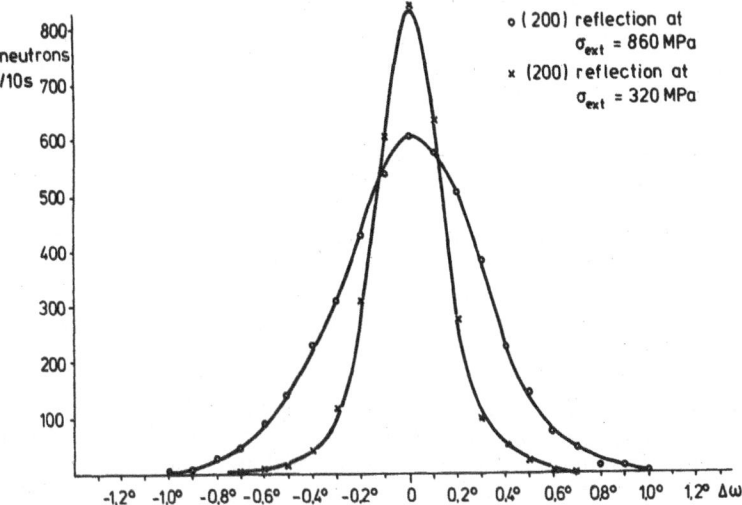

Figure 7. Comparison of the ω-profile of the (200)-reflection in the elastic and plastic deformation region (grain 1, specimen 1)

The measurements of the rocking curves were performed on the same reflections used for the elastic deformation analysis. Additional rocking curves using the q-configuration were measured on grain 1, 2 and 3 for specimen 1. The corresponding profiles were taken on the {200} reflections with <200> directions as axes of rotation. For the evaluation of the data the broadening of the reflection profiles in ω has been calculated by comparing the profiles in the elastic and plastic deformation region. The resulting tensor $\underline{B}$ describes the reflection broadening with respect to the crystal axes system. Thus, the diagonal tensor components B_{ii} give a measure for the broadening of reflections when rotating the crystal around the corresponding main crystal axis. It was found that the off-diagonal elements vanish within the experimental accuracy. The results obtained from the least-squares refinement of the ω-reflection broadenings measured in the symmetrical mode (ω = 2θ/2) were checked by measurements in the q-configuration which gives directly the values for the crystal main axes directions. Good agreement was found which demonstrates that the anisotropy of the plastic deformation can be analysed quantitatively. The increase in the full width half maximum values after plastic deformation is due to gliding processes. Following the Taylor-theory, each deformation of the crystal embedded in the matrix can be realized by 5 activated gliding systems. The corresponding combination out of the 12 gliding systems in the fcc-lattice is characterized by its minimal gliding sum. Usually, the calculation of the gliding sum is based on the macroscopical deformation of the sample material. In the present investigation, however, the more reliable experimental data for the microscopical deformations of single grains could be used. Thus, the inhomogeneous deformation behaviour of the grains is taken into account.

The calculations were carried out for all grains of specimen 1. Since the choice of the combination of gliding-systems is not unique, the gliding sums of all energetically equivalent combinations were added. The experimental reflection profile data always refer to observations in a plane perpendicular to a selected axis of rotation. The glidings, however, always take place in a special direction. Thus, for the comparison of the experimental data with the theoretically calculated glidings in the crystal, all glidings in the plane perpendicular to a main crystal axis had to be added. For crystallites with general orientations the so obtained gliding sums are in good agreement with the observed reflection broadenings . Only crystallites with a special orientation make an exception since here 6 or 8 gliding systems can be activated simultaneously. So, in most cases the procedure presented here offers the possibility to predict the plastic deformation behaviour of individual grains at small plastic deformation levels on the basis of experiments performed in the elastic deformation region.

Besides the analysis at a defined external stress level, also the onset and the development of the plastic deformation were followed. Fig. 8 gives evidence for a non-continuous development of the plastic deformation.

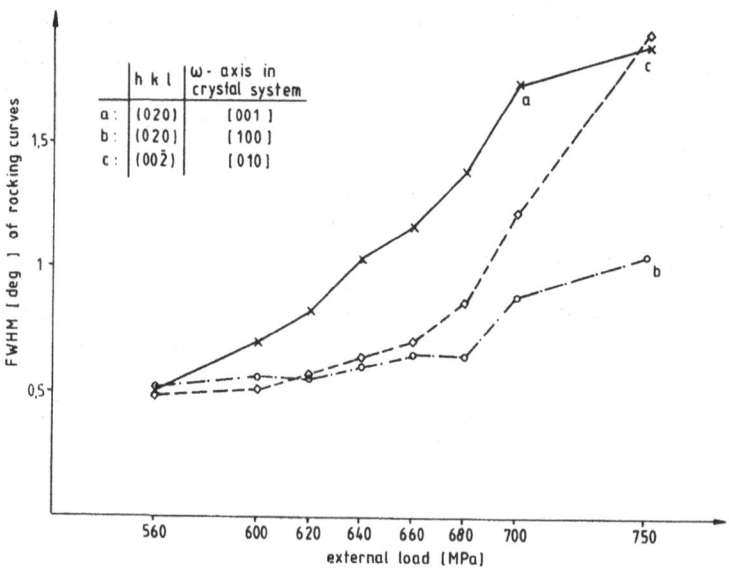

Figure 8. Reflection broadening in different crystal directions as a function of external load.

The reason for the observed plastic deformation behaviour has to be seen in relaxation processes in the crystallite and in its neighbourhood, so that the void where the crystallite is adapted to is not changing steadily. Fig. 9 compares the rocking curves of different crystallites.

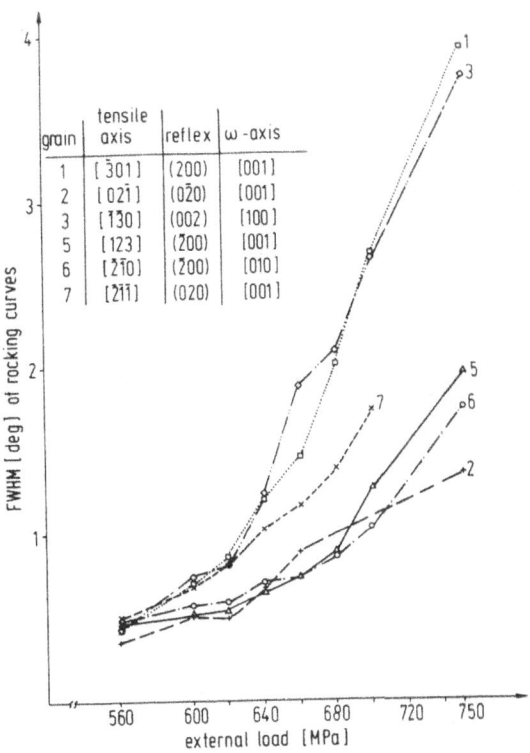

Figure 9. Rocking curves of different crystallites as a function of external load (specimen 2)

Whereas the plastic deformation in grain 1 and 3 (specimen 2) sets in at about 620 MPa, a significant increase in the reflection broadening for the grains 5 and 6 is delayed up to 650 MPa. Also, the inhomogeneity of the onset of the plastic deformation should be due as well to the orientations of the crystallites as to their coupling to neighboured crystallites and more detailed evaluations are under way.

First synchrotron measurements were made over the grain diameter in the plastic deformation region (fig. 10).

276

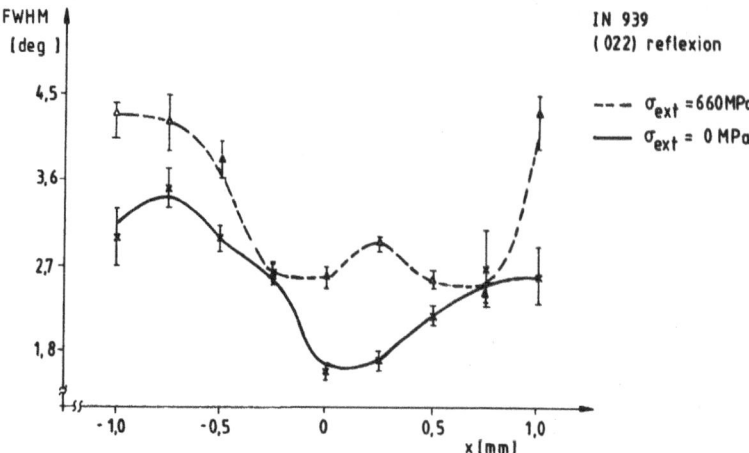

Figure 10. FWHM over the grain diameter. Synchrotron measurement at the surface.

The data, measured at LURE, Orsay, give evidence for a non-homogeneous deformation along the crystal diameter. Reasons herefor might be due to inhomogeneous elastic deformation states in the grain. Therefore, further experiments including elastic and plastic deformation analysis at the same measuring spots are preseen.

4. SUMMARY

The single grain measuring and evaluation technique can be used for residual stress analysis in large grained materials. Hereby, the definition of large grains depends on the intensity of source and, hence, on the radiation used. Using conventional X-ray generators, grain sizes down to $\emptyset \approx 40 \ \mu m$ can be measured whereas for neutron diffraction grain sizes of $\emptyset \approx 1$ mm are needed. Since the deformation state in individual crystallites is obtained, microstresses and, hence, stress maxima can be obtained. The macrostress state is then reproduced by averaging over several crystallites. The measuring technique is a tool for analysing the polycrystalline state on a microstructural level. In this framework information as well on the elastic as on the plastic deformation behaviour is accessible.

THE STRESS-FREE REFERENCE SAMPLE: ALLOY COMPOSITION INFORMATION FROM NEUTRON CAPTURE.

HANS G. PRIESMEYER
Institut für Reine und Angewandte Kernphysik
Universität Kiel c/o GKSS Research Center
POBox 1160
D-2054 Geesthacht FRG

ABSTRACT. Lattice constants often depend on the concentration of solutes in materials. The analysis of the gamma radiation following neutron capture can be used to detect variations of chemical composition. The measurements can be performed simultaneously with the strain measurements. Prompt gamma analysis (PGA) can detect hydrogen in the 30 to 50 ppm range. Electron-positron pairs are formed in the bulk material and positron annihilation radiation can be used to characterize plastic deformation.

1. Introduction

It is well recognized that three subfields in neutron diffraction for stress measurements need further investigation: the influence of texture, the treatment of grain size effects and the demand for accurate reference lattice spacings of the stress-free state. This paper deals with prompt gamma radiation following neutron capture to improve the reliability of the d_0-determination, which is in general affected by gradients in chemical composition of an alloy or solid solution. Neutron capture is a nuclear process that competes with neutron scattering. Therefore almost any sample investigated for strain in a neutron beam will at the same time absorb neutrons and subsequently emit gamma radiation whose energy characterises the capturing isotope and whose intensity varies according to the isotope's concentration. Neutron capture can lead to radioactivation of the sample, where the gamma radiation is emitted after a certain half-life of the nuclear decay or it can generate prompt gamma radiation which appears immediately after the capture process.

2. Correlation between lattice parameters and chemical composition

To determine the stress-free reference lattice spacing, it is not only necessary to identify a suitable volume element in the material, but also to make sure that its chemical composition will not be different from that in the volume sampled for stress. Generally in alloys and solid solutions the lattice parameters depend on the chemical composition and may often change considerably with changing concentration of any solute.

M. T. Hutchings and A. D. Krawitz (ed.),
Measurement of Residual and Applied Stress Using Neutron Diffraction, 277–284.
© 1992 *Kluwer Academic Publishers.*

278

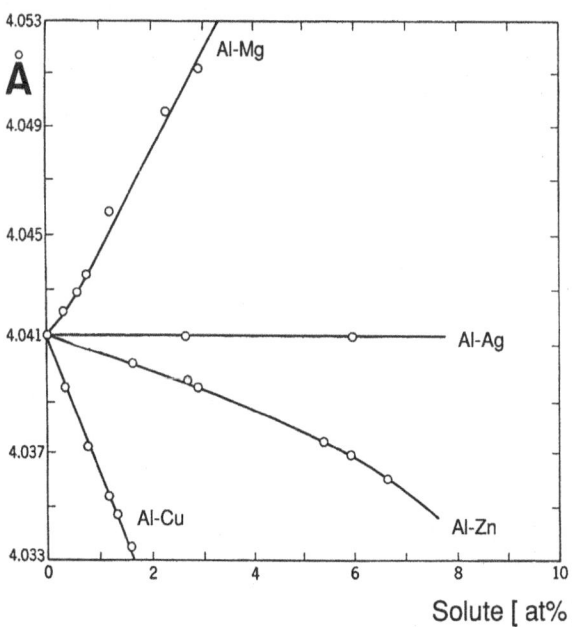

Fig.1: Changes of lattice spacings in solid solutions based on aluminum (adopted from H.J. Axon, W. Hume-Rothery (1948)).

Two examples shall illustrate this behaviour: The changes of lattice parameters in solid solutions based on aluminum have been investigated many years ago by Axon and Hume-Rothery (1948)(cf.Fig.1). For low concentrations of solutes (< 2-3 at%) the following relations can be derived:

$$a[nm] = 0.4041 + 0.00033 \times p \text{ at\%} \text{ for Al-Mg alloys}$$
$$a[nm] = 0.4041 - 0.00048 \times p \text{ at\%} \text{ for Al-Cu alloys.}$$

This implies that a 0.5 at% increase in solute concentration will change the lattice parameter (and thereby create a „virtual" strain) by a relative amount of + 4 x 10^{-4} for Mg and -6 x 10^{-4} for Cu.

The lattice sizes of steel are known to depend on the carbon concentration. They obey approximately the following relations:

Austenite: (fcc) $a[nm] = 0.3548 + 0.044 \times \text{wt\%}$
Martensite: (bc tetrag.) $a[nm] = 0.2861 - 0.0013 \times \text{wt\%}$
                       $c[nm] = 0.2861 + 0.0116 \times \text{wt\%}$

For the tetragonal-structured martensite a difference in carbon content will act on both the a and c lattice parameter. The linear relationship used to estimate the variation of lattice parameters with concentration of individual solutes is called "Vegard's Law" (cf.Barrett & Massalski 1980).

3. Experimental setup for prompt gamma analysis (PGA)(Bittorf 1990)

The high-resolution neutron time-of-flight spectrometer FSS at the FRG-1 was additionally equipped with a gamma spectrometer, using a 100 cm³ p-type closed-end HPGe detector (ORTEC GEM series) and ORTEC Type 672 spectroscopy amplifier. Data were taken using a stand-alone multichannel buffer (ORTEC 918 A) connected to an HP Vectra personal computer. The ADC resolution was 8192 channels in order to cover the whole range of expected gamma ray energies. The neutron beam was confined to 5 x 5 mm². The detector covered a solid angle of approx. 4 x 10⁻³ sr. The neutron flux at the sample position was measured by goldfoil activation and resulted in 1.3 x 10⁷ n/cm² s. Data evaluation was done using the model 918 A software after corrections for neutron self-absorption and gamma ray absorption had been made. To determine the concentrations, two methods have been applied: the peak-sum method (Heurtebise, Lubkowitz 1976) and the use of standards.

4. Results

A broad spectrum of different materials has been investigated in order to study the sensitivity of the method. To keep the error on the stress-free lattice spacing low, it may be necessary to know concentration gradients within 0.1- 0.2 at%. Fig. 2 shows the high-energy part of a typical gamma spectrum. The AlCu60 sample was a wire with a diameter of 0.8 mm in this case. Besides the full energy peaks single and double escape peaks can be seen. The number of peaks and the background may be reduced by pair spectrometry and coincidence methods.

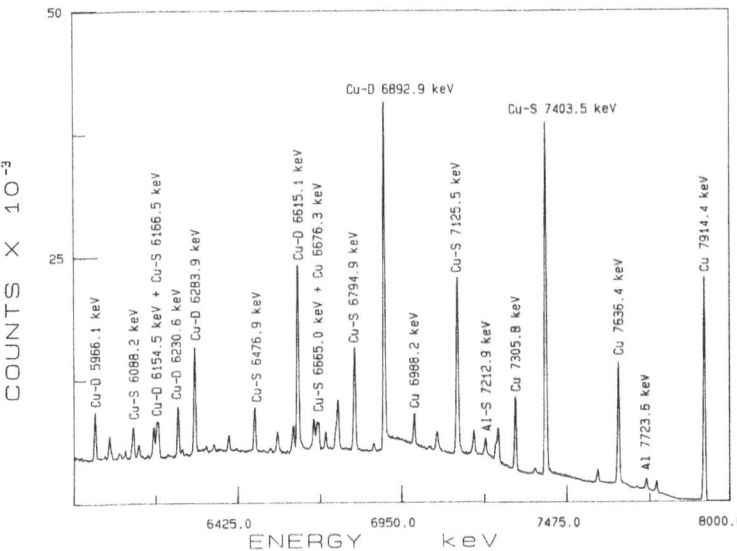

Fig. 2: Neutron capture prompt-gamma spectrum of a AlCu60 sample. The concentration of Cu was measured to be 55.7 at% instead of 60 at%.

The following tables show the composition of the Ni-base superalloys INCONEL 718 and WASPALOY compared to the values given by the manufacturer. The relative errors in PGA can be kept in the 1 % range for abundant solutes but will be an order of magnitude higher for the less abundant. Effects of target shape and neutron scattering on element sensitivities have to be considered (Mackey et al, 1991).

Table 1: Composition of INCONEL 718

Element	PGA (at%)	chemical analysis (at%)
Ni	49.4 ± 0.5	52.2 ± 4.2
Cr	21.7 ± 0.2	21.3 ± 2.6
Fe	20.6 ± 0.3	20.2 ± 5.6
Nb	3.9 ± 0.8	3.3 ± 0.3
Ti	1.8 ± 0.1	1.1 ± 0.3
Mo	2.6 ± 0.3	1.9 ± 0.2

Table 2: Composition of WASPALOY

Element	PGA (at%)	chemical analysis (at%)
Ni	58.0 ± 0.4	58.9 ± 4.8
Cr	24.6 ± 0.3	21.7 ± 1.9
Co	9.9 ± 0.2	13.3 ± 1.6
Ti	6.1 ± 0.2	3.6 ± 0.3
Mo	1.3 ± 0.4	2.6 ± 0.5

To calculate the concentrations of elements the most intense high-energy lines were taken, since the errors depend on the number of counts in a peak and the degree of interference with other lines.

Table 3: Sensitivity for different materials (in gammacounts/sec at% sr at a neutron flux of $10^6 n/cm^2$ sec)

Element	energy of prominent gamma ray used for PGA [keV]	sensitivity [cts/sec at% sr]
H	2223.1	1.09
C	4945.3	0.014
Al	7724.5	0.039
Ti	1381.7	9.50
V	7163.9	1.61
Cr	8884.8	1.18
Mn	7243.9	3.68
Fe	7631.1	1.26
Co	7212.5	3.10
Ni	8999.4	0.52
Cu	7914.4	0.62
Y	6081.0	0.58
Zr	934.1	0.35
Nb	294.3	1.32
Mo	778.1	3.10
W	6190.5	0.66

5. Hydrogen determination

Among the isotopes which can be characterized by PGA, hydrogen is of special interest to materials science. The presence of hydrogen in carbon steels may favor the growth characteristics of pores by methane formation. Molecular hydrogen at grain bounderies will lead to embrittlement. The reference lattice spacing may be affected by these influences. Total hydrogen contents in a number of samples have been determined relatively to a Zircaloy standard containing 400 ppm hydrogen. The sensitivity of PGA depends on the geometry of the sample, the neutron flux and the matrix material. For a zirconium matrix the detection limit attained was 30 ppm or 0.26 at%. Fig. 3 shows the characteristic 2223.3 keV capture gamma ray of the $1H(n,\gamma)2H$ reaction.

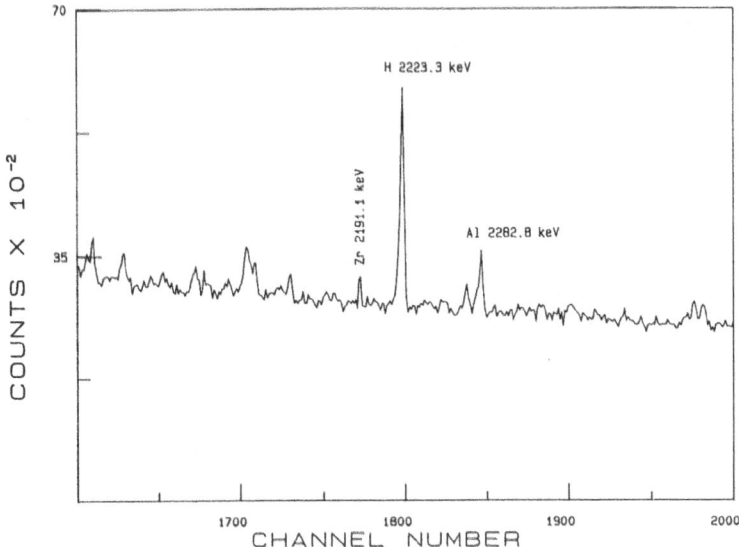

Fig. 3: Gamma ray spectrum of Zircaloy containing 400 ppm of hydrogen.

6. Pair production and annihilation

It has been shown (Allen et al.,1988) that the line width of the positron annihilation radiation is correlated to the plastic deformation of a sample, because of its sensitivity to defect concentrations. The energy of prompt radiation following neutron capture in most cases is above the threshold for electron/positron pair production. The cross section of this process depends on the atomic number Z of the target material and the gamma ray energy E: $\sigma \sim Z^2 \cdot \ln E$

The PGA experiments have shown that positrons are produced even in the lightest alloys (H.G. Priesmeyer et al. 1991). The energetic positrons are slowed down within a distance of $10 - 50\,\mu m$, so that the annihilation radiation is generated in almost the same sample volume where the antiparticle was created. Analysis of the Doppler broadening of the annihilation line nondestructively yields information about the bulk of the material. A pilot experiment using ARMCO iron samples with different plastic deformations has shown that in agreement with the general picture of positron annihilation the line shape can be approximated by two Gaussians (Fig.4). This corresponds to annihilation with both conduction band – and core-bound electrons and is equivalent to angular correlation results.

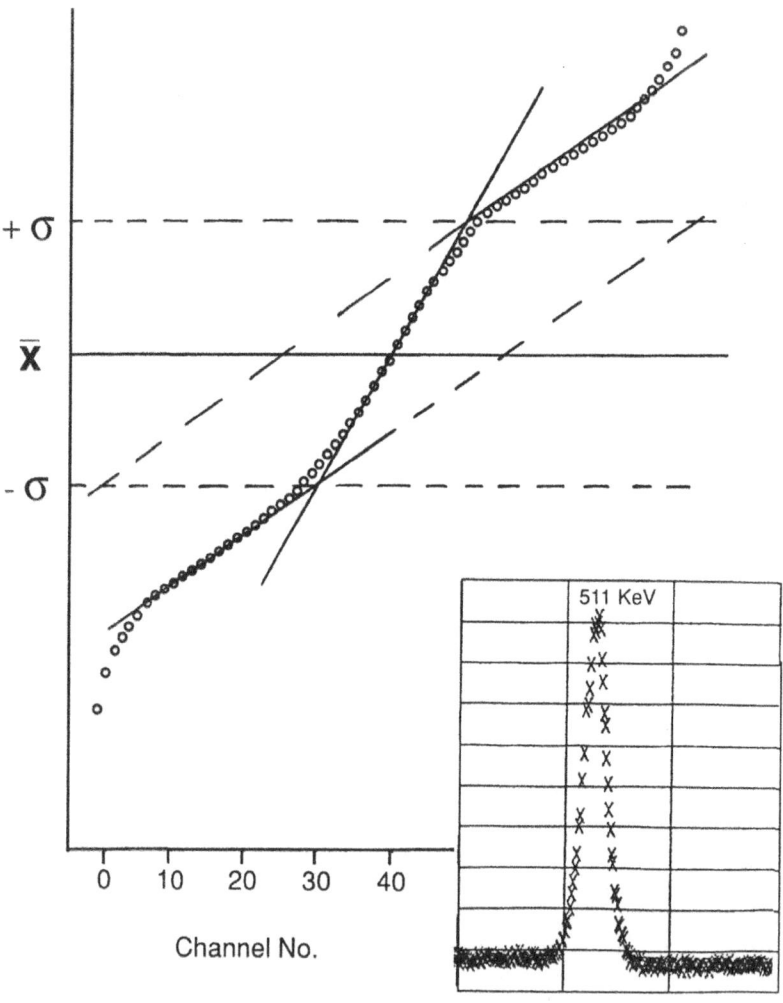

Fig. 4: Normalized integral distribution of annihilation peak, showing the presence of two overlapping Gaussian distributions.

As can be seen from first results presented in Fig. 5, the variation of the S-parameter with plastic deformation found with in-situ positron production (lower part of the figure) shows a similar behaviour as with positrons from an external source (upper part of the figure). Using external sources, only surface information can be gained, while in-situ produced positrons sample the bulk material non-destructively.

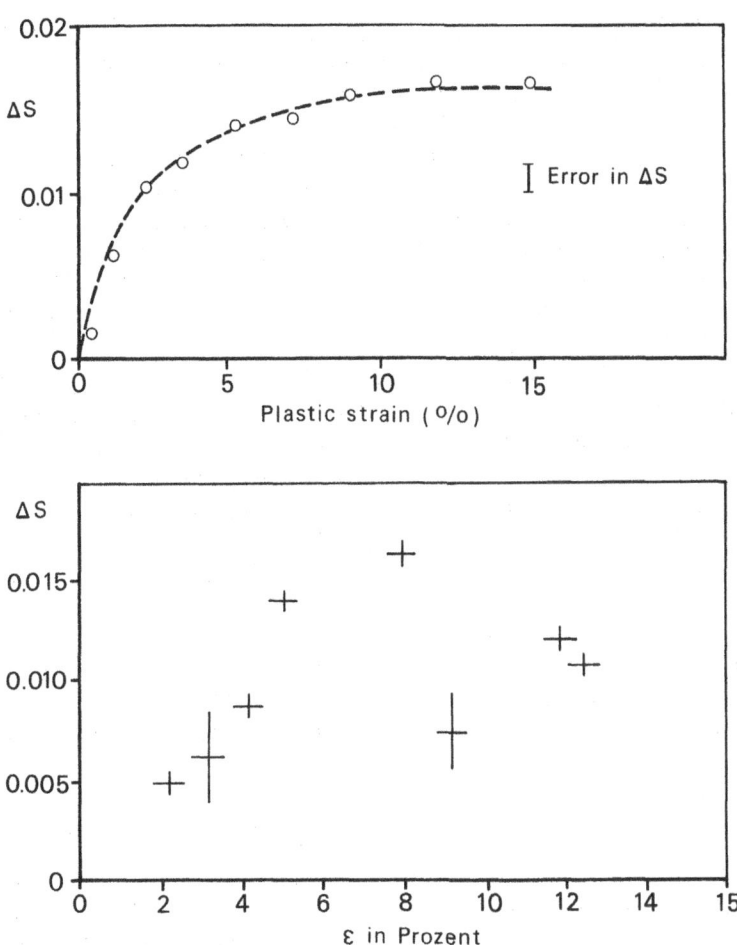

Fig. 5: S versus plastic deformation (upper part Allen at al. (1988), lower part Wiener (1991))

7. References

Allen, A.J., Coleman, C.F., Conchie, S.J., Smith, F.A. (1989), 'Applications of positron annihilation to the monitoring of fatigue damage and creep in technological components', Mat. Res. Soc. Symp. Proc. 142, pp. 131-142

Axon, H.J., Hume-Rothery, W. (1948), 'Proc.Roy.Soc. A193', referred to in 'Structure of Metals', Ch. Barrett, T.B. Massalski (1980)

Barrett, Ch., Massalski, T.B. (1980), 'Structure of Metals', 3rd edition, Pergamon Press

Bittorf, Ch. (1990), 'Prompte Neutronen-Einfang-Gamma-Analyse der Zusammensetzung von Werkstoffen'(Prompt neutron capture gamma analysis of materials with special emphasis on hydrogen, in German), Thesis University of Kiel, Germany 1990

Heurtebise, M., Lubkowitz, J.A. (1976), 'Determination of Metals in Alloys by Neutron Capture Gamma Ray Spectrometry', J. Radioanalyt. Chemistry 31, p. 503

Mackey, E.A., Gordon, G.E., Lindström, R.M., Anderson, D.L. (1991), 'Effects of Target Shape and Neutron Scattering on Element Sensitivities for Neutron-Capture Gamma-Ray Activation Analysis', Anal. Chem 63, (1991), pp. 288-292

Priesmeyer, H.G., Wiener, G., Bittorf, Ch., Mai Van Nhon (1991), 'Combined neutron scattering, neutron capture gamma ray and positron annihilation studies on materials under elastic and plastic deformation', Mechanical Behaviour of Materials-VI, Vol. 4 (Conf. Proc. ICM-6, Kyoto Japan), M. Jono, T. Inoue, ed., pp. 723-728

Wiener, G. (1991), 'Untersuchungen an plastisch deformierten Eisenproben mit Neutronenbeugung und Positronenvernichtung'(Investigations on plastically deformed iron samples using neutron diffraction and positron annihilation,in German), Thesis, Kiel University

THE PRECISION OF PEAK POSITION DETERMINATION IN DIFFRACTION MEASUREMENTS OF STRESS

COLIN G. WINDSOR
National Non-Destructive Testing Centre
AEA Industrial Technology,
B521, Harwell Laboratory, OX11 ORA, UK.

ABSTRACT. An analytic study is made of some of the various factors which lead to errors in the determination of diffraction peak positions. In particular the effects of scan width, of peak shifts, of resolution, of point spacing, of background, and of run time, are investigated. So as to be of use in residual strain determination, the results are presented in terms of the changes in macrostrain error when the scan parameters are changed at a constant overall scan time. All results are given in relation to a typical neutron diffraction scan for stress determination. Several conclusions are drawn. If peak positions are roughly known in advance, the run time may be reduced by a factor of 2 by a peak scan of only 1.5 peak widths (FWHH). Negligible error results from miscentring, as long as the peaks are centered in the scan to within 0.7 of the full width. Perhaps most importantly, the balance between count rate and resolution should often be drawn more towards increased intensity.

1. Introduction

This paper attempts to provide some answers to questions which were raised during discussion during the workshop. Our common objective was to obtain the most accurate possible strain values in a given total measuring time. A typical scan was shown by Dr Andrew Allen at the meeting, and is copied in figure 1[1]. Is such a curve the best compromise between resolution and intensity? To what degree should the scan be concentrated only on the vicinity of the peak? If the scan is limited in this way, does it matter if the peak centre is shifted away from the centre of the scan? Should an optimised instrument have high resolution, or high intensity? Does this level of background matter? How important are the number of individual points used to make up the scan?

There is no need to speculate on these matters, for precise answers are available. What is needed is to pose the questions correctly, and to evaluate the answers in such a way that firm conclusions can be drawn. In fact there is no need to perform actual experiments in order to obtain the necessary results, nor even to perform actual fits to simulated data. Both these courses of action introduce statistical errors into the results that tend to swamp the small trends being investigated. Here a semi-analytic method is used which gives the important parameter, the statistical error on the line position, with essentially the same result as a real fit to real data, but without the statistical fluctuations inherent in real fits.

M. T. Hutchings and A. D. Krawitz (ed.),
Measurement of Residual and Applied Stress Using Neutron Diffraction, 285–296.
© 1992 *UKAEA.*

2. The typical strain scan and its simulation

The neutron diffraction spectrum in figure 1 represents a typical scan used in strain determination. It has a signal to noise ratio of about 5. Its full width at half height (FWHH) is about 0.6 degrees. The scan extends to about 2 full widths in each direction. The point spacing was 0.1 degrees, so that the scan contained 31 individual points. As with most diffraction profiles on steady state sources it is well fitted by a Gaussian distribution lying on a flat background level. Quantitatively the peak itself may be described as an intensity y as a function of angle x, by 3 parameters p_1 to p_3. The peak will be assumed to be measured over a scan which can be defined by a further 3 parameters p_4 to p_6. The six parameters assumed to describe the peak and scan are therefore:

p_1 The peak centre position
p_2 The constant background
p_3 The full width at half height
p_4 The overall run time of the scan
p_5 The number of points in the scan
p_6 The total width of the scan

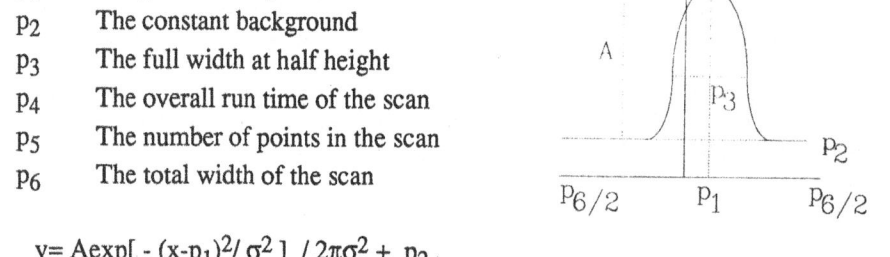

$$y = A\exp[-(x-p_1)^2/\sigma^2]\,/\,2\pi\sigma^2 + p_2. \tag{1}$$

σ represents the standard deviation of the Gaussian distribution. The assumed width parameter will be the full width at half height (FWHH) given by

$$p_3 = 2\,(2\log_e 2)^{1/2}\,\sigma = 2.35482\sigma. \tag{2}$$

The scan over the peak will be assumed to be composed of p_5 equally spaced points going from an initial angle of $-p_6/2$ to a final angle $p_6/2$.

The peak intensity, A, is proportional to the total run time divided by the number of points in the scan p_5. An assumption must be made on how the count rate varies with the resolution. The theory of Cagliotti[2] extended by Cooper and Nathans[3] enables the count rate of a diffractometer to be estimated for given values of the collimations and monochromator mosaic spreads. However the question remains as to which values give the optimal count rate for a given resolution. This question was addressed at the conference by Margaca[4]. The standard discussion was by Kalus and Dorner[5] who showed that the optimal configuration, with maximum intensity for a given resolution, was with the incident and scattered collimations equal and the mosaic spread large. Over 80% of this maximum is given when the collimations are both equal to the mosaic. Under either of these conditions the final count rate is inversely proportional to the square of the

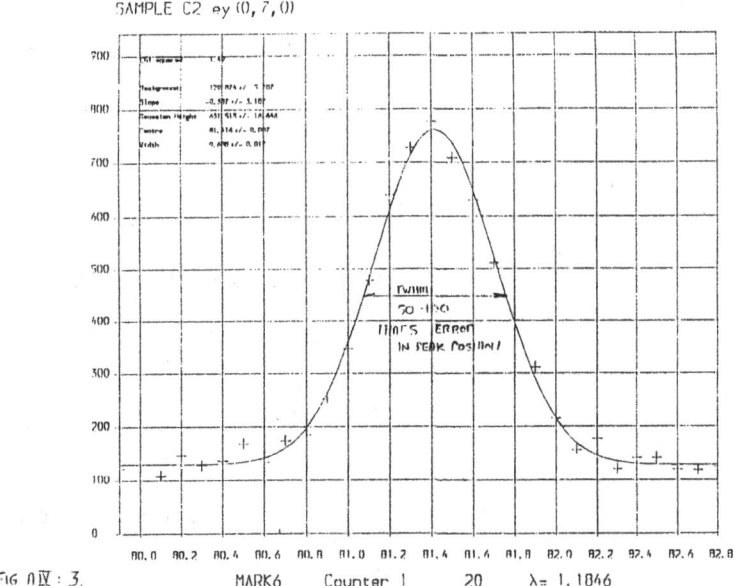

Figure 1 The neutron diffraction scan used as an illustration of a typical stress measurement by A J Allen in his talk at the Workshop[1]. The experimental points are denoted by crosses equal to the error bars. The peak has a full width at half height of 0.68 degrees. The fitted Gaussian profile shown by the continuous line gives an error in the peak position of 0.007 degrees, around 1/100th of the full width. The fitted accuracy corresponds to a typical macrostrain error $\delta d/d = \cot\theta \delta\theta$ of 71×10^{-6}.

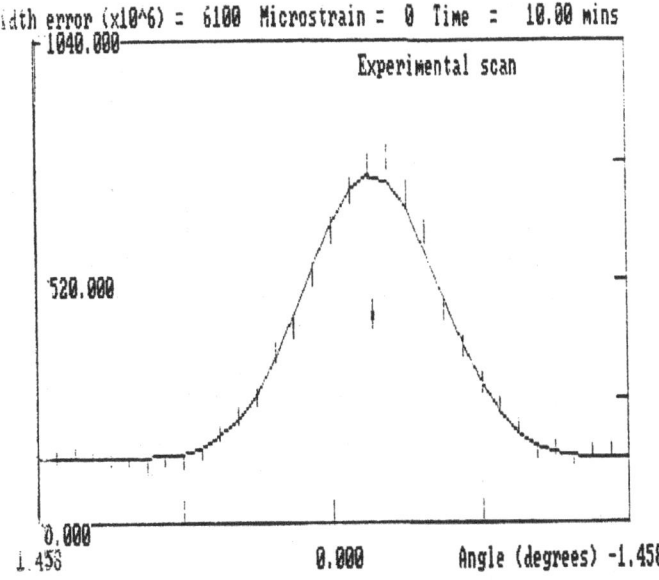

Figure 2 The simulated neutron diffraction scan to be compared with figure 1. Simulated experimental points are shown by the error bars. The peak has a full width at half height of 0.61 degrees. The fitted Gaussian profile shown by the continuous line gives an error in the peak position of 0.007 degrees, again around 1/100th of the full width. The fitted accuracy corresponds to a typical macrostrain error $\delta d/d = \cot\theta \delta\theta$ of 71×10^{-6}.

resolution[6]. With this assumption the expression for the intensity, A, in terms of the resolution p_3 is

$$A = p_4 / (p_5 \cdot p_3{}^2). \tag{3}$$

The full line in figure 2 shows a simulated diffraction peak with a full width at half height of 0.61 degrees, compared to 0.68 degrees in the experimental run. The intensity is 747 compared with 760, and the background 136 compared with 130. The simulated scan contained 33 points with a spacing of 0.091 degrees, compared with the experimental scan of 31 points with a spacing of 0.1 degrees. Thus in all important aspects the simulated run resembles the typical experimental run quoted by Allen[1].

Our method is able to evaluate the mean positional accuracy of such a scan having given parameters without the statistical fluctuations inherent in assuming any one set of points making up the scan. The points with error bars in figure 2 show such a set consistent with the parameters of the full line figure 2. They are evaluated from the smooth calculated curves by evaluating for each abscissa point the standard deviation equal to the square root of the ordinate. A deviation consistent with this error can then be evaluated. Such simulated points give a valuable visual impression of the effects of the parameters, and appear as insets in the figures. However they have no effect on the calculated positional accuracy which is a function only of the line parameters, as represented by the smooth curve in figure 2.

The error in any fitted parameter is related to the change in that parameter which doubles the chi squared parameter which defines the goodness of the fit. For a typical scan of measured experimental points $y_i{}^{exp}$ fitted to a calculated profile $y_i{}^{cal}$, the chi squared parameter χ^2 is given by

$$\chi^2 = (1/n) \, \Sigma_{i=1,n} \, (y_i{}^{exp} - y_i{}^{cal})^2 / (\Delta y_i{}^{exp})^2 \tag{4}$$

where $\Delta y_i{}^{exp}$ is the error in the experimental intensity[7]. This experimental error is simply related to the square root of the number of counts $\Delta y_i{}^{exp} = (y_i{}^{exp})^{1/2}$. In the present analytic calculation, the "experimental" intensity will be simulated by defining $y_i{}^{sim} = y_i{}^{cal} + \Delta y_i{}^{cal}$, where $\Delta y_i{}^{cal} = (y_i{}^{cal})^{1/2}$, and so equal to the calculated statistical error. It follows that for any form of the calculated profile the χ^2 function is equal to unity,

$$\chi^2 = (1/n) \, \Sigma_{i=1,n} \, (y_i{}^{cal} - y_i{}^{cal} + \Delta y_i{}^{cal})^2 / (\Delta y_i{}^{cal})^2 = 1. \tag{5}$$

However when the calculated peak centre is displaced by some small displacement dx, there will be two contributions to the deviation between y_i^{sim} and y^{cal}, a statistical error as before and an extra deviation caused by the change in the calculated profile. This second term is proportional to the absolute value of the gradient of the calculated peak shape

$$y_i^{sim} = y_i^{cal} + (y_i^{cal})^{1/2} + (\delta y_i^{cal}/\delta x)\delta x, \tag{6}$$

so that

$$\chi^2 = 1 + (1/n) \Sigma_{i=1,n} (\delta y_i^{cal}/\delta x)\delta x)^2 / (\Delta y_i^{cal})^2. \tag{7}$$

The χ^2 function doubles at the displacement δx where

$$\delta x = 1 / \{(1/n) \Sigma_{i=1,n} (\delta y_i^{cal}/\delta x)^2 / y_i^{cal}\}^{1/2}. \tag{8}$$

This formula is that used in all the subsequent analysis. It contains no non-analytic terms and so can be evaluated exactly for any calculated functional form whose differential is known. It is readily checked against the standard numerical fitting procedure. In the example of figure 1 the experimental profile was fitted with the least squares minimisation program PKFIT[8]. The analytically calculated error in the peak position parameter was $\delta x = 0.0068$ degrees compared to the standard error given by the fitting program of 0.007. Thus the analytic method is able to reproduce the results of the standard fitting procedure. This error corresponds to a strain error $\delta d/d = cot\theta d\theta = cot(81.4/2).(0.007.\pi/360) = 71.10^{-6}$. Thus the typical run has a macrostrain error of 71. In this study all results will be expressed in terms of this minimum macrostrain error.

3. What is the optimum width of the scan over the peak?

Typically neutron diffraction users have recorded scans, as in figure 1, which extend well beyond the region of the peak into the region of background. In contrast many X-ray users have only measured across the top of the peak itself. In the case of stress measurement, the actual background value is not important. Time spent measuring background is wasted compared with time spent close to the peak where the count rate is highest. But how large is this effect? How much does it matter?

Figure 3 is one of several figures which show the change in the accuracy of the macrostrain measurement error as a function of change in a parameter - in this case in the width of the scan at a constant run time. The abscissa in each case we consider will be the power of 2 by which the parameter in question is changed from its value for the standard scan. Thus an abscissa value of 0 will always correspond to the standard run and to a zero change in accuracy. The value of -2 corresponds to a scan width 1/4 of that of the

power of 2 by which the parameter in question is changed from its value for the standard scan. Thus an abscissa value of 0 will always correspond to the standard run and to a zero change in accuracy. The value of -2 corresponds to a scan width 1/4 of that of the standard scan and the left hand inset shows a typical scan at that width. It is seen that it covers only the region of the peak above the half height value. From the figure it is seen that the possible improvement in the macrostrain precision above its value in the standard run is around 30, or a 40%. The right hand inset shows a typical scan with 4 times the scan width. Some 80% if the scan is now in the background region and the error in macrostrain is increased by some 110 units. The full curve is seen to have a minimum at around -1.5 where the peak is measured between its half height positions. The error increases for still more restricted scans as the curvature of the measured region decreases.

The conclusion is that the scan width should be rather less than is typical in neutron diffraction - with a criterion that it certainly covers the positions of the peak where the intensity is half the maximum. The actual scan width will depend on the likely magnitude of the peak shifts discussed below.

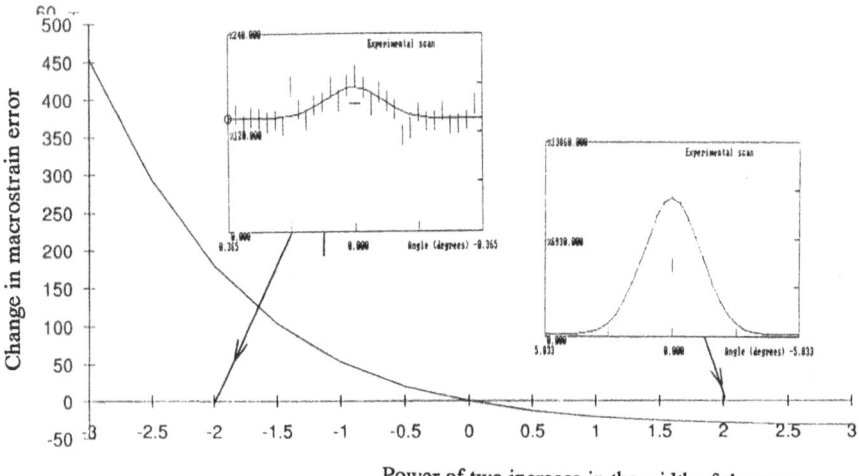

Figure 3. A graph of the change in the error in the measured macrostrain as a function of the total width of the scan. All macrostrain values are relative to the value 71 obtained using the standard scan of figure 2. The scan widths are also measured relative to the standard scan, and are shown as a function of the power of 2 increase over the standard run. Inset are shown typical scan corresponding to the values shown. The figure shows a minimum around where the scan extends to the half intensity position.

4. What are the effects of peak shifts on the position accuracy?

The argument often given against reducing the scan widths in diffraction strain measurements is that peak shifts necessarily occur during the measurement so that the peak is no longer centred in the scan. Figure 4 shows a set of results investigating the accuracy of the strain measurement as the standard scan is moved sideways across the measurement window. There is essentially no change in macrostrain accuracy until the truncated wing of the peak has risen to about 20% of the peak height. The macrostrain error is increased by an acceptable 10 (or when the peak shift is such that the truncated wing intensity is at the half height position). However the error rises rapidly at greater shifts.

The conclusion is that peak shifts have neglible effect on positional accuracy as long as the scan covers at least the portion of the peak above the half height intensity.

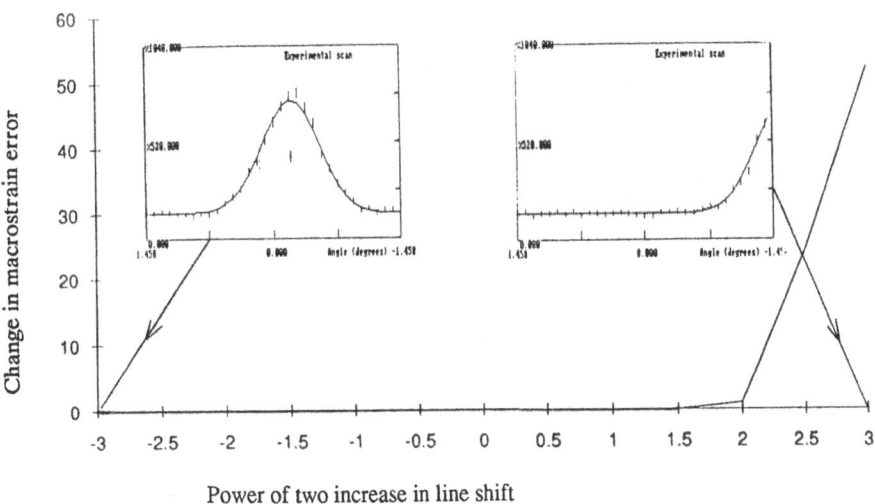

Figure 4. A graph similar to figure 3 of the change in the macrostrain error as a function of shifts in the position of the peak in the scan. The peak shifts increase in power of 2 intervals. The figure shows that peak shifts have essentially no effect until the peak is truncated above the trailing side half height position.

5. What is the optimal experimental resolution?

In any diffraction strain experiment there is always a compromise to be made between intensity and resolution. Resolution can generally be increased by installing angular collimators of lower divergence, and by reducing the monochromator mosaic spread, but these actions inevitably decrease the count rate and lead to an increased statistical error per point over a given run time. When is the statistical error in the macrostrain measurement a minimum for a given source and total run time? Figure 5 shows the change in macrostrain accuracy for this condition. As the resolution is increased the

292

accuracy actually decreases - improvement in statistics is more effective than improvement in resolution. However this analysis assumes a constant background, and the improvement is nullified if a background proportional to the count rate is assumed.

No clear conclusion can therefore be drawn. Improvement in resolution does not always improve the macrostrain accuracy, and the degree to which resolution can be relaxed depends on the variation of the background with resolution. Other instrumental effects such as peak overlap may make the accuracy decrease if the resolution is too coarse.

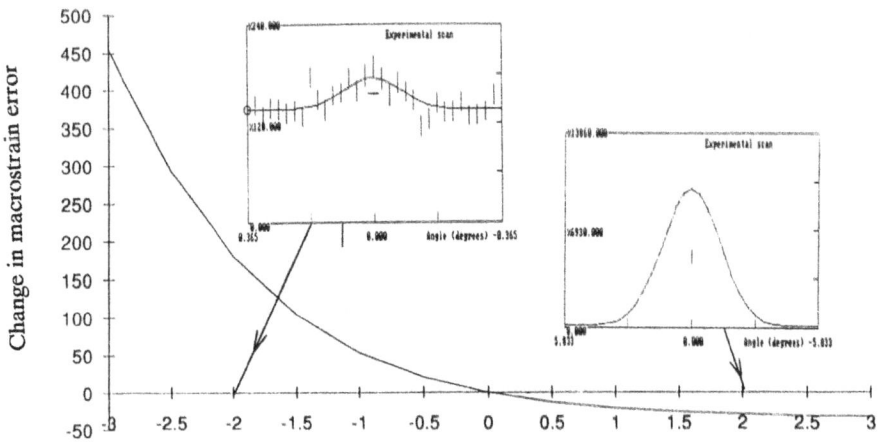

Power of two increase in line width at constant scantime

Figure 5. A graph similar to figure 3 of the change in the macrostrain error as a function of the experimental resolution. All curves are for a constant total run time, and the intensity of the incident neutron beam has been adjusted to be proportional to the square of the linewidth, as predicted for an optimised configuration. The figure shows that small linewidths are by no means always desirable in this type of experiment, and that greater precision in the line position may often be obtained by relaxing the resolution.

6. What is the optimum number of points in the scan?

Most spectrometers can be readily set up to include any number of points over a given scan. For a scan of given total run time, there is always a compromise to be made. Too many points and the statistical error for each point will be high, too few points and the shape of the profile will be poorly defined. In addition there will be generally be a time lost in setting up each point, proportional for the scan to the total number of points. although this is neglected in this analysis. Figure 6 shows the effects on the macrostrain accuracy of changing the number of points at constant run time. The graph is essentially flat, with no change being seen until the number of points has been reduced below 9. If setting up time is taken into account, then the scans with fewer points will be distinctly favoured. However if line assymetries may occur the definition of the line profile may then not be possible. The possibility of noisy points should also be considered. If a brief period of electronic or real noise spoils a given point, this will more easily be identified and removed if the counting time is short.

The conclusion is that the typical diffraction scan usually contains more than the optimum number of points.

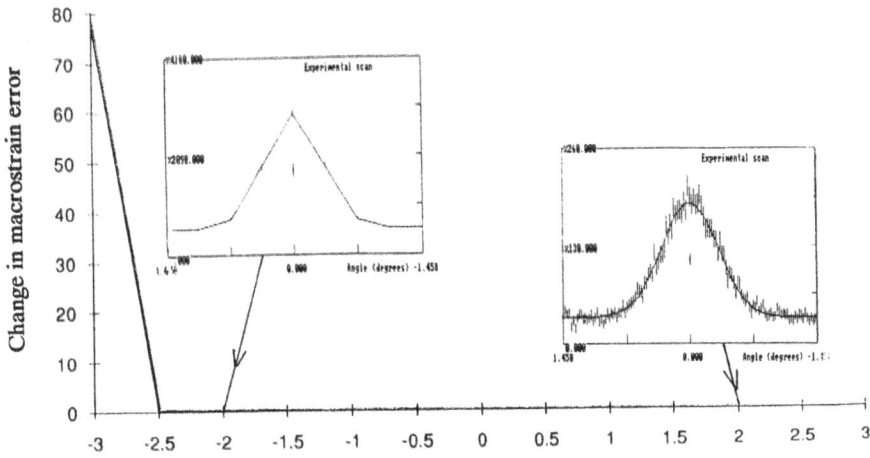

Power of two increase in the number of points at constant total width

Figure 6. A graph similar to figure 3 of the change in the macrostrain error as a function of the number of points in the scan. All curves are for a constant total run time. The figure shows that the precision of the line position is independent of the number of points in the scan over a wide range.

7. How important is the background level?

In most diffraction experiments the background level appears to be a small fraction of the peak height, but are its effects really negligible? It is always possible to reduce the background by extra shielding. Is this process worth the time and effort?

In some experiments, particularly those observing stress changes in minority phases, the signal intensity is low, so that the background effects become large. What values of the signal to noise ratio give acceptably accurate macrostrain accuracies? Figure 7 shows the effect on the macrostrain accuracy of changing the background level. The standard run had a signal to noise ratio of 4.5. Reducing the background by a factor 2 gives an improvement in macrostrain accuracy of 14 or 20%. A further reduction by a factor 4 is needed to improve it by 27 or 38%. Reducing it to zero improves it only to 35 or 50%.

In the other direction, the signal to noise ratio may be dropped to 0.45 before the macrostrain accuracy deteriorates beyond 100.

The conclusion is that in typical stress measurement diffraction it is worth ensuring that the signal to background is at least 10. Further reductions give increasingly less improvement and may not be worth while. Worthwhile stress measurements with macrostrain accuracies no worse than 50, or 70% of that of the standard run, can be achieved with signal to noise ratios of order unity. For better accuracy, longer run times must be used.

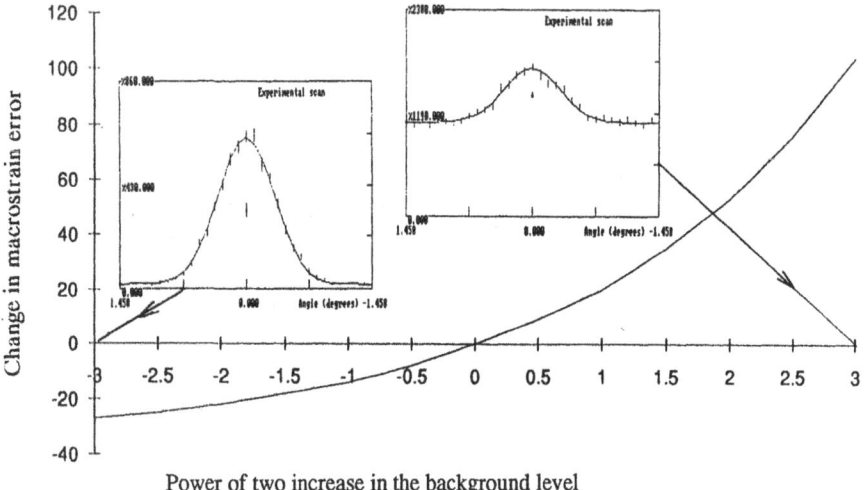

Figure 7. A graph similar to figure 3 of the change in the macrostrain error as a function of the background level. The figure shows that the precision of the line position can be improved significantly by decreasing the background level to around a signal to noise level of 10. Further improvement is marginal.

8. How long should the run be?

This is a mathematically trivial question since all statistical accuracies scale as the square root of the run time. However figure 8 serves as a calibration between macrostrain accuracy and run time for all the other figures. For example suppose it is desired to measure the macrostrain from a reflection with a 1.1 signal to noise ratio to the same accuracy that of the standard run. From figure 7 it can be read off that a macrostrain decrease of 53 is needed. From figure 8 it is seen that this improvement in accuracy corresponds to a run time increase of order 3 to around 30 minutes.

9. General conclusions

It has been shown that a careful consideration of the length and point spacing in stress diffraction experiment scans can lead to much improved accuracy, or to reduced run time for a given accuracy. In particular an optimum scan covering just the peak region above its quarter height intensity can reduce run times by a factor 2. Precise centering of scans about the peak centre is not necessary as long as the half intensity points are included. A preliminary scan to determine the centre of each scan may well be cost effective. Scans containing fewer points save time, space, and analysis time with negligible decrease in peak precision.

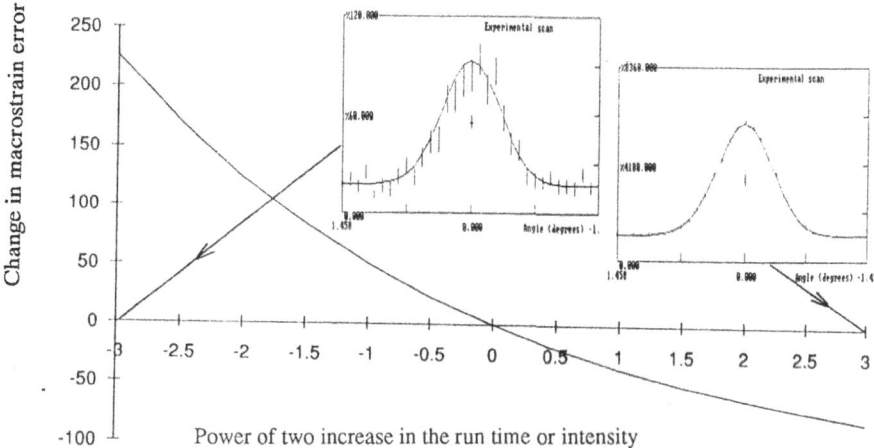

Figure 8. A graph similar to figure 3 of the change in the macrostrain error as a function of the total length of the run time. All errors decrease proportionally to the square root of the run time, as would be expected. The figure is useful in giving a calibration curve to figure 3 to 6 in terms of the change of run time which corresponds to a given change in macrostrain accuracy.

Since this paper was written, I have been able to view the thesis of Brant[9], who has made an investigation into certain of the scan parameters considered here by the numerical method of actual fitting to simulated data. The results presented are consistent with the present work.

This research was supported by the Corporate Research Programme of the UK Atomic Energy Authority.

References

[1]   Allen, A. J. (1991) "Errors in Analysis", This proceedings.
[2]   Cagliotti G., and Ricci, F. P. (1962) "Resolution and Luminosity of Crystal Diffractometers for Neutron Diffraction", Nucl. Inst. Methods **15**, 155-163.
[3]   Cooper, M. J, and Nathans, R. (1967) "The Resolution Function in Neutron Diffractometry", Acta Crystallogr. **23**, 357-367.
[4]   Margaca, F. M. A. (1991) "Optimised Geometry for a Stress Measurement Two Axis Diffractometer at a Reactor", This proceedings.
[5]   Kalus, J. and Dorner, B. (1973) "On the Use of In-pile collimation in Inelastic Neutron Scattering", Acta Crystallogr. **A29**, 526-528.
[6]   Windsor, C. G. (1981) Pulsed Neutron Scattering, Taylor and Francis, London.
[7]   Goodman, R. (1956) Teach Yourself Statistics, English Universities Press, London.
[8]   Hutchings, M. T. Lowde, R. D. and Tindle, G. L. (1986) "The Determination of S(Q,w) from Triple-axis Spectrometer Data", Neutron Scattering Data Analysis Workshop, Rutherford Appleton Laboratory. IOP Conference Series 81, 151-159.
[9]   Brand, P. C. (1991) "Stress Measurements by Means of Neutron Diffraction", Thesis, University of Twente, Report ECN-R--91-006, Netherlands Energy Research Foundation.

ERRORS IN ANALYSIS

A. J. ALLEN
National NDT Centre
AEA Technology
Harwell Laboratory
Didcot, Oxon., OX11 ORA, UK

ABSTRACT

The sources of error in the determination of residual stresses using neutron diffraction arise both from normal experimental uncertainties and from some inherent uncertainties in the nature of the measurements themselves. This presentation is aimed at provoking discussion firstly of errors and uncertainties in strain determination including:-

1. Errors on peak and multi-peak analysis;
2. The effects of the Ikeda-Carpenter function on the errors in pulsed source measurements;
3. Thermal, grain size, texture and compositional effects;
4. Sampling volume uncertainties in strain tensor determination (vs strain sensitivity);
5. Problems in establishing the true "zero strain" condition.

Secondly, the uncertainties introduced in transforming the strains into stresses must be considered. While use of Poisson's ratio and Young's modulus may suffice in many engineering applications, elastic anisotropy can complicate the transformation from lattice strain to true or engineering stresses. At best, there is uncertainty in the correct modulus and Poisson ratio values to be used. At worst, the full single-crystal elastic constant tensor must be considered. Grain-boundary interactions and slip mechanisms may also have an impact. Such effects are likely to be particularly important when residual stresses are to be determined from lattice strains for a material which is well into the plastic regime.

M. T. Hutchings and A. D. Krawitz (ed.),
Measurement of Residual and Applied Stress Using Neutron Diffraction, 297.
© 1992 *Kluwer Academic Publishers.*

5. INSTRUMENTATION
5a INSTRUMENTATION : STEADY STATE REACTOR

OPTIMIZED GEOMETRY FOR A STRESS MEASUREMENT TWO–AXIS DIFFRACTOMETER AT A REACTOR

F.M.A. MARGAÇA
LNETI/ICEN
Physics Department
EN 10
2685 Sacavém
Portugal

ABSTRACT. This paper reports a study on the design optimization of a two–axis diffractometer installed at a steady neutron source, dedicated to stress measurements. First, the requirements of both angular and spatial resolution are considered and their bearing on the design is discussed. Then, the general formulae for the instrumental broadening and luminosity of a Bragg peak is reviewed. The choice of the angular divergences of collimators and of the monochromator mosaic spread, which leads to the highest luminosity for a given resolution, is discussed for two different situations. Namely, for the setup at a neutron guide tube and at a beam tube.

1. Introduction

A powder diffraction pattern obtained with a two–axis neutron instrument, exhibits several peaks due to Bragg reflections, whose full width at half maximum (fwhm) and luminosity depend on the angular divergence of the collimators, on the monochromator mosaic spread and on the dispersion parameter which is determined by the relative Bragg angles at the monochromator and sample.

It is the purpose of this paper to present the most suitable choice for the values of these geometrical parameters, in order to optimize the luminosity for a given resolution, for a diffractometer dedicated to stress measurements.

First, the particular requirements imposed by stress measurements and their bearing on the diffractometer design are discussed. Then, the general fwhm and luminosity of the Bragg peaks is reviewed. Finally the optimized geometry for a stress measurement two–axis diffractometer is discussed for two different cases, namely, for installation at a neutron guide tube and at a neutron beam tube.

2. Stress Measurements by Neutron Diffraction

2.1. THE PRINCIPLE

The measurement of stress by neutron diffraction is based on the principle that small variations in the internal stress in the sample may be determined

M. T. Hutchings and A. D. Krawitz (ed.),
Measurement of Residual and Applied Stress Using Neutron Diffraction, 301–311.
© 1992 *Kluwer Academic Publishers.*

by measurement of the associated variations in the crystal lattice spacings, d. The strain is defined as the ratio between the small change in d and the stress free value d_o

$$e = \frac{d - d_o}{d_o} \tag{1}$$

Measurement of d can be made using Bragg's law

$$d = \frac{\lambda}{2 \sin \theta} \tag{2}$$

where λ is the neutron wavelength and θ is the sample Bragg diffraction angle.

The experimental method is to scan the counter angle and record the scattered neutron intensity as a function of scattering angle 2θ, for the Bragg peaks of interest, using a monochromatic incident beam. This should be done for both the stressed and stress–free states of the crystalline planes under study.

2.2. THE ANGULAR RESOLUTION

The Bragg peak position should be determined with an accuracy such that strains up to the yield point can be measured. This, for mild steel is 3×10^{-4}. In practice, it is, possible (Windsor (1990)) to determine the centre of a peak to an accuracy of $1/10$ of its full width at half height, F_ω. Thus, the resolution aimed for the stress measurements is that the Bragg peak fwhm should be

$$F_\omega \leq 3 \times 10^{-3} \, \text{rad} \cong 10' \tag{3}$$

In practice however, stress measurements have been performed at diffractometers with F_ω values ranging from 12' to 30', in D1A and Panda, respectively, at the ILL, Grenoble and Harwell (Hewat (1975), Hewat and Bailey (1976)).

2.3. THE SPATIAL RESOLUTION

In some experiments the aim is to measure the response of the whole sample to the stress. However in most experiments the objective is the measurement of the stress distribution as a function of position over the sample. For this type of work, well collimated "pencil" beams produced by cadmium masks, are required, to define a "gauge volume" within the specimen (Windsor (1990)). The sample is then translated in space, keeping the gauge volume constant, to measure the spatial changes in stress.

As the volume resolution is optimized when the incident and scattered beams intersect perpendicularly, a scattering angle at the sample of $2\theta = 90^\circ$ should be used.

A beam of angular divergence α projects a mask aperture of width ω, on a plane normal to the incident beam at the measuring point, with a size s given by

$$s = \omega + 2\ell \tan \alpha \qquad (4)$$

where ℓ is the distance from the mask to the measuring point.

Consider the usual situation where a mask of square aperture of 1 mm side is placed at the exit of a Soller collimator with 10' horizontal and 6° vertical divergences. If the mask is at a distance $\ell = 200$ mm, the illumined sample area at the measuring position will be 2×43 mm². However, for most experiments the sampled volume aimed for is approximately a cube of 2 mm side.

To be able to measure stress within a sample with a spatial resolution of $2 \times 2 \times 2$ mm³, the horizontal and vertical angular divergences should be equal to some value α for both the incident and scattered beams. A typical Soller collimator should not therefore be placed in the incident beam path. To define the incident beam collimation two cadmium (or boron) masks should be used instead. One fixed and positioned in front of the monochromator, with square aperture of side ω_M and at a distance ℓ_1 from the measuring point. The other should be placed as close as possible to the specimen, with square aperture side ω_s and at a distance ℓ_0 from the measuring point. This arrangement is shown in figure 1.

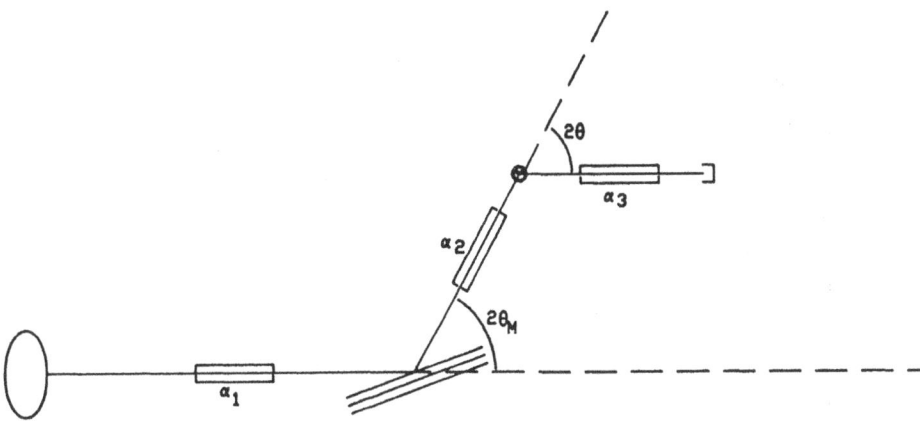

Figure 1. Incident beam collimator assembly.

To define the scattered beam collimation, a similar collimator should be used. However, a Soller collimator can be used in the scattered path as its large vertical divergence does not affect any longer the spatial resolution.

The beam collimation by such a system is

$$\tan \alpha = \frac{\omega_M + \omega_s}{2(\ell_1 - \ell_0)} = \frac{\omega_D + \omega_s'}{2(\ell_1' - \ell_0')} \qquad (5)$$

where ω_D, ω_s', ℓ_1' and ℓ_0' are the sides of the square apertures and the distances of the detector and sample cadmium masks, respectively, placed in the scattered beam. The square cross–section side s of the sampled illumined region is given by

$$s = 2\ell_1 \tan \alpha - \omega_M = \omega_s + 2\ell_0 \tan \alpha \qquad (6)$$

The full lines in Figure 1, which intersect at point P, represent the largest divergence allowed by the collimator. The sample mask position is therefore limited by the position of this point for a given collimation.

The distance from point P to the measuring point is given by simple geometrical optics as

$$\ell_P = \frac{s}{2 \tan \alpha} \qquad (7)$$

To obtain $s = 2$ mm using an incident beam of 10′ angular divergence, P will be at about 345 mm from the measuring point. However a beam with $\alpha = 5'$ would lead to $\ell_P \simeq 690$ mm. This means that a spatial resolution of 2 mm side can actually be obtained for relatively large samples if care is taken in the definition of the collimation, sample mask aperture and its position. This is illustrated in Table 1.

TABLE 1. Calculated values that should be used for ω_M and ω_s to obtain $s = 2$ mm for different collimation values α and different (specimen sizes) ℓ_0 distances, with $\ell_1 = 1$m.

α	5′	10′	12′	15′	18′
ω_M (mm)	0.9	3.8	5.0	6.7	8.5
ℓ_P (cm)	69	35	29	23	19
ℓ_0 (cm)	ω_s (mm) for $s = 2$ mm				
10	1.7	1.4	1.3	1.2	0.9
20	1.4	0.8	0.6	0.3	0
25	1.3	0.6	0.2	0	
30	1.1	0.3	0		
35	1.0	0			
40	0.8				
45	0.7				
50	0.6				

Table 1 shows that for a sampled volume resolution of $2 \times 2 \times 2$ mm³ to be achieved, the beam angular divergence should be $\alpha \leq 12'$ if engineering samples with linear dimensions greater than 25 cm are to be studied.

An identical situation occurs in the scattering path of the beam. The distance from to the measuring point being given by (7), then equation (6) holds provided ω_M, ω_s, ℓ_1 and ℓ_0 are replaced by ω'_D, ω'_s, ℓ'_1 and ℓ'_0, respectively.

3. Instrumental Broadening and Luminosity

3.1. GENERAL FORMULAE

Neutrons producing a Bragg peak in a powder diffraction pattern from a two–axis instrument must satisfy the Bragg law both at the monochromating crystal and at the sample. This leads to a correlation between the neutron wavelengths and the angular parameters defining the trajectories of the neutron.

Caglioti et al. (1958) developed general expressions for the full widths at half height of the Bragg reflections at any scattering angle and their luminosity in terms of the angular divergences α of collimators and of the mosaic spread β of the monochromating crystal. Later, these have been successfuly tested by experimental measurements (Caglioti et al. (1962) and Loopstra (1966)). The Bragg peak is characterized by a Gaussian profile whose full width at half maximum (fwhm) is

$$F_\omega = N/(\alpha_1^2 + \alpha_2^2 + 4\beta^2)^{1/2} \tag{8}$$

with

$$N = [(\alpha_1\alpha_2)^2 + (\alpha_1\alpha_3)^2 + (\alpha_2\alpha_3)^2 + 4\beta^2(\alpha_2^2 + \alpha_3^2)$$

$$- 4a\alpha_2^2(\alpha_1^2 + 2\beta^2) + 4a^2(\alpha_1^2\alpha_2^2 + \alpha_1^2\beta^2 + \alpha_2^2\beta^2)]^{1/2}$$

where $a = \tan\theta/\tan\theta_M$ with θ_M and θ being the Bragg angles at the monochromator and at sample respectively, appropriate to the wavelength λ. The dispersion parameter, a, is a measure of the relative dispersion undergone by the twice–reflected neutron beam for a given wavelength spread of the monochromatic beam. The collimator horizontal divergence is represented by α_i where $i = 1$, 2, 3 stands for the collimators installed, respectively, in–pile, between the monochromator and the sample and in front of the detector as shown in figure 2. The horizontal plane is that where the neutron beam axis lie and the sample Bragg angle is defined. Furthermore for a scattering angle of 90°, the angular spread of the detected neutrons is not dependent on the vertical divergences of the collimators. Thus, only the horizontal divergence of the collimators is considered.

The instrument luminosity, which is proportional to the total area of the Bragg peak, is given by (Caglioti et al. (1958)):

$$I = k \frac{\alpha_1\alpha_2\alpha_3\beta}{(\alpha_1^2 + \alpha_2^2 + 4\beta^2)^{1/2}} \; . \tag{9}$$

where k is a constant with respect to the instrumental parameters, α, β and a.

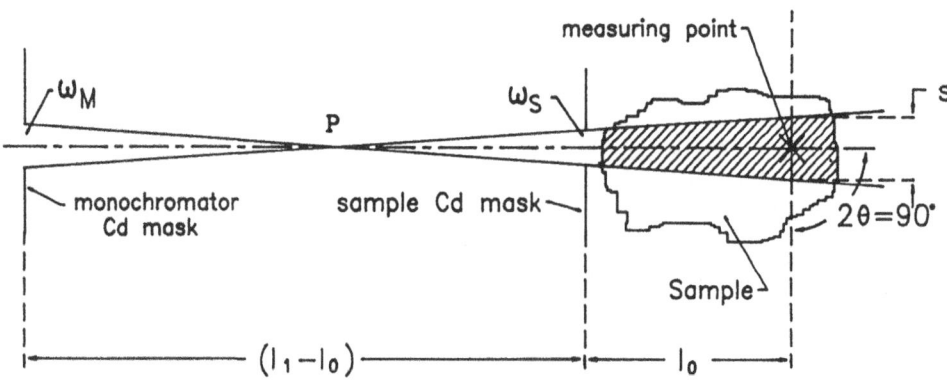

Figure 2. Powder two–axis diffractometer schematic lay–out.

Equations (8) and (9) show that both the resolution and the luminosity of a neutron diffractometer depend in a rather complicated way on the angular divergences α, of the collimators and on the mosaic spread β, of the monochromating crystal. The dispersion parameter a affects only the resolution. So, changing the value of a only changes the width of the Bragg peak, not its total area. Care must be taken therefore to choose the value of a which makes the peak sharper, i.e., that associated with the minimum value in F_ω.

3.2. INSTALATION AT A GUIDE TUBE

Consider a diffractometer installed at a neutron guide tube. The angular divergence of the transmitted beam is wavelength dependent. However, for the shorter transmitted wavelengths which are the most interesting for diffraction work, the beam divergence is approximately $\alpha_1 = 10'$ (Hewat (1976)). Taking into account that for good spatial resolution $\alpha_2 = \alpha_3 \cong 10'$ then, the installation of a stress neutron diffractometer at a guide tube is a particular case for which $\alpha_1 = \alpha_2 = \alpha_3 \equiv \alpha$. If $\beta = x\alpha$ where x is a real positive number, then equations (8) and (9) give

$$F_\omega = \alpha \left[\frac{3}{2} + 2a(a-1) + \frac{x^2}{1+2x^2} \right]^{1/2} \tag{10}$$

and

$$I = k \frac{\alpha^3 x}{(2+4x^2)^{1/2}} \tag{11}$$

The fwhm of the Bragg peak is minimum for $a = 1/2$. This corresponds to a monochromator take–off–angle $2\ \theta_M$ equal to $127°$ for a sample scattering angle of $2\ \theta = 90°$. Table 2 shows F_ω and I for different values of $2\ \theta_M$ and β in the case of the scattering angle of interest for stress measurements ($2\ \theta = 90°$).

The importance of the choice of the monochromator take–off–angle $2\theta_M$ is evident from table 2. In fact, for a given mosaic spread the Bragg peak fwhm decreases as the value $2\theta_M$ tends to $127°$. The small take–off–angle at the monochromator $2\theta = 50°$ produces a very broad peak indeed, in comparison with the larger angles shown in table 2. For a collimation of $\alpha = 10'$, the Bragg peak fwhm is $12'$ if $2\theta_M = 127°$ and $\beta = 20'$, whereas for the same collimation and mosaic spread the fwhm is $13.9'$ if $2\theta_M = 90°$. Using the latter take–off–angle the fwhm value could be reduced to $12.9'$ using $\beta = 5'$ however the peak luminosity is now only 62% of that in the first case where the fwhm is $12'$.

TABLE 2. Bragg peak fwhm and luminosity for a stress diffractometer in-staled at a guide tube for different values of the monochromator take–off–angle $2\ \theta_M$ and mosaic spread β.

a	$2\theta_M$	$\beta = \alpha/2$		$\beta = \alpha$		$\beta = 2\alpha$	
		F_ω	I	F_ω	I	F_ω	I
2.14	50°	2.56 α		2.59 α		2.61 α	
1.0	90°	1.29 α	0.29 $k\alpha^3$	1.35 α	0.41 $k\alpha^3$	1.39 α	0.47 $k\alpha^3$
0.5	127°	1.08 α		1.15 α		1.20 α	
0.27	150°	1.13 α		1.20 α		1.25 α	

It can be concluded that whatever the value required for F_ω, the highest luminosity is obtained for the monochromator take–off–angle which leads to a minimum in F_ω given by (10). In stress diffractometers installed at neutron guide tubes $2\theta_M$ should be about $127°$.

3.3 INSTALLATION AT A BEAM TUBE

For a stress diffractometer installed at a neutron beam tube there is no special limitation on the value of α_1. For the sake of good gauge volume definition the constraint $\alpha_2 = \alpha_3 \equiv \alpha$ still holds. Suppose $\alpha_1 = y\alpha$ and $\beta = x\alpha$ where y, and x are real positive numbers. Then, the fwhm and the luminosity of the Bragg peak can be written, from (8) and (9), respectively:

$$F_\omega = \alpha \left[\frac{1 + 2y^2 + 8x^2 - 4a(y^2 + 2x^2) + 4a^2(y^2 + y^2x^2 + x^2)}{y^2 + 1 + 4x^2} \right]^{1/2} \tag{12}$$

and

$$I = k \frac{\alpha^3 yx}{(y^2 + 1 + 4x^2)^{1/2}} \qquad \qquad (13)$$

The minimum value of F_ω is obtained, for the particular values of the dispersion parameter:

$$a_o = \frac{1}{2}\left(\frac{y^2 + 2}{2y^2 + 1}\right) \qquad \text{if} \quad \alpha = \beta \quad (x = 1)$$

and

$$a_o = \frac{1}{2}\left(\frac{y^2 + 8}{5y^2 + 4}\right) \qquad \text{if} \quad \alpha = 2\beta \quad (x = 2)$$

The value of a_o is always equal or smaller than $1/2$ for $y \geq 1$. Table 3 shows the influence of the various geometrical parameters in the fwhm and luminosity of the Bragg peak obtained at a sample scattering angle of $90°$ in a diffractometer installed at a beam tube.

TABLE 3. The fwhm and luminosity of a Bragg peak for $\alpha_2 = \alpha_3 \equiv \alpha$ for different values of α_1 in the two cases $\beta = \alpha$ and $\beta = 2\alpha$, for three different monochromator take–off–angles $2\theta_M = 90°$, $127°$ and $148°$.

		$\beta = \alpha$				$\beta = 2\alpha$		
		$\alpha^{-1}F_\omega$				$\alpha^{-1}F_\omega$		
α_1/α	$\dfrac{I}{k\alpha^3}$	$2\theta_M = 90°$	$2\theta_M = 127°$	$2\theta_M = 148°$	$\dfrac{I}{k\alpha^3}$	$2\theta_M = 90°$	$2\theta_M = 127°$	$2\theta_M = 148°$
1	0.41	1.35	1.15	1.19	0.47	1.39	1.20	1.24
2	0.67	1.80	1.25	1.20	0.87	2.06	1.40	1.29
3	0.80	2.05	1.31	1.21	1.18	2.62	1.59	1.35
4	0.87	2.19	1.35	1.21	1.38	3.04	1.75	1.40
5	0.91	2.27	1.37	1.22	1.54	3.33	1.86	1.44
6	0.94	2.32	1.38	1.22	1.65	3.54	1.95	1.47
7	0.95	2.35	1.39	1.22	1.72	3.69	2.01	1.49
8	0.96	2.37	1.39	1.22	1.78	3.80	2.05	1.51
9	0.97	2.39	1.40	1.22	1.82	3.88	2.08	1.52
10	0.98	2.40	1.40	1.22	1.85	3.94	2.11	1.53

The Bragg peak luminosity increases with α_1 and β, whereas the peak fwhm also increases with these values but it strongly depends on the monochromator take–off–angle. If a high value for $2\theta_M$ is set then the α_1 collimation has little influence on the peak fwhm but it does increase its luminosity in a significant way.

In fact, when $\beta = \alpha$, for the highest monochromator take–off–angle considered in table 3, $2\theta_M = 148°$, the peak fwhm is pratically constant (varies from 1.19 α to 1.22 α), although the collimation in the in–pile beam is changed from $\alpha_1 = \alpha$ to $\alpha_1 = 10\alpha$. However, for $2\theta_M = 90°$ this same α_1 variation gives a much broader peak width, F_ω varying from 1.4 α to 2.4 α. This means that the required fwhm, can be obtained using quite different parameters as is shown in Table 4 for $F_\omega = 12'$.

TABLE 4. Comparision of peak luminosities obtained with $2\theta_M = 148°$ and $90°$, for the same fwhm, $F_\omega = 12'$.

$2\theta_M$	α	α_1	Ik^{-1}	$Ik^{-1}(\times 10^8)$
148	10'	1.7°	0.98 α^3	2.41
90	9'	9'	0.41 α^3	0.73

Table 4 shows that an intensity gain amounting to a factor of 3 can be achieved for a 12' peak fwhm, if $2\theta_M = 148°$ is used instead of the conventional $2\theta_M = 90°$ setting.

The choice of the neutrons with both the right direction and correct wavelength is made through the combined action of the monochromating crystal and the polycrystalline sample where these neutrons are Bragg reflected. This combined action is very selective for $2\theta_M = 90°$ and even more for higher $2\theta_M$ values. In the latter case increasing α_1 only provides a larger neutron source area from which neutrons with both the suitable wavelength and direction can reach the detector. For lower values of $2\theta_M$ neutrons within a larger range of wavelengths and directions are twice Bragg reflected and reach the detector, and a relatively narrow Bragg peak is obtained only via stringent collimation α, including that in–pile, α_1.

In conclusion the optimized diffractometer installed at a neutron beam tube, should have a large monochromator take–off–angle of about 150° and the in–pile collimation α_1 value should be such as to allow the monochromator to receive neutrons from the largest portion of the beam tube end close to the reactor core.

4. The Optimized Stress Measurement Diffractometer

The diffractometers which have been used for stress measurements show quite different resolutions, their Bragg peak fwhm for the scattering angle $2\theta = 90°$

varying from about 0.2° to 0.6° in D1A at the ILL, Grenoble and Panda at Harwell, respectively. Consider the optimized setup for a good resolution diffractometer with $F_\omega = 0.2°$.

For installation of this instrument at a neutron guide tube, table 2 shows that this value of F_ω can be obtained using $\alpha = 10'$ e $\beta = 20'$ for $2\theta_M = 127°$ or $\alpha \cong 9'$ and $\beta = 18'$ for $2\theta_M = 90°$. In the latter case the peak luminosity is 70% of the former case. If the take–off–angle, for some reson, has to be smaller, e.g., $2\theta_M = 50°$, $F_\omega = 0.2°$ is achieved using $\alpha = 5' = \beta$ and the luminosity is only 10% of that obtained in the former situation.

For installation of the same instrument at a beam tube a good choice, according to table 4, is that where $2\theta_M = 148°$, $\alpha = \beta = 10'$ and $\alpha_1 \simeq 1.7°$. In comparision, using a take–off–angle of 90°, this resolution would require $\alpha = 9' = \beta$, with a peak luminosity equal to 30% of that obtained in the former case.

In any case a low monochromator take–off–angle should not be used if an optimum setup is desired. Even $2\theta_M = 90°$ is too low as this leads to a considerable instrumental broadening unless stringent in–pile collimation is used thus wasting beam intensity.

5. Conclusions

The measurement of stresses by neutron diffraction demands a scattering angle at the sample of 90° and that the incident and scattered beams have approximately the same angular horizontal and vertical divergence α, in order to define accurately a small cubic sampled volume within the specimen.

It was shown that 1 m radius free space available for sample positioning together with a sampled volume resolution of $2 \times 2 \times 2$ mm³ prevent the use of angular divergences larger than about 12'. Even so there may be limitations on the positioning of the sample cadmium mask for large specimens. These can only be overcome if the constraint on such a small sampled volume is relaxed.

The instrumental broadening of the Bragg peak is shown to strongly depend on the dispersion parameter which, for a fixed scattering angle, reduces to a strong dependence on the monochromator take–off–angle. It was proved that this sould be about 127° for a diffractometer installed at a neutron guide tube. Its value should be higher, about 150° for installation at a beam tube, in which case the peak broadening is almost insensitive to the in–pile beam collimation, which can thus be relaxed, to increase considerably the peak luminosity. In both installations significant luminosity gains are obtained in comparision with the setup using 90° monochromator take–off–angle, which has been usually considered as a good choice.

Acknowledgements

The author is grateful to Frederico Carvalho whose cooperation made this work possible and to José Salgado for helpful discussion of the manuscript.

References

Caglioti, G., Paoletti, A. and Ricci, F.P. (1958) 'Choice of collimators for a crystal spectrometer for neutron diffraction', Nucl. Instr. & Meth 3, 223–228.

Caglioti, G. and Ricci, F.P. (1962) 'Resolution and luminosity of crystal spectrometers for neutron diffraction', Nucl. Instr. & Meth. 15, 155–163.

Hewat, A.W. (1975) 'Design for a conventional high–resolution neutron powder diffractometer', Nucl. Instr. & Meth. 127, 361–370.

Hewat, A.W. and Baley, I. (1976) 'D1A, a high resolution neutron powder diffractometer with a bank of mylar collimators', Nucl. Instr. & Meth. 137, 463–471.

Loopstra, B.O. (1966) 'Neutron powder diffractometry using a wavelength of 2.6 Å', Nucl. Instr. & Meth. 44, 181–187.

Windsor, Colin G. (1990) 'The use of neutron diffraction for the solution of industrial problems', International School on 'Industrial and Technical Applications of Neutrons', Lerici, 19–29th June.

IMPLEMENTATION AND APPLICATION OF A PSD SET-UP FOR NEUTRON DIFFRACTION STRAIN MEASUREMENTS

T. LORENTZEN, T. LEFFERS AND D. JUUL JENSEN
Materials Department
Risø National Laboratory
DK-4000 Roskilde
Denmark

ABSTRACT. A new experimental configuration for residual stress studies by neutron diffraction has been implemented at Risø . The set-up is based on a linear position sensitive detector. This paper describes some of the technical implications in operating this set-up. Examples are given on the application of the instrument. These are studies of residual strains in a welded plate and in a welded tube assembly.

1. Introduction

Neutron diffraction is becoming a well established technique for internal stress determination. It is a valuable probe for studies of bulk material properties such as internal stress/strain distributions. The neutron diffraction experiments can provide insight into internal stresses which can be difficult to calculate, and hence the technique can support studies of the processes which generate residual stresses such as plastic deformation. However, neutron diffraction strain measurements can be rather time consuming; as neutron beam time is scarce, and the equipment is expensive to operate, any development towards a minimization of the required beam time is of great interest.

Many of the first residual stress studies by neutron diffraction were conducted on standard instruments used in physics research [1,2,3]. Though some of the studies include experiments on instruments utilizing multi detector banks or position sensitive detectors, none of these instruments were built specifically with internal stress studies in mind. As more and more research were focussed on the field of residual stress studies by neutron diffraction, scientists began to design experimental configurations dedicated to internal stress studies, and some of these are now being realized at facilities around the world.

Risø National Laboratory has been actively engaged in the development and implementation of such an experimental configuration for residual stress studies since 1986. The approach has been to develop a set-up based on a linear position sensitive detector (PSD). This set-up is designed to provide great flexibility, and it has been the aim to reduce the required measuring time. Further the instrument is implemented at a location allowing the monochromator take off angle to be set as high as $2\theta_M \approx 150°$ in order to enhance the instrumental resolution. The instrument is used in three different configurations which will

313

M. T. Hutchings and A. D. Krawitz (ed.),
Measurement of Residual and Applied Stress Using Neutron Diffraction, 313–327.
© 1992 *Kluwer Academic Publishers.*

314

be described together with the technical implications these set-ups have on the calibration and operation of the detector system. Further, some examples are given on the application of the experimental configurations.

2. Basic operation of the PSD based detector system

The PSD based detector system is operated as a so called delay line circuit, also named an RC-line (see figure 1.).

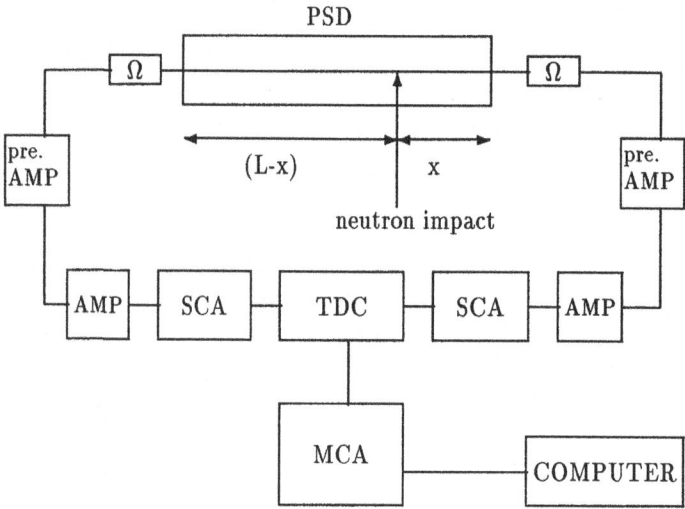

- SCA (single channel analyzer)
- TDC (time to digital converter)
- MCA (multi-channel analyzer)
- Ω (circuit resistance)
- AMP/pre. AMP (amplifiers)

Figure 1: The electronic delay line circuit used for the PSD.

The He^3 detector is 350mm long and 25mm in diameter. In this detector the neutron impact ionizies the gas and produces charged particles which will drift in the electric field between the anode wire and the outer tube lining (cathode). This generates current pulses which travel both ways in the anode wire, and the essence of determining where the neutron impact happened along the detector, is to measure the difference in the arrival time of these two pulses. One might think that this time decoding must be very accurate with currents traveling with the speed of light, which would give a maximum time difference between the signals of only ≈ 1 nano second, however, the signals <u>do not</u> travel with the speed of light in such an RC-line. In fact the travel speed can be orders of magnitude slower than that of light.

The uni-polar pulses from the preamplifiers are double differentiated by the main amplifiers, hence producing a bi-polar sine shaped signal which is fed to the timing single channel analyzers. These single channel analyzers detect the zero crossing of the bi-polar pulses, and each sends out a 5V pulse. The localization of the neutron impact is done by measuring the time difference between these two 5V pulses. The fact that the signal is double differentiated and that the zero crossing is used as the characteristic time of the signal, corresponds to a detection of a sign change on the curvature of the uni-polar amplifier signals. Hence, this technique is often described as a rise-time measurement, because the signal processing in effect determines a characteristic rise-time for the pulses. Further the uni-polar pulses from the amplifiers change shape depending on how far away from the pre-amplifier the neutron impact happened, and hence the position where the change in curvature sign appears is shifted. This shift is in opposite directions for the two pulses, hence this gives an inherent amplification of the time difference. The described shift in fact means that one could operate the PSD circuit by studying only the signal from one end of the detector with the free end grounded. However, as described, this prevents a utilization of the inherent time difference amplification, and time measurements would need to be more accurate.

In discussing the resolution of the PSD detector, the accuracy in measuring the described time difference is crucial. This resolution is not the strain resolution of the instrument, but is merely a measure of how accurately the neutron impact can be located. Hence this is rather a measure of the spatial resolution along the detector. However, the individual neutron impacts must obviously be located very accurately in order to achieve an accurate definition of a Bragg peak, hence this spatial resolution will affect the strain sensitivity of the instrument.

3. Experimental configurations

The PSD based detector system is normally operated in three different experimental configurations, two of which closely resemble one another, and one is a rather special set-up developed at Risø. The most simple application is when the PSD detector system is used to record complete Bragg peaks in single exposure type measurements. This type of set-up is shown in figure 2.

In this set-up the spatial resolution along the PSD directly affects the instrumental resolution hence affecting the ability of the instrument to resolve multiple adjacent peaks, as well as the ability to quantitatively analyze peak shapes and widths.

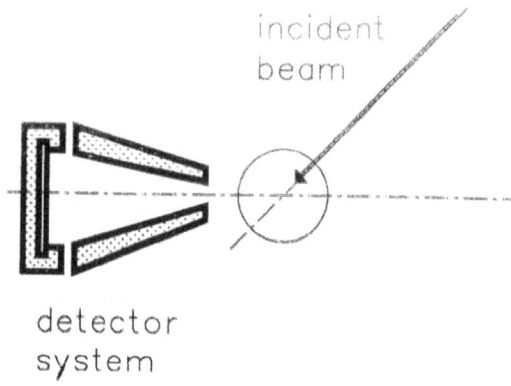

Figure 2: The single exposure set-up.

The set-up is utilized in two slightly different configurations depending on the sample size and the required spatial resolution in the sample. In many investigations of internal stresses, like internal stresses in metal matrix composites, test samples are often small blocks of the material or small uniaxial tensile specimens, and the aim of the neutron diffraction strain measurements is to evaluate the average internal stresses over an entire cross section of the sample. In such investigations a large part of the sample is covered by the incident neutron beam, and there is no need for accurate aperture arangements to define the gauge volume. The front end of the detector shielding is kept open to allow an appropriate range of diffraction angles to pass through (often $\approx \pm 2°$ off the specific Bragg angle $2\theta_S$). A similar set-up is also used in depth profiling, where a small gauge volume must be defined inside the sample. For this application the front end of the detector shielding can be equipped with a nose-type aperture mounting stage. This allows small fixed cadmium apertures to be positioned immediately above the sample surface, hence providing an optimum definition of the gauge volume inside the sample. If required, a similar nose type aperture stage can also be mounted on the incident beam line. In this approach to depth profiling all attention is focussed on a single gauge volume at a time, and the Bragg peak is again recorded in a single exposure type measurement.

Both applications described above rely on the same set of calibration measurements, which must be completed whenever the PSD is installed or moved on the spectrometer diffraction arm. One of the necessary calibrations serves to relate the angular position along the detector to the channel numbers into which the recorded intensities are stored depending on the observed time difference between signals. In the second calibration run the variation in the detector efficiency along the detector is recorded, and this gives an intensity calibration which will be used on all subsequent measurements. Typical examples of these calibration runs are given in figure 3.

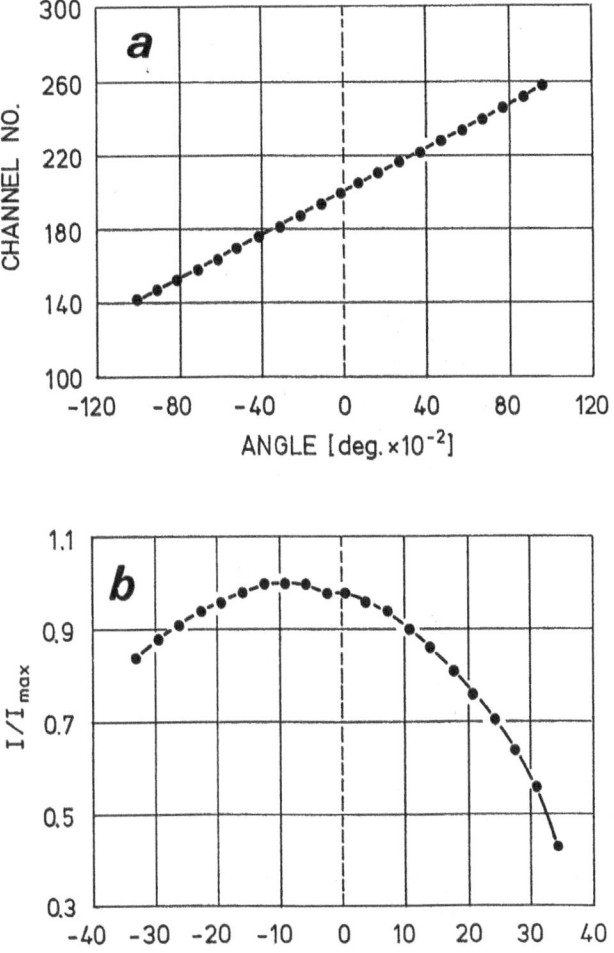

Figure 3: Calibration of the PSD is done by scanning the detector across an attenuated direct beam, and recording a) channel position versus angle, and b) intensity versus channel position.

In the set-up for depth profiling described above, all attention is focussed on a single gauge volume at a time, however, an alternative approach to depth profiling is to study scattered intensity from all gauge volumes along the profile simultaneously. This experimental configuration is developed at Risø , and consists of the PSD equipped with a neutron collimator in front of the detector. In this set-up the incident beam defines the line of interest through the sample, and the collimator in front of the PSD serves as a multi-slit system, assuring that recorded intensity at specific locations along the PSD can only come from a well defined gauge volume inside the sample. Using this set-up the sample and detector are kept fixed relative to each other, and the incident neutron wavelength is scanned in order to record Bragg peaks from all gauge volumes. Clearly the gauge volume with the longest beam path, will determine the appropriate counting time, and gauge volumes with decreasing path lengths will be recorded with an increasing improvement of counting statistics. However, all gauge volumes will be inspected in the time it takes to record the one which is the most time consuming. The set-up is shown in figure 4.

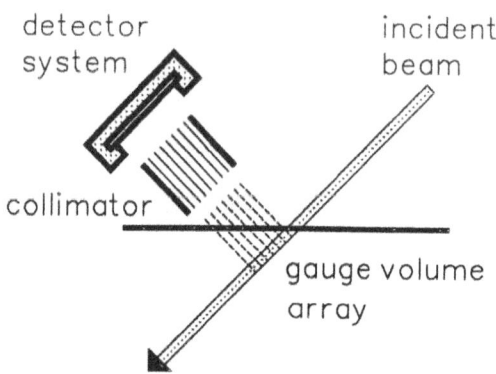

Figure 4: Experimental configuration for depth profiling focussing the attention on an array of gauge volumes simultaneously.

4. Optimization of signal shape and spatial resolution

The described experimental configurations have different requirements to the optimization of the PSD detector system. As mentioned the level of time difference between the two pulses identifying the location of the neutron impact is a crucial parameter for the spatial resolution along the detector. The larger the shift is in this time difference for a given shift of the neutron impact along the detector, the more accurately we can determine the location. It turns out that for a specific area of the detector this change in time difference can be optimized. The time difference between the two signals can be hundreds of nanoseconds, with the maximum change in time difference (the change in difference between two neutron impacts occuring at each end of the 350mm long detector) beeing fixed by the detector design. However, the change in time difference between two neighboring neutron impacts at two specific locations along the PSD is not fixed. By inserting resistors in series with the detector the change in time difference along the detector can be linearized, which in fact means that for a specific area of the PSD, e.g. the centre 7cm, such a linearization will increase the time difference change with increasing resistance inserted. The more the time difference changes with position, the easier it becomes to accurately determine the position of the neutron impact, hence this is one way of improving the spatial resolution for a specific section of the PSD. To illustrate this a series of test measurements were done using a direct beam of less than 1mm width, and with the detector placed on a translation stage allowing movements of ±35mm perpendicular to the beam. Resistors in the range of 18.2 to 85.4$k\Omega$ were tested by exposing the detector to the direct beam at +35 and −35 mm, respectively, and for each measurement the time difference ΔT given in figure 5 was determined. This time difference ΔT was seen to vary with curcuit resistance according to figure 6.

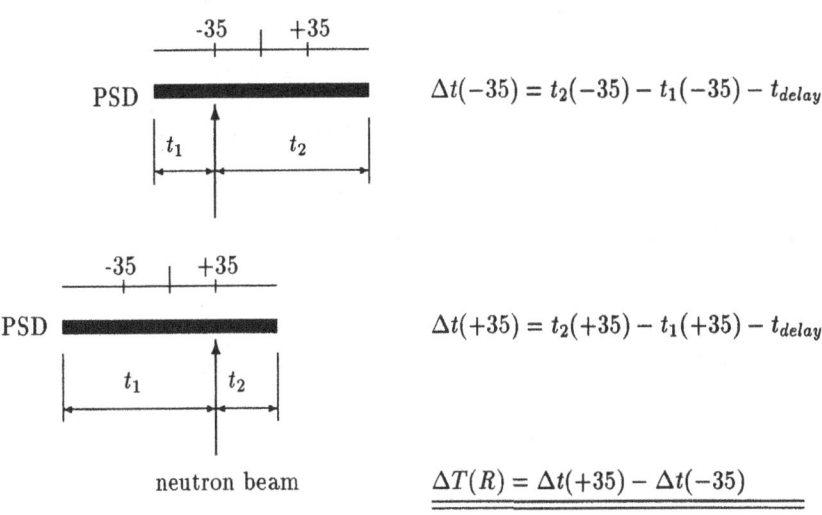

$$\Delta t(-35) = t_2(-35) - t_1(-35) - t_{delay}$$

$$\Delta t(+35) = t_2(+35) - t_1(+35) - t_{delay}$$

$$\Delta T(R) = \Delta t(+35) - \Delta t(-35)$$

Figure 5: The time difference ΔT determined as a function of circuit resistance. The t_{delay} is a fixed time delay which is deliberately imposed on one of the pulses to assure that this pulse can always be treated as the stop pulse having the longest arrival time.

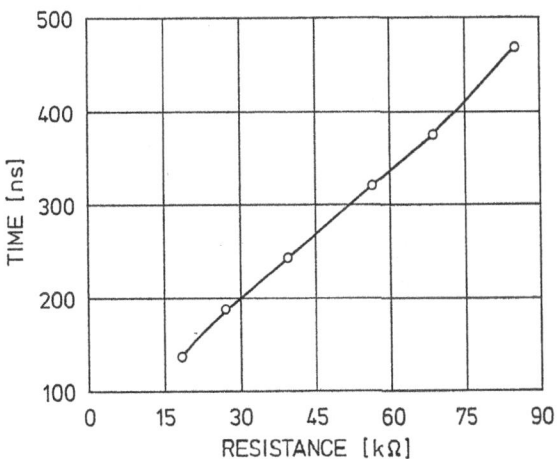

Figure 6: ΔT as a function of circuit resistance.

As seen the time difference is almost linearly increasing with the resistance, hence improving the accuracy in localizing the neutron impact. From the described test measurements alone it appears that we keep gaining time/spatial resolution with increasing resistance. However; there are other detrimental effects of the increase of circuit resistance.

By studying the signal on the detector coming from a very narrow incident beam it is seen that this narrow beam will generate a signal which is spread out along the detector, and consequently the signal appears to be broader than the width of the beam. Such signal broadening affects the resolution of the instrument in the sense that Bragg peaks become broader and the ability to quantitatively study Bragg peak shapes and the ability to resolve adjacent diffraction peaks deteriorate. Further, when using the configuration for depth profiling using a collimator in front of the PSD the aim is to record a band of parallel neutron beams simultaneously. In this set-up the signal width becoms a source of "pollution" as the signal from one gauge volume will be overlapped by signals originating from neighboring volumes. Hence the width of signals on the PSD becomes an issue for all the described configurations.

It turns out that this signal width is also affected by the increase in resistance, and for the above series of resistances, the signal shapes were observed as shown in figure 7, where the channel numbers along the abscissa corresponds to measures of position along the PSD.

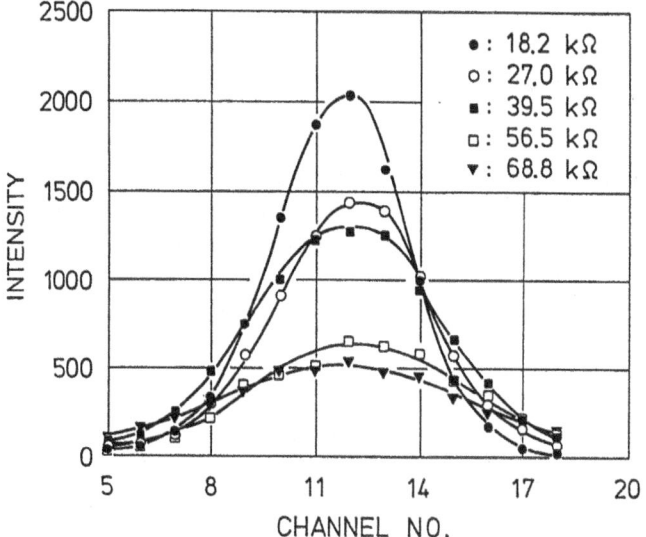

Figure 7: Alterations in peak intensity and in Full Width at Half Maximum (FWHM) as a function of load resistance.

As ΔT increases with resistance it is observed that the peak intensity decreases dramatically, and at for instance $56k\Omega$ it is reduced by a factor of 3 though the integrated intensity under the peak is constant. This reduction in peak intensity does not affect the data collection if it is acceptable simply to take the integrated intensity as a measure of the scattered intensity at that specific location of the PSD. However, this broadening of individual signals may not be acceptable in experimental configurations where the aim is to record several neighbouring neutron impacts simultaneously like in the collimator-based set-up described previously. Using that approach to depth profiling, the collimator consists of 1mm wide parallel channels, hence the PSD is seeing a band of parallel 1mm wide neutron beams coming from the array of gauge volumes, and as described previously such 1mm wide beams will appear broader on the PSD. This means that there will be a certain overlap of signals on the PSD depending on the width of the signals. Hence the effect of circuit resistance making signals broader and lower is a critical effect for this experimental configuration. As a simple means of weighting the described beneficial time difference increase with resistance and the detrimental peak broadening effect, a resolution factor S is defined as the ratio between these two quantities.

$$S = \frac{FWHM[\text{ns}]}{(\Delta T(R)/distance)[\text{ns/mm.}]} \tag{1}$$

Figure 8. shows this factor S as a function of the circuit resistance R, and it is evident that for the given set-up, the optimum spatial resolution is found for a circuit resistance of approximately $56k\Omega$, where the FWHM was 3.8mm.

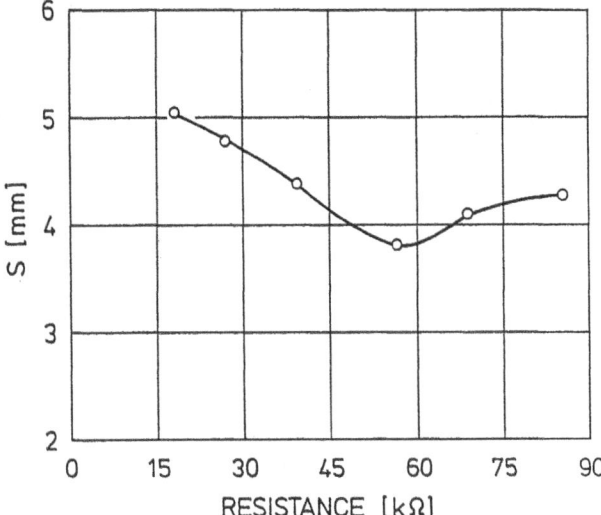

Figure 8: Resolution factor S as a function of load resistance. S is here defined as the ratio of the peak width (FWHM) to the signal time difference for a fixed distance between two neutron impacts.

Realizing that signals do appear broader on the PSD than the actual width of the neutron beam there are two options when using the depth profiling concept. The signal width can be minimized increasing the circuit resistance as described above, and this will allow a simultaneous recording of parallel neutron beams which are very close to each other. The optimum would be to simultaneously record beams coming through every collimator channel, hence allowing the strain profile to be recorded with a spatial resolution of ≈ 1mm. However, as seen in figure 6, this also leads to a dramatic reduction in the peak intensity, and this may for some applications render a prohibitively low signal to noise ratio. In case

322

signals are too weak to allow such a reduction of the intensity, the circuit resistance must be reduced, and the degree of signal overlap increases. To illustrate this signal overlap, a 1mm wide neutron beam was scanned across the detector/collimator arrangement in steps of 2,3 and 4mm. This simulates scattered intensity coming from gauge volumes which are separated by 1, 2 and 3mm respectively, and the degree of signal overlap is evident from figure 9.

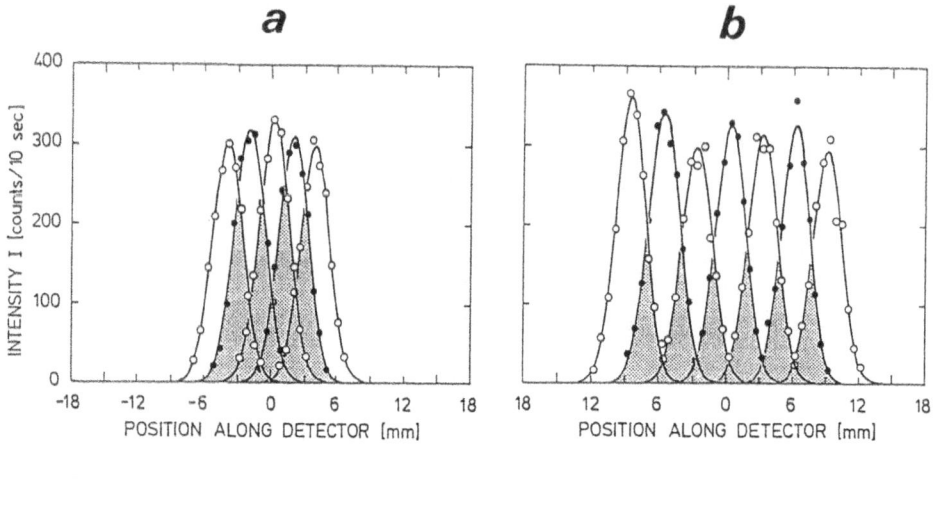

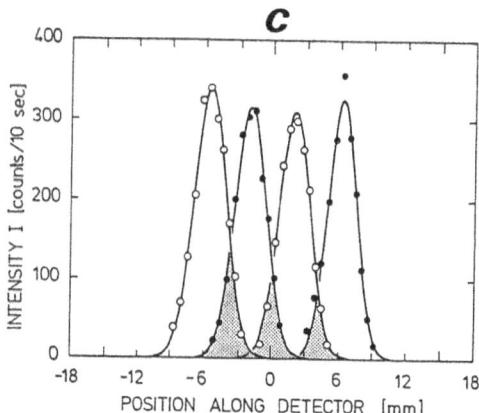

Figure 9: The detector signals generated by a 1mm wide beam. Moving the beam in increments of 2,3 and 4mm simulates diffracted intensity originating from sample volumes separated by 1, 2 and 3mm respectively.

From figure 9a it is obvious that the degree of overlap is rather large when gauge volumes are only separated by 1mm, and that the signals are almost completely separated when gauge volumes are separated by 3mm (see figure 9c). This implies that in order to accept every single one of these signals as coming from a specific gauge volume and not being polluted by scattered intensity from neighboring volumes, gauge volumes must be separated by 3mm, which is brought about by blocking three collimator channels for every open one. However, an alternative is to take only the peak intensity of each signal in figure 9a as the recorded intensity from each gauge volume. This will allow us to study all gauge volumes along the incident beam simultaneously without the need for blocking collimator channels. This approach, however, is rather costly intensity wise, and it may in fact be as efficient to block some collimator channels and then shift the sample parallel to the beam in order to complete the strain profile.

From the above discussion on resolution and peak widths the optimum choice of circuit resistance is $56k\Omega$ in order to provide a good compromise between the issues of signal time difference and that of peak overlap. This is a reasonable compromise as long as the signal to noise ratio is acceptably large. However, it is our experience that in many depth profiling investigations, beam paths are relatively long and hence the signal to noise ratio rapidly deteriorates to an unacceptably low level. For such investigations we cannot accept the inherent decrease of peak intensity due to the inserted circuit resistance (see figure 7), and these have to be dismounted. Normally the circuit resistances are mounted directly in the preamplifiers, and it is not recommendable that soldering components in the preamplifiers becomes a regular part of setting up the instrument. Preamplifiers are very sensitive components, and bad soldering may be a source of noise, or even worse the preamplifiers are easily permanently damaged during the soldering operation. Considering the range of experimental investigations which most commonly occur at the facility, it was decided that operating the PSD without the circuit resistances would be the most versatile configuration, and hence it will be necessary to block some collimator channels when doing depth profiling by focussing the attention on an array of gauge volumes simultaneously. However, the discussion on circuit resistors provides the basis for a possible adjustment of the spatial resolution, which is beneficial for specific investigations, where the signal to noise ratio is relatively high.

5. Depth profiling with a PSD based set-up

The application of the described PSD set-up, where the detector is utilized to record complete Bragg peaks in single exposure type experiments, is straightforward. Either the sample dimension itself helps defining the gauge volume, or alternatively for depth profiling the nose type aperture is inserted, and the sample is positioned with the direction of the strain measurement parallel to the scattering vector. However, depth profiling using the PSD/collimator set-up requires careful planning of the experimental configuration. Here the incident beam defines a line of irradiated material, and the collimator splits this line of material into an array of small gauge volumes. This set-up also fixes the direction of the scattering vector, which will be identical for all gauge volumes. Hence the array of gauge volumes and the direction of the strain measurement are rigidly connected in this set-up. The challenge is to arrange the sample to give the strain component of interest

324

along the line of interest through the sample. This may cause problems for some sample geometries, and often compromises must be made. However, for many common sample geometries acceptable solutions can be found. One typical sample geometry is rectangular plate assemblies as in T-butt weldings, where the aim often is to determine the variation of three orthogonal strain components with position in the plate. For such a sample geometry the experimental configuration could be like sketched in figure 10, and an example of such an investigation is a typical strain profile of the ϵ_x strain component as shown in figure 11 (for further details see [4]).

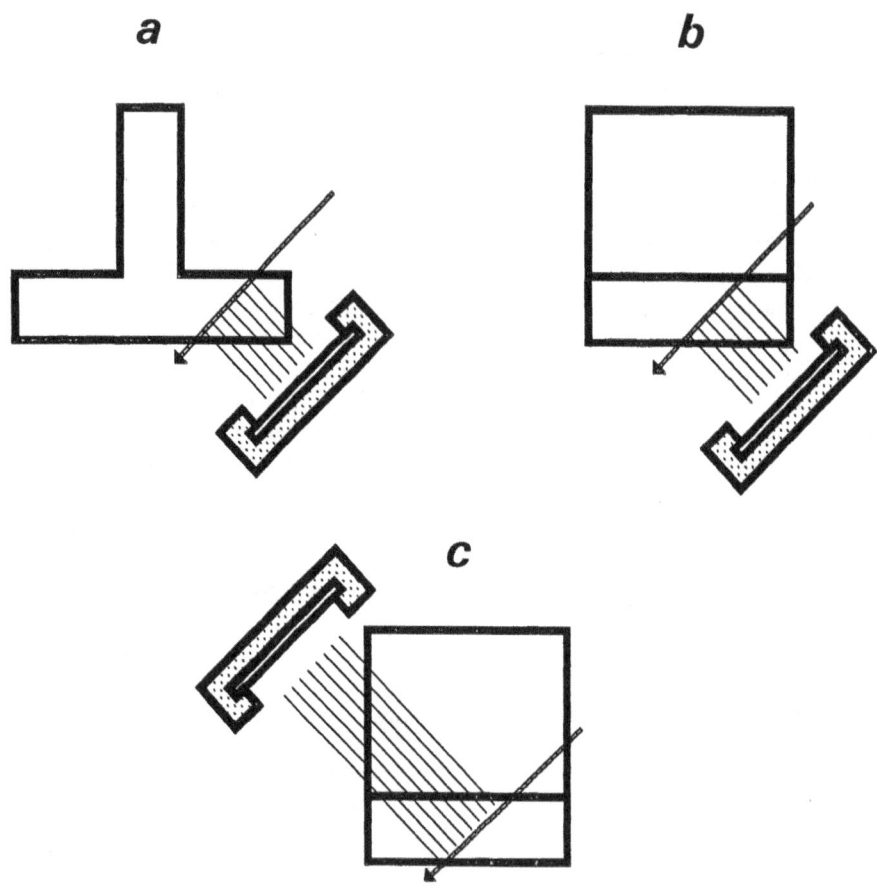

Figure 10: Typical experimental configurations for studies of residual strains in welded plate assemblies using the PSD/collimator based set-up.

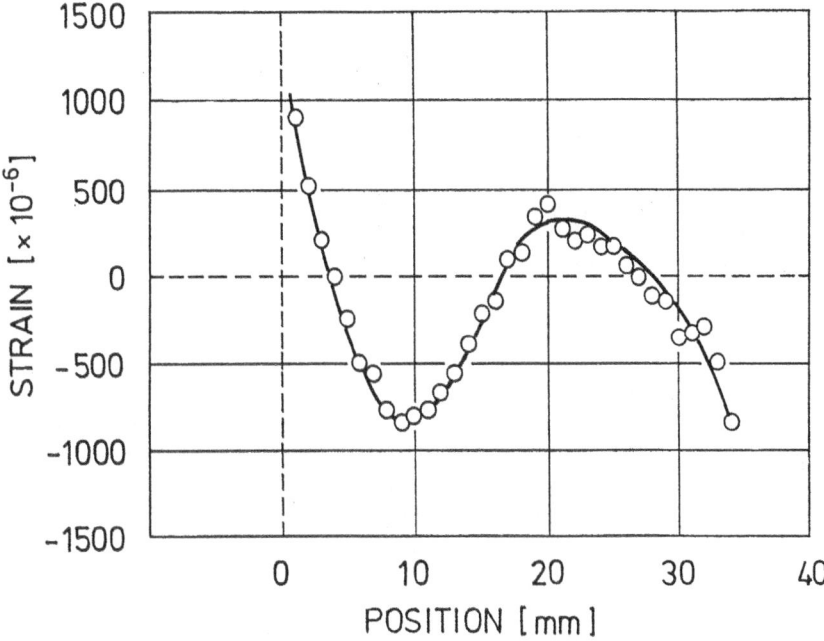

Figure 11: Strain profile of the ϵ_x component through the thickness of a 25mm thick plate in a T-butt welded plate assembly. The strain profile is recorded using the set-up shown in figure 10a.

Another typical sample geometry for residual strain investigations by neutron diffraction is thick walled tubes or welded tube assemblies, where the aim normally is to evaluate the axial, radial and tangential strain components as a function of position through the tube wall. For this sample geometry typical experimental configurations can be as shown in figure 12. One example of such an investigation is the strain profile of the ϵ_A component recorded in a thick walled tube using the experimental configuration shown in figure 13 (for further details see [5]).

From the above examples it is evident that some times compromises must be made on the exact direction of the strain measurement, and often compromises must be made regarding the exact location of the gauge volumes used for different components. However, such compromises are often consistent with commonly applied assumptions on sample symmetry and axes of rotational symmetry, and despite these compromises valuable information of the residual strain state can be recorded rapidly using the PSD/collimator based set-up.

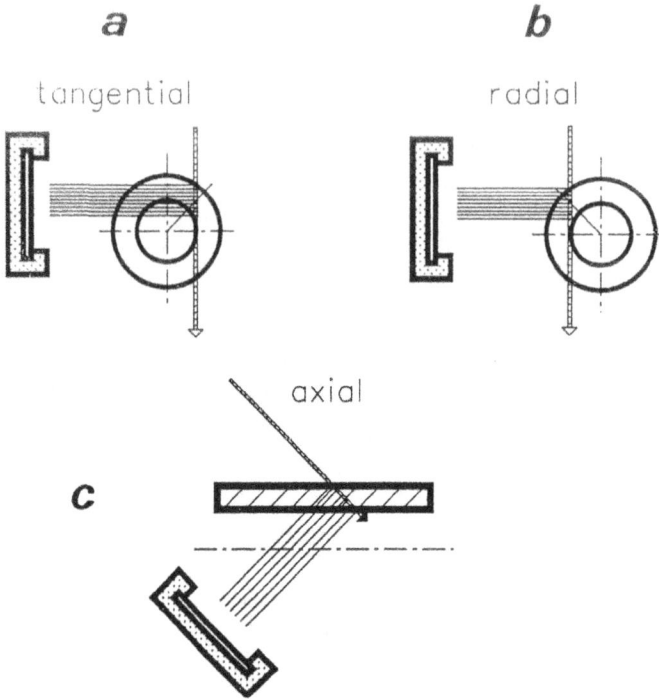

Figure 12: Typical experimental configurations for studies of residual strains in thick walled tubes using the PSD/collimator based set-up.

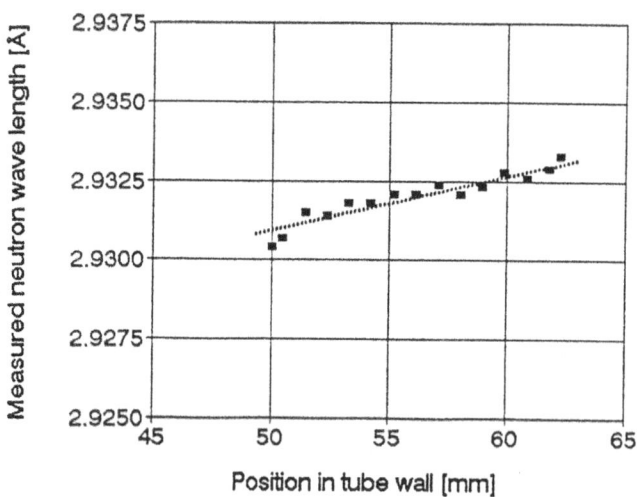

Figure 13: Strain profile of the ϵ_A component through the thickness of a thick walled tube. The strain profile is recorded using the set-up shown in figure 12c.

6. Summary

PSD based experimental configurations for residual stress studies by neutron diffraction have been presented. The technical implementation of the instrument has been described, and optimization in terms of spatial resolution, strain sensitivity and efficiency has been discussed. A potential way of optimizing the spatial resolution was presented. However, it is inherent in this optimization that the detector efficiency is lowered, and for some applications this may cause a prohibitively low signal to noise ratio. The application of the instrument was illustrated by examples of a residual strain measurements on a welded plate assembly and on a welded assembly of two thick walled tubes.

References

[1] Allen, A., Andreani, C., Hutchings, M.T. and Windsor, C.G. (1981) 'Measurements of internal stresses within bulk materials using neutron diffraction'. NDT International, October, pp.249-254.

[2] Krawitz, A.D., Brune, J.E. and Schmank, M.J. (1981) 'Measurements of stress in the interior of solids with neutrons'. Proceedings of the 28th Sagamore Army Materials Conference, Sagamore, USA, pp.139-155.

[3] Pintschovius, L., Jung, V., Macherauch, E., Schäfer, R. and Vöhringer, O. (1981) 'Determination of residual stress distributions in the interior of technical parts by means of neutron diffraction', Proceedings of the 28th Sagamore Army Materials Conference, Sagamore, USA, pp.467-482.

[4] Lorentzen, T. (1990) 'Bulk residual stress studies by neutron diffraction', Ph.D.thesis, Aalborg University, Aalborg, Denmark.

[5] Lorentzen, T., Leffers, T. and Juul Jensen, D. 'Residual stress evaluation in a welded tube assembly using neutron diffraction', to be published in proceedings of the third International Conference on Residual Stresses, ICRS-3, Japan, 1991.

REVERSE TIME-OF-FLIGHT FOURIER TECHNIQUE FOR STRAIN MEASUREMENTS

HANS G. PRIESMEYER
Institut für Reine und Angewandte Kernphysik
Universität Kiel c/o GKSS Research Center
POBox 1160
D-2054 Geesthacht FRG

ABSTRACT. Fourier type correlation spectrometers at reactors allow strain measurements by neutron time-of-flight diffractometry. For instruments of this type intensity and resolution are independent. Data acquisition is done by on-line correlation, which is possible in the reverse-time-of-flight Fourier (RTOF) mode. A spectrometer designed for strain measurements is now operating at the 5 MW FRG-1 research reactor at GKSS Geesthacht.

1. Introduction

The neutron time-of-flight method is not necessarily limited to pulsed neutron sources, but it is not economic on steady-state neutron sources when used in the conventional way. In order to exploit the advantages of the method for neutron diffraction, correlation spectrometry has been developed some time ago. Among these techniques the mechanical analog to harmonic analysis has become widely known as the Fourier chopper spectrometer. The inherent deficiencies of the early Fourier method - successive measurements at ground frequency and higher harmonics - have been elegantly overcome by the reverse-time-of-flight Fourier (RTOF) technique (P. Hiismäki, 1972). An RTOF spectrometer has been designed and commissioned at the 5 MW research reactor FRG-1. It has been optimized for the measurement of internal stresses.

2. The Neutron Time-of-Flight Method in Nondestructive Strain Measurements

The well-established advantages of the time-of-flight method have been decisive for the choice of spectrometer:

the possibility to use a fixed sample-detector geometry,

utilisation of the white neutron spectrum transmitted by the guide tube,

the simultaneous investigation of many reflexions from the same volume in consideration of orientation-dependent elastic constants,

high instrumental resolution to investigate peak shapes,

simpler sample positioning to meet special environmental requirements as for heavy samples,

M. T. Hutchings and A. D. Krawitz (ed.),
Measurement of Residual and Applied Stress Using Neutron Diffraction, 329–334.

330

samples under operating conditions, temperature and load variations,

and the possibility to use several 90-degree detectors to reduce measuring times for strain tensor determinations,

have been considered as credit points of the method. It is a special feature of correlation spectrometry, that unlike conventional TOF intensity and resolution are uncorrelated. This means that the instrumental resolution can be made very high without loss in neutron intensity.

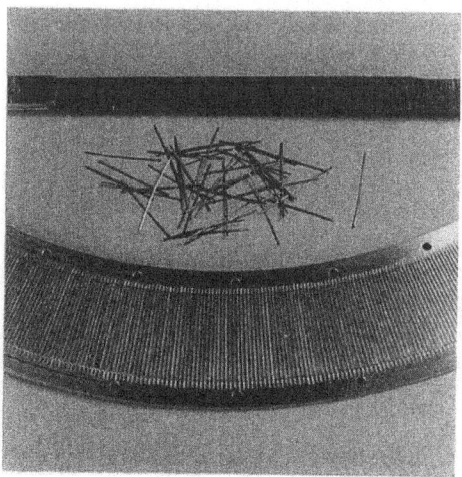

Fig. 1: Grooved disk with gadolinium strips.

Fig. 2: Rotor/stator unit.

3. The RTOF spectrometer at Geesthacht

Fourier methods require a mechanical modulator for the thermal neutron beam, which can vary the intensity in a sinusoidal way from zero to more than 150 kHz. The period of the highest modulation frequency achieved will determine the instrumental resolution, if all other incertainties from flight-path length and scattering angle are chosen in such a way that they may be neglected.

The modulator consists of a combelike rotor/stator system (Fig.1). The rotor is a titanium disk of 60 cm o.d., containing along its outer edge 1024 equally spaced strips of gadolinium metal as an efficient neutron absorber. These strips are 35 mm long, 0.7 mm wide and 0.25 mm thick. A stator having the same neutron absorber pattern is positioned at 6 mm behind the rotor. The rotor speed chosen for routine experiments with 5 x 5 mm² neutron beam size is 6000 rpm, resulting in a modulation frequency of 102.4 kHz. Figure 3 shows the outline of the experiment.

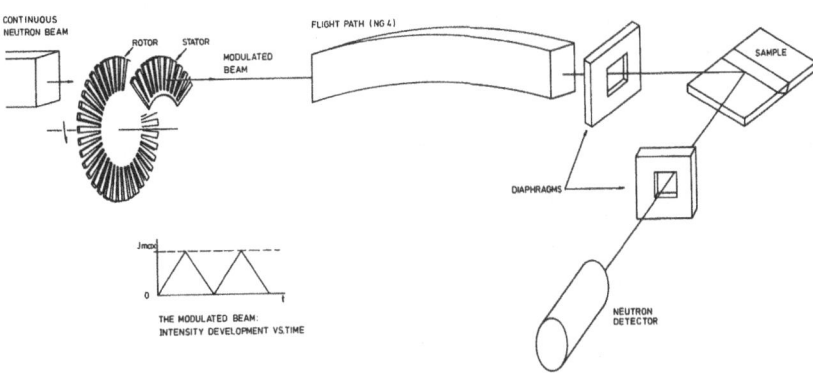

Fig. 3: The Geesthacht RTOF spectrometer.

The flight path is essentially a neutron guide with a characteristic wavelength of 0.183 nm. Its cross section is 15 x 108 mm² and can be used in full or sectioned. The length of the guide is 19.8 m, it protects the sample from the direct view to the reactor core. The neutron flux measured by gold foil activation is 1.6×10^6 n/cm² sec with a maximum intensity near 0.165 nm. Up to 16 Li-6 glass-scintillation detectors will be set up in time-focussing geometry to cover most of the solid angle into which the neutrons are scattered. The correlation calculation necessary to determine the neutron diffraction spectrum is done on-line and easy to explain: An optical pickup system which is fixed to the rotor axis, produces binary signals, by which the chopper status „open" and „closed" can be determined. This signal chain is fed into a 2048 channel electronic shift register according to a clock pulse frequency which defines the timing resolution (cf.Fig. 4). Let us consider now the case of a single monoenergetic neutron, which will need a single flight-time to traverse the distance from the chopper to the detector. Obviously it can only have started from the chopping point during the „open"-status.

332

The detected neutron triggers the parallel transfer of the shift register containing the „phase information" (open/closed status of the rotor/stator system) to an accumulating register. Therefore on the average 1024 channels are incremented, one of which corresponds to the true neutron flight-time. The rotor speed is controlled according to a predetermined frequency distribution, so that it is different, when the next monoenergetic neutron appears, adding a different binary pattern to the accumulating register. It is the channel corresponding to the correct neutron time-of-flight, that discriminates against the rest of the others in the course of the experiment, as is indicated in Fig. 5. This is the principle of on-line correlation by which diffraction spectra containing many Bragg lines have been measured.

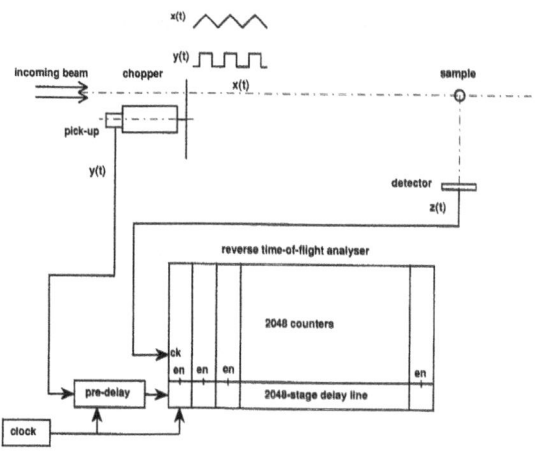

Fig. 4: Scheme of the RTOF method.

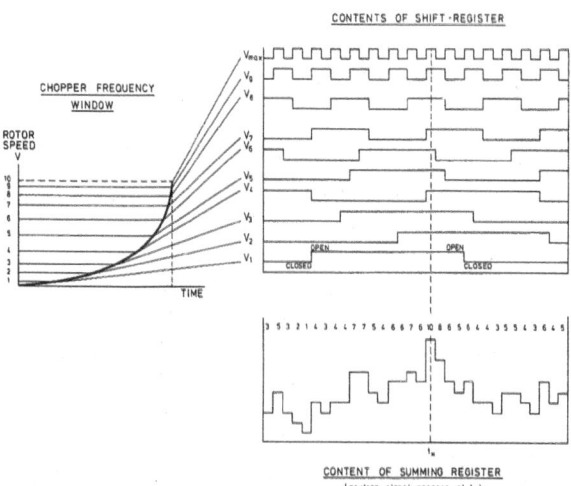

Fig. 5: Principle of signal accumulation.

The following Fig. 6 shows time-of-flight spectra collected with the system. Measuring times are of the order of 2 hours for 5 x 5 x 5 mm³ scattering volume in steel and can be reduced by implementation of further neutron detectors (including at a later stage detectors in different directions for simultaneous measurement of different angles for strain tensor or texture determination).

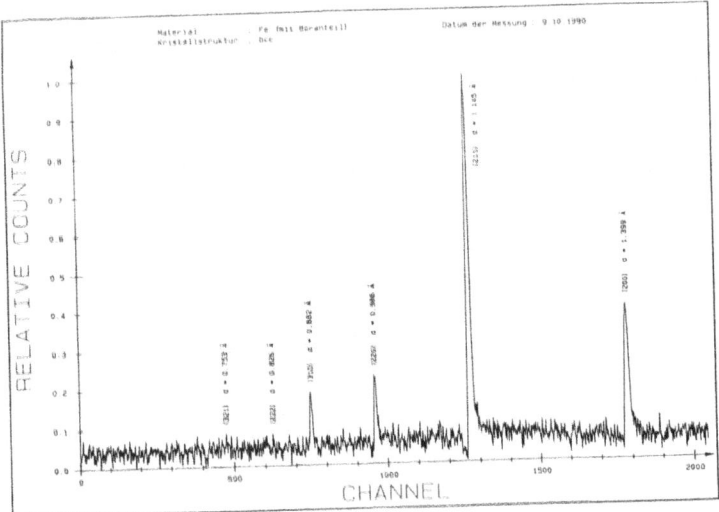

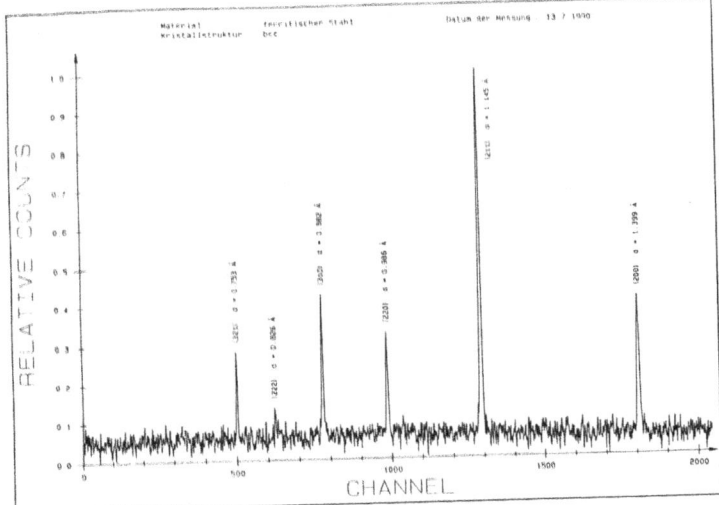

Fig. 6: Time-of-flight spectra of two different steel samples (with and without texture).

4. References

Keuter, J. (1990), 'Neutronographische und röntgenographische Spannungsanalyse geschweißter T-Stöße' (Neutron and x-ray stress analysis of T-butt weldments, in German), Thesis, Kiel University

Hiismäki, P. (1972), 'Inverse time-of-flight method', IAEA Conference Neutron Inelastic Scattering Grenoble 1972, IAEA Vienna, pp. 803-807

Hiismäki, P., Trunov, V.A. et al. (1985), 'Experience of the Fourier time-of-flight (TOF) neutron techniques for high-resolution neutron diffractometry', Neutron Scattering in the 'Nineties' (Conf. Proc. Jülich 1985), IAEA Vienna, pp. 453-459

Priesmeyer, H.G., Schröder, J. (1990), 'Residual stress analysis by neutron time-of-flight at a reactor source', Neutron scattering for materials science (Mat. Res. Soc. Symp. Proc., No. 166), S.M. Shapiro, S.-C. Moss, J.D. Jorgensen ed., pp. 299-304

THE ALIGNMENT OF INSTRUMENTATION AND THE POSITIONING OF SPECIMENS FOR STRESS MEASUREMENTS BY MEANS OF NEUTRON DIFFRACTION

P. C. BRAND
Netherlands Energy Research Foundation ECN
P.O. Box 1
1755 ZG Petten
The Netherlands.

ABSTRACT. Adequate alignment for sample and instrument is a very important aspect of residual stress measurements by means of neutron diffraction. In this paper a sequence of alignment procedures is presented. The first purpose of this is to align transportable aperture systems along the primary and diffracted beams. The second purpose is to find the location of a specimen in the diffractometer geometry. Where possible, the power of the specimen to diffract neutrons is used to perform the alignment steps.

1. Introduction

The alignment requirements for a neutron diffractometer that is used for stress measurements are more strict than for a normal neutron powder diffractometer.

A neutron diffractometer for stress measurements generally differs from a normal diffractometer in two major ways:

1. The presence of gauge volume defining apertures in the primary and secondary beams. Both can be translated along the radius of the diffractometer.
2. The presence of a specimen co-ordinate table on the diffractometer. Usually, this table has three translation axes (x, y and z) and an independent rotation axis (ψ), which is co-axial with the neutron counter movement axis (2θ).

In this paper, a method for the alignment of both aperture movements and the sample in the neutron beam geometry will be presented [1].

2. Alignment requirements

The alignment requirements for a diffractometer that holds a specimen are the following:

M. T. Hutchings and A. D. Krawitz (ed.),
Measurement of Residual and Applied Stress Using Neutron Diffraction, 335–346.
© 1992 *Kluwer Academic Publishers.*

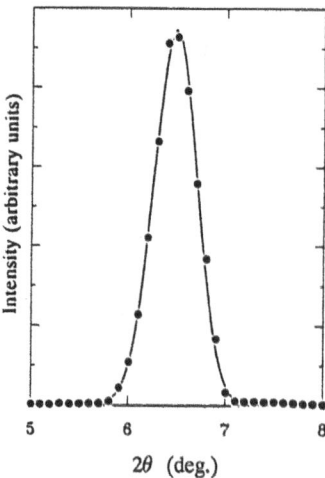

Fig. 1 An example of a scan around the direct beam. The solid line represents the result of a Gaussian fit to the measured data points.

1. The centre line of the primary and diffracted beams should always point to the centre of the diffractometer, regardless of the position of the beam apertures.
2. The sliding direction of the diffracted beam aperture should be along an angle 2θ with the primary beam.
3. The co-ordinates x and y at which the specimen surface coincides with the central axis of the diffractometer should be known.

3. Alignment procedures

In order to achieve adequate alignment, *i.e.* satisfying the above stated conditions, a series of procedures has been developed. These procedures will be treated in a logical order.

3.1. THE CENTRAL AXIS OF THE DIFFRACTOMETER

After levelling the diffractometer using a spirit level, the central axis of the diffractometer should be found. For this purpose a precision cylinder, which is mounted on the specimen table is used. On top of the cylinder a co-axial steel pin is mounted. Using the ψ-motor, the cylinder and pin are rotated around the (vertical) central axis. Using both the x and y movements and a dial gauge the pin is centred. This has proved to be accurate within 0.01 mm. Both apertures can now be adjusted such that they point towards the pin. Establishing electrical contact between the pin and the apertures may be of help here.

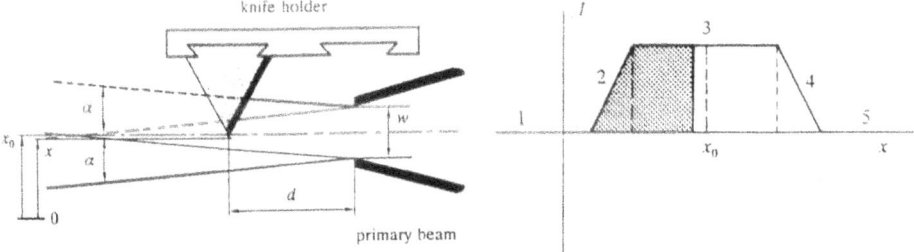

Fig. 2 The single knife edge assembly in the primary beam geometry.

Fig. 3 The intensity profile seen at a distance d from the primary beam aperture as a function of the co-ordinate perpendicular to the beam direction. The numbers in the figure refer to the functions given in the text. The shaded area symbolizes the intensity that is currently 'seen' by the detector.

3.2. THE DIRECT BEAM POSITION

In the not aligned condition, the position of the neutron detector is given relative to an arbitrary zero position of the 2θ-motor. However, the position of the 2θ-motor must be known relative to the position of the primary beam. This can be achieved by a scan of the primary beam. This results in a more or less Gaussian shaped intensity profile of which the maximum intensity position: $2\theta_0$ can be obtained by fitting a Gaussian. The thus found $2\theta_0$ is the offset value for the detector angle. An example of a successful fit is given in figure 1. The statistic error in $2\theta_0$ is usually less than $0.005°$.

3.3. THE SINGLE KNIFE EDGE SCAN

This part of the alignment procedure is meant to obtain the first out of two points that define the direction of the neutron beam.

The counter is positioned at $2\theta_0$. On the specimen table a device called the single knife edge is mounted. The single knife edge is a wedge covered with cadmium, which is connected to a holding piece as is shown in figure 2. The instrument angle ψ is chosen such that the x-movement transports the knife edge more or less perpendicularly to the primary beam. A scan that gives the direct beam intensity I as a function of the x-position of the knife edge is performed.

The form of the beam cross-section at the site of the single knife edge is defined by the beam divergence α, the aperture width w and the aperture distance from the knife edge d. The sought after parameter is x_0, the knife edge position at which half of the primary beam is covered.

338

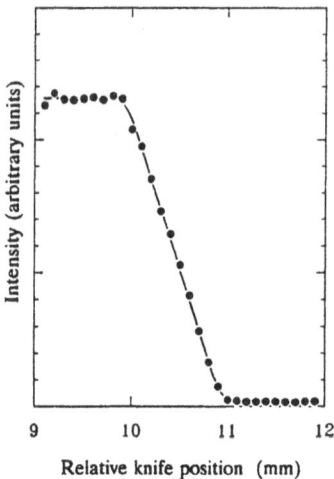

Fig 4. Result of the single knife edge scan. The solid line is the final result of a Levenberg-Marquardt fitting procedure using the single knife edge function described in the text (equation 1).

The scan starts with the single knife edge fully covering the primary neutron beam and the neutron detector will only detect the background radiation I_b. The detected intensity I as the knife edge uncovers the beam can be deduced from the intensity profile shown in figure 3.

The value of this function is proportional to the shaded surface of figure 3 (I_0 is the constant of proportionality). It is a continuous function which is divided into 5 x-regions.

1. $x < x_0 - \frac{1}{2}w - \frac{1}{2}\alpha d$

$$I = I_b \tag{1a}$$

2. $x_0 - \frac{1}{2}w - \frac{1}{2}\alpha d \leq x < x_0 - \frac{1}{2}w + \frac{1}{2}\alpha d$

$$I = \frac{(x - x_0 + \frac{1}{2}w + \frac{1}{2}\alpha d)^2}{2\alpha d} I_0 + I_b \tag{1b}$$

3. $x_0 - \frac{1}{2}w + \frac{1}{2}\alpha d \leq x < x_0 + \frac{1}{2}w - \frac{1}{2}\alpha d$

$$I = (x - x_0 + \frac{1}{2}w)I_0 + I_b \tag{1c}$$

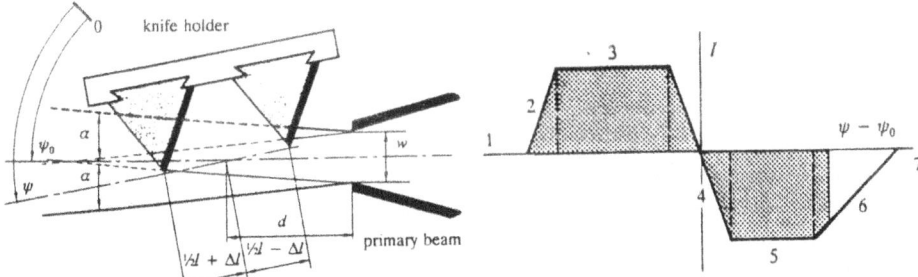

Fig. 5 Construction and beam geometry of the double knife edge.

Fig. 6. Two profiles of the intensity as a function of the co-ordinate perpendicular to the beam direction, which has been transformed int terms of $\psi - \psi_0$. These profiles are active during the double knife edge scan. The numbers correspond to the functions derived in the text.

4. $x_0 + \frac{1}{2}w - \frac{1}{2}\alpha d \leq x < x_0 + \frac{1}{2}w + \frac{1}{2}\alpha d$

$$I = I_0 - \frac{(x_0 - x + \frac{1}{2}w + \frac{1}{2}\alpha d)^2}{2\alpha d} I_0 + I_b \qquad (1d)$$

5. $x \geq x_0 + \frac{1}{2}w + \frac{1}{2}\alpha d$

$$I = I_0 + I_b \qquad (1e)$$

In order to obtain the value of x_0 from the equations 1a to 1e they have to serve as the model function in a Levenberg-Marquardt fitting procedure [2]. Input to this fitting procedure are the scan data. The free parameters are: x_0, I_0, I_b, w, and the product αd. An example of a successful fit is given in figure 4

Now one point of the line that defines the direction of the primary beam is obtained. At this position, the double knife edge scan which gives a second point on this line, will be performed. The error in the thus found x_0 is usually less than 0.01 mm.

3.4.   THE DOUBLE KNIFE EDGE SCAN

Next to the single knife edge a second − identical − knife is mounted. The construction of the knives and the holder (figure 5) is such that the connection line between the two knife edges and the backside of the knife holder is parallel. The co-ordinate parallel to the beam direction will be chosen such that the point between

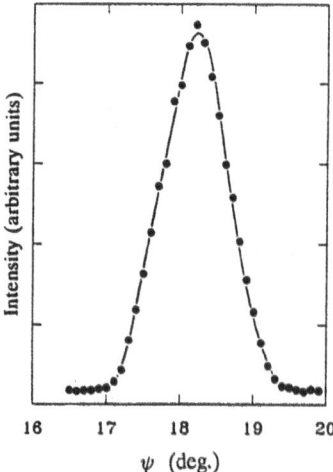

Fig. 7 Result of the double knife edge scan. The solid line is the result of a Levenberg-Marquardt fitting procedure of the function that was given in the text (equation 3).

the two knives will approximately coincide with the central axis (misplacement Δl). At this point a ψ-scan is carried out from the situation that one of the knives is completely covering the primary beam to the situation where the other knife is fully covering the beam. Between these extremes a maximum, corresponding to the situation that the two knife edges are in line with the beam, exists.

The scan variable of the double knife edge scan is ψ, so the purpose of the scan is to determine the ψ-value at which the line connecting the two knives is parallel to the primary beam centre line. This ψ-value will be called ψ_0.

The beam geometry is defined by: α and d and w. In the following it is assumed that the scan starts at the situation where the knife closest to the primary beam defining aperture, covers the beam, so that only background radiation I_b is detected. The direction of ψ is chosen to be counterclockwise.

In figure 6 the intensity profile of the beam at the counter side of the double knife edge is given. The detected intensity $I(\psi - \psi_0)$ is proportional to the hatched surface under the curve. It should be noted that the surface under the right hand part of the curve has a negative contribution. For the derivation of the next equations we have assumed the following geometrical condition:

$$\tfrac{1}{2}\alpha < \frac{\tfrac{1}{2}w - \tfrac{1}{2}\alpha(d + \tfrac{1}{2}l - \Delta l)}{\tfrac{1}{2}l - \Delta l} \qquad (2)$$

This corresponds to the situation that the divergence spread out by the first knife seen from the beam aperture (*c.f.* figure 5) is covered by the second knife before it enters the lower divergence wedge of the primary beam aperture. The intensity function can now be divided into 7 ψ-regions:

1. $\psi - \psi_0 < \dfrac{-\frac{1}{2}w - (d - \frac{1}{2}l - \Delta l)\frac{1}{2}\alpha}{\frac{1}{2}l + \Delta l}$

$$I = I_b \tag{3a}$$

2. $\dfrac{-\frac{1}{2}w - (d - \frac{1}{2}l - \Delta l)\frac{1}{2}\alpha}{\frac{1}{2}l + \Delta l} \leq \psi - \psi_0 < \dfrac{-\frac{1}{2}w + (d - \frac{1}{2}l - \Delta l)\frac{1}{2}\alpha}{\frac{1}{2}l + \Delta l}$

$$I = \dfrac{\left\{(\psi - \psi_0)(\frac{1}{2}l + \Delta l) + \frac{1}{2}w + (d - \frac{1}{2}l - \Delta l)\frac{1}{2}\alpha\right\}^2}{2(d - \frac{1}{2}l - \Delta l)\alpha} I_0 + I_b \tag{3b}$$

3. $\dfrac{-\frac{1}{2}w + (d - \frac{1}{2}l - \Delta l)\frac{1}{2}\alpha}{\frac{1}{2}l + \Delta l} \leq \psi - \psi_0 < -\frac{1}{2}\alpha$

$$I = \left\{(\psi - \psi_0)(\frac{1}{2}l + \Delta l) + \frac{1}{2}w\right\} I_0 + I_b \tag{3c}$$

4. $-\frac{1}{2}\alpha \leq \psi - \psi_0 < \frac{1}{2}\alpha$

$$I = \left\{\frac{1}{2}w + (\psi - \psi_0)(\frac{1}{2}l + \Delta l)\right\} I_0 - \dfrac{l\left\{(\psi - \psi_0) + \frac{1}{2}\alpha\right\}^2}{2\alpha} I_0 + I_b \tag{3d}$$

5. $\frac{1}{2}\alpha \leq \psi - \psi_0 < \dfrac{\frac{1}{2}w - (d + \frac{1}{2}l - \Delta l)\frac{1}{2}\alpha}{\frac{1}{2}l - \Delta l}$

$$I = \left\{\frac{1}{2}w - (\psi - \psi_0)(\frac{1}{2}l - \Delta l)\right\} I_0 + I_b \tag{3e}$$

6. $\dfrac{\frac{1}{2}w - (d + \frac{1}{2}l - \Delta l)\frac{1}{2}\alpha}{\frac{1}{2}l - \Delta l} \leq \psi - \psi_0 < \dfrac{\frac{1}{2}w + (d + \frac{1}{2}l - \Delta l)\frac{1}{2}\alpha}{\frac{1}{2}l - \Delta l}$

$$I = \dfrac{\left\{\frac{1}{2}w + (d + \frac{1}{2}l - \Delta l)\frac{1}{2}\alpha - (\psi - \psi_0)(\frac{1}{2}l - \Delta l)\right\}^2}{2(d + \frac{1}{2}l - \Delta l)\alpha} I_0 + I_b \tag{3f}$$

7. $\psi - \psi_0 \geq \dfrac{\frac{1}{2}w + (d + \frac{1}{2}l - \Delta l)\frac{1}{2}\alpha}{\frac{1}{2}l - \Delta l}$

$$I = I_b \tag{3g}$$

These equations serve to obtain ψ_0 from the data of a double knife edge scan, using a fitting procedure. The free parameters are Δl, w, I_0, I_b, α, and ψ_0.

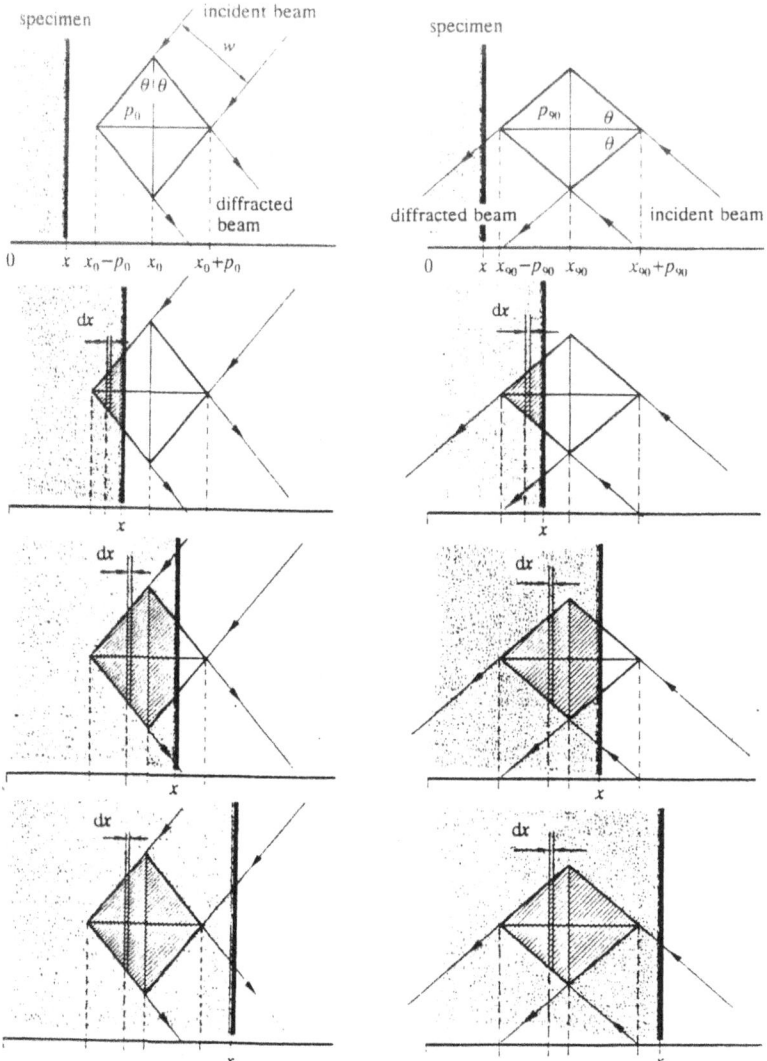

Fig. 8 The beam geometry during four distinct stages of the entering curve scans at $\psi = 0°$ (left sequence) and at $\psi = 90°$ (right sequence).

An example of a successful fit is given in figure 7. The statistic error in ψ_0 is usually less than 0.01°.

While maintaining the specimen table at ψ_0, a steel ruler is connected to the back side of the knife holder. Using a dial gauge connected to one of the moving parts of the primary aperture, the ruler is used as a physical reference to the beam direction.

Adjustment bolts on the primary aperture assembly serve to adjust the slide system in an orientation parallel to the primary beam.

In order to adjust the secondary aperture sliding system, the neutron counter is moved to an angle 2θ and the specimen table is rotated towards $\psi_0 + 2\theta$. Again a dial gauge serves to align the secondary aperture system.

3.5. CHECKING THE ALIGNMENT OF THE SYSTEM

Having performed the actions described in the foregoing sections, the diffractometer should be well aligned. This should, however, be checked before starting any measurements.

In this section a method will be treated that uses diffraction as the main tool to determine the state of alignment of the diffractometer. For this purpose a special test specimen is used. It consists of a rectangular box made of thin (0.1 mm thick) vanadium foil. The dimensions of the box are $200 \times 50 \times 10$ mm^3. The box is filled with iron powder *p.a.* and closed with epoxy resin in order to prevent oxidation.

On this sample successive 2θ-scans are performed. After each scan, the specimen is moved a small distance into the direction of its surface normal. This process proceeds from the situation where the gauge volume is located completely outside the specimen to the situation where the gauge volume is located completely inside the specimen. The measurement results are presented as the integrated intensity ΣI of diffraction peaks as a function of the relative specimen position. Qualitatively, at any ψ, the result is an s-curve starting at a background value I_b, in order to arrive at a maximum where the rhombus is completely embraced by the specimen. We will now model entering curves for $\psi=0°$ and $\psi=90°$.

3.5.1. *Entering curve at $\psi = 0°$* The primary and secondary beam apertures, both of width w define a rhombic prism in space. The rhombus is defined by w and 2θ (see figure 8). In the well aligned condition, the point of intersection of both diagonals of the rhombus coincides with the centre of rotation of the diffractometer. At specimen position x_0, the intersection of the diagonals of the rhombus lies in the specimen surface.

The measured integrated intensity ΣI from a partly occupied region (characterized by x) is proportional to the occupied part of gauge volume. One aspect of the $\psi = 0°$ case is that the intensity, obtained from a co-ordinate somewhere below the specimen surface, is attenuated along the total travelling path for a neutron. This is taken into account by an exponential absorption factor, which includes the absorption coefficient μ.

In the left part of figure 8 the cross-section of the stationary rhombic prism is shown together with the moving specimen. The function that describes the integrated intensity as a function of the relative specimen position is divided into 4 x-regions. In the next equations ΣI_0 is a proportionality constant, ΣI_b is the background intensity and p_0 is half the rhombus diagonal which is oriented perpendicular to the specimen. It can be shown that

344

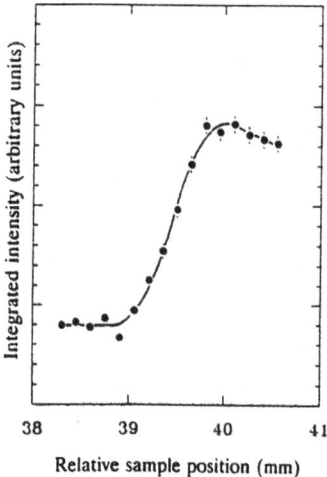

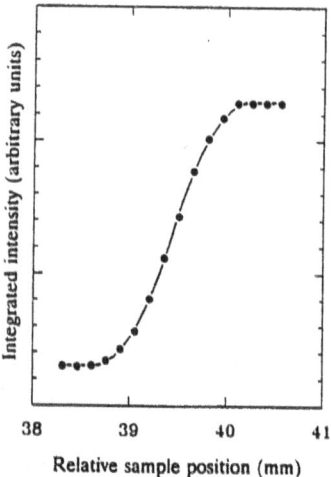

Fig. 9 Result for the entering curve at $\psi = 0°$, The solid line belongs to the function that was given in the text (equation 5).

Fig. 10 Result for the entering curve at $\psi = 90°$, The solid line belongs to the function that was given in the text (equation 7).

$$P_0 = \frac{\sin\theta}{\sin 2\theta} w \qquad (4)$$

The function that gives the integrated intensity as a function of the relative position x is:

1. $x < x_0 - P_0$

$$\Sigma I = \Sigma I_b \qquad (5a)$$

2. $x_0 - P_0 \leq x < x_0$

$$\Sigma I = \Sigma I_0 \frac{x - x_0 + P_0}{\mu} - \Sigma I_0 \frac{\sin\theta}{2\mu^2}\left\{ 1 - e^{\frac{-2\mu}{\sin\theta}(x - x_0 + P_0)} \right\} + \Sigma I_b \qquad (5b)$$

3. $x_0 \leq x < x_0 + P_0$

$$\Sigma I = \Sigma I_0 \frac{\sin\theta}{2\mu^2}\left\{ e^{\frac{-2\mu}{\sin\theta}(x - x_0 + P_0)} - 2e^{\frac{-2\mu}{\sin\theta}(x - x_0)} + 1 \right\} + \Sigma I_0 \frac{x_0 + P_0 - x}{\mu} + \Sigma I_b (5c)$$

4. $x \geq x_0 + p_0$

$$\Sigma I = \Sigma I_0 \frac{\sin\theta}{2\mu^2} \left\{ e^{\frac{-2\mu}{\sin\theta}(x - x_0 + p_0)} - 2e^{\frac{-2\mu}{\sin\theta}(x - x_0)} + e^{\frac{-2\mu}{\sin\theta}(x - x_0 - p_0)} \right\} + \Sigma I_b \quad (5d)$$

Equations 5a to 5d give the expected integrated intensity as a function of x. From the measurement of ΣI at a series of x-values, the values of the unknown parameters can be obtained by a fitting process. The free parameters are: x_0, w, ΣI_0, ΣI_b and μ. An example of a successful fit is given in figure 9. The statistic error found in the result for x_0 is less than 0.02 mm.

3.5.2. *Entering curve at* $\psi = 90°$ For the situation where $\psi = 90°$, *i.e.* when the diffraction vector q is parallel to the specimen surface, the derivation of the integrated intensity as a function of the relative sample position is less complicated as compared to the $\psi = 0°$ case. In this case, the total travelling path for neutrons in the specimen material does not depend on the position in the specimen. Therefore all absorption information will be hidden in the parameter ΣI_0.

The geometry for this case is given in the right part of figure 8. The parameter that symbolizes the situation where the intersection point of the two rhombus diagonals meets the specimen surface is called x_{90}. The length of the diagonal that is now perpendicular to the specimen surface is called p_{90}, its value is

$$p_{90} = \frac{\cos\theta}{\cos 2\theta} w \quad (6)$$

The model that gives the integrated intensity as a function of relative specimen position is divided into 4 x-regions:

1. $x < x_{90} - p_{90}$

$$\Sigma I = \Sigma I_b \quad (7a)$$

2. $x_{90} - p_{90} \leq x < x_{90}$

$$\Sigma I = \Sigma I_0(x - x_{90} + p_{90})^2 + \Sigma I_b \quad (7b)$$

3. $x_{90} \leq x < x_{90} + p_{90}$

$$\Sigma I = \Sigma I_0 \left\{ 2p_{90}^2 - (x_{90} + p_{90} - x)^2 \right\} + \Sigma I_b \quad (7c)$$

4. $x \geq x_{90} + p_{90}$

$$\Sigma I = 2\Sigma I_0 p_{90}^2 + \Sigma I_b \tag{7d}$$

Equations 7a to 7d give the expected integrated intensity as a function of x and a number of parameters. From the measurement of ΣI at a series of x-values, the values of the unknown parameters can be obtained by a fitting process. The free parameters are: x_{90}, w, ΣI_0, and ΣI_b. An example of the entering curve at $\psi = 90°$ is given in figure 10. The statistic error in x_{90} is less than 0.03 mm. x_0 and x_{90} are usually found to be within their respective error limits.

4. Conclusion

Using fitting functions which result from geometrical considerations of the primary and diffracted neutron beams, an adequate state of alignment can be achieved. It is beyond the scope of this paper to give general equations that apply for arbitrary specimen shapes, however, these can be found along the same line of thoughts as the ones presented here.

5. References

1   P.C. Brand, *Stress measurements by means of neutron diffraction*, PhD-thesis, Enschede, pp 131-148, (1991).

2   W.H. Press; B.P. Flannery; S.A. Teukolsky and W.T. Vetterling, *Numerical Recipes*, Cambridge University Press, 509ff (1986).

PROPOSAL FOR A NEUTRON STRAIN MEASUREMENT APPARATUS

M. KOCSIS and J. KULDA
Institut Laue Langevin
156X
38042 Grenoble CEDEX
France

ABSTRACT. The design of a dedicated diffractometer for strain measurements is considered. The proposed setup makes use of bent perfect silicon crystals both for incident beam monochromatization and for the analysis of the Bragg angle variations in the diffracted beam. The implementation of an asymmetrically cut analyzer facilitates good gauge volume definition and makes possible the use of a PSD for simultaneous recording of the examined reflection profile.

1. Introduction

Neutron diffraction is a unique technique for non-destructive measurement of internal strain field in bulk samples. The strain (which is a relative change in lattice spacing, d) is determined using Bragg's law of diffraction. Differentiating Bragg's law for a given reflection,

$$\varepsilon = \Delta d/d = -\cot\theta \cdot \Delta\theta \qquad (1)$$

indicating, that lattice strain ε causes change in Bragg angle θ. The measured variation in scattering angle is $\Delta\Phi = 2\Delta\theta$. The direction of the measured component of strain in the sample is parallel to the scattering vector.

The probe region (volume) for mapping residual stress distribution in the sample is defined by the use of apertures in the incident and diffracted beams. The optimum scattering angle for definition of the sampled volume is $\Phi = 2\theta$

347

M. T. Hutchings and A. D. Krawitz (ed.),
Measurement of Residual and Applied Stress Using Neutron Diffraction, 347–352.
© 1992 *Kluwer Academic Publishers*.

= 90°. Because internal strains are small, typically $\Delta d/d \approx 10^{-3}$-$10^{-4}$, a high resolution powder diffractometer (HRPD) is required. The instrumental resolution (FWHM) of the fixed wavelength λ diffractometer is given by [1]

$$(\Delta d/d)^2 = (\Delta\lambda/\lambda)^2 + (\cot g\theta . \Delta\theta)^2 \quad . \tag{2}$$

Usually peaks of interest are about 2° wide at the base and step scans are made in 0.1° increments so that conventional detectors require 20 points per scan. Application of a position sensitive detector (PSD) can greatly enhance the measurement speed, but the receiving slit geometry is complicated by the need to place the diffracted beam slit (for the best observed volume definition) close to the sample. The most important consequence of this geometry is that whenever the observed peak is not precisely in the center of PSD, any displacement of the diffraction slit will lead to shift in peak position.

To overcome the above mentioned problems - long measuring time and the artificial peak shift - we propose a new diffractometer configuration. The base point of the proposal is the fact that we analyse only one peak in the diffraction pattern. It seems, that by using a bent perfect Si single crystal as an analyser and a linear PSD (cf. fig.1) we get an arrangement, which has the advantages of high resolution and short measuring time.

2. The resolution function.

In what follows we shall assume that the reflection by the monochromator as well as by the analyzer can be described by a Gaussian profile

$$\sigma_M(\omega_1, \lambda) = \exp\left[-\left(\omega_1 - \frac{\lambda - \lambda_0}{2d_M \cos\theta_M}\right)^2 \eta_M^{-2}\right] \tag{3}$$

with $\omega_1 = \theta - \theta_M$, $2d_M \sin\theta_M = \lambda_0$ and mosaic spread η_M. For a more accurate analysis one may use more exact analytical formulas for the rocking curves of bent crystals [2], however, their integrals and convolutions have to be calculated numerically.

The reflectivity of the sample, neglecting any kind of broadening due to grain size etc., will be represented by a delta-function

$$\sigma_S(\omega_2, \lambda) = \delta\left[\omega_2 + 2\left(\theta_S - \theta_M + \frac{\lambda - \lambda_0}{2d_S \cos\theta_S}\right)\right] \tag{4}$$

where similarly to the monochromator case $2d_s \sin\theta_s = \lambda_o$ and $\omega_2 = \theta - \theta_s$. The angular profile of the beam as seen by a detector is obtained by the convolution

$$\sigma_{MS}(\omega_2,\lambda) = \int_{-\infty}^{\infty} \exp\left[-(\omega_1 - \xi_M)^2 \eta_M^{-2}\right] \cdot \delta\left[\omega_2 + 2(\theta_s - \theta_M + \xi_s) - \omega_1\right] d\omega_1 =$$

$$= \exp\left[-(\omega_2 + 2(\theta_s - \theta_M) + 2\xi_s - \xi_M)^2 \eta_M^{-2}\right] \qquad (5)$$

where ξ_M (ξ_s) is the dispersion at monochromator (sample). We see that for the nominal wavelength λ_o the center of the profile is shifted by $2(\theta_s - \theta_M)$ from the position for a parallel setup with identical crystals. For stress measurements the magnitude of this shift or better its variation $\delta\theta$ due to variations of d_s of eq.(1) is of primary interest. In this double-axis setup, however, the shift is blurred by the wavelength dependent term

$$2\xi_s - \xi_M = (\lambda - \lambda_o)\left(\frac{2}{2d_s \cos\theta_s} - \frac{1}{2d_M \cos\theta_M}\right) \qquad (6)$$

which leads to a significant broadening of the reflection profile as for any given angular position ω_2 of the detector an integral over the whole available wavelength spread $\Delta\lambda$ is registered.

When an analyzer crystal, identical to the monochromator, is placed after the sample, the resulting reflection profile changes to

$$\sigma_{MSA}(\omega_3,\lambda) = \int_{-\infty}^{\infty} \exp\left[-(\omega_2 + 2(\theta_s - \theta_M) + 2\xi_s - \xi_M)^2 \eta_M^{-2}\right]$$

$$\exp\left[-(\omega_3 - \xi_M - \omega_2)^2 \eta_M^{-2}\right] d\omega_2 = \qquad (7)$$

$$= \exp\left[-2(\theta_s - \theta_M + \xi_s - \xi_M - \omega_3)^2 \eta_M^{-2}\right]$$

Now the dispersion-broadening term instead of (6) becomes just $(\xi_s - \xi_M)$ and it can be significantly suppressed by choosing $d_s = d_M$. For identical interplanar distances the wavelength spread does not give rise to any increase in profile width, the resolution is only related to the misorientation spread η_M of the monochromator and analyzer crystals. It has, however, to be emphasized that both of them have to provide peak reflectivities very close to unity (as bent perfect crystals do) otherwise severe losses in intensity are encountered.

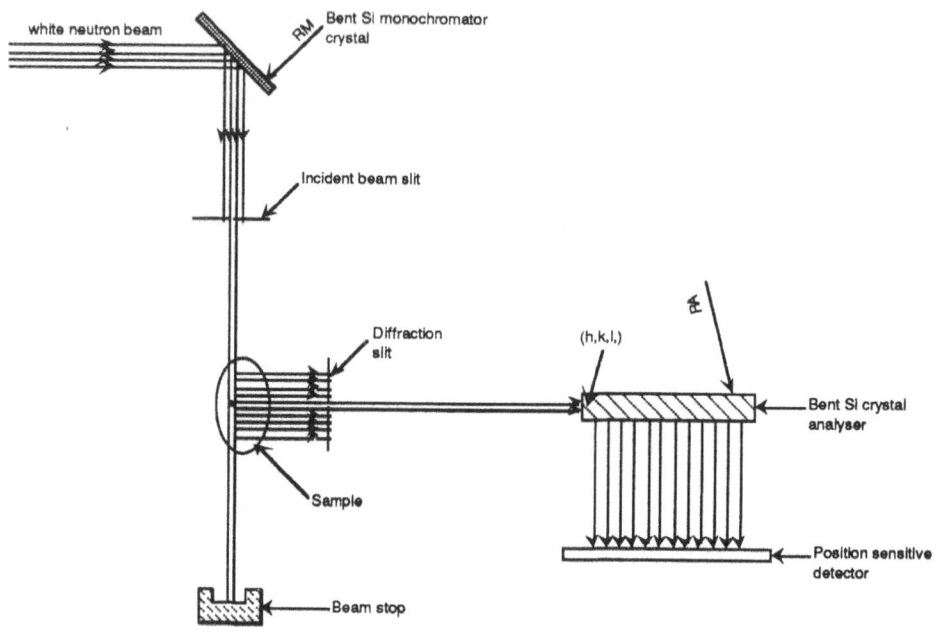

Fig.1. Schematic arrangement of the proposed apparatus

3. Diffraction by bent crystals.

The rocking curve of a perfect single crystal plate of thickness D elastically bent with radius R can be approximated by a rectangle of width

$$\delta\theta = \frac{D}{R}\cot g\theta \tag{8}$$

θ being the Bragg angle (symmetric reflection geometry is assumed) and height

$$r = 1 - \exp(-QR / \cos\theta) \qquad (9)$$

with

$$Q = \frac{F_G^2 \lambda^3}{\Omega^2 \sin 2\theta} \qquad (10)$$

denoting the kinematical reflectivity [3]. For a more involved analysis an analytical expression describing correctly the tails of the rocking curve is available [2]. The integrated reflecting power is limited by the mechanical properties of silicon; the minimum bending radius of a plate of thickness 5 mm is about 10 m; for strong low order reflections and wavelengths about 2 Å such deformation still does not give rise to any important decrease of the reflectivity r. The corresponding angular width is of the order of a few minutes of arc matching well the required resolution of $\Delta d/d = 10^{-3} - 10^{-4}$.

4. Experimental setup.

In order to benefit from the improved resolution due to the use of a bent analyser crystal we propose the following experimental arrangement, schematically presented in fig.1. The neutron beam from a reactor beam hole or from a neutron guide is monochromated by a perfect silicon crystal bent to a radius about 10 m; the reflection and wavelength are to be chosen so as to provide a take-off angle about 90°. A combination of input and output slits near to the sample position defines the diffracting volume. The sample reflection is chosen to produce a scattering angle as close as possible to 90°. As an example may serve the combination of silicon (004) at $\lambda = 1.92$ Å ($2\theta = 90°$) with iron (002) ($2\theta = 84.3°$ at the same wavelength).

The analyzer parameters are the same as those for the monochromator. However, an asymmetrically cut crystal is used which makes possible to employ a PSD to collect and analyze the whole reflection profile simultaneously, making use of the 1:1 relation between position and local misorientation along the beam path in the analyzer crystal [4]. The analyser is cut asymmetrically with incident beam passing along the plate (see fig.1). The local misorientation of the crystal lattice is proportional to the path S_o travelled by the incident beam, $\Delta\theta = \Delta\theta_o + S_o/R$. Hence from the position, at which reflection of a given neutron took place its angular deviation according to eq.(7) can be deduced. A spatial uncertainty of 1-2 mm, limited both by properties of a PSD and by the beam divergence, corresponds to an angular resolution of the order of 10^{-4}rad. The total length L of the crystal plate determines the angular range $\Delta\theta_{max}$ = L/R which can be analysed simultaneously, in practice L is limited by transmission losses of the material. With silicon, featuring a total cross section of the order of 0.2 barn at neutron wavelengths about 2 Å [5], a reasonable choice is L $\approx$ 200 mm. We then have $\Delta\theta_{max} \approx 0.02$ rad for R = 10 m, a value which corresponds to maximum measurable strain.

5. Concluding remarks.

According to preliminary calculations the proposed setup should provide not only a strongly improved resolution as compared to present HRPDs but should at the same time offer a gain in luminosity by decoupling of the $\Delta d/d$ resolution from the beam divergence (and hence from the wavelength spread) and permitting to record simultaneously the whole profile of the diffraction line. The latter conclusion will be subject to careful experimental checks, the real situation may be also strongly influenced by the particular sample properties. The gain will obviously be the greatest for samples providing sharp diffraction lines, permitting to make use of the inherent high angular resolution of the setup.

References.

1. Allen, A. J., Hutchings, M. T., Windsor, C. G. & Andreani, C. (1985), "Neutron Diffraction Methods for the Study of Residual Stress Fields", Adv. Phys., 34, 445 - 473.
2. Kulda, J.,&Lukas, P., (1989), "Plane Wav e Diffraction by a Deformed Crystal", Phys. Stat. Sol. vol.(b), 153, 435 - 442.
3. Kulda, J., (1984), "A Novel Approach to Dynamical Neutron Diffraction by a Deformed Crystal", Acta Cryst., A 40,120 - 124.
4. Mikula, P., Lukas, P., & Eichhorn, F., (1988), " A New Version of a Medium - Resolution Double - Crystal Diffractometer for the Study of Small - Angle Neutron Scattering (SANS)", J. Appl.Cryst., 21, 33 - 37.
5. Freund, A., (1983), "Cross Sections of Materials Used as Neutron Monochromators and Filters", Nucl. Instr. Meth. 213, 495 - 501

MICROBEAM TECHNIQUES IN DIFFRACTION: A THEORETICAL TREATMENT

I. C. NOYAN
IBM Research Division
Yorktown Heights
NY 10566, USA

ABSTRACT

This talk will focus on two main topics. In the first half, we will discuss the idea of the "representative" volume for deformation and its determination with diffraction. In the second half, we will treat the kind of deformation fields sampled by microbeam techniques and the criteria for selecting the "correct" equations linking the stress to strains in such cases.

M. T. Hutchings and A. D. Krawitz (ed.),
Measurement of Residual and Applied Stress Using Neutron Diffraction, 353.
© 1992 *Kluwer Academic Publishers.*

SUMMARY OF THE PANEL DISCUSSION ON INSTRUMENTATION AT STEADY STATE SOURCES

T. LORENTZEN
Materials Department, Risø National Laboratory,
P.O. Box 49, DK-4000, Roskilde, Denmark.

P. C. BRAND
Netherlands Energy Research Foundation ECN,
P.O. Box 1, 1755 ZG, Petten, The Netherlands.

1. Introduction

The evolution of the neutron diffraction technique for internal stress/strain determination has initiated a new branch of instrumental developments at many steady state reactors around the world. The technique has some instrumental requirements which are usually not encountered in other neutron diffraction applications. One of the requirements is the application of apertures that define the size and the position of the primary and diffracted beams, which in turn define the location and the size of the gauge volume within the specimen. Another − related − requirement is that the specimen, which may have unusually large and awkward dimensions, must be accurately positioned with respect to the location of the gauge volume. Furthermore the instrumentation should provide accurate movement possibilities for the specimen once it is installed. As measurements are often extremely time consuming due to the use of very small gauge volumes and relatively long through-specimen beam path lengths, the experimental configuration must be optimized such that the measurement time consumption is minimum.

2. Gauge volume size and definition

An instrumental issue of major concern is the gauge volume definition. In most other fields of neutron scattering, the specimen is simply embedded in the neutron beam. On the other hand, for the determination of internal strains in structural components, the aim is to record a strain value from a given (*hkl*)-lattice spacing at a specific position in the

Present adress: National Institute of Standards and Technology (NIST)
                      Gaithersburg, MD 20899, U.S.A.

M. T. Hutchings and A. D. Krawitz (ed.),
Measurement of Residual and Applied Stress Using Neutron Diffraction, 355–359.
© 1992 *Kluwer Academic Publishers.*

specimen. This calls for an accurate definition of a small gauge volume, which can be located anywhere inside the component. The gauge volume, which is usually discussed in terms of square boxes of say 1 mm^3 to 10 mm^3, is defined by the beam apertures, the scattering angle, the beam collimation and the distance between the beam apertures and the specimen.

The appropriate sizes of the beam apertures which define the gauge volume depend on the strain gradient in the direction of a co-ordinate of interest. In general it is desirable to make the gauge volume dimension as small as possible in the direction of that gradient. It is the general view that the instrumentation should allow for aperture sizes as small as 0.5 mm, although for most applications 1 - 2 mm is believed to be sufficient. For exceptional cases, the apertures should be adjustable to dimensions as large as 10 - 20 mm. Whenever possible one or two of the gauge volume dimensions should be enlarged in a direction where the stress gradient is zero, in order to improve the scattered intensity.

In terms of how to define gauge volumes inside a given component it was the general view that a crucial parameter is the specimen to aperture distance and the collimation for both the incident and diffracted beams. Therefore, in order to achieve an optimum gauge volume definition, the instrumentation should be equipped with apertures that can be moved to a position immediately next to the specimen surface.

A counter angle of $2\theta = 90°$ is usually found to provide the optimum gauge volume definition. In fact most reported gauge volumes are of nearly rectangular shape. For some applications $e.g.$ the study of steep strain gradients near surfaces, it will actually be beneficial for the spatial resolution to deliberately move towards a back scattering configuration ($2\theta \approx 180°$). This will enlarge the gauge volume dimensions perpendicular to the strain gradient and minimize the dimensions parallel to the gradient. Also a configuration where 2θ is very small is a possibility in some steep gradient cases. However, this situation results in low sensitivity to changes in interplanar spacing because of the nature of Bragg's law.

When the gauge volume dimensions are as small as a few cubic millimetres, the initial specimen installation and alignment procedures become of great importance. Many ideas have been presented on adequate ways of aligning specimens. The most popular ones appear to be those based on laser beams and on theodolite systems. In the laser based systems, a well defined laser beam can be made to coincide with the neutron beam, providing a visual inspection of the incident beam path as well as an inspection of where the beam enters the specimen surface. Through the use of mirrors or by moving the diffraction arm to $2\theta = 0°$ before specimen installation, inspection of the position of the beam on the sample and the path of the diffracted beam can be made. A theodolite based system is either a fixed or a portable system like the ones commonly used for the levelling of buildings. This system also allows for a visual inspection of the specimen positioning on the spectrometer table. These light based systems provide a comfortable way of visual inspection of the alignment. The final alignment procedure involves step-scanning through the surface, either manually or under computer control, in order to determine the precise position of the surface as reckoned by the actual neutron optics.

3. Instrumental Resolution

Measurement of internal strains by means of neutron diffraction requires a relatively good instrumental resolution. Therefore, some discussion was focused on the relation between instrumental resolution and what was eventually defined as the strain sensitivity. It was concluded that most high resolution powder instruments around the world provide adequate instrumental resolution for determination of internal strains. It should be mentioned, however, that the $\Delta d/d$ reported as the Full Width at Half Maximum (FWHM) of the instrumental resolution function, is orders of magnitude larger than the strain sensitivity required for many internal strain investigations. In general it is acceptable to record internal strains to an accuracy between $\pm 5 \times 10^{-5}$ and $\pm 1 \times 10^{-4}$ which is usually 50 - 100 times smaller than the FWHM of the Bragg peak. We are thus looking for extremely small peak shifts in terms of FWHM.

The instrumental resolution is first of all a measure of the instrument's capability to resolve variations in peak shapes and to resolve overlapping peaks. As such, a good instrumental resolution is crucial in the evaluation of whether the peak fitting function (which is often supposed to be of Gaussian shape) is a good description of the measured diffraction peak profile. If the instrumental resolution is poor, deviations from the expected peak shape can, through the numerical peak fitting, mistakenly be interpreted as a peak shift due to internal strains.

Even for a perfectly Gaussian profile, the statistical error in the peak position of any applied numerical peak fitting procedure strongly depends on the FWHM of the measured peak. So although the instrumental resolution seems adequate at most high resolution powder diffractometers, as a figure of merit it can not be neglected.

It was also pointed out that whereas a strain sensitivity between $\pm 5 \times 10^{-5}$ and $\pm 1 \times 10^{-4}$ is adequate for the determination of internal strains in metal based components, this sensitivity might not be sufficient for the internal strain determination in ceramic based components. Ceramics are usually very stiff, so the internal strain level is often expected to be much lower than in the metal based systems.

The conclusion is hence that the present resolution of most high resolution powder diffractometers is adequate for many investigations of internal strains in metal based structural components. However, in search of new areas of application − such as structural ceramics − the instrument developers should keep pressing the level of instrumental resolution to the lowest possible values.

4. Experimental configuration

Measurement of internal stresses in gauge volumes as small as a few cubic millimetres, at locations deep inside the material, requires high neutron flux and/or very efficient data acquisition systems. Furthermore, as the acceptable strain sensitivity is relatively small as compared to the spatial variation of the internal strains, specimen alignment and positioning becomes of outmost importance. Therefore, much effort is focused on developing new and efficient experimental configurations.

More than a dozen different configurations were presented during the workshop, each presenting new and advantageous aspects. However, none of them is regarded as the most efficient set-up for all types of problems. It was the general view that instrument developers should focus their attention on optimizing the detector systems in order to make more use of the diffracted neutrons, which are currently wasted during the experiments. This turns the discussion away from the use of standard double and triple axis instruments that are equipped with single detectors and towards the use of multi-detectors or position sensitive detectors (PSD's).

Most of the presented experimental configurations involved the use of multi detectors or PSD's. In general the detector systems utilizing PSD's are either used to record diffracted neutrons from different gauge volumes simultaneously, or alternatively these detectors are used to record complete diffraction peaks in single exposure type of measurements. As such the application resembles the multi detector bank systems, which often must be scanned through a limited range of angles in order to record a complete peak.

In conjunction with the discussion on the use of position sensitive detectors, some attention has been focused on the development of radial Soller-collimators. These will provide a PSD based system with a good spatial resolution for recording complete Bragg peaks from a single well defined gauge volume.

The above discussion has emphasized detector and collimation systems. However, some attention was focused on monochromator systems. Most instruments utilize a single monochromator crystal, which can be both tilted and vertically focused. Besides this, the option of a double monochromator system was also discussed. The double monochromator system has the advantage that the wavelength can be altered without any movement of spectrometer components outside the monochromator system. This in turn should be beneficial for the accuracy, as the movement of the larger spectrometer components, which must be positioned within fractions of a degree, is omitted. Double monochromators offer the possibility to perform wavelength scans as an alternative to scanning the diffraction angle. This is favourable in view of beam path attenuation effects and diffraction cone curvature errors which both depend on the scattering angle and can potentially cause errors in the determination of peak positions. It should be mentioned that, while some laboratories are currently establishing such double monochromator facilities, one has recently been dismantled at ILL in Grenoble. This was presumably due to the relatively high intensity loss in the double monochromator system.

5. Concluding Remarks

One of the aims of the panel discussion was to shed some light on what should be incorporated in an ideal instrument for internal stress determination by means of neutron diffraction. Many ideas were discussed and the above only highlights some of the topics which are believed to be of major importance. Though much attention has been given to PSD's and multi detector banks, there will still be much use of the standard single detector system as found on double and triple axis instruments.

The key word seems to be *flexibility*, and it was generally felt that such a thing as 'the ideal instrument' simply does not exist. The appropriate experimental configuration is to a great extent application dependent. Some applications may not work with PSD based systems and some systems may greatly benefit from using a PSD or a detector bank based system. For some applications it may seem appropriate to record diffracted intensity from an array of gauge volumes simultaneously and for others it may seem beneficial to focus all attention on a single gauge volume at a time. Also the specimen shape and dimensions may influence the choice of optimum apertures, collimators and detector systems.

Realizing that none of the mentioned experimental configurations will be the optimum choice for all applications, the advice to new instrument developers is to build a very flexible and versatile instrumental base *e.g.* such that both the incident and diffracted beam line is equipped with an optical rail system, which will allow for easy installation of a single detector, a position sensitive detector or a multi detector bank. Also the apertures and collimators, either parallel or converging, should be installed on the rail system, such that all items can be freely moved in and out on the spectrometer arms in order to optimize the distance between the specimen and these items. As specimen dimensions may vary enormously, the distance to the centre of the spectrometer should be adjustable between 0 and 1.5 - 2 m. Especially for a PSD based system with a converging collimator, the specimen to detector distance must be relatively large and also in this case 1.5 - 2 m seems appropriate.

For initial specimen installation and alignment, laser beams or systems based on a theodolite are adequate and should be incorporated in new developments. However, the traditional way of locating the specimen surface from measurements of diffracted intensity as the specimen is step by step shifted towards the gauge volume, will always be necessary for validation of specimen alignment.

5b INSTRUMENTATION : PULSED NEUTRON SOURCE

RESIDUAL AND APPLIED STRESS MEASUREMENTS AT IPNS

J. W. RICHARDSON, JR.
IPNS Division
Argonne National Laboratory
Argonne, IL 60439
USA

ABSTRACT. The Intense Pulsed Neutron Source (IPNS) at Argonne National Laboratory has operated as a user facility since 1981. From the early days on, highly precise measurement of structural parameters including lattice constants has been a major objective of developments on the powder diffractometers at IPNS. The General Purpose Powder Diffractometer (GPPD) combines excellent instrumental resolution - roughly independent of d-spacing - with good neutron flux to provide an exceptional tool for the measurement of residual and applied stresses.

1. Introduction

IPNS is a spallation neutron source, where diffracted intensities are measured as a function of Time-of-Flight (TOF) - related to neutron wavelength - at a series of fixed diffraction angles.   Whereas all TOF diffractometers share the characteristic of simultaneously collecting data over an extensive range of d-spacing, the actual range depends on the wavelength spectrum of the source and the complement of detectors in use. These factors may vary from one instrument to another, so it is important to recognize and exploit the unique characteristics of each.

As the user program at IPNS has matured, the types of residual stress experiments performed have progressed from direct collaborations with individuals (MacEwen et al. (1983,1984)) to routine interactions with external researchers (Krawitz et al. (1986,1987), Kupperman et al. (1989), Majumdar et al. (1988, 1989)). As the tremendous opportunities offered by neutron diffraction have become more apparent to the general community, we have worked to enhance our capabilities and strengthen interactions with scientists active in the field. This is most visibly illustrated by increased acquisition of beamtime by a number of industrial and governmental concerns.

2. Instrumentation

A detailed description of the design characteristics and performance of the two powder diffractometers at IPNS, the SEPD and GPPD, is given by Jorgensen et al. (1989). On the SEPD and GPPD scattering is approximately in the horizontal plane (±7°). Both instruments have eight banks of detectors positioned at 2θ angles ranging from ±15° to ±150°, with varying numbers of detectors in each bank. On the GPPD the banks are centered at scattering angles ±148° (31 detectors), ±90° (20 det.), ±60° (12 det.), 30° (8 det.) and -20° (8 det.). Each detector is 15" high and spans a 2θ

M. T. Hutchings and A. D. Krawitz (ed.),
Measurement of Residual and Applied Stress Using Neutron Diffraction, 363–367.
© 1992 *Kluwer Academic Publishers.*

range of 0.5°. Beam dimensions are adjustable via in-line collimation or additional slits, with a maximum allowable size of 2" high by 1/2" wide.

In addition to providing a nicely shaped gauge volume for all data, having banks of detectors at ±90° offers us the opportunity of measuring diffraction from the axial and radial directions of an oriented sample simultaneously, when the sample is positioned at 45° to the beam. Furthermore, the ±148° and ±60° banks provide additional data for orientations between axial and radial, such that roughly 1/3 of the entire range is covered. The instrumental resolution ($\Delta d/d$) for ±148° is 0.25%, for ±90° it is 0.40% and for ±60° 0.77%.

Diffraction profiles typically cover the time-of-flight range 3-30 milliseconds which corresponds to roughly 0.4-4.0 Å d-spacing. Individual diffraction peaks on the GPPD have a well-defined asymmetry (Von Dreele et al. (1982)), due to the pulsed nature of the source, fittable by convoluting leading and lagging exponentials with a Gaussian. Broadening effects such as microstrains alter only the Gaussian full-width.

In order to develop our ability to routinely handle a wide variety of residual stress samples, we have emphasized ancillary equipment capabilities. On the GPPD, data can be taken at essentially all temperatures from 10K to 1600K in a variety of sample configurations. We have a specially-designed translator-rotator which can be utilized for computer-controlled depth-profiling and 3D orientation analysis.

3. Data Analysis

Residual strains are typically calculated from d-spacings and full-width-half-maximum values obtained from singlet or multiplet peak fitting of key reflections. One such fit is illustrated in Figure 1 for data from a 30%Si_3N_4-Al_2O_3 ceramic composite.

Inherent to a TOF experiment, however, is the acquisition of many orders of those reflections defining unique crystallographic directions (e.g., (111), (222), etc.). These additional data can be used for confirmation of residual stress measurements from lowest order reflections, if systematic factors such as peak shape variation with TOF are understood. The Rietveld profile refinement technique (Rietveld (1968), Rotella

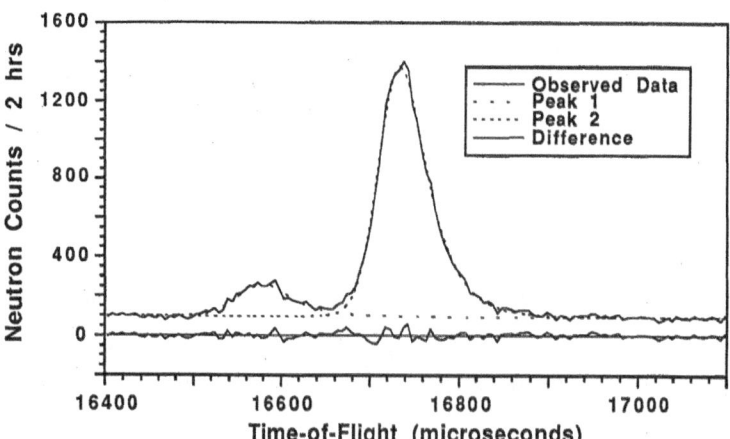

Figure 1. Typical least squares fit of two overlapping peaks (d~1.58 Å) from a Si_3N_4- Al_2O_3 composite. Note the slight asymmetry of the peaks which is inherent to the source.

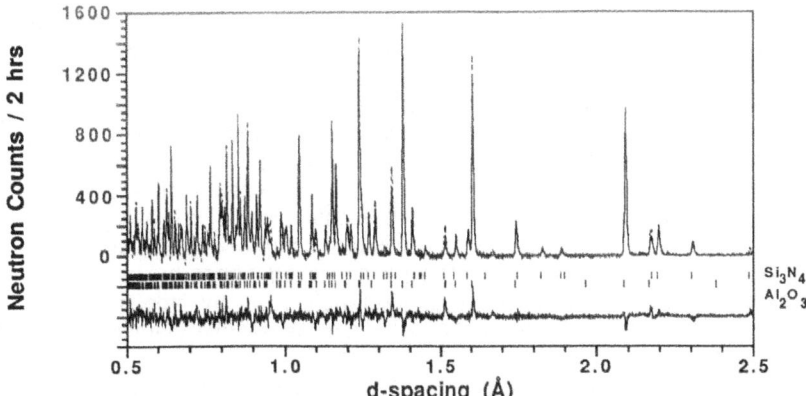

Figure 2. Rietveld profile fit for Si_3N_4-Al_2O_3 showing that: (a) both phases retain their structures, (b) peak breadths for both phases are modellable and (c) slight preferred orientation exists particularly in the second (Al_2O_3) phase.

(1989)) can be used to great advantage for this. A significant number of characteristics such as preferred orientation, anisotropic strain (peak breadths dependent on crystallographic direction) and strain-induced phase transitions can also be modelled within the framework of the Rietveld formalism. A Rietveld profile fit for the Si_3N_4-Al_2O_3 composite is shown in Figure 2.

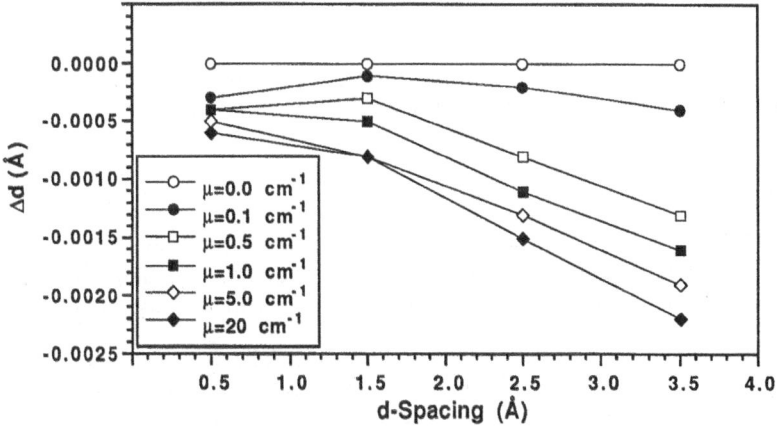

Figure 3. Monte Carlo calculations of the effect of absorption on apparent peak position on a time-of-flight diffractometer.

4. Analytical Concerns

The demand for very high precision and accuracy in residual stress analyses requires us to understand the absorption-weighted centroid of scattering for each sample. This is particularly critical when calculating stress-free parameters from standards with dramatically different bulk densities than the materials of interest. It is equally important to understand variations in sample position for different environments, e.g., a furnace versus a displex refrigerator. The importance of this can be calculated using Monte Carlo simulations of the instrumental characteristics. Shown in Figure 3 are plots of calculated Δd - representing shifts in peak positions - as a function of d-spacing for a series of possible sample linear neutron absorption coefficients. Such potential errors can be corrected either by using a stress-free standard with the same absorption characteristics or by applying an empirical correction. A similar effect can be expected when the sample diameter is changed as shown in Figure 4.

Although these errors are potentially severe, the Monte Carlo calculations appear to accurately reproduce the experimental behavior and empirical corrections appear to be sufficient when needed.

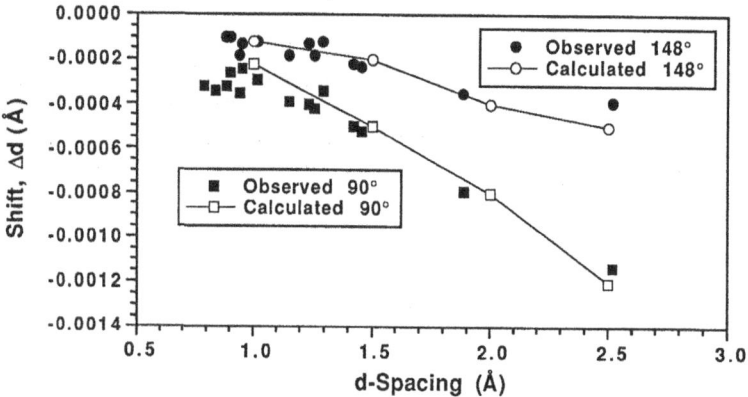

Figure 4. Apparent peak shifts due to changing from a $1/4$" diameter WC powder sample to a $7/16$" sample on the GPPD.

5. Conclusions

The GPPD powder neutron diffrctometer at IPNS combines good resolution with reasonable neutron flux to accomodate routine measurement of residual and applied stresses. Time-of-flight diffraction at fixed angle allows measurements from a geometrically fixed gauge volume throughout the d-spacing range. Lower time-average neutron fluxes relative to reactor sources can be partially overcome by examining changes in many reflections to obtain more complete strain information. This will be especially important as new mixed phase materials such as composites are studied in detail. Successfully exploiting some of the unique aspects of time-of-flight diffraction for stress measurements has broadened the scope of neutron diffraction as an non-destructive examination technique.

6. Acknowledgments

This manuscript has been authored by a contractor of the U.S. Government under contract No. W-31-ENG-38.

References

Jorgensen, J. D., Faber, Jr., J., Carpenter, J. M., Crawford, R. K., Haumann, J. R., Hitterman, R. L., Kleb, R., Ostrowski, G. E., Rotella, F. J. and Worlton, T. G. (1989) 'Electronically Focused Time-of-Flight Powder Diffractometers at the Intense Pulsed Neutron Source', *J. Appl. Cryst.*, **22**, 321.

Krawitz, A. D., Roberts, R. and Faber, Jr., J (1986) 'Residual Stress Relaxation in Cemented Carbide Composites', Science of Hard Materials, Institute of Physics Conference Series No. 75, pp. 577-589.

Krawitz, A. D., Reichel, D. G. and Hitterman, R. L. (1987) 'Residual Stress and Stress Distribution in a WC-Ni Composite', *J. of Mater. Sci. Eng.*, **A119**, 127-134.

Kupperman, D. S., Singh, J. P., Faber, Jr., J. and Hitterman, R. L. (1989) 'Application of Neutron Diffraction to the Characterization of Residual Strains in $YBa_2 Cu_3 O_{7-x}$ / Ag', *J. Appl. Phys.*, **66**, 3396-3398.

MacEwen, S. R., Faber, Jr., J. and Turner, A.P.L. (1983) 'The Use of Time-of-Flight Neutron Diffraction to Study Grain Interaction Stresses', *Acta Metall.*, **31**, 657-67.

MacEwen, S. R., Faber, Jr., J. and Turner, A.P.L. (1984) 'The Influence of Texture on the Interpretation of Diffraction Data to Determine Residual Stress', *Scripta Metall.*, **18**, 629-33.

Majumdar, S., Kupperman, D. S. and Singh, J. P. (1988) 'Determination of Residual Thermal Stresses in a SiC/Al_2O_3 Composite Using Neutron Diffraction', *J. Am. Ceram. Soc.*, **71**, 858-863.

Majumdar, S. and Kupperman, D. S. (1989) 'Effects of Temperature and Whisker Volume Fraction on Average Residual Thermal Strains in a SiC/Al_2O_3 Composite', *J. Am. Ceram. Soc.*, **72**, 312-313.

Rietveld, H. M. (1968) 'A Profile Refinement Method for Nuclear and Magnetic Structures', *J. Appl. Cryst.*, **2**, 65-71.

Rotella, F. J. (1989) 'Users Manual for Rietveld Analysis of Time-of-Flight Neutron Powder Diffraction Data at IPNS', Argonne National Laboratory, USA.

Von Dreele, R. B., Jorgensen, J. D. and Windsor, C. G. (1982) 'Rietveld Refinement with Spallation Neutron Powder Diffraction Data', *J. Appl. Cryst.*, **15**, 581-589.

RESIDUAL STRESS MEASUREMENT USING THE PULSED NEUTRON SOURCE AT LANSCE

M. A. M. BOURKE & J. A. GOLDSTONE
LANSCE
Los Alamos National Laboratory
New Mexico, 87545, USA

T.M. HOLDEN
AECL Research, Chalk River, Ontario, K0J 1JO

ABSTRACT. The presence of residual stress in engineering components can effect their mechanical properties and structural integrity. Neutron diffraction is the only technique which can make spatially resolved non-destructive strain measurements in the interior of components. By recording the change in the crystalline interplanar spacings, elastic strains can be measured for individual lattice reflections. Using a pulsed neutron source, all the lattice reflections are recorded in each measurement which allows anisotropic effects to be studied. Measurements made at the Manuel Lujan Jr. Neutron Scattering Centre (LANSCE) demonstrate the potential for stress measurements on a pulsed source and indicate the advantages and disadvantages over measurements made on a reactor.

1. Introduction

Residual stress measurements by neutron diffraction with spatial resolutions of a few millimeters in engineering components were first made in the early eighties on reactor sources [1,2]. However as the flux and reliability is improved at pulsed sources like LANSCE and IPNS in the United States and ISIS in the United Kingdom, their original limitations are diminishing and it is now becoming feasible to define and examine small sampling volumes in components. The demand for beam time at reactors remains high so pulsed sources can provide a useful and in some cases superior alternative.

In a standard diffraction experiment all the lattice reflections of a powder specimen are usually required. Typically reactors provide a continuous source of thermal neutrons which can be monochromated to provide relatively intense beams of collimated neutrons. By moving a detector around a specimen an intensity (2θ) scan can be obtained at a monochromatic wavelength. Monochromation of a neutron beam from a reactor uses only a small fraction of the available neutrons to produce the diffracted spectrum, however the intensity of even a monochromated beam is considerable.

By contrast at a pulsed source the integrated neutron flux over any wavelength range is much lower than on a reactor. However it is inherent in the operation of a pulsed source that all of the neutrons in each pulse can contribute to the measured spectrum making efficient use of the available neutrons. Each pulse represents a wavelength scan which is analogous to scanning in 2θ from 0 to 180°. When the complete diffraction pattern is required the timescale for measurement on pulsed sources is comparable with reactor sources.

Efficiency of measurement is crucial for examining components which involve small sampling volumes and large path lengths. The feature that works against a pulsed

369

M. T. Hutchings and A. D. Krawitz (ed.),
Measurement of Residual and Applied Stress Using Neutron Diffraction, 369–382.
© 1992 *Kluwer Academic Publishers.*

source in achieving comparable count times with steady state sources is that macroscopic strain measurements can often be inferred from just one or two lattice reflections. Thus only a small 2θ range is scanned on a reactor source and the pulsed source count times are long in comparison. Nevertheless measurements at LANSCE made in 1990 indicate that the intensity of pulsed sources has reached a level where it is feasible to define small (<100mm^3) sampling volumes in plate-like specimens and to make measurements in hours rather than days. A number of features have made this possible including the summation over all of the angle subtended by the 90° detectors, measurement in more than one scattering geometry, and analysis by the use of profile refinement.

The principle of measurement on a pulsed neutron source is identical to a steady-state source but practical differences and difficulties exist because of the time of flight analysis of the diffracted neutrons. Apart from the experimental difficulties, pulsed source measurements conveniently provide a comprehensive sample of the strain for all the measurable lattice directions. This information can elucidate both the grain interaction stresses and help to validate the use and selection of single reflections at reactor sources. The data can also be used to test the different models of material deformation. Stress rig and pressure cell experiments are easily accommodated and allow a range of lattice responses to be examined simultaneously.

Measurements at LANSCE demonstrate that a sampling volume of less than 40mm^3 could be defined in a flat specimen while maintaining sufficient neutron intensity to make measurements even after attenuation through 20mm of steel. Changes in lattice spacing in steel of 50 μstrain are resolvable giving a stress discrimination of ≈10MPa. The work described in this paper was performed using the Neutron Powder Diffractometer (NPD) at LANSCE although preliminary measurements on a 5mm thick steel ring were attempted at ISIS [3,4].

2. Anisotropic strain effects

Even in the elastic regime the manner in which solid polycrystalline materials deform is not well understood. Measurements using a stress rig have shown that the strain state in a single phase steel is far from the homogeneous assumption invariably used in engineering design [5]. When intergranular effects, elastic anisotropy and the anisotropic yield in grains of different orientations are considered, it is clear that the theoretical situation is very complicated.

Theoretical limits were proposed by Reuss [6] and Voigt [7]. Reuss proposed that each crystallite experienced the same stress field while Voigt suggested that each crystallite experienced the same strain field. The Voigt model implies that no strain anisotropy will be observed for different lattice reflections while the Reuss model predicts that strains in the polycrystalline aggregate will depend on the orientation of the crystallites with respect to the applied stress. A more realistic model was proposed by Kroner [8] in which crystallites are assumed to be embedded in a homogeneous elastic medium with elastic properties equal to those of the bulk material . The Kroner model has been applied to measurements made at LANSCE and is described elsewhere in this volume [9]. The implications of the three models are discussed by Sayers [10]. The other model which is receiving considerable attention particularly in the context of composite materials is that of Eshelby [11]. Checks and validation of the different models require the measurement of the strain response of many different lattice reflections.

Diffraction measurements of residual lattice strains examine the elastic strain present in the grains contributing to the diffracted maxima. However this may result both from the macroscopic strain effects or from grain interaction effects which may be loosely associated with type 1 and type 2 residual stresses [12]. It is important to make the distinction since it is usually the type 1 stresses which concern the engineer. The benefits of a pulsed source in the measurement of grain interaction stresses [13] were demonstrated in 1983 but hitherto we are unaware of any spatially resolved strain measurements in solid components at pulsed neutron sources.

On a reactor source when one lattice reflection is examined only a small number of grains are examined with the risk that the selection may not be representative of the bulk material or may represent a special case. Although in many cases the use of a single reflection to infer macroscopic (type 1) residual stresses has been justified, if there is any reason to suppose that grain interaction stresses may be significant then there is a case for making a more detailed survey of the lattice reflections which may beneficially be performed at a pulsed source.

3. Operation of a pulsed source

At a pulsed source neutrons are produced by spallation, which occurs when energetic particles interact with target nuclei. Heavy element targets offer the best efficiency for emitted neutrons per unit energy of the incident pulse. At LANSCE proton bunches are accelerated to 800MeV then are directed at a tungsten target. The target nuclei are excited and "boil off" neutrons and fragments of the target nuclei in an evaporation process. Each incident proton bunch produces a pulse of highly energetic neutrons. For diffraction experiments the fast neutrons from the initial pulse must be thermalised using a moderator.

The difference between a pulsed and steady state source concerns the manner in which the scattered radiation is detected. If a neutron is created at a known time and position (i.e. when the proton pulse interacts with the target) its wavelength on detection can be determined from the distance and time it took to travel to the detector. Thus no collimation between the detector and specimen is needed because the wavelength is inferred from the geometry of the scattering process and the time of flight (TOF). Each pulse contains a continuous spectrum of energies thus the Bragg equation for all the lattice planes will always be satisfied by all directions. By correcting for slightly different scattering geometries a final spectrum can be constituted from the summation of the many individual spectra from individual pulses and from individual detector tubes. Thus lattice spacings, d_{hkl}, are determined by maintaining a constant diffraction angle and scanning the wavelength. The wavelength is inversely proportional to the velocity thus the TOF, t, is proportional to the wavelength, λ, and the lattice strain is given by:

$$\frac{\Delta d_{hkl}}{d_{hkl}} = \frac{\Delta \lambda_{hkl}}{\lambda_{hkl}} = \frac{\Delta t_{hkl}}{t_{hkl}}$$

The diffracted maximum intensity for each lattice spacing occurs at a discrete wavelength. The strain is determined from the change in TOF between the measured value and unstrained material.

The easiest way of improving the resolution of a spectrometer on a pulsed source is to increase the path length or the time of flight between the target and the detector. However the improved resolution is often compromised by beam losses along the flight path and ultimately for long guides "frame overlap" occurs when the slow neutrons from one pulse are overtaken by fast neutrons of its successor. Since the measurement of engineering strains in components is usually limited by the intensity of the neutron beam the use of the highest resolution spectrometers at pulsed sources are probably not warranted. On the other hand medium resolution spectrometers like the NPD at LANSCE or POLARIS at ISIS have good intensity while still having adequate resolution.

4. Definition of a sampling volume

A detailed description of the manipulation and collimation system (MACS) employed at LANSCE is included elsewhere in this volume. However a brief discussion of the collimation is included here.

One of the advantages of the NPD is that it has 4 detector banks at $\pm 90°$ and $\pm 148°$ which permit 4 simultaneous strain measurements. The NPD is 32m from the target and the divergence of the incident beam at the sample position is small. Consequently the distance between the incident aperture and the specimen is not crucial. The penumbra produced by a square incident aperture with edges 2mm placed 100mm from the specimen is ≈ 0.2mm in each dimension.

By contrast the position of the exit aperture is critical to the dimension of the sampling volume along the incident beam because there is no inherent collimation between the sample and the detectors. In principle soller collimation could be installed but it is essential to maintain a large solid angle of detector visible at the specimen if the count times are to be kept reasonable. The requirement that most or all of the NPD 90° bank is accessible to the sampled volume combined with the requirement of millimeter spatial resolution along the incident beam constrain the exit collimation to be placed as close to the incident beam direction as possible (typically less than 30mm).

The general situation for an aperture of width d, thickness t and distance L from the incident beam path (ignoring the finite size of the incident beam) is shown in figure 1. The drawing is not to scale and typically L will be 20mm, d ≈ 3mm and t≈ 13mm. The aperture width, d, and to some extent its thickness, t, are not easily changed however the distance, L should always be minimized. The spatial resolution along the incident beam decreases with increasing distance of the exit aperture from the incident beam.

For the common case of plate specimens inclined at 45° to the incident beam direction masks have been fabricated with the aperture at 45° to their surface. These masks are supported from rails can be moved adjacent to the specimen surface which minimizes the dimension L with a commensurate improvement in figure 2.

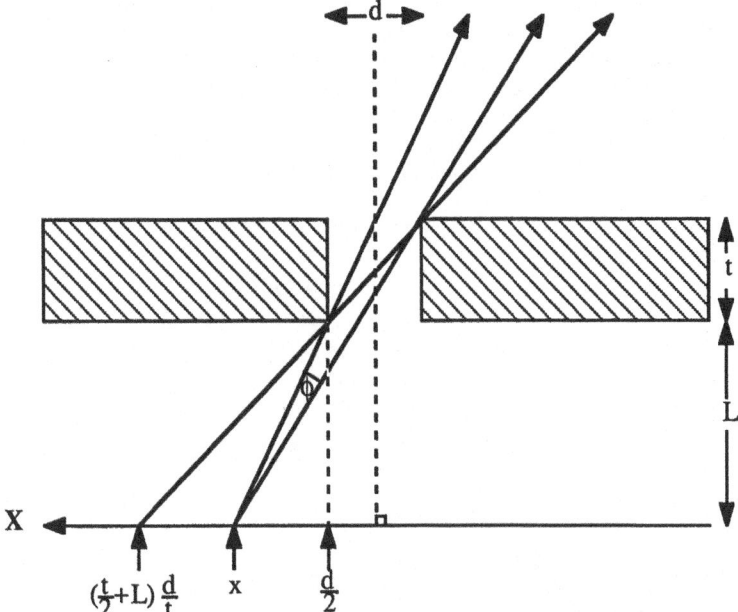

Figure 1: **Resolution along the incident beam**

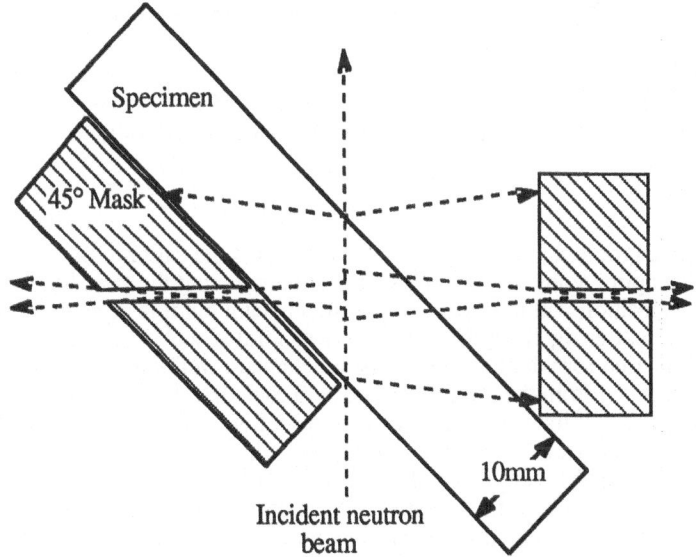

Figure 2: **Improvement in resolution for 45° apertures**

The two 90° banksconsist of 16 helium tubes which subtend ≈11° at the sample position. The spectra obtained in individual tubes correspond to slightly different strain directions but accumulate too slowly to give adequate statistics. Consequently the spectra from all of the detectors are summed to give an integrated spectrum corresponding to a measurement with a spread in scattering vector (and thus of the strain measuring directions) of ≈5.5° (fig 3). This average is not large when considering macroscopic engineering strains but is necessary to obtain favourable count times.

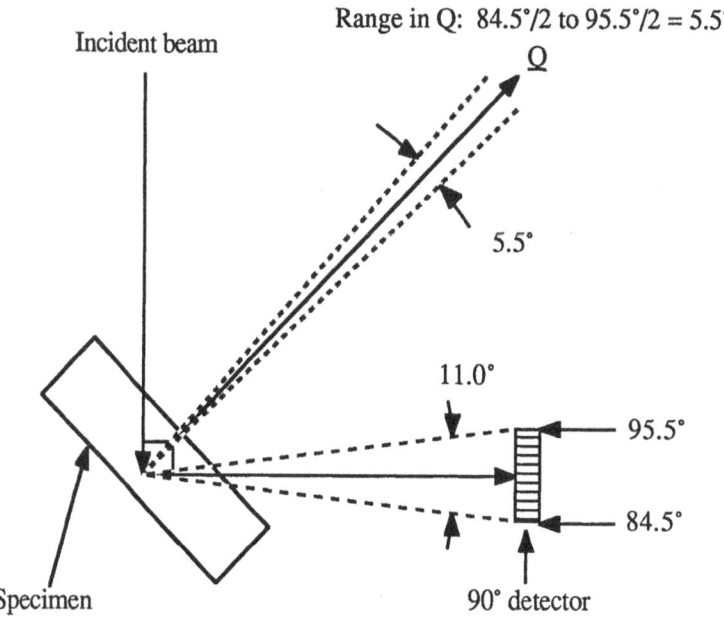

Figure 3: Average in **Q** by summing over 90° detector

5.   Calibration

One unique problem to the pulsed sources concerns the calibration of the detectors for a particular sampling volume. On a time of flight instrument the sampling volume can be defined at any position within or along the incident beam but it is necessary to calibrate the detectors for scattering from that point. In general a well characterised calcium fluoride powder is used for which the lattice parameters are well known and thus the path lengths can be predicted from the sampling volume to each detector bank .

The sampling volume is defined by the intersection of the incident and diffracted neutron beams. If an aperture is moved in a direction which is not parallel to the incident or 90° diffracted beams the sampling volume will be displaced in space. Not only does this introduce an ambiguity into the position of the sampling volume in the specimen but because the detectors are calibrated for a sampling volume centered on one specific point a constant error is introduced into the time of flight spectrum. On the NPD errors associated with ≈1mm displacements of the sampling volume are equivalent to constant strain errors of $\approx 30\mu$strain. If the sampling volume is unknowingly displaced by more than a few millimeters, then a significant strain error can be introduced. Since the apertures are frequently moved or replaced over the course of an experiment, care must be taken if repeat calibrations which would take up valuable beam time are to be avoided.

For the 1991 run cycle at LANSCE an optical alignment system will be installed for the MACS equipment. It will consist of two alignment telescopes mounted on a dummy sample chamber which is identical to the real sample chamber on the instrument which has no optical access. The intersection of the directions defined by the telescopes mounted at $90°$ to one another will identify the centre of the sampling volume and will ensure that the apertures have not been displaced. For correct placement of the apertures the telescope directions will be placed accurately parallel and normal to the beam direction. When the system is implemented it is expected that the specimen and aperture positions will be reproducible to within 0.1mm.

6. Data analysis by profile refinement

Bragg reflections in each spectrum can be fitted individually to give strains for grains in different orientations relative to the direction of the scattering vector enabling the examination of anisotropic effects. This is analogous to individual measurements using a steady state sources except that no reorientation of the component is necessary to obtain data for all of the lattice reflections in the same direction in the component. At LANSCE a program is available which fits, calculates and tabulates the strains of all the reflections in each spectrum (after the initial positions have been identified).

An alternative for analyzing the data is profile refinement. If the crystal structure is known then the intensities and positions of the observed lattice reflections can be predicted using the Rietveld method of profile analysis [14]. By making a least squares fit between the observed and predicted profiles, the atomic positions and lattice parameters for the material can be determined. In regions of compression or tension the lattice parameter will change and can be used to infer the strain.

The use of profile refinement is advantageous for several reasons. For cubic materials (which have one lattice parameter) the accuracy with which the lattice parameter is determined in a profile refinement is better than the accuracy for the fit of an individual reflection because all of the measured reflections contribute to the refinement. This is true despite the anisotropic elastic strain effects that occur for different reflections. Thus the profile refinement does not take into account the difference in compliance exhibited by different directions in a crystal lattice. It ignores any deviation from perfect crystalline behaviour and fits the best possible model to the data. In the absence of preferred orientation, the lattice parameter can (arguably) be assumed to offer the best estimate of an isotropic strain for engineering calculations.

A second advantage of profile refinement relates to the issue of count times. Since refinements use the data from all the lattice reflections, reasonable refinements are obtained long before the accuracies of fits of individual reflections approach usable values. If the anisotropic strain information for individual reflections is not required then count times can be substantially reduced permitting measurements from small sampling volumes to be made in times approaching those on steady state sources. When the count times are long enough to give reasonable accuracies for the fits of individual peaks the accuracy specified for the lattice parameter is often better than 25μstrain. One future possibility would be to modify the Rietveld refinement to accommodate the elastic anisotropy but at this time the analysis is better served by ignoring it.

7. Examples

7.1 Deformed Austenitic Ring

Calibration measurements were made on a plastically deformed ring to assess the count rates and feasibility of defining a small sampling volume using the NPD. The specimen was previously examined at a steady-state source as part of a program to validate a finite element calculation [15]. A detailed description of the measurements is given elsewhere where they are compared with values predicted by the Kroner model [9]. The experimental parameters are described here to illustrate the count times that are possible for flat specimens and to show the agreement obtained between the strains obtained from the lattice parameter and from the fits of individual reflections.

The ring dimensions were 76mm internal diameter, 127mm outer diameter and 13mm thick. It was plastically deformed by diametral loading in compression which introduced a residual stress pattern similar to a bent beam. Measurements were made along a radius in a section at 90° to the previously applied stress direction. The ring was placed at 45° to the incident beam so that normal strain directions were measured in the plane of the ring and normal to it. Depending on the orientation of the ring either the hoop and axial or the radial and axial strains were measured simultaneously in opposing 90° banks. Using a sampling volume of 40 mm³ at a beam current of ≈75μA two strains were measured every 4 hours. Additional information was recorded on the back scattering detectors although no collimation was used. A spectrum from one of the 90° banks is shown in figure 4.

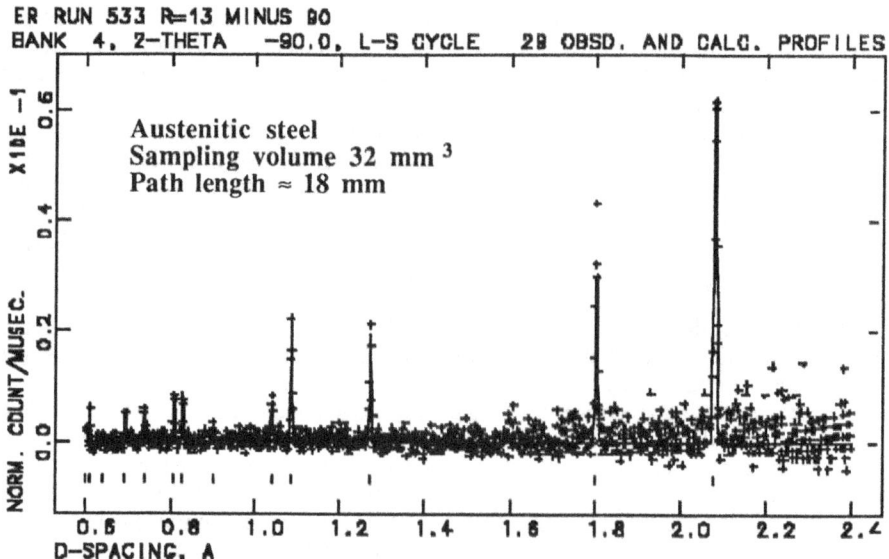

Figure 4: Spectrum at 90° obtained in 4 hours

With count times of 4 hours the accuracy for the peak fits of strong individual reflections was ≈ 80μstrain. This compares to an accuracy specified for the lattice parameter from the profile refinement of ≈ 30μstrain. In figure 5 the tangential (hoop) strains are plotted for 5 reflections of differing elastic compliance together with the strains predicted by the profile refinement. The strains by lattice parameter are connected by a solid line. The solid line is bounded by the stiff (220,111) and compliant (200) directions which reflects the averaging effect of the profile refinement on the elastic anisotropic effects. The smooth variation of the tangential strain predicted by the lattice parameter gives credence to its validity as a bulk average and to the comparatively small errors predicted by the refinement.

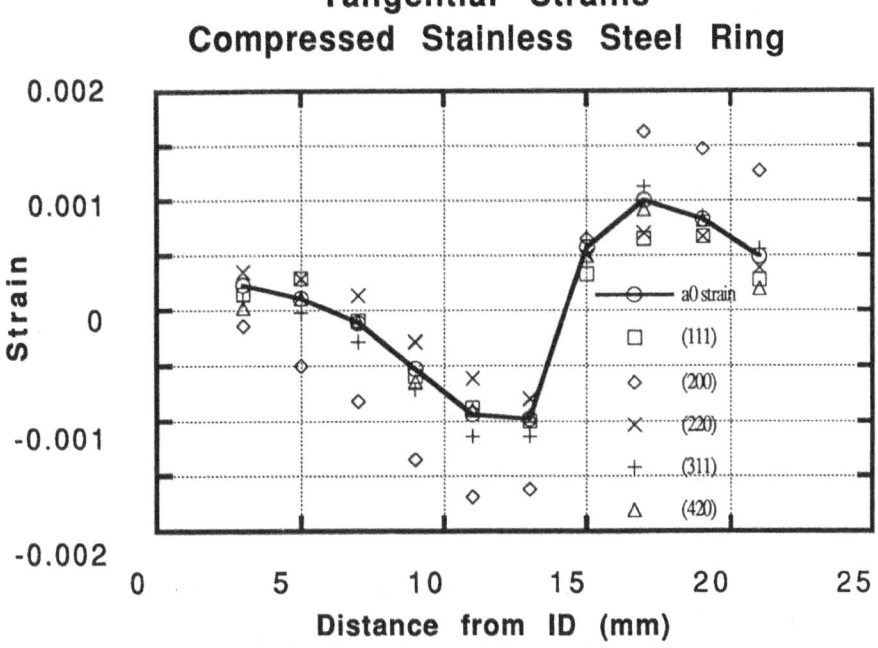

Figure 5: Hoop strains by individual reflection and by lattice parameter

The strains shown above were obtained from the combined spectrum at each position obtained by summing all 16 individual detectors in each 90° bank. To assess whether profile refinement can be used to infer the strain in shorter periods we summed progressively fewer detectors and examined the errors cited for the lattice parameters from repeated profile refinements. The results are shown in figure 6. By summing only 4 of the available 16 detectors a 1 hour counting period was simulated. The error at 1 hour for the lattice parameter of ≈ 50 μstrain which is still adequate. Of course the peak fits on individual lattice reflections are poorer and the error for the strongest reflection was ≈120 μstrain. Nevertheless if the lattice parameter is sufficient then for specimens where advantage can be taken of the opposing 90° banks count times for a 32mm^3 sampling volume can approach 30 minutes per strain direction.

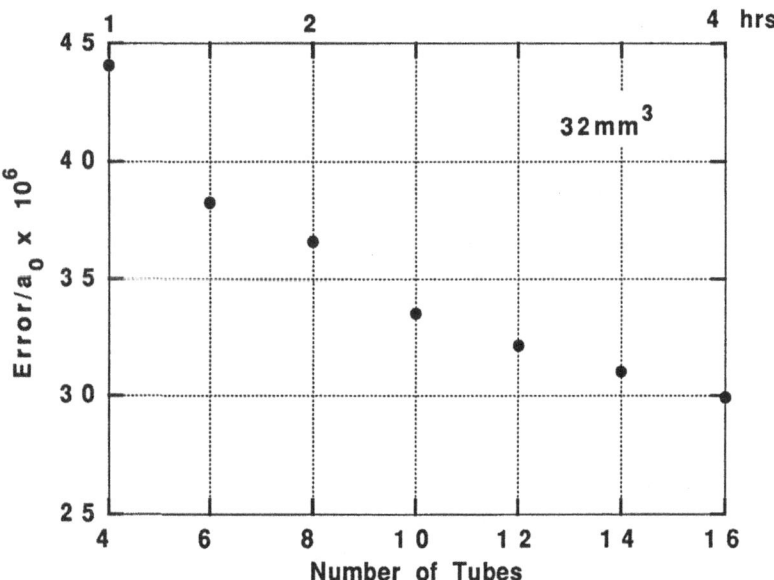

Figure 6: Error in lattice parameter vs # of detector tubes i.e. count time

7.2 Other examples

Another experiment, which took advantage of the characteristics of a pulsed source, addressed the effectiveness of vibratory stress relief in reducing the residual stresses near welds. Two multipass welds were examined in 300mm by 300mm by 25mm thick austenitic steel plates one of which was subsequently stress relieved. Strong texture changes occurred as the sampling volume was moved through distances as small as 4mm through the weld and heat affected zones because of epitaxial growth of crystals from the melt and because of the differences produced in sequential weld beads. However because all the lattice reflections were examined in each spectrum the irregular absence of some reflections was not a major problem.

By taking advantage of the symmetry along the welding direction an elongated "matchstick" sampling volume (70*4*4 mm) was used with usable spectra being obtained in 3-4 hours. The preferred orientation was too severe to make refinements of the data so the strain information was deduced solely from the fits of individual reflections. The errors obtained for individual peak fits were typically ≈ 100 μstrain. However by calculating the lattice parameter from a least squares fit <u>just</u> of the peak positions (ignoring the intensities) it is expected that more accurate strain information can be obtained from the data.

Other preliminary studies addressed whisker and particulate reinforced Al-SiC composites , compacted powders and an autofrettaged ring. In the experiments on the metal-matrix composites uniaxial tension and compression specimens were deformed to differing levels of plastic deformation. On release of the applied loading the residual strain state of the two components in the composite were examined. These experiments did not require definition of a small sampling volume because the grain interaction strains of interest were assumed to be uniform throughout the samples. Neutron diffraction is valuable for studies of this nature because the matrix and reinforcement can be examined simultaneously and non-destructively. In addition relatively large specimens can be examined making count times comparatively fast.

8. Discussion

The advantages of the time of flight method are: that all the lattice reflections are recorded with the same resolution in each spectrum, that simultaneous strain measurements in different directions are possible, that a spectrum can be recorded at any angle (subject to the geometry) of the component and that multiphase materials can easily be examined. The disadvantages compared to monochromatic measurements are: that good volume resolution is only possible by placing the diffracted beam apertures close to the specimen, that calibration is harder and more time consuming, and that count times are generally longer.

Studies in materials exhibiting strong texture variations or in multi-phase materials are particularly suitable for a pulsed source because all the lattice reflections are collected in each detector bank. Examples include the study of grain interaction strains in metal matrix composites and the unexpected appearance of a weak ferritic second phase in the stainless steel weld material. Using monochromatic neutrons to follow texture changes or to examine extra phases necessitates changes in wavelength and orientation to keep the scattering vector correctly aligned relative to the specimen.

Sampling volumes in the range 30-50mm^3 were routinely defined on the NPD and in one case as small as 8mm^3 (in a 5mm thick ring). However in all cases the specimens had flat and parallel surfaces against which the exit apertures could be placed. If the principal axes are not assumed specification of the stress tensor requires a minimum of 6 independent strain measurements. This will require the rotation of specimens and may compromise the sampling volume if the apertures must not be moved. The only solution to this problem is to use radial soller collimation which would permit the definition of a sampling volume at a distance but this remains unproven technology.

The uncertainty concerning the count times to give specified strain accuracies for different materials, specimen geometries and aperture geometries reflects the paucity of experience in defining small (< 100mm^3) sampling volumes on the NPD. However if strains can be inferred from the lattice parameters of Rietveld refinements and if simultaneous strain measurements can be made in different detectors the count times are not dissimilar to measurements on steady state sources. Of course the option of simultaneous strain measurements in different detectors is dependent on the specimen geometry and may not always be possible.

Determination of the unstrained material response remains a problem both in general and because changes in path length cause shifts in the TOF spectrum which can be misinterpreted as strain variations. Its measurement on a pulsed source in small off-cuts or annealed specimens has all the same uncertainties which apply to a steady state source. One extra option for estimating the unstrained value is possible on a pulsed source if the assumption is made that there is no hydrostatic stress in a region where the stress changes from tension to compression. In the deformed ring described above an elastic core exists approximately between 11 and 17mm from the bore of the ring. In this region the stress in the hoop direction varies from compression to tension. Strains can be calculated using an arbitrary unstrained value and then the data for different reflections can be plotted against the stiffness anisotropy factor [1], A_{hkl} where

$$A_{hkl} = \frac{(h^2k^2 + k^2l^2 + l^2h^2)}{(h^2 + k^2 + l^2)}$$

The strains differ due to the anisotropy of the elastic compliance in different directions within the unit cell. When the strains for different reflections in one measuring position are plotted against A_{hkl} the slope is approximately linear but changes in magnitude with the stress and in sign between the tensile region and the compressive region. An example is shown in figure 7.

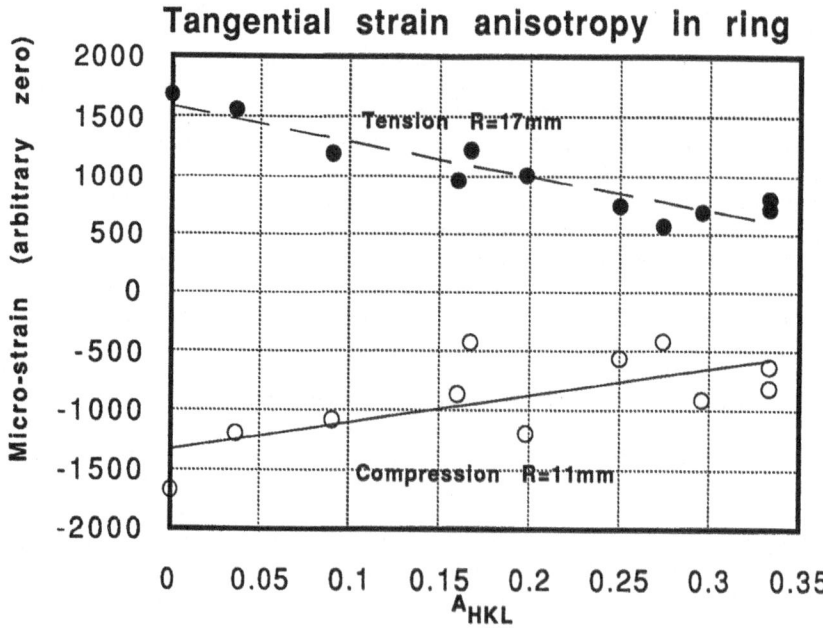

Figure 7: Strain anisotropy at two positions in compression and tension

The details of the gradient of the slope are complicated but if the change in sign is assumed to indicate the change from compression to tension then if a measuring point is identified in the ring where the gradient is found to be zero then it is reasonable to assume that the material at that point is strain free. The approach is limited in its application but may occasionally prove useful and is easy to apply for pulsed neutron data where all the reflections are available.

Much calibration work remains to be performed using the new MACS set-up on the NPD. In particular the sensitivity of the CaF_2 calibrations to the size and shape of the sampling volume. The definition of the sampling volume using boron nitride apertures is predictable by geometric considerations however the effects of strongly absorbing materials in weighting the result towards the incident side of the sampling volume need to be considered. If radial soller collimation can be engineered then the biggest limitation of the measurements on a pulsed source which requires the proximity of the diffracted beam apertures to the incident beam path could be circumvented. This needs consideration.

9. Conclusions

Incident and exit beam collimation have been installed on the NPD at LANSCE permitting spatially resolved measurements to be made in solid components. Good spatial resolution can only be achieved in plate specimens but where this is the case the NPD provides strain information from all the lattice reflections in count times which are not dissimilar to a steady state source. Situations where grain interaction strains are being investigated in multiphase materials are particularly appropriate for spectrometers at pulsed neutron sources. All three of the main pulsed sources, LANSCE, ISIS and IPNS have spectrometers with adequate resolution and flux for residual strain measurements.

On the NPD diffracted beam collimation has only been installed on the ±90° detectors but strains are also recorded on the back scattering detectors providing extra information. For plate specimens using the two opposing 90° banks two strain measurements (normal to each other) with defined sampling volumes are obtained in each measurement. The use of profile refinement can reduce the count time required to obtain acceptable strain accuracies.

The relationship between macroscopic residual stress and the strain response of a single lattice reflection depends on many features including orientation, elastic compliance, anisotropic yield and the constraint provided by surrounding grains. Currently the mechanics of material deformation are not well understood and there is often ambiguity concerning the nature and origin of the strain measured by a single lattice reflection. The current interest in microstrain effects and grain interaction strains can be well served by measurements on a pulsed source.

10. Acknowledgements

The Manuel Lujan, Jr., Neutron Scattering Centre is a national user facility funded by the United States Department of Energy, Office of Basic Energy Science. This work was supported in part by DOE contract 7405-ENG-36. The authors would like to thank Elane Flower of Lawrence Livermore, for providing the ring described in section 7.1, and Kathy Lovell who was closely involved in the design of the MACS equipment.

References

[1]   Allen, A.J., Hutchings, M.T., Windsor, C.G. & Andreani, C, (1985) "Neutron diffraction methods for the study of residual stress fields". Advances in Physics, **34**, No4, 445-473.

[2]   Stacey, A., MacGillivray, H.J., Webster, G.A., Webster, P.J. & Ziebeck,K.R.A. ,(1985a) "Measurement of residual stresses by neutron diffraction". Journal of Strain Analysis, **20**, No 2, 93-100.

[3]   Bourke, M.A.M., (1990) "Measurement of residual stress in engineering components by neutron diffraction", PhD Thesis, University of London .

[4]   Bourke, M.A.M., Ezeilo, A.N, (1989/90) " Measurement of residual stress in autofrettaged tubing", ISIS Annual report A63.

[5]   Allen, A.J., Bourke, M.A.M., David, W.I.F., Dawes, S., Hutchings, M.T., Krawitz, A., Windsor, C.G., (1989) "Effects of elastic anisotropy on the lattice strains in polycrystalline metals and composites measured by neutron diffraction", Proc of ICRS2, p78 Elsevier Applied Science.

[6]   Reuss, A., Z. (1929) "Berechnung der fliessgrenze von misch kristallen auf grund der plastizitatsbedingung fur einkristalle", Angew, Math. Mech, **9**, 49.

[7]   Voigt, W. (1928) "Lehrbuch der kristallphysik". Leipzig: Teubner, 962.

[8]   Kroner, (1958)E., Z. Phys, **151**, 504.

[9]   Holden, T.M, Hosbons, R.R, MacEwen, S.R, Flower, E.C, Bourke, M.A.M, Goldstone, J.A.. (1991) "Comparison between finite element calculations in complex parts and neutron diffraction" This publication.

[10]   Sayers, C.M., (1984) "The strain distribution in anisotropic polycrystalline aggregates subjected to an external stress field", Philosophical Magazine A, **49**, No.2, 243-262.

[11]   Eshelby, J.D., (1957) "The determination of the elastic field of an ellipsoidal inclusion, and related problems", Proceedings of the Royal Society, **241A**, 376-396.

[12]   Pintschovius, L., Macherauch, E., & Scholtes, B., (1986) "Determination of residual stresses in autofrettaged steel tubes by neutron and X-ray diffraction "Proceedings of the international conference on residual stresses, Garmisch-Partenkirchen (FRG), 159-165.DGM Informationsgesellschaft.

[13]   MacEwen, S.R., Faber, J. & Turner, P.L., (1983) "The use of time of flight neutron diffraction to study grain interaction stresses", Acta Metall, **31**, No 5, 657-676.

[14]   Von Dreele, R.B., Jorgensen, J.D. & Windsor, C.G., (1982) "Rietveld refinement with spallation neutron powder diffraction data", Journal of Applied Crystallography, **15**, 581-589.

[15]   Flower, E.C., MacEwen, S.R., Holden, T.M., Hosbons, R.R.,(1988) " A comparison of finite element analysis with neutron and X-ray measurements of residual strains in stainless steel", Proceedings of the international conference on residual stresses, Nancy pbld by Elsevier.

STRESS MEASUREMENT : EXPERIENCE AT ISIS

S.HULL, W.I.F.DAVID and M.W.JOHNSON
The ISIS Facility,
Rutherford Appleton Laboratory,
Chilton, Didcot,
Oxfordshire,
OX11 0QX,
United Kingdom.

Abstract

The ISIS facility, situated at the Rutherford Appleton Laboratory, is the world's most intense source of pulsed neutrons. The two powder diffractometers at ISIS are described, with particular reference to their use for measurements of applied and residual stress. A brief outline of an instrument optimised for enginnering applications of neutron scattering, currently being designed at ISIS, is also given. Finally, novel analysis software based on the traditional Rietveld technique allows (hkl) dependent peak shifts and broadening to be investigated.

1. Introduction

In the first stage of the neutron production process, H^- ions are accelerated to 665keV by a five stage Cockcroft-Walton multiplier and then to 70MeV by four linear r.f. accelerating cavities. On injection the two electrons are stripped from the H^- ions by passing the beam through a 0.25μm alumina foil. The remaining protons are accelerated inside a synchrotron to 750MeV by repeated passes through six r.f. accelerating cavities. Pulses of protons are extracted, 50 times a second, and directed into a heavy water cooled depleted uranium target. Fast neutrons are produced in the target by the spallation process and, to a lesser extent, by fission of uranium nuclei. The neutrons produced are slowed down to energies suitable for condensed matter research inside an array of four moderators containing hydrogeneous materials. These comprise two water moderators at 316K, plus one liquid methane (at ~100K) and one liquid hydrogen (at ~20K). The ISIS instruments are situated at the end of beamlines which view that moderator which provides the highest flux over the particular wavelength range of interest.

There are currently two powder diffraction instruments in operation at ISIS. The High Resolution Powder Diffractometer (H.R.P.D.) provides constant and extremely high resolution whilst the Polaris diffractometer offers more modest resolution but higher incident neutron flux. These instruments are discussed in more detail in the following two sections, with brief examples of measurements of applied and residual stress. A diffraction instrument specifically designed for engineering applications of neutron scattering is currently being designed at ISIS and is described in section 4. The pulsed nature of the incident neutron flux allows the whole diffraction pattern to be obtained at fixed scattering geometry, using the time-of-flight technique. The final section describes analysis software which has been developed at ISIS to include the effects of (hkl) dependent peak shifts due to elastic and plastic anisotropy.

M. T. Hutchings and A. D. Krawitz (ed.),
Measurement of Residual and Applied Stress Using Neutron Diffraction, 383–388.
© 1992 *Kluwer Academic Publishers*.

2. The High Resolution Powder Diffractometer (H.R.P.D.)

The High Resolution Powder Diffractometer at ISIS [1], shown schematically in figure 1, is situated at the end of a ~100m long curved guide and receives neutrons from a 100K methane moderator. A resolution of $\Delta d/d \approx 5 \times 10^{-4}$ is provided by the Li-doped glass scintillator detectors situated at backscattering angles ($160° < 2\theta < 176°$). A detector bank at $2\theta \approx 90°$ constructed of ZnS scintillator modules is currently being installed on H.R.P.D. and will provide a resolution $\Delta d/d \approx 2 \times 10^{-3}$. The fixed geometry allows simultaneous measurement of a large number of Bragg reflections, with essentially constant $\Delta d/d$ resolution, allowing many reflections to be measured to approximately the same accuracy. These features, coupled with the high intensity which allows rapid data collections, makes H.R.P.D. well suited to measurements of both macrostrain (peak shifts) and microstrain (peak broadening).

Diffraction measurements performed on H.R.P.D. have investigated the response of various metal specimens to applied compressive and tensile loads [2,3]. Under tensile loading the lattice strain of annealed mild steel exhibits surprisingly complex behaviour at the yield stress. With the load applied along the axis of a vertical cylindrical specimen (40mm long x 3mm diameter) the backscattering detectors measure in a direction perpendicular to stress direction. In the elastic regime these planes are under compressive stress and the peak shape is that given by the instrument resolution. The measured macrostrain is observed to increase linearly with applied stress and demonstrates the effects of elastic anisotropy [4], those planes with a high value of $A_{hkl} = (h^2k^2 + k^2l^2 + l^2h^2)/(h^2 + k^2 + l^2)^2$ being the most stiff. At ~250MPa there is significant relaxation in the strain and above this stress the peaks broaden rapidly showing an increase in microstrain. Different reflections exhibit different relaxations, which depend on the amount of slip allowed by neighbouring grains.

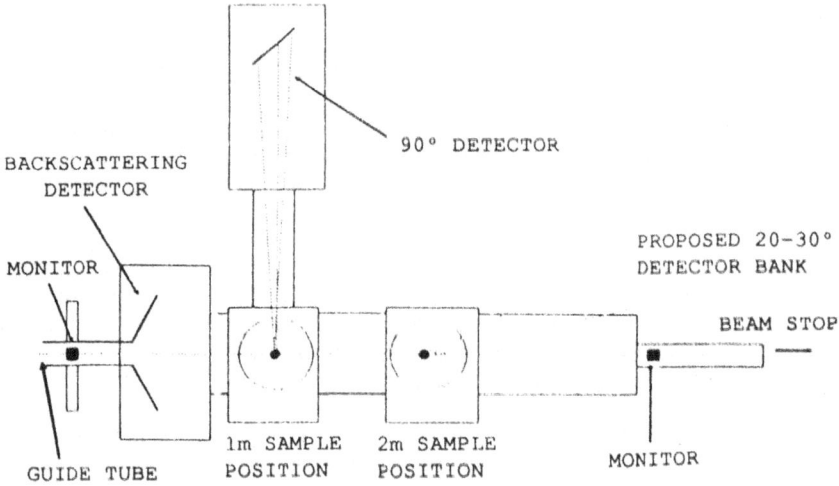

Figure 1. Schematic Diagram of the High Resolution Powder Diffractometer (H.R.P.D.) at ISIS.

3. The Polaris Diffractometer

The Polaris diffractometer [5], illustrated in figure 2, offers a complementary performance to that of H.R.P.D. by providing a higher incident flux with more modest resolution. This is achieved by situating the instrument only 12m from an ambient H_2O moderator. Three ^3He detector banks simultaneously collect data at backscattering angles ($135° < 2\theta < 158°$, $\Delta d/d \approx 5 \times 10^{-3}$), at $2\theta \approx 90°$ ($87° < 2\theta < 93°$,

$\Delta d/d\approx 6x10^{-3}$) and at low angles (14°<2θ<30°, $\Delta d/d\approx 2x10^{-2}$). The 2θ≈90° detector bank has, in particular, demonstrated the unique advantages of fixed scattering geometry offered by the time-of-flight technique. The planned installation of a second detector bank at 2θ≈-90° will extend this capability and allow simultaneous diffraction measurements to be performed in two perpendicular directions, such as parallel and perpendicular to an applied load, or to the fibre axis in composite materials.

Measurements of the residual stress distributions within an autofrettaged steel ring have been performed on Polaris, with the sample scanned remotely through the beam [6]. The spatial distribution of the hoop, radial and axial strains compare favourably with those obtained from the same sample measured on the D1A diffractometer at the I.L.L., as demonstrated in figure 3. However, the Polaris data for a_0 are obtained by time-of-flight Rietveld refinement of a large number of Bragg reflections whilst those from the fixed wavelength instrument are given by fitting only the (211) reflection, this being taken as representative of the bulk behaviour. Although comparable count times were used in each case the sampling volume required for the measurements on Polaris was ~5 times larger than that on D1A. However, future increases in neutron flux and solid angle of detector coverage are expected to reduce this difference.

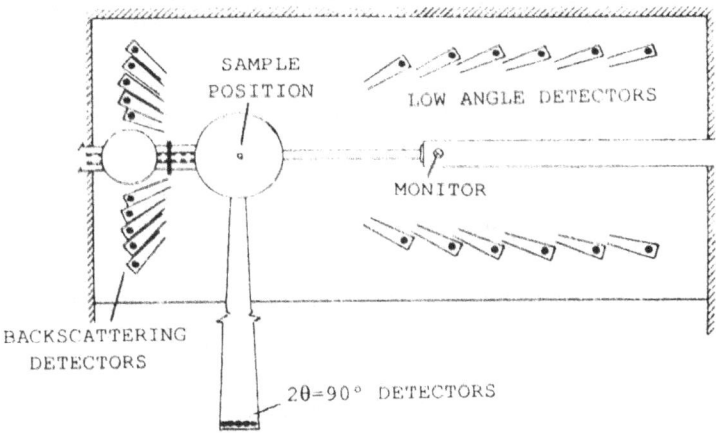

Figure 2. Schematic diagram of the Polaris medium resolution diffractometer at ISIS.

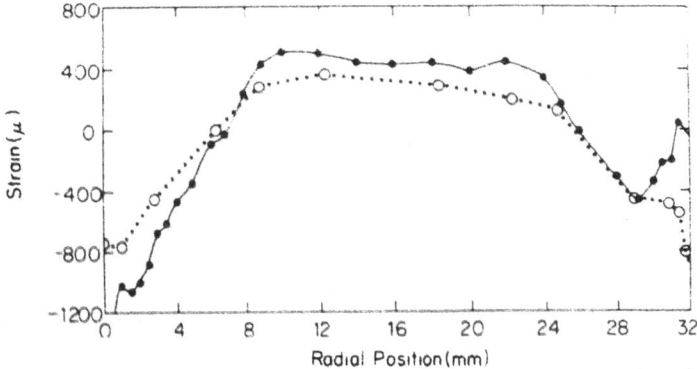

Figure 3. Comparison between the residual strain measured in the hoop direction across an autofrettaged sample on Polaris at ISIS (open circles) and D1A at the I.L.L. (closed circles). The discrepancies between the two sets of data near the edge probably result from the use of a larger sampling volume in the former.

The potential increase in count-rate available on a time-of-flight diffractometer by installing large area detectors requires that each detector element views the same sampling volume. To this end, the possibility of installing a radial collimator on Polaris is currently under investigation. This device will be situated between the sample and the detector bank and will allow a 2x2x2mm volume to be defined within a sample. The collimator will be constructed of converging vanes, which are only 0.5mm apart at the sample end. This allows the collimator to be positioned 0.5m from the specimen and gives clear advantages over a simple slit close to the sample when a large and irregularly shaped specimen is being scanned through the beam.

4. The Proposed Engineering Instrument

Recent experience at ISIS has demonstrated the potential of the time-of-flight diffraction technique for applied and residual stress measurements, particularly where the measurement of a large number of Bragg reflections is required. However, the studies undertaken to date have been performed on the H.R.P.D. and Polaris diffractometers which are optimised for diffraction studies of powder samples. A diffraction instrument specifically constructed for engineering applications of neutron scattering, provisionally called ENGIN, has been designed at ISIS (figure 4).

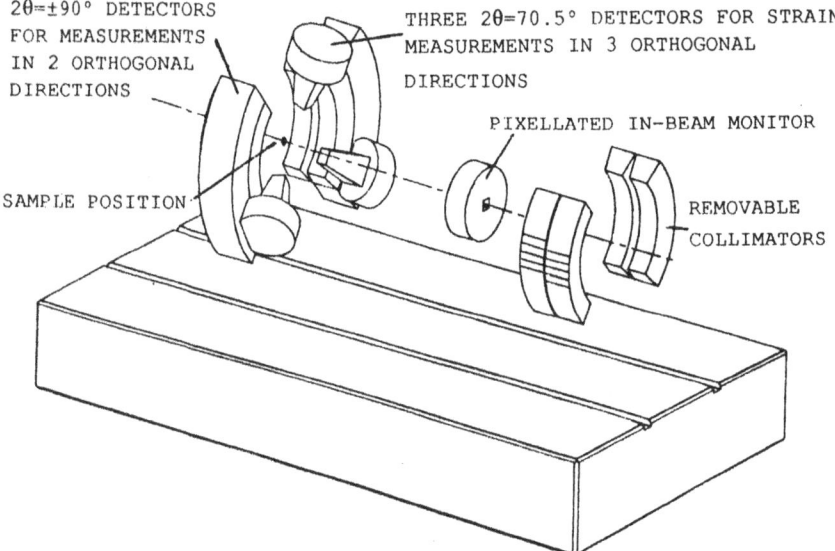

Figure 4. Schematic diagram of the proposed diffraction instrument dedicated to engineering applications of neutron scattering, currently being designed at ISIS.

A significant increase in incident neutron flux is obtained by situating the instrument only 7.5m from the ambient H_2O moderator. The $2\theta \sim 90°$ detectors are situated on both sides of the specimen at a distance 0.75m and subtend scattering angles between 80° and 100° and azimuthal angles up to ±45° from the horizontal. The detector banks are constructed of individual ZnS scintillator elements, each 3mm wide and 5cm high, which provide a resolution of $\Delta d/d \approx 3 \times 10^{-3}$. Removable radial collimators between the sample and detectors permit small sample volumes to be defined within bulk specimens. Three further detectors are to be situated at 120° intervals round a cone at $2\theta \sim 70.5°$. This unique geometry permits simultaneous measurements in three orthogonal directions. If the principal stress directions are known, this allows the complete three dimensional stress tensor to be determined in a single measurement,

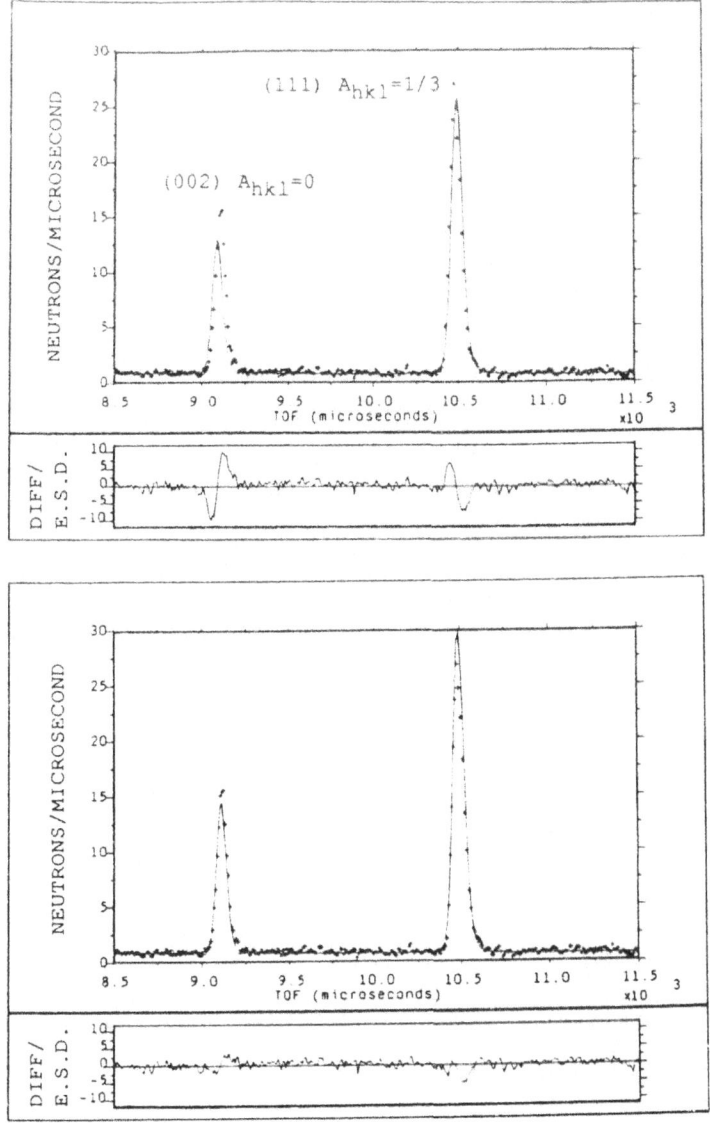

Figure 5. Diffraction data (crosses) collected from a 5mm diameter sample of an Ni-alloy tensile specimen, under a load of 900MPa. The diffraction peaks correspond to planes perpendicular to the applied load. The fitted profile (full line) is shown for the cases of excluding (top) and including (bottom) the (hkl) dependent peak shifts due to elastic anisotropy.

without having to reposition the sample. Finally, a two dimensional detector with a spatial resolution of ~1mm will be installed in the transmitted beam to allow radiography measurements of residual stress distributions and imperfections [7].

5. Analysis Software

The major advantage of the time-of-flight technique lies in its ability to investigate the (hkl) dependence of diffraction peak shifts and line broadening. Although the peak shape in time-of-flight powder diffraction is complex, with both epithermal and thermal components, the resultant asymmetrical peak shape can be well modelled [8-10]. In simple terms, the measured d-spacing (in Å) in a time-of-flight diffraction experiment is given by $d=1.977 \times 10^{-3} t/L \sin\theta$ where t is the measured neutron time-of-flight (in μsec.), L is the overall flightpath (in m.) and 2θ is the scattering angle. However, recent experience at ISIS indicates that the complex behaviour of the target/moderator assembly leads to a small, though significant, t dependence of L.

Analysis software has recently been developed at ISIS which is a modification of the traditional Rietveld method [11-12] and allows for model independent refinements of both peak widths and peak shifts. For the latter case, diffraction measurements of high purity NBS standard Si powder has been used to determine the intrinsic instrumental peak shifts due to the dependence of L on t. These can then be fitted to a quadratic polynomial and fixed in future analysis. The effects of including the detailed behaviour of L(t) is to increase the precision of lattice parameters measured on Polaris from $\sim 4 \times 10^{-5}$ to $\sim 5 \times 10^{-6}$ and on H.R.P.D. to $\sim 1 \times 10^{-6}$.

The analysis software discussed above has recently been applied to the analysis of diffraction data from metal samples under applied uniaxial stress where (hkl) dependent peak shifts result from the cubic anisotropy A_{hkl} (figure 5). Using this procedure it has proved possible to measure lattice parameters to an accuracy corresponding to ~ 30 microstrain. Furthermore, his technique has important applications for the determination of the 'stress-free' lattice parameter d_0, any (hkl) dependent peak shifts giving a direct indication of the presence of lattice strain.

Acknowledgements

We are grateful to M.T.Hutchings, P.J.Webster, G.A.Webster and M.Bourke for their permission to include experimental data collected at ISIS.

References

[1] W.I.F.David, D.E.Akporiaye, R.M.Ibberson and C.C.Wilson, Rutherford Appleton Laboratory Report, RAL-88-103 (1988).
[2] A.J.Allen, M.Bourke, W.I.F.David, S.Dawes, M.T.Hutchings, A.D.Krawitz and C.G.Windsor, International Conf. on Residual Stress, ICRS2, eds. G.Beck, S.Denis and A.Simon (Elsevier,London), 78 (1989).
[3] M.T.Hutchings, Nondestr. Test. Eval., 5, 395 (1990).
[4] C.G.Windsor, paper at this meeting.
[5] S.Hull and J.Mayers, Rutherford Appleton Laboratory Report, RAL-89-118 (1989).
[6] M.Bourke, PhD Thesis, Imperial Coll., London (1990).
[7] H.G.Priesmeyer, paper at this meeting
[8] R.B.Von Dreele, J.D.Jorgensen and C.G.Windsor, J. Appl. Crystallogr., 15, 581 (1982).
[9] J.M.Carpenter, R.A.Robinson, A.D.Taylor and D.J.Picton, Rutherford Appleton Laboratory Report, RAL-84-082 (1985).
[10] S.Ikeda and J.M.Carpenter, Nucl. Inst. Meth., A235, 553 (1985).
[11] H.M.Rietveld, Acta Cryst., 22, 151 (1967)
[12] H.M.Rietveld, J. Appl. Crystallogr., 2, 65 (1969)

TRANSMISSION BRAGG-EDGE MEASUREMENTS

HANS G. PRIESMEYER
Institut für Reine und Angewandte Kernphysik
Universität Kiel c/o GKSS Research Center
POBox 1160
D-2054 Geesthacht FRG

ABSTRACT. The transmission of polycrystalline materials in a white neutron spectrum shows sharp edges as a function of neutron wavelength or time-of-flight. They correspond to the breakdown of the Bragg interference condition for those sets of lattice planes which scatter by 180 degrees. Materials stresses as well as texture will influence the position and shape of the edges. Transmission measurements may be favourable for a number of experimental tasks.

1. Introduction

Low-energy neutron time-of-flight transmission measurements of polycrystalline materials show diffraction edges, which correspond to wavelength cut-offs of the individual lattice spacings for 180 degree backscattering. The angular resolution contribution in this case is minimized and there is no contribution from beam or sample size to the flight path resolution. Johnson and Bowman (1982) have shown the high resolution of transmission measurements for an iron sample, where they have been able to distiguish diffraction edges up to $n = 196$ and to assign them below $n = 90$ (n being the sum of squares of the Miller indices). Transmission measurements can be favorable for a number of new experiments. In this paper we shall briefly discuss d0-determination, nondestructive strain measurements in a spinning wheel and the Fast-Transient Diffraction experiment, which is being set up at LANSCE.

2. Neutron transmission through polycristalline material

The position of a diffraction edge in a time-of-flight spectrum is characterized by
$d[nm] = \lambda /2 = 1.978.10^{-4} \cdot t/L$, t being the flight time in μsec and L the flight path in m. It is not affected by the position of the sample in the neutron beam or its size. Since the transmission measurement integrates over the sample thickness, strain determination is easiest for two-dimensional stress fields. A stress-free untextured sample will show diffraction edges with slopes representing the instrumental resolution and heights according to the orientation distribution of the crystallites. Once the flight-path from the pulsed neutron source to the detector is calibrated, absolute d-values can be found. Because of the high intensity in transmission geometry, the sample can be

M. T. Hutchings and A. D. Krawitz (ed.),
Measurement of Residual and Applied Stress Using Neutron Diffraction, 389–394.
© 1992 *Kluwer Academic Publishers.*

tilted in the beam and the stability of the edge position can easily be checked. Changes would than have to be attributed to strain. Fig. 1 shows a typical time-of-flight spectrum of steel taken with a simple Fermi-chopper at the FRG-1 reactor.

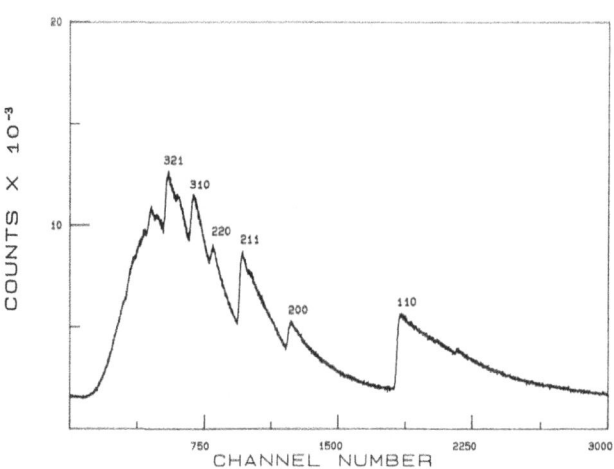

Fig. 1: Time-of-flight spectrum of a steel sample of 10 mm thickness taken with a Fermi-chopper at a neutron guide tube which cuts off the high-energy neutrons of the reactor spectrum.

3. Nondestructive strain measurements in a spinning gyroscope

A gyroscope rotor spinning at 18000 rpm has been investigated using the GELINA linear accelerator of CBNM/Geel-Belgium at low frequency as a pulsed neutron source. Diffraction edge positions have been compared to those in the stationary case. Detailed analysis of the experiment is under way (K. Meggers 1991). From the edge-shifts compressive strains in the direction of the rotor axis can be calculated, after the specimen motion effects have been taken into account. (Shull et al. (1964) & (1968)).

The following Fig. 2 shows the time-of-flight spectrum, for which the shapes of the diffraction edges look different from Fig.1 because they represent the moderator spectrum of the linac.

Fig. 3 shows qualitatively how the 211 - Bragg edge is shifted from the stationary case when rotating at 18000 rpm.

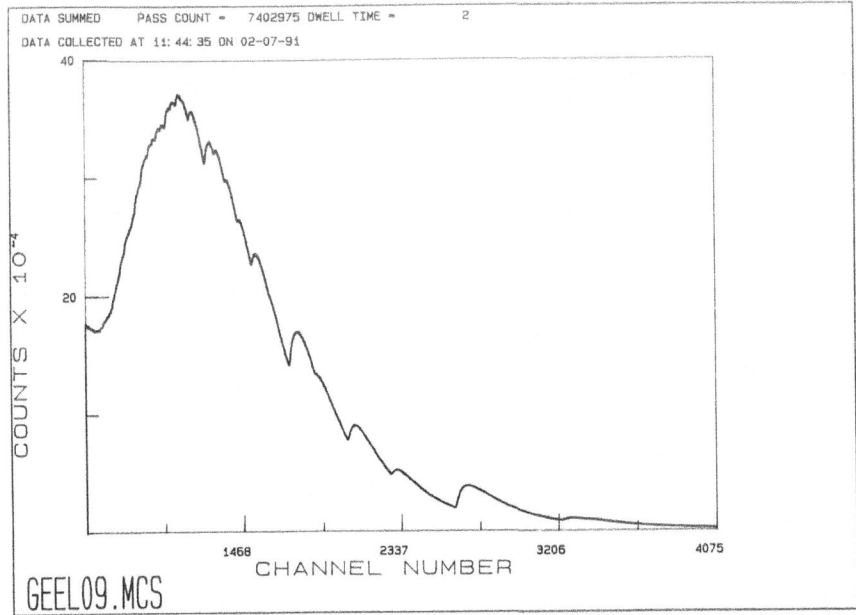

Fig. 2: Time-of-flight spectrum of steel rotor using neutrons from a linear accelerator neutron source.

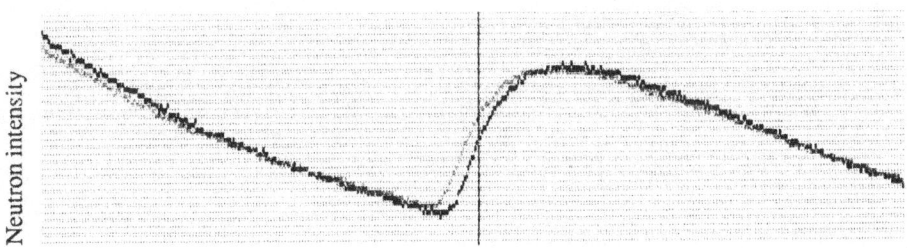

Fig. 3: Example for diffraction edge shifted by centrifugal force.

392

4. Fast-Transient Diffraction

In 1989 it could be shown at the intense spallation neutron source in Los Alamos, using the current-mode scintillation detector method, that it is possible to measure the transmission diffraction spectrum with a single neutron burst and thereby perform time-resolved measurements in the millisecond or even microsecond range (R. Pynn (1990)). Figure 4 shows the original data for an iron sample of 1" thickness.

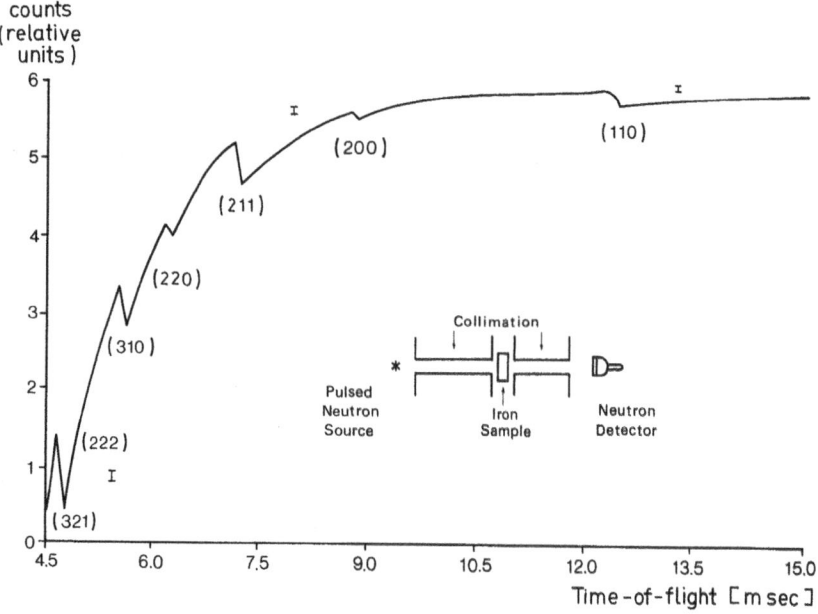

Fig. 4: One-shot transmission spectrum of iron, 1" thick.

The FTD method will allow interesting dynamic experiments, which cannot be done by other means. The experiments at Los Alamos are financially supported by Deutsche Forschungsgemeinschaft and will continue this year.

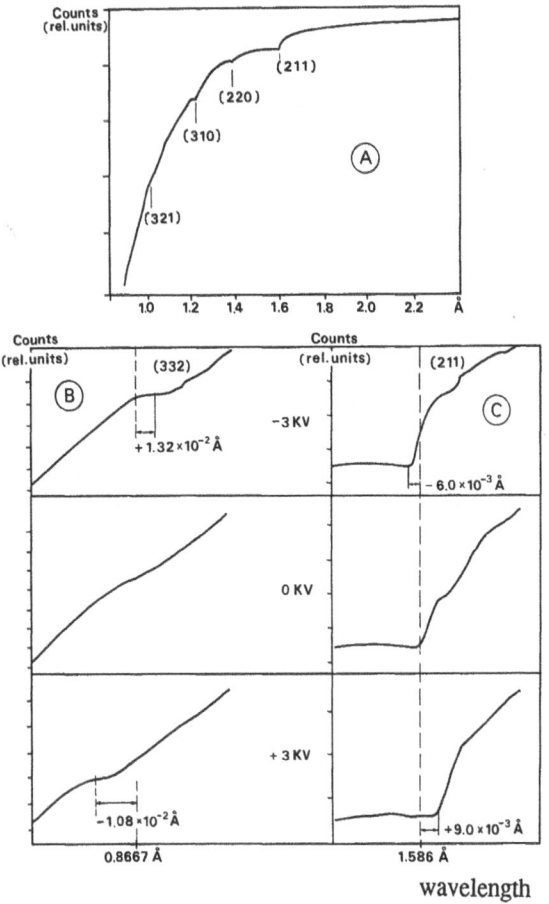

Fig. 5: Transmission spectra of piezoelectric material PZT5A with different applied high-voltages.

5. Outlook for Bragg-edge white source transmission measurements

The most promising outlook for future experimental work is with the FTD method, with its capability to perform neutron stroboscopic investigations of fast transients like high external loads, which can be held only for short times, fast phase changes and/or unstable phases during high pressure, melting or solidification processes, or annealing for stress relief. Any time-of-flight spectrometer should have a transmission detector at the location of the beam dump, which could be useful to identify strain free areas of the sample or regions having large strain gradients. The high neutron intensity in transmission geometry will allow to confine the beam to less than 1 mm². It should also be mentioned that the method may be extended to strain radiography (similar to photoelastic imaging), once the suitable position sensitive detectors with good timing resolution are available. Finally the method can be useful if transportable pulsed neutron sources should be developed.

394

6. References

Johnson, R.G., Bowman, C.D. (1981), 'High-resolution powder diffraction by white source transmission measurements' Neutron Scattering 1981 (AIP Conf. Proc. No. 89), J. Faber. ed., pp. 53-55

Shull, C.G., Morash, K.R., Rogers, J.G. (1968), 'Specimen motion effects in neutron diffraction', Acta Cryst. A24, pp. 160-163

Shull, C.G., Gingrich, N.S. (1964), ' Neutron diffraction effects with moving lattices', J. Appl. Phys. 35, pp. 678-682

Pynn, R. (1990), 'Neutron scattering methods for materials science' Neutron scattering methods for materials science (Mat. Res. Soc. Symp. Proc., No. 166), S.M. Shapiro, S.-C. Moss, J.D. Jorgensen ed., pp. 15-26

Meggers, K. (1991), 'Neutronendiffraktometrische Untersuchungen von Gitterdehnungen unter Einflußzentrifugaler Kräfte' (Neutrondiffraction measurements of lattice strains caused by centrifugal forces, in german) Thesis, Kiel University

MACS, THE MANIPULATION AND COLLIMATION SYSTEM ON THE NPD AT LANSCE

M. A. M. BOURKE , J. A. GOLDSTONE & K.J. LOVELL.
LANSCE
Los Alamos National Laboratory
New Mexico, 87545, USA

ABSTRACT. The practical problems associated with beam collimation and specimen manipulation at a pulsed neutron source are identical to those on a steady state source. However extra constraints result from the limited space available and from the time of flight analysis of the diffracted neutrons. A manipulation and collimation system (MACS) has been designed for the neutron powder diffractometer (NPD) at the Manuel Lujan Jr. Neutron Scattering Centre (LANSCE). It provides specimen motion and aperture positioning with accuracies of better than 0.1mm and is constructed as a rigid unit. For flat sided specimens sampling volumes less than 30mm^3 have been obtained demonstrating the viability of making spatially resolved strain measurements at a pulsed neutron source.

1. Introduction

Spatially resolved measurements by neutron diffraction of residual lattice strains are made at different positions in a component by translating it through a defined sampling volume. Strains in different directions are measured by reorienting the specimen with respect to the detectors. Thus the basic requirements for a measuring system are multi-axis specimen manipulation, precise collimation of the neutron beams, and an ability to place a specimen accurately relative to a reference point. In engineering applications the sampling volumes of interest are often less than 50mm^3, except when a symmetry can be assumed or when spatially resolved measurements are not required. In situations involving steep stress gradients sampling volumes of less than 10mm^3 are sometimes necessary. Despite the small sampling volumes the mass of specimens can vary from a few hundred grammes to tens of kilograms. A final requirement is that the sampling volumes can be rapidly and accurately changed for different experiments.

The requirements are necessary both for pulsed and steady-state sources. On POLARIS at ISIS in the U.K. and on the NPD at LANSCE in the U.S., the sample position lies within aluminium vessels which are normally evacuated during experiments to minimize air scattering. On the NPD the vacuum vessel is surrounded by bulk shielding which cannot be removed (unlike POLARIS) and precludes any optical access to the specimen when it is in position for measuring a spectrum. The sample position is on the axis of the circular opening in the top of the vacuum vessel at a specified height below the top flange corresponding to the centre of the neutron beam. When the specimen is in position it is surrounded by four banks of detector at $\pm$ 90° and $\pm$ 148° which provide 4 simultaneous strain measurements, although the best volume definition is achieved with the 90° detectors.

M. T. Hutchings and A. D. Krawitz (ed.),
Measurement of Residual and Applied Stress Using Neutron Diffraction, 395–400.
© 1992 *Kluwer Academic Publishers.*

2. Translators

All collimation and manipulators must pass through the circular opening (74cm diameter) in the vacuum vessel. The approach on the NPD has been to mount the manipulators and beam apertures into a rigid frame attached to a top plate which can be placed precisely in the instrument using cone locators. Prior alignment is possible in a copy of the instrument sample can to which optical access is possible.

The minimum requirement for the specimen movement system was XYZ orthogonal translation and a rotation about a vertical axis. However the necessity that the system passed through the round aperture constrained the design and limited the extent of the movements particularly in the two horizontal directions. Lowering specimens into the beam would require a clamp which can accommodate many different specimen sizes and masses. For light specimens this is not a problem, however massive specimens of unusual shape can more conveniently be supported on a flat surface and raised into the neutron beam. Consequently vertical motion is provided by raising or lowering a flat table on 4 shafts which are interconnected by a drive chain. This provides 300mm of vertical motion between the beam height and the top of the rotary stage.

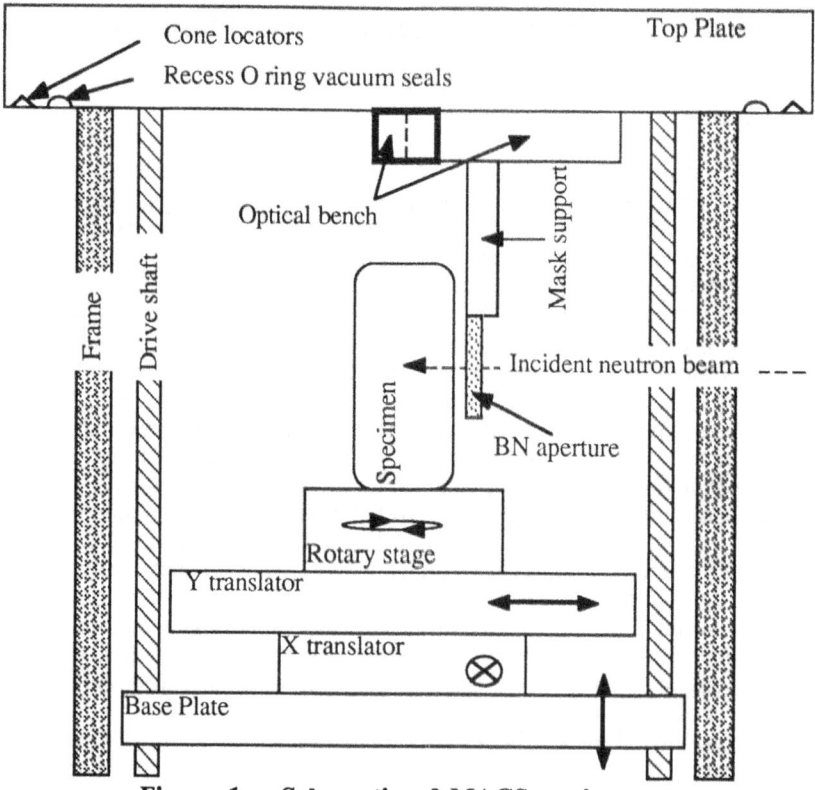

Figure 1:. Schematic of MACS equipment

Horizontal translators are mounted parallel and normal to the incident beam direction. To move a specimen along the scattering vector for one of the 90° banks , i.e. at 45° to the incident beam (which is frequently necessary), the translators are operated sequentially to give a zig-zag motion. Smooth motion along the scattering vector is advantageous and can be obtained by simultaneous operation of the motors. The horizontal slides each provide a total travel of 25 cm. Together with the vertical motion, the MACS system can accommodate specimens up to 50kg. A schematic of the manipulation system is included in figure 1. Figure 2 shows a $(300mm)^2$, 25mm thick weld specimen (RS-1 in the photo) in position in the MACS system.

Figure 2: **Photograph of MACS equipment with $(300mm)^2$ 25mm thick weld (RS1) in position for measurement**

3. Collimation

On monochromatic diffractometers which scan the diffracted intensity as a function of angle, the direction of the diffracted beam must be accurately defined at the detector. Consequently collimation is inherent between the specimen and the detector in the form of soller slits. This facilitates the definition of a sampling volume because a slit can be installed anywhere between the sample and the detector. On a pulsed source there is no inherent collimation between the specimen and detectors and diffracted neutrons can reach the detector from any scattering geometry. Care must be taken in masking the specimen so that only neutrons from the sampling volume reach the detector.

The incident beam entering the NPD vacuum vessel is normally 50mm high and 10mm across. The sample position is 32m from the tungsten target (where the neutrons are produced) and consequently the divergence of the incident neutron beam at the sample position is small. Hence the distance between the incident aperture and the specimen does not affect the definition of a sampling volume. The penumbra at the sample position around a beam defined by a 2mm square placed 50mm from the specimen is less than 0.2mm both vertically and horizontally. Thus collimation of the incident beam is not strongly related to the distance of the incident aperture from the specimen.

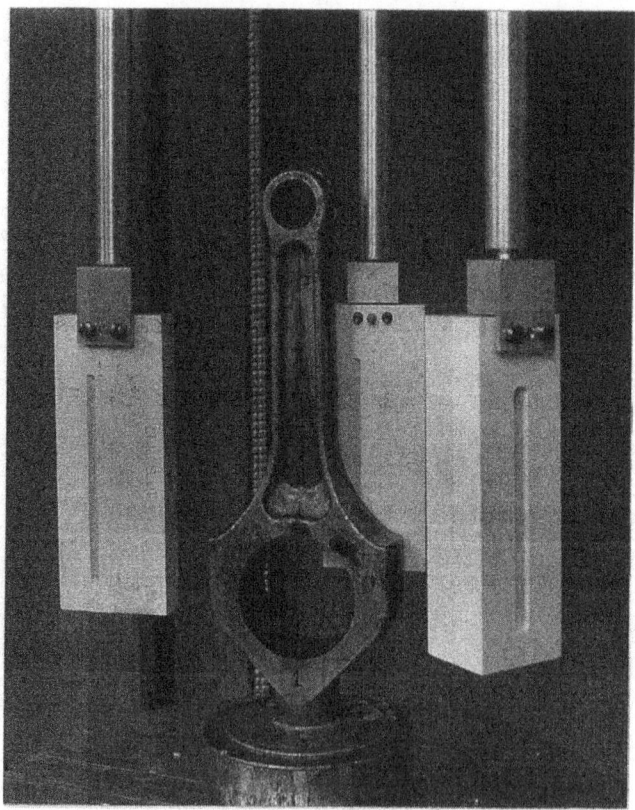

Figure 3: BN apertures by connecting rod.

The situation for the diffracted beam apertures differs because close proximity between the aperture and the incident beam direction is necessary. Although a thick aperture placed 10cm from the specimen could give millimeter spatial definition along the incident beam, it would necessarily obscure a large solid angle of the 90° detectors which subtend $\approx 11°$. Count rates from sampling volumes less than 100mm^3 necessitate that most or all of the detector solid angle is employed. Good spatial resolution along the incident beam can only be obtained, while maintaining a large solid angle of detector, by having close proximity of the diffracted beam apertures to the incident beam. The best spatial resolution is achieved for flat plates where the apertures can be placed directly adjacent to the surfaces.

Both the incident and diffracted apertures are supported from slides on optical benches which are attached to the underside of the top plate (fig 1). The rails are aligned parallel with and normal to the incident beam direction. This allows the apertures to be moved to different radii from the centre of the sampling volume without any lateral displacement. Fine positional adjustment is performed using manual microcontrol adjusters.

At a spallation source the wavelength spectrum in each pulse is continuous and boron, not cadmium, is used as a mask because it remains opaque to neutrons of wavelength less than 0.5Å. Boron nitride proved to be preferable to boron carbide because it was easier to machine to precise dimensions. A variety of apertures were fabricated from a 13mm thick boron nitride plate. Incident apertures are tailored to specific requirements by combining vertical and horizontal slits. Figure 3 shows a photo of a connecting rod surrounded by the white BN apertures. For clarity the 90° masks have been withdrawn from the specimen but in practice they would be adjacent to the specimen surface.

4. Calibration

The sampling volume is defined at a point in space at the intersection of the incident and diffracted beams. The position is arbitrary thus calibration of the detectors is important if the absolute values of the lattice spacings obtained from different banks are to be consistent. To calibrate the detectors the diffracted spectrum for a well characterised stress free powder is measured by the defined sampling volume. From knowledge of the structure of the powder, the calibration constants (which relate the time of flight to the wavelength and thus the lattice spacing) for each detector can be determined. If the centre of the sampling volume is unmoved for different aperture combinations, the calibration remains effective. However if an aperture is displaced and the centre of the sampling volume moved then recalibration will be necessary. For this reason the apertures must be accurately replaceable and movable along directions which are either parallel or normal to the incident beam direction to avoid repeated recalibration.

Intensity checks were made by translating a vertical steel pin through the sampling volume defined by different aperture sizes and positions. The diffracted intensity over a specific wavelength range was noted as the pin was moved parallel and normal to the incident beam. When the pin was moved normal to the incident beam direction the intensity profiles showed sharp reductions at the limits of the aperture size indicating the parallel nature of the incident beam. Intensity scans were also recorded in the 90° detectors as the pin was moved parallel to the incident beam. Knowledge of the angle subtended by the detectors combined with the width, thickness and distance of the apertures from the incident beam allowed the resolution along the incident beam to be predicted.

5. Future development

A displacement of the sampling volume by only a few millimeters may be misinterpreted as a strain variation due to the change in path length and commensurate error in calibration. Previously alignment procedures were performed by dead reckoning but we are installing two alignment telescopes on the dummy sample can which will ensure that apertures and specimens are replaceable to within 0.1mm for concurrent measurements. A transmitted beam monitor will also be installed to aid in aligning specimens by noting the change in the ratio of the transmitted to the incident beam intensity as the specimen is moved into the beam.

6. Conclusions

A manipulation and collimation system has been installed on the NPD at LANSCE. Components up to 50kg can be moved within the confines of the sample tank. Sampling volumes as small as $10mm^3$ have been achieved for thin flat samples. However for specimens where the exit collimation cannot be placed within 30mm of the incident beam it is impossible to get get good spatial resolution along the incident beam without obscuring part of the $90°$ detector bank and compromising the count times. Optimum spatial resolution is obtained by installing the diffracted beam apertures adjacent to the specimen surface or as close as allowed by the geometry of a particular measurement. This makes flat sided specimens where strains are required at different positions in the plane of the specimen most suitable for measurements on a pulsed instrument.

7. Acknowledgements

The Manuel Lujan, Jr., Neutron Scattering Centre is a national user facility funded by the United States Department of Energy, Office of Basic Energy Science. This work was supported in part by DOE contract 7405-ENG-36.

THE DESIGN OF A PULSED SOURCE INSTRUMENT:
SUMMARY OF DISCUSSION SESSION

S. HULL,
The ISIS Facility,
Rutherford Appleton Laboratory,
Chilton, Didcot,
Oxfordshire,
OX11 0QX,
United Kingdom.

1. Introduction

The discussion group considered a number of design features which were considered important for a pulsed source diffraction instrument optimised for engineering applications. The discussion time available proved insufficient to develop a full specification, though a number of important points were raised and are detailed below. The relative merits of pulsed and steady state sources were also discussed though it is clear that the ability of a pulsed source to measure a large number of Bragg reflections simultaneously must be exploited in order to be competitive in terms of count-rate. Pulsed sources have advantages for the study of the mean stress over a large volume in multicomponent systems, such as composites. The choice between pulsed and steady state sources depends critically upon the particular problem under investigation and there is clearly a degree of complementarity between the two types of source.

Current time-of-flight diffractometers have been designed for conventional structural investigations using powder samples. At ISIS a preliminary design of an instrument optimised for engineering use of neutron scattering has been developed. Discussions were, therefore, centred on the features contained in this design.

2. Proposed Design

A large sample space is required to accommodate bulk specimens and to allow accurate optical alignment of the sampling volume. However, the attenuation of neutrons through steel limits the maximum size of sample that can be investigated and a distance of ~2m in each direction was considered a reasonable compromise.

An estimate of the accuracy required for strain measurements was ~50-100 microstrain. This is routinely obtained on even modest resolution pulsed source diffractometers and does not impose any significant design problems. However, for measurements of microstrain it is essential to investigate the detailed peak broadening. This requires significantly better instrumental resolution, though this may be better suited to a different instrument.

The optimal scattering geometry for definition of well defined gauge volumes within bulk specimens has scattering angles of $2\theta \approx 90°$. The installation of two detector banks, at $2\theta = \pm90°$ has important advantages since this allows the diffraction pattern to be measured simultaneously in two orthogonal directions, such as parallel and perpendicular to the fibre axis in composite materials. At a time-of-flight source it is possible to extend this to three orthogonal directions, by arranging three detector banks at 120°

M. T. Hutchings and A. D. Krawitz (ed.),
Measurement of Residual and Applied Stress Using Neutron Diffraction, 401–402.

intervals around a cone at $2\theta \approx 70.5°$. However, this requires measurement of the diffraction pattern in transmission geometry, with the inherent problems of beam attenuation. Furthermore, alignment of a large specimen in the appropriate orientation may not always be possible. This possibility requires further consideration and perhaps test experiments using existing instrumentation.

The count-rate on a pulsed source diffractometer can be improved by installing large area detectors, covering a range of scattering angles, 2θ. The resulting measurements include a distribution of scattering vectors $\underline{Q}$, though $\pm \sim 5°$ was not felt to present a significant disadvantage. A more severe problem concerns the need to ensure that each element of the detector bank views the same gauge volume. This can be achieved using either a radial soller collimator or a defining slit close to the sample. Following a lengthy discussion, the ability to position a radial collimator at a fixed distance from the specimen and define a fixed volume was felt to be advantageous. The reduction in measured background was also considered important. Potential problems concerning the 'triangular' distribution of intensity measured over the gauge volume and the transmission of the radial collimators were not felt to be significant. The major drawback concerns the difficulties in constructing a radial collimator, with vanes separated by perhaps as little as 0.5mm. Furthermore, a different collimator is required for each size of gauge volume. The use of parallel collimators to define multiple gauge volumes simultaneously, of the type used at Risø, was not considered to be appropriate for a pulsed source instrument since the count-rate for each element appears prohibitively small.

6. COMPOSITES

STRESS MEASUREMENTS IN COMPOSITES USING NEUTRON DIFFRACTION

AARON D. KRAWITZ
Department of Mechanical and Aerospace Engineering
and Missouri University Research Reactor
University of Missouri
Columbia, Missouri 65211
USA

ABSTRACT. The measurement of residual microstresses in composites using neutrons is discussed. The basic theory of the measurements is presented and the advantages of neutrons noted, including a discussion of the limitations of x-rays. A number of experimental issues are considered, including grain size, stress-free standards, sample alignment, and the use of steady state versus pulsed sources. Studies illustrating these issues are presented for which the techniques, problems and potential of the method are emphasized. Measurements of Al-SiC metal matrix composites deal with the anisotropy of the stress state in the presence of aligned fibers, particle shape effects, and changes in residual stress due to uniaxial compressive and tensile plastic deformation. Problems encountered with grain size and composition in the stress-free standards are noted. Studies of WC-Ni cemented carbide composites as a function of temperature deal with effects of microstructure and the interpretation of peak broadening in terms of stress distributions.

1. Introduction

The use of neutrons for the study of residual stresses in composite materials is complementary to the use of x-rays [1-7]. The advantages derive from the greater penetration of neutrons in most engineering materials [8]. This makes surface preparation less critical, enables the study of systems containing heavy elements and/or large diameter fibers or particles, and enables the $\psi = 90°$ orientation to be readily reached using transmission geometry and the ψ-goniometer configuration. Neutrons also provide bulk volume averaging that enables sampling of, for example, entire cross-sections of tension/compression test specimens. Thus, neutrons are well-suited to the study of the inherently volumetric microstress states found in composites. Furthermore, subsurface micro- or macrostress gradients can be studied, by stopping down the beam, and macro- and microstresses may be separated [9].

This paper will deal exclusively with the determination of residual stresses of the microstress kind (Type 2) in composite materials. Discussions of theory, experimental aspects, and problems precede the presentation of examples intended to illustrate the possibilities and problems. The cases considered include whisker and particle reinforced metal matrix composites (SiC reinforced Al alloys) and cemented carbide particulate composites containing heavy metal carbide particles (WC reinforced nickel).

405

M. T. Hutchings and A. D. Krawitz (ed.),
Measurement of Residual and Applied Stress Using Neutron Diffraction, 405–420.
© 1992 *Kluwer Academic Publishers.*

2. Theory

The coordinate systems used herein are shown in Fig. 1 [1]. The S_i define the specimen system, with S_1 and S_2 in the surface of the specimen. The L_i define the laboratory system such that L_3 is in the direction of the normal to the diffracting planes. L_2 is in the S_1-S_2 plane and makes an angle ϕ with S_2. The angle ψ is between S_3 and L_3. For the case of a cylindrical sample, the ϕ rotation is around the cylinder axis.

The strain measured in a direction specified by the angles ϕ and ψ in terms of the residual stress components σ_{ij}, is [1]

$$<\varepsilon_{\phi\psi}> = \frac{S_2}{2}[<\sigma_{11}>\cos^2\phi + <\sigma_{12}>\sin2\phi + <\sigma_{22}>\sin^2\phi]\sin^2\psi +$$

$$\frac{S_2}{2}<\sigma_{33}>\cos^2\psi + S_1[<\sigma_{11}> + <\sigma_{22}> + <\sigma_{33}>] +$$

$$\frac{S_2}{2}[<\sigma_{13}>\cos\phi + <\sigma_{23}>\sin\phi]\sin2\psi , \qquad (1)$$

where $S_2/2$ is $(1+v)/E$, S_1 is $-v/E$, and E and v are Young's modulus and Poisson's ratio, respectively, for the diffracting planes. For a cylindrical sample, the direction 3 is parallel to the cylinder axis, with 1 and 2 in radial directions. The ϕ axis is parallel to 3 so that the ϕ rotation is around the cylinder axis.

Two cases are of particular interest for microstresses in composites, cylindrical and hydrostatic stress states. We assume for both that no shear stress is measured, i.e. $\sigma_{ij} = 0$ for $i \neq j$. For the cylindrical case, $\sigma_{11} = \sigma_{22}$, and the only other non-zero stress component is σ_{33}. Since there is no ϕ dependence, Eqn. (1) becomes

$$<\varepsilon_\psi> = \frac{1+v}{E}(<\sigma_{11}> - <\sigma_{33}>)\sin^2\psi + \frac{1}{E}(<\sigma_{33}> -2v<\sigma_{11}>) . \qquad (2)$$

Now there are only two unknowns and the stresses can be determined provided that a stress-free standard is available in order to calculate the strains. This is the case for many aligned fiber or whisker composites. Note that the stresses in the axial (σ_{33}) and radial ($\sigma_{11} = \sigma_{22}$) directions are, in general, different.

For the hydrostatic case, all the principal stresses are equal, i.e. $\sigma = \sigma_{11} = \sigma_{22} = \sigma_{33}$. Since there is neither ϕ nor ψ dependence, Eqn. (1) becomes

$$<\varepsilon_{00}> = <\varepsilon> = \frac{1-2v}{E}<\sigma> . \qquad (3)$$

The volume strain, Δ, is defined as sum of the strains in the three axial directions, so

$$\Delta = 3<\varepsilon> = \frac{3(1-2v)}{E}<\sigma> = \frac{<\sigma>}{K} , \qquad (4)$$

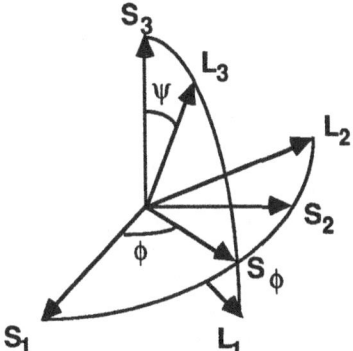

Figure 1. Axial convention used in stress measurements. S_i, L_i are the sample and laboratory systems, respectively, and are related by ϕ and ψ. The diffracting planes are normal to L_3.

where K is the bulk modulus. In this case, which is usually applicable to as-produced particulate and randomly oriented whisker composites, a single measurement in any direction is, in principle, sufficient. Note that the strain measured in any direction is one-third of the volume strain.

   The carats on the strain and stress components in the above equations indicate volume-averaged values. Locally, considerable variation and anisotropy may exist.

3. Experimental

3.1. SOURCES AND SPECTROMETERS

Both steady state [10-15] and pulsed sources [16-20] have been used to measure residual microstresses in composites . Steady state sources are, at this time, more convenient for triaxial measurements in, for example, aligned fiber composites. This is because four-circle sample stages are easier to use at such sources. Reasonably well optimized double-axis powder spectrometers are generally the choice and provide adequate resolution for many studies. Optimization generally means narrow α_1 (pre-monochromator), wide α_2 (between the monochromator and sample), and narrow α_3 (between the sample and detector) collimation, with a high monochomator take-off angle, e.g. 90° and ideally above 120° [21 and Margacza in this volume]. As dedicated instruments are developed the utility of steady state instruments should increase. Another advantage is that peak shapes are substantially Gaussian, enabling full peak fitting with high statistical accuracy.
   Pulsed sources offer the ability to collect many diffraction peaks at high resolution and diffraction angle [22]. The full control of sample orientation is more complex because the sample chambers are enclosed spaces, although such control is certainly possible. Another disadvantage is that the peak shapes are more complex. However, fitting functions have been developed and advantage can be taken of the high resolution of such instruments. It should be noted that high resolution instruments on steady state sources are becoming competitive in this regard. Pulsed sources are very powerful for hydrostatic stress states, particularly when strain anisotropy within a phase is present and many different

crystallographic orientations are required for full analysis. Since all peaks are collected at diffraction angles as high as 150° 2θ, the strain sensitivity is very great. In addition, the use of low and high temperature stages is convenient.

3.2. ALIGNMENT, SLITS AND PEAK POSITIONS

Although these issues are covered elsewhere in this volume, a few remarks specific to composites are in order. When triaxial measurements are made (employing φ and ψ tilts), the center of scattering should not change. This can be a problem even with a cylindrical sample as one goes from ψ = 0 to 90° since the paths change and the specimen may not be precisely centered. One way to deal with this is to stop down the incident beam so that it is always entirely within the specimen. In this way any sample displacements during the measurement will not affect peak positions as long as the beam never comes out of the sample. There are still possible problems with changes in path length but they won't matter as long as all data collection is done in the same way. However, even if this condition is satisfied, misaligned instruments can generate spurious effects such as ψ-splitting.

For steady state sources, determination of peak positions involves individual peak fits as for stress measurement with other kinds of materials. For pulsed sources, single fits have generally been used but another possibility is Rietveld profile refinement [23]. This method, developed for structural modelling of powder diffraction data, can be used to determine cell parameters using all peaks collected, and has been employed with success. If directional mechanical loading of the sample is involved, either in situ or prior to the measurement, or aligned fibers are involved, the symmetry of the cell may be lowered, complicating the analysis.

Finally, when the position of a peak in a composite is compared to that in a stress-free standard, it is critical that both samples be positioned at the same place, that the same slits are used, and that the sizes and shapes of each sample are similar so that absorption does not affect the result.

3.3. USE OF X-RAYS

The penetration of neutrons is particularly useful when heavy elements and volumetric stress states are involved. For x-rays the beam penetration can be sufficiently low to be sensitive to surface relaxation effects. Using a formulation of Hanabusa and coworkers [24], it is possible to model such effects. If a $\sin^2\psi$ analysis is used with a WC-8 wt.% Ni composite, for various interparticle separations, a, of 1, 2, 4, and 10 μm, the results are shown in Fig. 2, for chromium radiation [25]. The data is modelled using a diffraction angle of about 154° 2θ for WC and a hydrostatic compressive stress of -400 MPa. The actual stress state is represented by the horizontal line labelled "hydrostatic value." The shallow beam penetration senses the relaxation of the stress state as it becomes biaxial at the surface. This gives rise to lines with slopes that depend on the interparticle separation, as shown. If these lines are analyzed using the assumption of biaxiality, i.e. the conventional x-ray analysis, stress values of -59, -95, -147, and -177 MPa will result for interparticle separations of 1, 2, 4, and 10 μm, respectively. In the biaxial limit, for essentially no beam penetration, the heavy line will yield a value of -120 MPa. This analysis assumes that other factors, such as surface preparation stresses, play no role. In reality, it would be difficult to avoid such effects within a few microns of the surface.

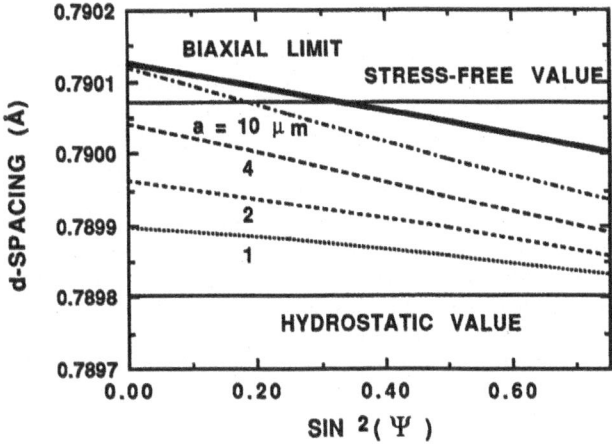

Figure 2. Calculated d vs. sin$^2\psi$ plots for WC-8 wt.% Ni using chromium radiation and interparticle spacings of 1, 2, 4, and 10 μm.

The significance of this analysis is twofold. First, if a triaxial stress state is present and properly measured with x-rays, it must be analyzed as such. Thus, if the hydrostatic line at the bottom of the figure were, in fact, measured, it must be used in conjunction with the stress-free line to reckon the stress values present. Biaxial analysis, i.e. use of the slope, can cause significant error. Second, the x-ray beam penetration may be sufficiently shallow to cause relaxation or surface preparation effects that will lead to incorrect values of the actual microstresses.

3.4. PROBLEMS

3.4.1. *Grain Size.* As for x-rays, coarse grain size can cause erratic results in peak position determinations. Although the large irradiated volumes relative to x-rays help, it still appears that grain sizes greater than 50-100 μm can cause problems. The presence of a large grain size can be detected using rocking curve type measurements. The detector position is set at the top of a peak and changes in the peak intensity are recorded as a function of changes in the sample angle θ (ω). Measurements over ± 5° in steps of 0.2° are generally sufficient. Should minor grain size problems be present, sample motion during data collection can improve the situation. However, it is desirable to metallurgically refine the grain size. In actual samples, as opposed to standards or model material, this is not in general possible and one must either live with the scatter or abandon the experiment.

3.4.2. *Stress-Free Standards.* Neutron stress measurements in composites are always triaxial in nature and thus stress-free reference standards are required, unless the hydrostatic component is not of interest [9]. The most common source of such standards is the starting material used to produce the composite. This is particularly true for the reinforcing phase, which is typically a stable refractory compound that is not alloyed during production of the composite. Materials such as SiC fibers, can be heavily faulted or, like TiC_x, have variable stoichiometry.

The situation for the matrix phase may be complicated by alloying during production of the composite due to fiber dissolution. This is the case for WC-based cemented carbides, in which the metal binder takes up W (and a small amount of C) during liquid phase sintering. In this case it is possible to use diffraction or other means to measure the amount of W in solution, then to make stress measurements at elevated temperature in order to determine where the stresses completely relax. This enables, in principle, a correction for the W solubility. Another problem, for the case of alloy matrices, is that the precipitation kinetics can be affected by the presence of the reinforcement, particularly when it is very fine scale. Thus, the metallurgical state of a 6061 Al alloy differs when it is cooled in the presence of SiC whiskers. Even if an unreinforced standard is produced in a manner identical to the composite, the result will not be the same. In this case, one approach is to grind some of the composite and lightly anneal it to sharpen the peaks. If the ground particles are sufficiently fine, most or all of the stresses will be relaxed even though the matrix and reinforcement are not completely separated.

As long as one phase can be unambiguously measured, the force balance relationship

$$V_r\sigma_r + V_m\sigma_m = 0 \tag{5}$$

can be used, where r and m refer to the reinforcement and matrix phases, respectively, V is volume fraction, and σ is stress. This relationship may be generalized to any number of phases but for n phases, n-1 must be unambiguously determined.

4. Examples

4.1. Al-SiC

4.1.1. *1100 Al Reinforced with SiC Whiskers, Angular and Spherical Particles.* Composites with 20 vol.% SiC in the form of whiskers, angular and spherical particles in an 1100 Al matrix were studied [26]. The whiskers had an average l/d ratio of 3 to 4 with diameters of about 0.5 μm. The particulate was 3 to 5 μm in the largest dimension and the spherical particles were approximately 0.5 μm in diameter. The samples were produced by hot extrusion in the form of cylinders. The stress-free standard was 1100 Al produced in the identical manner. Steady state measurements were made at the Missouri University Research Reactor (MURR) using a wavelength of 1.20 Å. The Al 311 peak, at about 60° 2θ, was measured for φ = 0, 45 and 90° at ψ = ± 0, 30, 45, 60 and 90°. The as-produced pure 1100 Al reference standard was initially found to give erratic peak positions. Rocking curves, using the 311 peak, were measured over ±5° θ (ω) for the as-produced state and after swaging and annealing at 500°C for 1 h; see Fig. 3. The large as-produced grain size caused dramatic variation in peak intensity which was reduced upon the recrystallization treatment. Results for the whisker-reinforced sample for (a) 2θ and (b) strain vs. $\sin^2\psi$ are shown in Fig. 4. The range of data points from the measurements at ±ψ and three φ values for each ψ leads to a spread in values at each $\sin^2\psi$ that is much greater than the statistical fitting errors for each peak. This presumably arises from grain size effects and perhaps real variation within the material. It is typical and indicates that several points should be measured for good statistics. Comparative results for all reinforcements are shown in Fig. 5. The resultant stresses are shown in Table 1. Whisker alignment leads to greater axial stresses

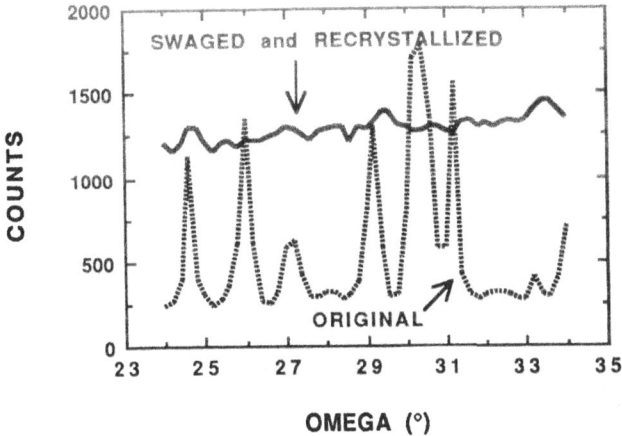

Figure 3. Rocking curves of pure 1100 Al alloy stress-free standards showing coarse grain size in as-produced sample and refined grain size after swaging and recrystallizing at 500° C for 1 h.

TABLE 1. Radial (σ_{11}) and axial (σ_{33}) stresses in 1100 Al alloy reinforced with 20 vol.% SiC of various morphologies

SiC Morphology	σ_{11} (MPa)	σ_{33} (MPa)
Whiskers	58±20	110±19
Particulate	32±16	43±16
Spherical	94±20	100±19

whereas the results for the angular and spherical particulates are essentially the same in both directions. The greater stress in the case of spherical particles presumably arises due to the significantly shorter mean-free-path in the matrix phase for the very fine particles, resulting in a greater amount of cold-worked matrix material.

4.1.2. Influence of Plastic Deformation on 6061 Al Reinforced with SiC Whiskers.
Measurements were made on a series of mechanical test specimens containing 20 vol.% SiC whiskers in a 6061 Al matrix for the purpose of observing changes in the as-produced residual stress state [27]. Specimens were measured before and after tensile or compressive plastic deformation. The data was collected in the same manner as above except that only one ϕ orientation was used. Problems were again encountered with the stress-free standard. In this case, the precipitation kinetics in the unreinforced standard differed from that in the presence of reinforcement. In particular, upon solutionizing at 530°C for 12 h and furnace cooling, more solute precipitates in the presence of the SiC. Air-cooling from 530°C reduces the solute in solution by 80% compared to quenching in ice brine in the presence of SiC and only 30% in unreinforced material [28]. A series of experiments involving various cooling

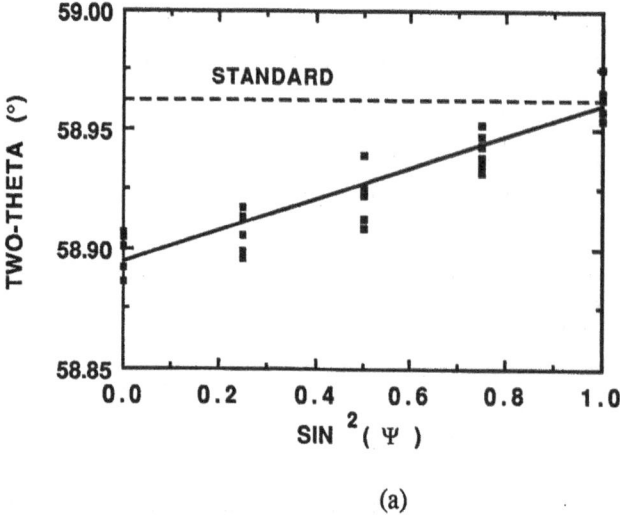

(a)

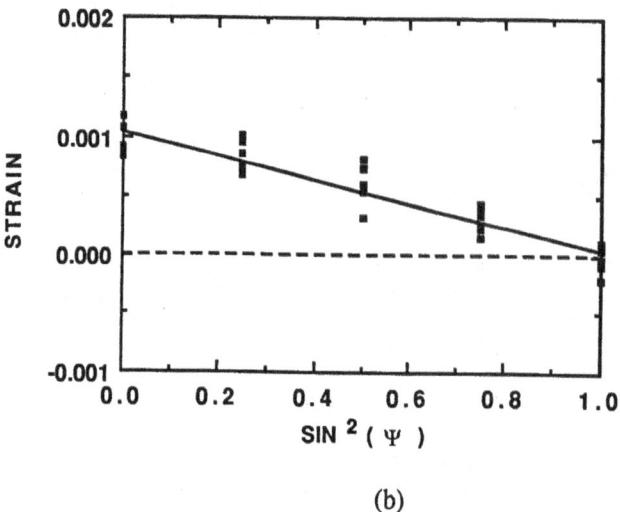

(b)

Figure 4. Results for 1100 Al reinforced with 20 vol.% SiC whiskers showing (a) 2θ vs. $\sin^2\psi$ and (b) strain (ε) vs. $\sin^2\psi$.

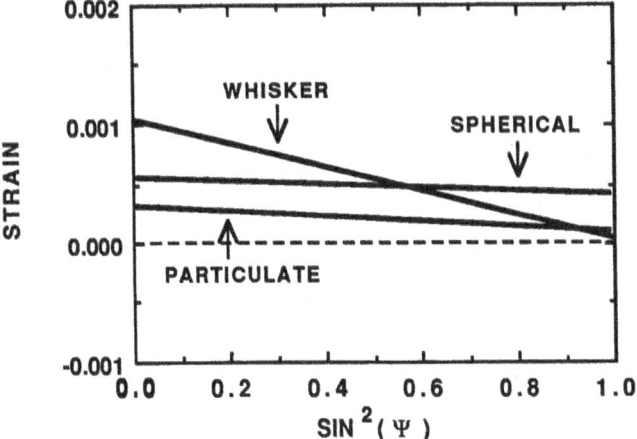

Figure 5. Results for 1100 Al reinforced with 20 vol.% SiC in the form of whiskers, particulate, and spherical particles showing ε vs. $\sin^2\psi$.

treatments, plus filing a composite, indicated that less solute was indeed being removed from solution. However, for the purpose of observing relative changes in residual stress with plastic deformation, the furnace cooled value for the unreinforced material was used as a reference point from which relative change could be reckoned.

The results are shown in Fig. 6 for the axial direction. A minimum is observed in the stress relaxation due to compressive plastic strain while tensile deformation results in a monotonic decrease. This behavior is in agreement with finite element and volume mismatch models developed by Shi and Arsenault [29].

4.2. WC-Ni

4.2.1. *Absolute Stresses.* Samples of WC-Ni containing 6 and 26 wt.% Ni were studied on the General Purpose Powder Diffractometer (GPPD) at the Argonne Intense Pulsed Neutron Source (IPNS) [30]. Only the 26 wt.% data will be discussed herein. Two WC particle sizes were employed, fine (0.8 μm) and coarse (6 μm). The samples were produced by liquid phase sintering at 1440°C. The WC particles are hexagonal, with an angular plate morphology for which the c-axis is normal to and the a-axes lie in the plane of the plates. WC is both thermally and elastically anisotropic, with linear coefficients of thermal expansion, α_a and α_c, of 5.4 and 4.8×10^{-6} K^{-1}, respectively, at 700°K. Measurements were made on bar specimens at 100, 200, 300, 500, 700, and 900°K. Preliminary measurements at MURR confirmed that there was no directionality in the stress states of the samples, i.e. that the stress state is hydrostatic. The WC and Ni cell parameters were obtained from Rietveld analysis of the powder spectra, which were obtained using the high angle ($\approx 148°$ 2θ) detector bank of the GPPD. Loose powder in a vanadium can was used for stress-free

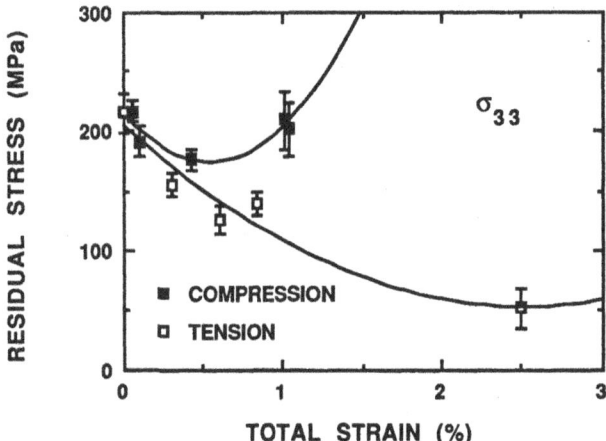

Figure 6. Results for 6061 Al reinforced with 20 vol.% SiC whiskers showing change in residual stress state with tensile and compressive plastic deformation in the axial (σ_{33}) direction.

standards for both the WC and Ni phases. This is ideal for the WC but the Ni in the composite takes up a considerable amount of W in solution during liquid phase sintering, causing a substantial change in cell parameter as will be discussed below.

The a and c cell parameters for the WC phase of the 26 wt.% Ni specimen containing coarse particles are shown in Fig. 7, as a function of temperature. The strain in the a-axis direction is greater than that in c. This is a result of the thermal and elastic anisotropy of the WC cell. The sense and degree of strain anisotropy, even in a hydrostatic stress field, is a function of the shape of the particles, i.e. varies with the aspect ratio L_a/L_c. The result of a calculation using the Eshelby formulation for the WC particles, modelled as oblate spheroids, is shown in Fig. 8 [31]. The data is best fit for $L_a/L_c \approx 1.4$. The corresponding variation in lattice parameter for the FCC Ni phase, and in a pure Ni powder, is shown in Fig. 9. The difference in the two curves is primarily due to the diffusion of W into solution in the Ni matrix during liquid phase sintering, as the position of the corrected curve indicates. The W content in the binder phase was independently determined to be about 6 at. %.

The stress in the WC phase was used to calculate the stress in the Ni via the force balance relationship, Eqn. (5). The stress values for WC and Ni in the 26 wt.% Ni samples containing coarse and fine WC particles are shown in Fig. 10. The results indicate that a difference exists due to particle size, i.e. binder content is constant for the samples. The stress state is greater in the presence of fine particles due to the reduced mean free path which creates stronger interparticle effects, a more rigid matrix, and perhaps greater dislocation activity, and therefore strengthening, in the binder phase.

4.2.2. *Stress Distribution.* In addition to large thermal residual stresses, striking peak breadth effects are observed which can, in principle, yield information about the distribution of local thermal stresses. Individual peak fits from seven WC and five Ni peaks all yielded

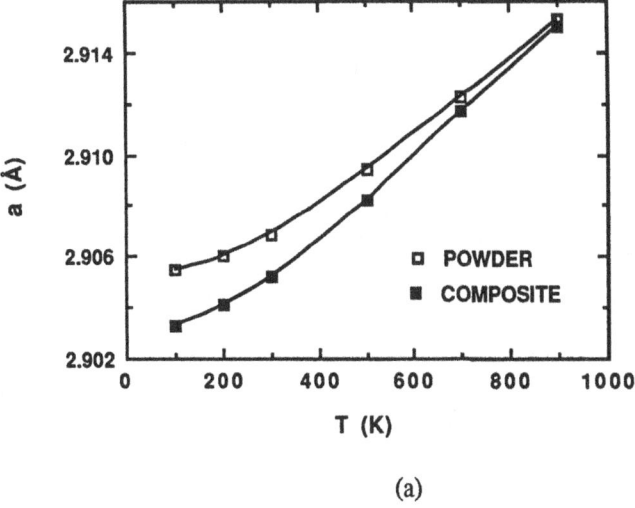

(a)

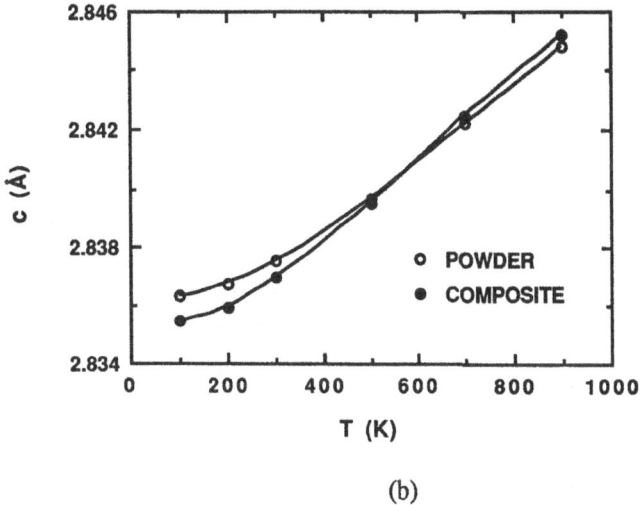

(b)

Figure 7. Values of (a) *a* and (b) *c* cell parameters for the WC phase of WC-26 wt.% Ni containing coarse (6 μm) particles in the composite, and a stress-free standard, as a function of temperature.

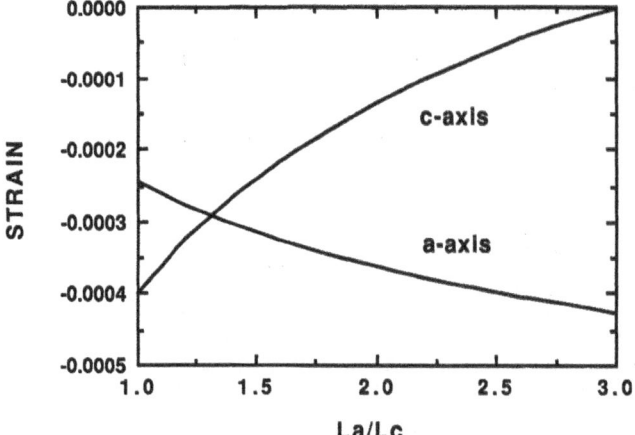

Figure 8. Calculated strain values in the a -and c -directions for a WC single crystal in the shape of an oblate spheroid in a hydrostatic compressive stress field using the Eshelby model.

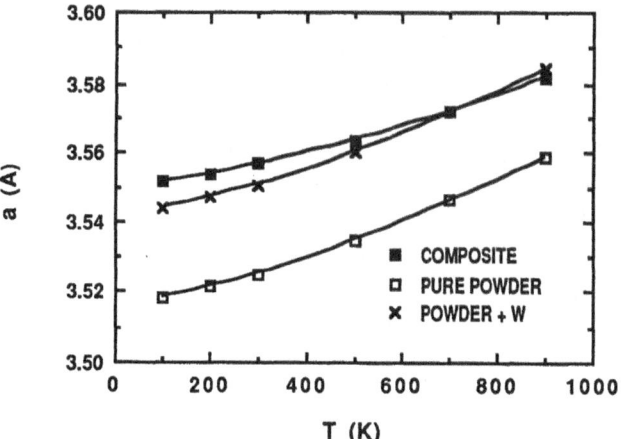

Figure 9. Values of cell parameters for Ni phase in the WC-26 wt.% Ni containing coarse (6 μm) particles in the composite, and a stress-free standard, as a function of temperature. Approximate correction for the presence of W in solution is also shown.

similar behavior with temperature; see Fig. 11. This behavior is reversible with temperature, suggesting that it is not due to plastic deformation. Separation of particle size and strain effects using an approach based on the the Wagner-Aqua integral peak breadth analysis [16, 32] also indicates that the effects are almost entirely due to elastic strain, i.e. the mosaic size due to dislocation activity appears to be quite large, on the order of hundreds of nanometers, and therefore does not contribute much to the broadening. Since the WC phase is not subject to compositional variation, this cannot be a source of the broadening in that phase, nor would it lead to reversibility with temperature. It is believed, therefore, that there are two primary sources of the broadening: a true stress distribution arising from spatial variation within the angular WC plates and in the binder, particularly the regions near interfaces, and a strain distribution arising from a range of WC particle aspect ratios in the microstructure. Modelling studies [18,30] support this conclusion.

5. Conclusions

Neutron diffraction is an effective tool for the study and characterization of residual microstress states in composite materials. The ability to sample large volumes, study systems containing heavy elements, make measurements at low and high temperature, repeatedly measure the same sample before and after thermal or mechanical treatments, and avoid complications due to surface condition and/or macrostresses means that the method extends, and is complementary to, x-ray methods. Useful data may be obtained at both steady state and pulsed sources. At the present time, triaxial measurements are more conveniently done at steady state sources while higher resolution can be obtained at pulsed sources. Both peak position and breadth may be studied.

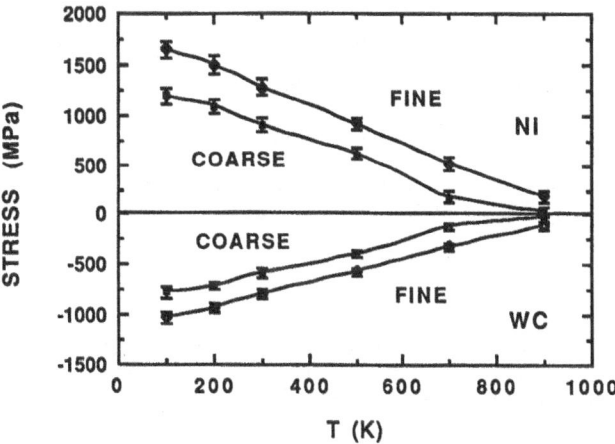

Fig. 10. Residual stress as a function of temperature for the WC and Ni phases of WC-26 wt.% Ni containing coarse (6 μm) and fine (0.8 μm) WC particles.

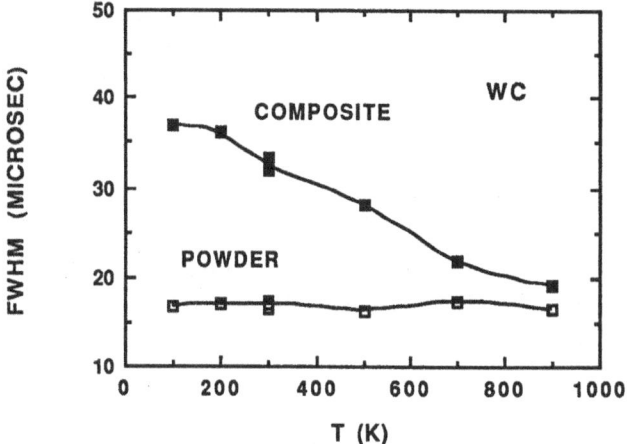

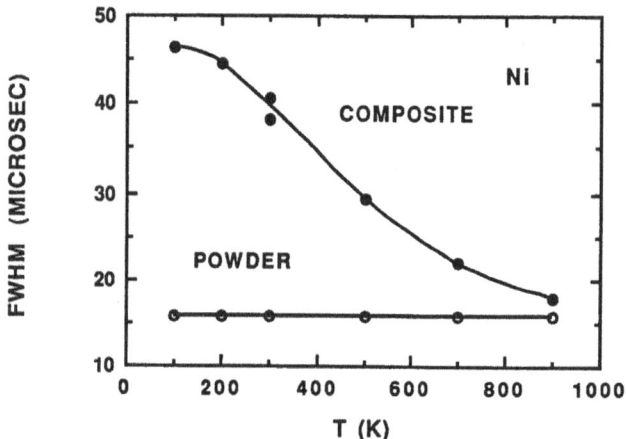

Figure 11. Full-width at half-maximum (FWHM) values for the WC 201 and Ni 311 peaks in WC-26 wt.% Ni, with coarse (6 μm) particles, and stress-free powders, as a function of temperature. Breadths given in pulsed source time-of-flight units (microseconds).

6. Acknowledgements

The author wishes to acknowledge Drs. F. Ross, R. Berliner and W. Yelon at MURR, Mr. L. Smith in the MAE Department at UMC, and Drs. J. Faber (now at AMOCO), J. Richardson and B. Brown, and Mr. R. Hitterman, at IPNS. Permission to present unpublished data from the work of Dr. K. Seol, UMC, and Prof. R. Arsenault and Mr. N. Shi, University of Maryland, is greatly appreciated. This work has benefitted from the use of the Intense Pulsed Neutron Source at Argonne National Laboratory which is funded by the U. S. Dept of Energy, BES-Materials Science, under Contract No. W-31-109-ENG-38.

7. References

1. Noyan, I. C. and Cohen, J. B. (1987) Residual Stress, Springer Verlag, New York.
2. Cohen, J. B. (1986) 'The measurement of stresses in composites', Powder Diffraction 1, 15-21.
3. Eigenmann, B. Scholtes, B. and Macherauch, E. (1989) 'Determination of residual stresses in ceramics and ceramic-metal composites by x-ray diffraction methods', Mater. Sci. Eng. A118, 1-17.
4. Allen, A. J., Hutchings, M. T., Windsor, C. G. and Andreani, C. (1985) 'Neutron diffraction methods for the study of residual stress fields', Adv. in Phys. 34, 445-473.
5. Krawitz, A. D. (1990) 'Residual stress analysis with neutrons' in S. M. Shapiro, S. C. Moss and J. Jorgenson (eds.), Neutron Scattering for Materials Science, Mat. Res. Soc. Symp. Proc. Vol. 166, Materials Research Society, 281-292.
6. Krawitz, A. D. and Holden, T. M. (1990) 'The measurement of residual stresses using neutron difraction', MRS Bull. XV, 57-64.
7. Pintschovius, L., Jung, V., Macherauch, E. and Vohringer, O. (1983) 'Residual stress measurements by means of neutron diffraction', Mater. Sci. Eng. 61, 43-50.
8. Bacon, G. E. (1975) Neutron Diffraction, 3rd ed., Oxford Univ. Press, Oxford.
9. Winholtz, R. A. and Cohen, J. B. (1989) 'Separation of the macro- and microstresses in plastically deformed 1080 steel', Adv. X-Ray Anal. 32, 341-353.
10. A J. Allen, M. Bourke, M. T. Hutchings, A. D. Krawitz and C. G. Windsor (1987) 'Neutron diffraction measurement of internal stress in bulk materials - metal matrix composites', in Macherauch, E. and Hauk, V. (eds.), Residual Stresses in Science & Technology, DGM Verlag, Vol. 1, 151-157.
11. Allen, A. J., Bourke, M., David, W. I. F., Dawes, S., Hutchings, M. T., Krawitz, A. D. and Windsor, C. G. (1989) 'Effects of elastic anisotropy on the lattice strains in polycrystalline metals and composites measured by neutron diffraction', in Beck, G., Denis, S. and Simon, S. (eds.), Proc. Int. Conf. on Residual Stresses (ICRS2), Elsevier Applied Science, 78-83.
12. Lilholt, H. and Juul Jensen, D. (1987) 'Internal stresses measured by neutron diffraction in metal matrix composites exposed to thermal treatments', in Herriot, J. (ed.), Composites Evaluation, Butterworths, Sevenoaks, 156-161.
13. Withers, P. J., Lilholt, H., Juul Jensen, D. and Stobbs, W. M. (1988) 'An examination of diffusional stress relief in metal matrix composites', in Andersen, S.I., Lilholt, H. and Pedersen, O.B. (eds.), Proc. of 9th Risø Symp. on Metallurgy and Materials Science, Risø, Roskilde, 503-510.

420

14. Withers, P. J., Lorentzen, T. and Stobbs, W. M. (1989) 'a study on the relation between internal stresses and the external loading response in Al/SiC composites', in Yunshu, W., Zhenlong, G. and Renjie, W. (eds.), Proc. 7th Int. Conf. on Composite Materials, International Academic Publishers, Beijing, 429-434.

15. Seol, K. and Krawitz, A. D. (1990) 'Anisotropic residual stress relaxation in cemented carbide composites', Mater. Sci. Eng. A127, 1-5.

16. Majumdar, S., Kupperman, D. and Singh, J. (1988) 'Determinations of residual thermal stresses in a SiC-Al$_2$O$_3$ composite using neutron diffraction', J. Am. Ceram. Soc. 71, 858-863.

17. Majumdar, S. and Kupperman, D. (1989) 'Effects of temperature and whisker volume fraction on average residual thermal strains in a SiC/Al$_2$O$_3$ composite', J. Am. Ceram. Soc. 72, 312-313.

18. Majumdar, S., Singh, J. P., Kupperman, D. and Krawitz, A. D. (1991) 'Application of neutron diffraction to measure residual strains in various engineering composite materials', J. Eng. Mater. Technol. 113, 51-59.

19. Krawitz, A. D. Roberts, R. and Faber, J. (1986) 'Residual stress relaxation in cemented carbide composites' in Almond, E., Brookes, C. and Warren, R. (eds.), Science of Hard Materials, Inst. of Physics Conf. Series No. 75, Adam Hilger Ltd, Bristol, pp. 577-589.

20. Krawitz, A. D., Reichel, D. G. and Hitterman, R. L. (1989) 'Residual stress and stress distribution in a WC-Ni composite', Mater. Sci. Eng. A119, 127-134.

21. Caglioti, C., Paoletti, A. and Ricci, F. P. (1958) 'Choice of collimators for a crystal spectrometer for neutron diffraction', Nucl. Instrum. 3, 223-228.

22. Jorgensen, J., Faber, J., Carpenter, J., Crawford, R., Haumann, J., Hitterman, R., Kleb, R., Ostrowski, G., Rotella, F. and Worlton, T. (1989) 'Electronically focused time-of-flight powder diffractometers at the Intense Pulsed Neutron Source', J. Appl. Cryst. 22, 321-333.

23. Rietveld, H.M. (1969) 'A profile refinement method for nuclear and magnetic structures', J. Appl. Crystallogr. 2, 65-71.

24. Hanabusa, T., Nishioka, K. and Fujiwara, H. (1983) 'Criterion for the triaxial x-ray residual stress analysis', Z. Metallkd. 74, 307-313.

25. Krawitz, A. D. (1985) 'The use of x-ray stress analysis for WC-base cermets', Mater. Sci. Eng. 75, 29-36.

26. Smith, L. F. (1990) 'Measurement of thermal residual stresses in Al/SiC composites', Report, Dept. of Mech. & Aero. Eng., Univ. of Missouri, Columbia.

27. Shi, N. (1991) 'Experimental study on the influence of external deformation on the thermal residual stress in short fiber reinforced metal matrix composites', Report, Dept. of Nuc. Eng., Univ. of Maryland, College Park.

28. Papazian, J. M. (1988) 'Effects of SiC whiskers and particles on precipitation in aluminum matrix composites', Metall. Trans. 19A, 2945-2953.

29. Shi, N. and Arsenault, R. J. (1991) 'The effect of thermal residual stresses on the assymetric constitutive behavior of metal-matrix composites', to be published in Composites Science and Technology.

30. Seol, K. (1990) 'Microstructural effects on the residual stress and stress distribution in WC-Ni composites', Ph.D. Dissertation, Univ. of Missouri, Columbia, MO, USA.

31. Majumdar, S. and Krawitz, A. D. (1990) 'Residual thermal stresses and strains in a WC-based cemented carbide composite', Mater. Sci. Eng. A123, L1-L3.

32. Wagner, C. N. J. and Aqua, E. N. (1964) 'Analysis of broadening of powder pattern peaks from cold-worked face-centered and body-centered cubic metals', Adv. X-Ray Anal. 7, 46-65.

THEORY AND MODELLING OF COMPOSITES

P. J. WITHERS
Department of Materials Science & Metallurgy
Cambridge University
Pembroke Street
Cambridge CB2 3QZ, UK.

ABSTRACT: If composites are to fulfil their design specifications it is important to be able to measure and to assess critically the performance of the constituent phases. In this context, the neutron diffraction technique has proven to be an invaluable tool, facilitating the monitoring of internal stress/strain development throughout thermal and/or applied loading. However, an analytical model is required before one can examine the extent to which the full potential for load transfer is realised. This paper examines three predictive models with an eye to interpreting neutron diffraction measurements. Given the difficulty of assessing point to point variations in microstrain by diffraction, attention is focussed on the prediction of average phase strains using an number of illustrative examples.

1. Introduction

Composites can achieve a portfolio of desirable mechanical properties, not to be found in a single material, through a blending of the complementary attributes from two or more phases. Since this usually requires that the properties of each phase be very different from those of the others, composites are characterised by inhomogeneous behaviour at the microstructural level. In service this leads to the development of microstrains which fluctuate according to the 'periodicity' of the microstructure. These stresses do not normally distribute evenly between the phases, so that the average stressing of each phase is different. In fact, the potential of composites as materials of improved specific stiffness, strength and elevated temperature properties derives from, and depends upon, a favourable partitioning of internal and applied stresses towards the reinforcing second phase.

While neutron diffraction data can give a unique insight into stress partitioning, some form of analytical model is required before one can make a quantitative appraisal of the effectiveness of the included phase. It is the aim of this paper to present a framework for the interpretation of neutron diffraction-derived microstrain measurements, both in terms of the internal stresses and the associated micromechanisms of load transfer. Attention is focused on discontinuous metal matrix composites, although many of the modelling procedures are equally applicable to continuous fibre composites as well as ceramic matrix composites.

M. T. Hutchings and A. D. Krawitz (ed.),
Measurement of Residual and Applied Stress Using Neutron Diffraction, 421–437.
© 1992 *Kluwer Academic Publishers.*

2. Fluctuating and Average Stress Fields

It is extremely rare to encounter a completely stress free composite system. Differences in elastic constants, yield stress, coefficients of thermal expansion, etc. between the phases inevitably lead to the development of internal stresses even in the absence of an externally applied force. These internal stresses vary from point to point on a scale approximately equal to the particle spacing (typically ~ 1 - 50 μm). This distance is much smaller than the spatial resolution of conventional diffraction techniques and so the associated lattice strain changes result in a broadening of the diffraction peaks (Figure 1). These internal stresses, which are unaffected by sectioning, have been referred to within this workshop as microstresses because their average over the composite is equal to zero (eqn. 1). Within this paper a more precise notation will be adopted.

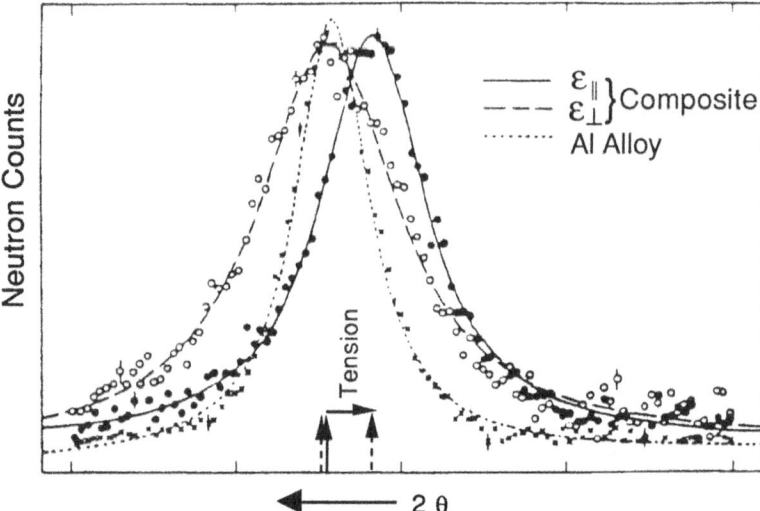

Figure 1. Aluminium (002) diffraction peaks observed from Al(2014) alloy, and in-plane and out-of-plane from an Al (2014)/20% SiC planar random whisker composite [1]. Note how the composite peaks are broadened by a factor of about two, chiefly as a result of the point to point variation in thermal stress, and shifted as a result of a net thermal mean matrix stress. The different magnitude of the peak shift in the in-plane and out-of-plane directions is related to the approximately planar random whisker alignment.

Consider an unloaded composite, the stresses internal to the composite, for example those arising from differential thermal contraction must average to zero whatever the fluctuations in either phase. However, as has been discussed in earlier papers, the average in either phase need not necessarily be equal to zero subject to the condition:

$$(1-f) <\sigma>_M + f <\sigma>_I = 0 \qquad (1)$$

where f denotes the volume fraction of inclusions (I) in a matrix (M). In the case of thermal stresses it is usually the case that the ceramic contracts more than the metal matrix on cooling from the fabrication temperature, so that the matrix is in net residual tension, the reinforcement in compression. I will call these average quantities *mean phase stresses* (<σ>). They are related directly via the phase stiffness tensors $C_{M,I}$ to the mean phase strains $<\varepsilon>_{M,I}$ which would be measured as a peak shift in a neutron diffraction residual strain experiment.

Except in the unlikely event of the elastic and plastic properties of the two phases being identical, external loading is also characterised by a markedly inhomogeneous stress state. In this case the stresses do not average to zero, but one can still make use of the concept of a mean phase stress which is superimposed upon the applied stress (σ^A) and for which eqn (1) still holds. The stress state in either phase could then be described as a mixture of a macrostress (σ^A) and an average microstress ($<\sigma>_{M,I}$). In this case, the *average phase stress* ($\overline{\sigma}$) sampled by diffraction is the sum of the macro and microstress fields:

$$\overline{\sigma}_M = \sigma^A + <\sigma>_M \qquad \text{(tensor notation)} \qquad (2)$$

with $\qquad (1\text{-}f)\,\overline{\sigma}_M + f\,\overline{\sigma}_I = \sigma^A \qquad$ and $\qquad (1\text{-}f)<\sigma>_M + f<\sigma>_I = 0$

As has been discussed in earlier papers, neutron diffraction peak profiles are usually approximately Gaussian in shape. This enables the definition of peak position, and hence average lattice strain ($\overline{\varepsilon}_{M,I}$), to very high accuracy. It should be remembered, however, that if the stress distribution within either phase is asymmetric, an analysis based on the fitting of symmetric Gaussian peaks could lead to misleading results. For example, the fibre strain distribution might become skewed by a suppression of the high strain population through fibre breakage occurring at high stresses.

3. Predictive Models

Composites are usually designed with specific thermomechanical properties in mind. It is thus important to be able to assess the extent to which the two phases work together under the relevant thermal or mechanical loading environment. Neutron diffraction is an invaluable tool in this respect because it is capable of providing dynamic information about the partitioning of stress between the phases under a variety of loading conditions. Consequently, my primary objective in this section is to present and assess models which make definite predictions about the partitioning of stress, and which allow an assessment of the relative importance of the potential load transfer mechanisms.

3.1. THE SHEAR LAG MODEL

Derived by Cox [2] in order to model the stiffness of fibrous paper, this model has since been used extensively to determine the transfer of load between the phases of nearly continuous fibre composites. The approach is simple, but is only applicable to elastic deformation. Load transfer to the reinforcement is visualised as taking place via shear strains generated in the matrix parallel to the fibre axis by the difference in stiffness between the phases. These shears build up the fibre stress from zero at the fibre end to a maximum at the centre. By calculating the form of these shear stresses it is possible to determine this build-up of stress along the fibre length, the average stressing of the phases, and the composite modulus.

$$\overline{\sigma}_I = f\,E_I\left(1 - \frac{\tanh\beta l/2}{\beta l/2}\right) \qquad \text{with } \beta^2 = \frac{4\,\mu_M}{r_0^2\,\ln(\frac{\pi}{f})\,E_I} \qquad (3)$$

where μ_M is the matrix shear stress, E_I the fibre modulus and l and r_0 the fibre length and radius. A number of simple modifications [eg. 3] have been made to the shear lag approach. These

424

introduce an element of load transfer across the fibre end, and thus extend the applicability of the model to lower aspect ratio composites.

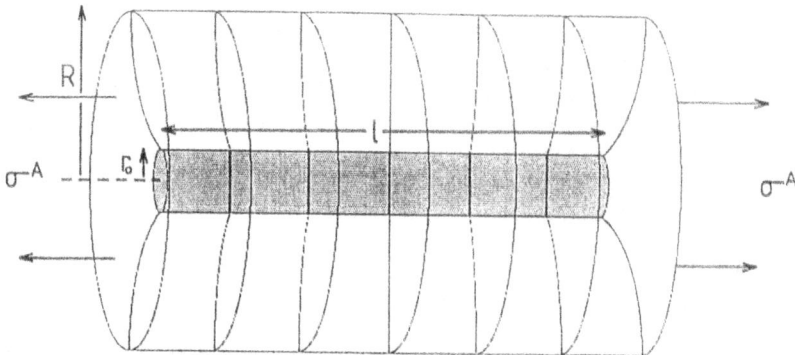

Figure 2. In Cox's model, load is transferred by matrix shear, as illustrated here, and so the fibre load builds up from zero at the fibre end to a maximum at the fibre centre.

The model has the following attributes:

√ it can be computed on a pocket calculator   × it is inaccurate for low stiffness ratios
√ it is accurate at large aspect ratios ($l/r_0 > 20$)   × it neglects load transfer across the fibre end
√ the modified version predicts the observed   × it underestimates the average fibre stress
   [4] form of stress variation within the fibre   because of the neglect of fibre end stress
                                                   × it is unsuitable for plastic deformation or
                                                   thermal loading

3.2. THE ESHELBY MODEL

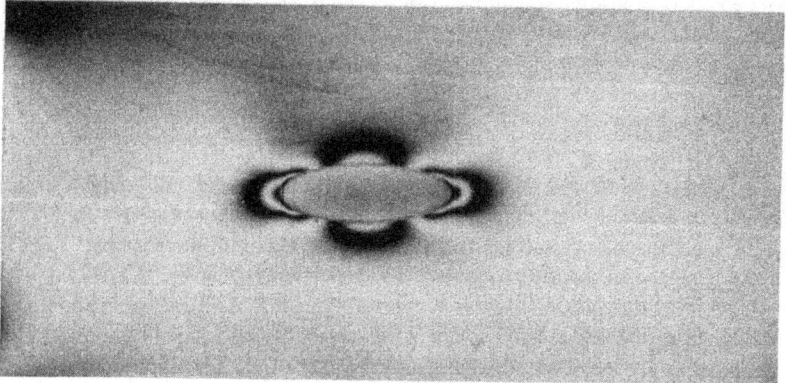

Figure 3. A slice from a three dimensional photoelastic model which had been loaded in axial compression showing the uniformity of stress within an ellipsoidal inclusion in contrast to the highly variable stress state locally within the matrix.

When a misfitting homogeneous ellipsoidal inclusion is constrained to lie inside an ellipsoidal hole in a matrix of the same material the stress and strain within it is always uniform (Figure 3). Eshelby [5] realised that because of the simple nature of the constrained inclusion shape and stress state, the solution to this one problem could be translated across to a wide variety of different situations. This is because whatever the stresses generated by the misfit, one can always generate the same stress state for another similarly constrained inclusion of a different material, given that one has complete freedom over the choice of the original shape of the second inclusion. Conversely, for a reinforcing inclusion of particular elastic constants, no matter what the matrix/inclusion misfit it is always possible to imagine an equivalent inclusion made of the matrix material which will generate the same inclusion stress when constrained to the same shape (Figure 4). Because the two inclusions have the same constrained shape and stress state they can be interchanged without disturbing the matrix, and thus solutions for the stress state of the elastically homogeneous problem can be used to solve the reinforcing problem. The trick, given that the elastically homogeneous problem is easily solved, is to calculate the elastically homogeneous inclusion/matrix shape misfit equivalent to the elastically inhomogeneous problem of interest (see later sections).

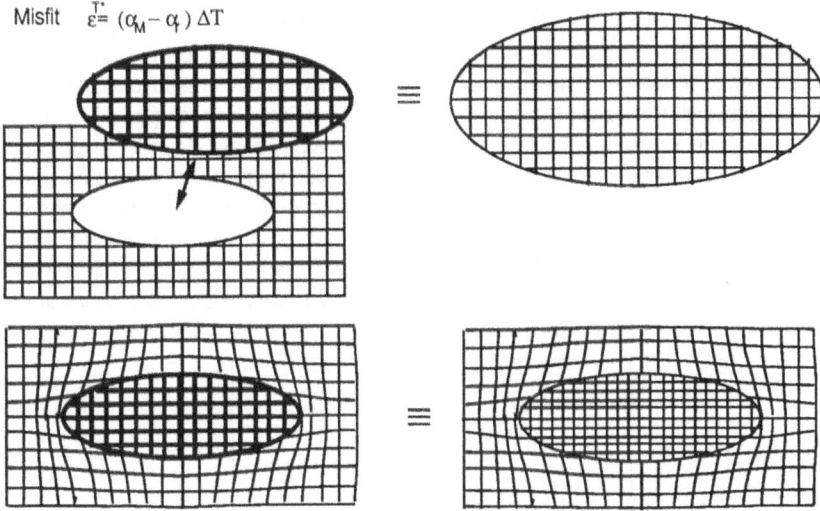

Figure 4. Eshelby's approach is most easily understood by considering a thermal residual stress state. Here the misfit between the matrix and the reinforcing particle is simply the difference in expansion coefficients multiplied by the temperature drop. Naturally, in this case the equivalent inclusion, having the same elastic constants as the matrix, is more oversized than the reinforcing one.

For a dilute composite the approach is mathematically rigorous. In order to satisfy the boundary conditions at the external surfaces of a non-dilute composite, Eshelby introduces the idea of an image stress (used in a similar way to the image techniques in electrostatics). In this form the model assumes that the inclusions are randomly distributed within the matrix. Because the stress in the inclusion is uniform, the mean inclusion stress is easily calculated (see below). Using the condition of microstress balance expressed in eqn (1) this leads directly to the mean matrix stress, without the need to explicitly average over the locally fluctuating matrix stress field.

426

The model has the following attributes:

√ *it has a rigorous analytical basis at low inclusion concentrations*
√ *it is valid at all aspect ratios*
√ *it can be computed on a programmable calculator or personal computer*
√ *it applies to thermal and load-induced stress fields*

× *it is rigorous only for ellipsoidal particles*
× *it is conceptually difficult to grasp*
× *additional assumptions are required to model plastic flow*

3.3. FINITE ELEMENT MODELS

The finite element method relies on the construction of a two or three dimensional mesh made up of connected geometrical shapes. The relationship between nodal forces and nodal displacements is then derived using the principle of virtual work for each element in the mesh. In this way the response of the elements can be calculated, and the complex behaviour of a continuum reduced to that of an equivalent discrete system.

Ideally one would like to be able to model a random array of more or less aligned short fibres. However, this would require a unit cell equal to the size of the composite, or at least big enough to contain hundreds of fibres in a psuedo-random array. This is quite impractical computationally and a much simpler matrix/fibre distribution must be used. Usually symmetry is invoked to reduce the size of the cell, however this can artificially increase the constraint. For example, it is common to assume circular symmetry so that only a single fibre need be considered. In order to tesselate the collection of unit cells it is then common to stipulate that the side walls of the cylinder must remain vertical throughout deformation. This sidewall constraint tends to overestimate the hydrostatic stress component, and thus also the ultimate tensile strength (see Figure 5). While other cell schemes may be more or less successful in avoiding symmetry related artefacts, it is a worrying fact that, given the random nature of most practical systems, finite element predictions are often very sensitive to the choice of the cell tesselation symmetry.

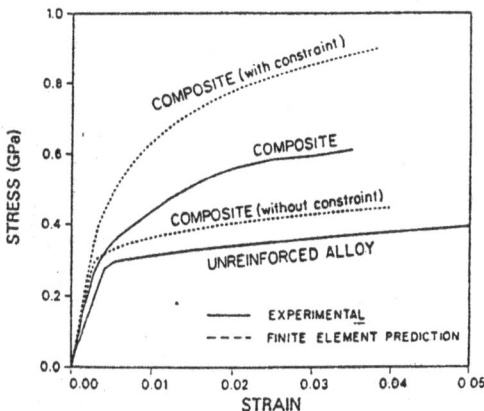

Figure 5. A comparison between finite modelling with and without sidewall constraint and experimental behaviour for a Al/SiC$_W$ composite[6]. The condition that the sidewalls must remain vertical increases the hydrostatic constraint and thus increases the yield stress.

To date, finite element analyses have been used primarily for the prediction of stress/strain curves and for the analysis of debonding and void initiation, however, there is no practical reason why they shouldn't be used for the prediction of the development of average internal strains in either phase for direct comparison with diffraction observations.

The method has the following attributes:

√ *it is very versatile*   × *computation requires a finite element program*
√ *allows great control over constuitive equations*   *and reasonable computing power*
   (eg. can use actual work hardening data from a   × *the choice of repeat cell is problematical*
   'representative' matrix)   × *not possible to express <ε> explicitly*
√ *it allows the introduction of effects like*
   interfacial debonding and breakage

3.4. EXPERIMENTAL VALIDATION OF THE MODELS

An important question is the degree to which these models accurately predict the average stresses which develop in each phase, as might be measured by diffraction. Three dimensional frozen stress photoelastic studies [6] have been undertaken to examine the form of the stress fields about and within cylindrical and ellipsoidal inclusions as a result of uniaxial compression.

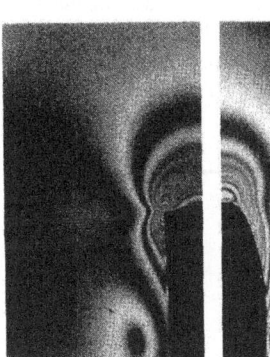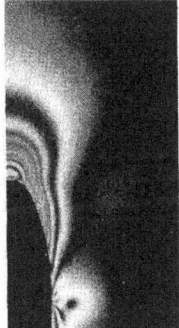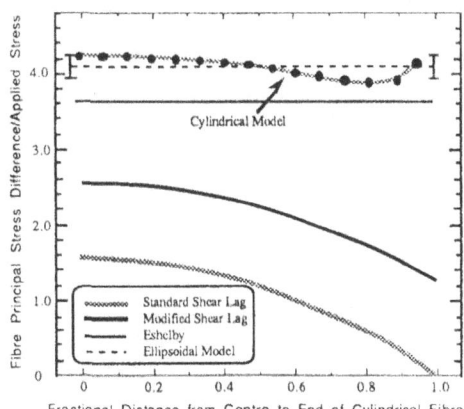

Figure 6. a) Slices from two three dimensional photoelastic models loaded in compression. While these highlight the differences in the stress fields local to the cylinder/ellipsoid 'corners', they are remarkably similar overall, in agreement with de Saint Venant's hypothesis. b) The axial variation in the principal stress difference along aspect ratio 3.3 ellipsoidal and cylindrical inclusions compared with the theoretical predictions [4].

These results have shown that:

● *the large differences in the matrix stress fields are confined to the immediate vicinity of the inclusion 'corners' (Figure 6)*

- these differences would affect the broadening of the diffraction peaks.

● *the stress in the cylindrical fibre is similar in form to that predicted by the modified shear-lag model, but the average value is closer to that predicted by the Eshelby model [6]*

- this suggests that for elastic deformation the Eshelby approach is adequate for the interpretation of the diffraction peaks from the reinforcing phase.

428

• *though difficult to assess directly, the mean value in the two matrices must be approximately the same, given the similarity of the average fibre stresses*

- this suggests that despite differences in the peak matrix stress the Eshelby model is useful for the correlation of average matrix and fibre strains.

These observations would seem to confirm the validity of the Eshelby method for the interpretation of the diffraction peak shifts recorded during elastic deformation. However, the utility of the Eshelby approach for plastic deformation has yet to be proven conclusively for short fibre composites. Certainly, future work should be directed at the correlation of average strains predicted by the Eshelby and finite element models with neutron diffraction strain data under a variety of loading conditions.

4. Fibre Aspect Ratio

It is common practice to extrude or roll composites after fabrication, in order to homogenise the microstructure, reduce porosity and, if short fibres are present, to increase the reinforcement alignment. The degree of alignment achieved, and the extent of fibre breakage which occurs during this process are of great importance in terms of predicting the subsequent composite behaviour. For highly aligned systems, fibre aspect ratio is predicted to have a strong effect on composite properties. In this context, work by Takao and Taya [8] has indicated that for all but the most skewed of aspect ratio distributions the volume averaged aspect ratio is sufficiently representative of the composite as a whole. Figure 7 shows the predicted dependence of Young's modulus on aspect ratio using the Eshelby method plotted alongside experimental data.

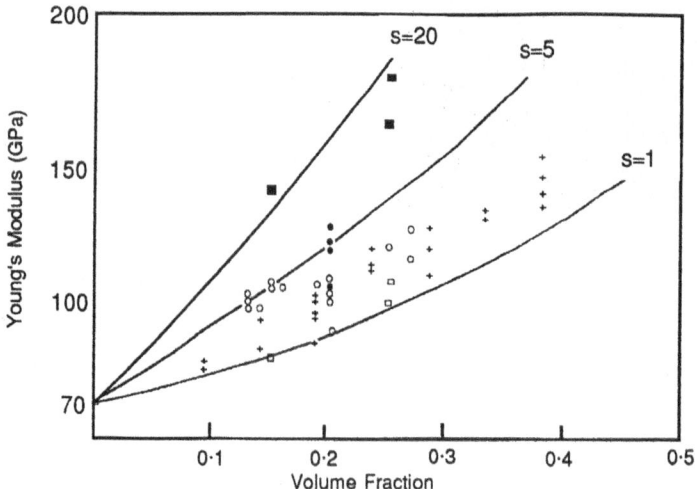

Figure 7. A comparison between experimental data for Al-SiC particle (+) and whisker (• extruded or o otherwise) composites and predictions of the Eshelby model. An aspect ratio of around 5 is typical of most whisker systems; much of the scatter in the data derives from the differing degree of alignment achieved by different processing routes. The whisker data of Lare, Ordway and Hahn [9] (parallel (■), transverse (□)) are the result of good alignment, and high aspect ratio.

5. Fibre and Matrix Crystallography

Much attention has already been spent within this workshop examining the difficulty of the interpretation of lattice strains measured from different lattice planes arising from their differing stiffness. Clearly composites are no less vulnerable to misinterpretation than are single phase materials, but since no extra degree of complication is introduced by the second phase I will leave the discussion to earlier papers. That phase anisotropy is important can be seen from the gradients (Young's moduli) shown in Table 1 which are taken from lattice strain measurements, such as those summarised in Figure 10, taken from systems under elastic loading.

TABLE 1. Gradients (in GPa) measured from lattice strain/applied load unloading curves, from the response of SiC planar random whisker and particulate composites, and unreinforced Al alloy [10]. High values correspond to high stiffness and/or low degrees of loading.

Reflection	Parallel to Loading			Perpendicular to Loading		
	Al or SiC	Al/SiCp	Al/SiCw	Al or SiC	Al/SiCp	Al/SiCw
Predicted Al	70	86	96	-212	-275	-355
Al(002)	71	78		-198	-271	-306
Al (311)		93	91	-196	-270	-283
Al (220)	70		83	-218	-239	-317
Al (111)	72	96			-347	-293
Predicted SiC	500	300	260	-2900	-1330	-925
SiC (311)		235	256		-1253	-1029
SiC (220)		235	373		-1524	-763
SiC (111)		225	238		-2058	-1953

Consistent with the anisotropy in elastic stiffness of the two phases, strains measured using different reflections are not directly comparable in terms of stress. For the unreinforced Al alloy, the strain response of the 002 reflection is seen to be the most compliant and that from 111 the least, as might be expected. A similar trend is observed for the particulate composite; the related gradients (Table 1 and Figure 10) for the two phases mirror the associated changes in stiffness with crystal direction. Thus differences in lattice strain between the different lattice reflections are *not* indicative of the experimental error of a single measurement, which is ~ ±150με, but arise from the anisotropy in crystal stiffness.

The gradients for the whisker phase show a larger and more sporadic variation. In this context it is important to consider the effect of the relationship between crystallography and reinforcement geometry (Figure 8). SiC whiskers comprise a mixture of hexagonal and cubic close packed polytypes but the growth direction is always parallel to the $<111>_{fcc}/<0001>_{hcp}$ direction. Thus measurements of the axial fibre strain can only be made using the (111) reflection. It is therefore not surprising that the gradient of the (111) reflection is the lowest (ie the largest strained) parallel to the load. Similarly, the (220) is the least stressed, so that the average strain taken over the range of planar random orientations would be predicted to lie somewhere between these extremes. This is consistent with the intermediate value of the predicted gradient.

430

Considering the strains measured perpendicular to the applied load; from a Poisson's contraction $(\upsilon(SiC)\sim 0.19)$ viewpoint one would expect the (220) strain to be the highest (ie. to have the lowest gradient). This simple argument is consistent with the observations, although a slight compressive stress in the inclusions (tensile stress in the matrix) transverse to the applied load also increases the compressive strain and lowers the gradient.

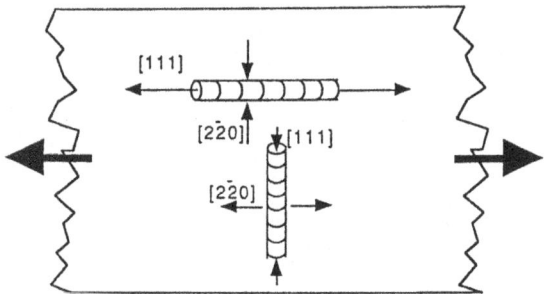

Figure 8. A schematic showing how the whisker geometry and crystallography combine to give the (111) and $(2\bar{2}0)$ lattice strains ‖ and ⊥ to the applied load for a hypothetical composite containing just two fibre orientations. In reality interpretation is further complicated by the multiplicity of the various lattice planes (eg. the three other {111} planes [11]) and the spectrum of fibre orientations.

The above analysis is oversimplistic, but the important point to note from this example is that in fibre composites geometry/crystallography relations must be considered when correlating diffraction measurements with predictive models, whether or not the model explicitly takes into account crystal stiffness anisotropy. For highly anisotropic fibres, it is possible to incorporate anisotropy into both finite element and Eshelby models using a transversely isotropic fibre.

6. Modelling Thermal Stresses

In most composite materials the individual phases have markedly different coefficients of thermal expansion (for example, in the Al/SiC system; Al: $\alpha_M = 23x10^{-6}$ /K, SiC: $\alpha_I = 4x10^{-6}$ /K). Fabrication at elevated temperature therefore results in the development of internal stresses on cooling. The evaluation of these stresses is of great importance in understanding the subsequent mechanical properties, and this is reflected in the large number of neutron diffraction experiments which have been undertaken to examine these stresses [12-14]. A useful parameter for the assessment of the thermal residual stresses is the equivalent temperature drop (ΔT), namely the temperature change necessary to generate the thermal residual stresses were the shape misfit to be elastically accommodated. In practice, stress relaxation, which is discussed elsewhere in this workshop, often occurs so that not all the strain misfit is accommodated elastically, ie the effective temperature is usually below the temperature at which stresses begin to build up on cooling.

Finite element and Eshelby models of the thermal residual stress state have been developed. Of these the Eshelby approach is perhaps the simplest, in that the mean phase stresses are easily expressed in terms of the shape misfit $(\varepsilon^{T*} = (\alpha_M - \alpha_I) \Delta T)$ between the reinforcing inclusions and the matrix [15] (Figure 4).

$$<\varepsilon>_M = f\,[S\text{-}I]\,[(C_M\text{-}C_I)[S\text{-}f(S\text{-}I)]\text{-}C_M\,]^{-1}\,C_I\,\varepsilon^{T*} \qquad (4)$$

where C_M, C_I are the matrix and inclusion stiffness tensors, I the indentity tensor and S is the Eshelby tensor, which is a simple function of the aspect ratio and the matrix Poisson's ratio. From this expression one can see that the elastic stresses and strains are linearly related both to the temperature drop (ΔT) and to the difference between the CTEs of the two phases ($\Delta \alpha$). The thermal strains evaluated by neutron diffraction for various Al/SiC composites are plotted in Figure 9, alongside the effective temperature drop given by eqn (4). It is perhaps significant that the data contained within the plot can be roughly divided into two groups: those with a $\Delta T \sim 200K$ and those with a $\Delta T \sim 300K$. This could be because the former group comprise aligned whiskers (large generation of stress/degree) in pure aluminium (low yield stress) while the latter group are either less well oriented whiskers in 2014 or particulate material (low generation of stress/degree). This explanation would suggest that relaxation governs the final residual stress state, probably by plastic means, such that the maximum mean matrix stresses are approximately independent of the reinforcing parameters (ie. the largest elastic strains are comparable for all the systems).

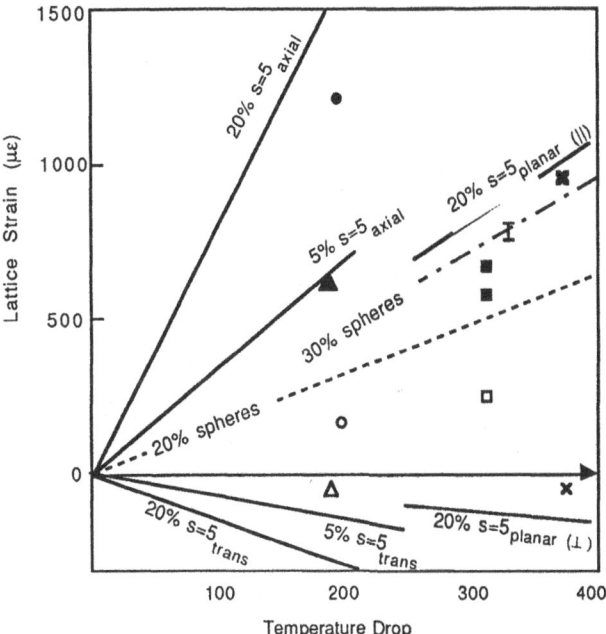

Figure 9. Neutron diffraction derived lattice strain data for (I) 30%SiCp/pure Al [12] (strains deduced from pub. stresses), (□ and x) 20% SiC_W rolled/2014 Al [10], (O) extruded 20%SiC_W/1100 Al [13], (▲) extruded 5% SiC_W/pure Al [14]. In the case of the whisker composites the bold symbols denote directions parallel to the whiskers.

7. Modelling Elastic Deformation

When a composite is subjected to an applied load, the stiffness differential between matrix and reinforcement generates an inhomogeneous stress field as a result of the strain misfit. For elastic loading this misfit is proportional to the applied load and thus disappears on unloading. Following Eshelby's analysis the mean matrix strain can be expressed explicitly [15]:

$$\langle\varepsilon\rangle_M = f \, [S\text{-}I] \, [(C_M\text{-}C_I)[S\text{-}f(S\text{-}I)]\text{-}C_M \,]^{-1} \, (C_I\text{-}C_M) \, C_M^{-1} \, \sigma^A \tag{5}$$

so that the average phase strain as measured by diffraction is given by:

$$\bar\varepsilon_M = \langle\varepsilon\rangle_M + \varepsilon^A = G\varepsilon^A + \varepsilon^A$$

where ε^A ($= C_M^{-1}\,\sigma^A$) is simply the elastic strain of the unreinforced matrix. It can be seen from Figure 10, that the observed lattice strain variation upon elastic loading is as well explained by eqn (5) as could be expected given the isotropic elastic constants assumed by the model.

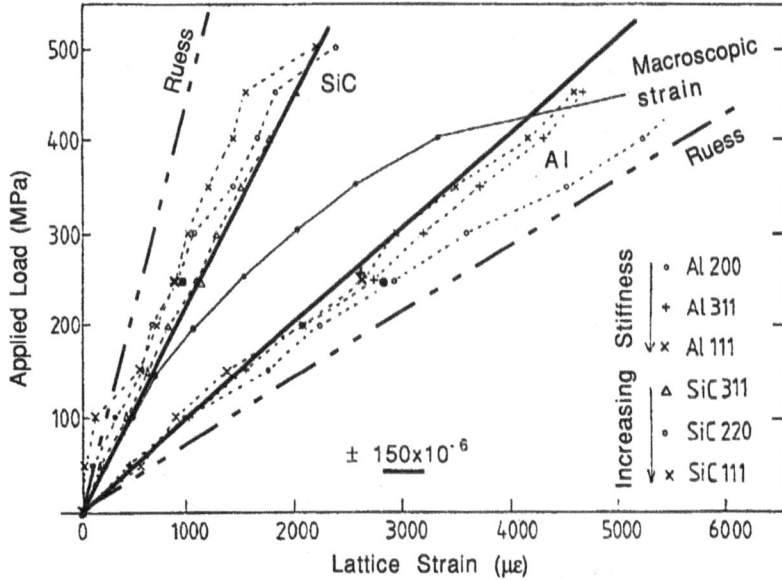

Figure 10. Plot of axial lattice strain response for 20%SiCp/2014Al [10] during elastic deformation showing fairly good correlation with Eshelby predictions. The gradients of the various curves are given in Table 1.

8. Modelling Plasticity

To all intents and purposes most ceramic reinforcements are entirely elastic over the temperature range pertinent to MMC behaviour, so that plastic flow takes place solely by plastic deformation of the matrix. The onset of composite yielding is thus governed by the onset of matrix yielding and the important question to ask is: 'At what point in space and time does matrix yielding occur?'.

Figure 3 demonstrates that even under elastic deformation, the partitioning of applied loads is always very uneven between the phases. As a natural consequence of this, the stress state in the matrix varies greatly from point to point. The position of highest stress is in the vicinity of the fibre end and intuitively one would expect flow to initiate there. Finite element models confirm such expectations (Figure 11). Initially flow is highly localised and is termed microplastic. It is responsible for the early departure of the stress strain curve from linearity. With increased straining, the plastic zone expands further into the matrix resulting in a continuous decrease in the flow stress with strain. As yet, finite element predictions of the average elastic strain response of the constituent phases to forward loading have not been published, although this is a reflection of the interests of finite element modellers themselves rather than of the inherent computational complexity of the task. In fact the relevant calculations must already have been made in order to calculate the macroscopic response of such composites.

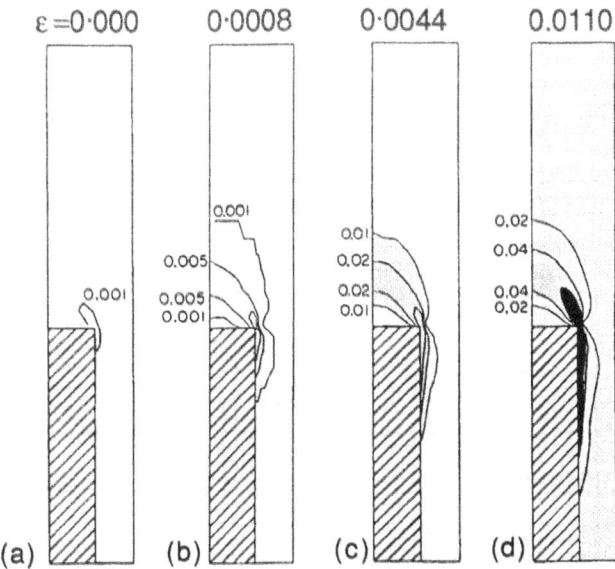

Figure 11. Finite element model [6] showing the growth of the plastic zone with increased straining.

The Eshelby approach can be modified to calculate the plastic straining response. The idealisation of flow which is necessary to facilitate this is somewhat oversimplistic, but at least the model does have the advantage that the average phase strains can be expressed analytically and that the transfer of stress from matrix to reinforcement observed by diffraction can be easily visualised.

Despite the premature onset of flow, the composite yield strength is usually higher than that for the unreinforced material. This observation suggests that the overall yield stress is governed not so much by localised yielding, but by the much later realisation of the yield stress by the average matrix stress. A microscopic explanation lies in the fact that for bulk deformation dislocations must travel large distances through the matrix, and that they are unable to do so until either the local areas of high stress link up (ie. when the whole cell in figure 11 is plastic), or the average matrix stress is sufficient to cause the movement of dislocations. At lower stresses only a local reduction of the matrix stress is possible, allowing more limited elastic and inelastic straining.

In common with the criterion derived for monolithic materials, it would seem natural to develop a yield criterion which takes into account the transverse stresses as well as the axial. Most popular is a Tresca-type yield criterion which can be modified by assuming that bulk flow occurs when the average matrix stress difference reaches that which would give rise to flow were it experienced by the matrix[†] (σ_{YM}) free from the reinforcement. This occurs when:

$$\sigma_{YM} = \overline{\sigma}_{Maxial} - \overline{\sigma}_{Mtrans} \qquad (6)$$

[†] the yield stress of the matrix in-situ is unlikely to be the same as that of the unreinforced matrix.

434

In order to model the load transfer which occurs subsequent to the onset of plastic flow consider an unstressed composite containing spherical particles before and after being subjected to a shear stress. Prior to loading, there are no shape misfits or dislocations (Figure 12). However, if the particle could be dissolved away *after* plastic shear loading, then the shape of the cavity within the matrix would be different from the original shape. Eshelby's continuum approach can be used to model the stresses arising from this shape change by smearing out the dislocation structure [16] and assuming an effectively homogeneous plastic strain in the matrix, i.e. the matrix is assumed to deform uniformly throughout in a manner similar to that of a matrix containing no reinforcing particle. In view of the highly variable state of stress around the particle prior to the onset of flow and the FEM predictions of Figure 11 of the plastic strain local to the fibre, this oversimplification should be regarded with suspicion. Nevertheless, this approach has proven useful for dispersion strengthened materials [17] and continuous fibre composites [18]. This assumption, which is similar to that required to carry over the method from ellipsoidal particles to cylindrical fibres despite differences in the local stress fields, has the advantage that the internal stress field can be derived in terms of the shape misfit (ε^{T*}) between the spherical particle and the ellipsoidal cavity.

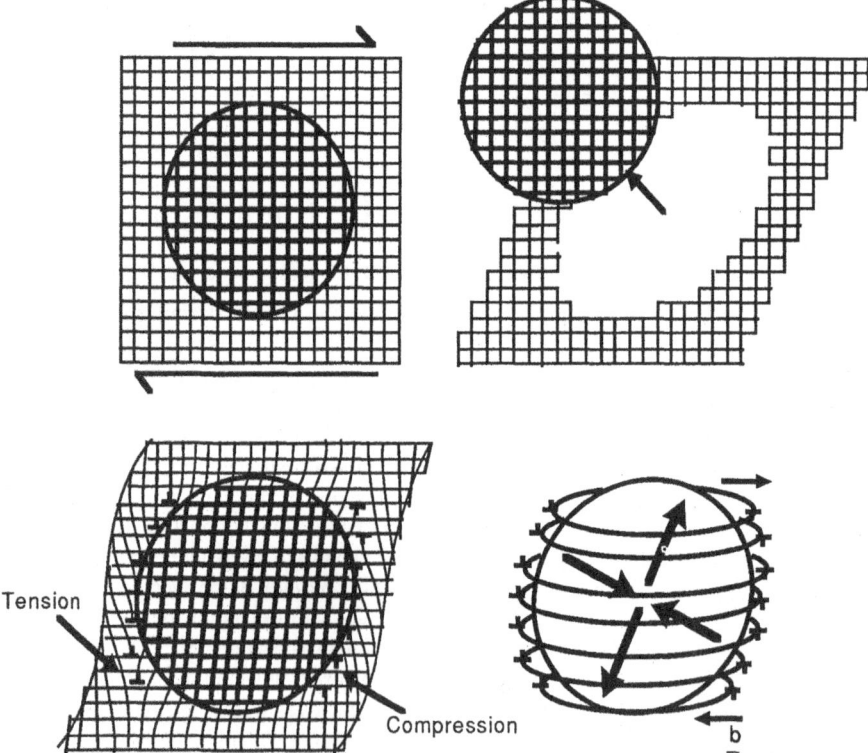

Figure 12. A dislocation picture showing how the interruption of flow within a crystal by a reinforcing inclusion necessarily brings about the generation of a uniform plastically induced inclusion/matrix misfit . In practice, secondary dislocation structures, not related to the external shape change but to the internal stress, can act so as to reduce the shape misfit.

From a dislocation viewpoint, the uniform shear depicted in Figure 12 is achieved by the passing of 9 dislocations through the matrix from one side to another. Because the particle is non-shearable these dislocations are held up at the particle and the only way they can reach the other side of the composite is by leaving behind an Orowan loop. It is the resulting dislocation array which distorts the inclusion and thus generates the mean inclusion stress. Provided a reasonable number of slip lines intersect the particles, these dislocations can be smeared out and the plastic distortion in the continuum approach related directly to the number of necessary dislocations [16].

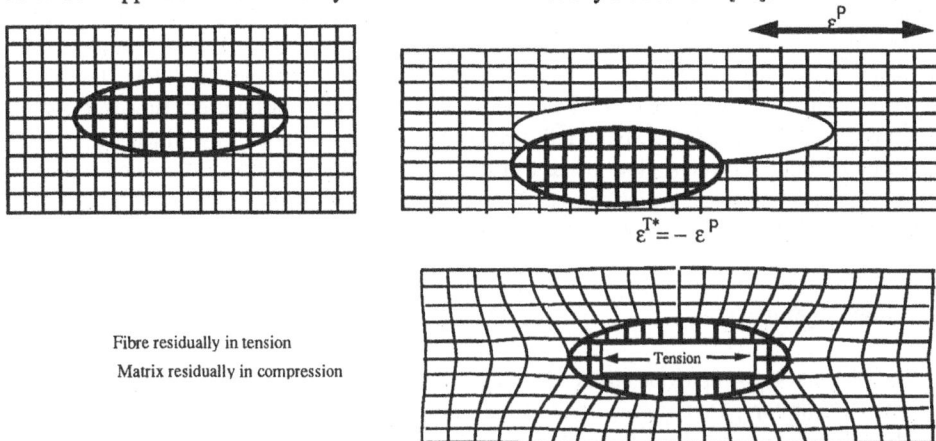

Fibre residually in tension

Matrix residually in compression

Figure 13. The plastic generation of an inclusion/matrix misfit. As can be seen plastic flow transfers load from the matrix to the inclusions so that subsequent after unloading the inclusion is residually in tension the matrix compression.

For uniaxial loading (Figure 13), the shape misfit between the plastically distorting matrix and the elastically distorting inclusion is given by:

$$\varepsilon^{T*} = -\varepsilon_M^P = -\varepsilon_M^P \ (-1/2, -1/2, 1, \ 0, 0, 0) \tag{7}$$

where ε_M^P is the global plastic strain in the matrix in the direction of loading (3). This is consistent with the volume conservation requirement for plastic flow ($\Sigma \varepsilon_j^P = 0$). The mean stress arising from plastic flow can then be calculated in terms of the misfit using an expression similar in form to that of eqn (4) [15]

$$<\sigma>_M = f \, C_M \, [S - I] \, [(C_M - C_I)[S - f(S - I)] - C_M]^{-1} \, C_I \, \varepsilon_M^P \tag{8}$$

This equation indicates that the transfer of load to the reinforcement with plastic straining is a linear function of matrix plastic strain (prior to the activation of relaxation), and that the rate of load transfer increases with increasing volume fraction and inclusion aspect ratio. This gives rise to a high rate of work-hardening, even when the matrix itself exhibits no inherent work hardening capacity.

In real systems the high initial rate of work hardening is short lived because relaxation mechanisms act, even at moderately low strains, to restrict the build up in reinforcement loading. This decrease in the rate of misfit growth between matrix and reinforcement can also be modelled by the Eshelby approach but this will not be discussed here, instead this is described in detail in a separate paper on relaxation within this workshop [19]. It is important to note, however, that the finite element and Eshelby approaches differ in that the Eshelby approach distinguishes between plasticity brought about through the motion of necessary dislocations which increases the reinforcement/matrix misfit, and plastic relaxation caused by the generation of secondary dislocations and structures which decreases the misfit. The finite element method, since it treats plasticity in a continuum manner, makes no such distinction.

9. Conclusions

Finite element and Eshelby methods enable a critical evaluation of neutron diffraction strain data. The finite element method has yet to fully circumvent tesselation problems, but offers scope for the investigation of effects such as shape, clustering and matrix work hardening. Though not yet applied to average phase stresses, on-going work suggests that such methods will have much to contribute to future diffraction experiments. The Eshelby method is also well suited to the prediction of average phase quantities, but is computationally much simpler than FEM. It can thus be used for an on-line evaluation of average phase strains, at the diffractometer. However, for the description of plastic flow behaviour it relies upon dislocation-based arguments which are probably not strictly valid for polycrystalline metals at a scale characteristic of MMCs. Both models clearly have their disadvantages: nevertheless, without such models it is not possible to begin to understand, and hence control, the mechanisms of load transfer central to composite behaviour.

Acknowledgements

The financial support provided by Alcan International and The Materials Department at the Risø National Laboratory, Denmark is gratefully acknowledged. Helpful discussions with W.M. Stobbs, T.W.Clyne and N.Hansen are appreciated.

References

1. Allen A.J., Bourke M., Hutchings M.T., Krawitz A.D. and Windsor C.G. (1987) Res. Stresses in Sci. & Tech., Ed. E. Macherauch, V. Hauk, DGM Verlag, 1, pp. 151.
2. Cox H.L. (1952), The elasticity and strength of paper and other fibrous materials', Brit J. Appl. Phys., 3, 72-79.
3. Clyne T.W. (1989) 'A simple development of the shear lag theory appropriate for composites with a relatively small modulus mismatch', Mat.Sci. & Eng., A122, 183-192.
4. Withers P.J., Chorley E.M., Clyne T.W. (1990),'Use of the frozen-stress photoelastic method to explore load partitioning in short fibre composites, to appear in Mat. Sci. & Eng., A132.
5. Eshelby J.D. (1957) 'The determination of the elastic field of an ellipsoidal inclusion, and related problems', Proc. R. Soc., 241A, 376-396.
6. T. Christman, A. Needleman, S.Suresh, (1989), An experimental and numerical study of deformation in MMCs, Acta Metall., 37, 3029-3050.
7. Withers P.J., Cecil G.J., Clyne T.W., (1991), 'Frozen stress photoelastic determination of the extent of fibre stressing in short fibre composites', To be published in Euromat '91, , Inst of Metals, Cambridge.
8. Takao Y. and Taya M. (1987), 'The effect of variable fibre aspect ratio on the stiffness & thermal exapansion coeifficients of a short fibre composite', J. Composite Materials, 21, 140-156.
9. Lare P.J., Ordway F. and Hahn H., (1971), Research on whisker-reinforced metal composites, Naval Air systems command, contract N00019-70-C-0204, Final Report, Dec.

10. Allen, A.J.,.Bourke, M, Dawes, S., Hutchings, M.T. and Withers, P.J, The analysis of internal strains measured by neutron diffraction in Al/SiC MMCs, (1991), to be submitted.

11. Majumdar, S., Kupperman, D. and Singh, J., (1988), Determination of residual thermal stresses in a SiC-Al$_2$O$_3$ composite using neutron diffraction, J. Am. Ceram. Soc., 71, 858-863.

12. Walker, J.R., Pickard, S.M., Derby B.,(1987), 'Internal thermal stresses and deformation of MMCs', Paper 21, Metal Matrix Composites: Strucutre and Property Assessment, 23-24th Nov., Inst. of Metals, London.

13. Krawitz, A.D, Holden, T.M., (1990) 'The measurement of residual stresses using neutron diffraction', MRS Bul, Nov. 57-64.

14. Withers P.J., Lorentzen T, (1989), 'A Study on the relation between the internal stresses & the external loading response in Al/SiC composites', ICCM VII, Beijing, 2, 429-434.

15 . Withers P.J., Stobbs W.M., Pedersen O.B., (1989) 'The application of the Eshelby method of internal stress determination to short fibre metal matrix composites', Acta Met., 37 3061-3084.

16. Ashby M.F.(1966), 'Work hardening of dispersion hardened crystals', Phil Mag.,14, 1157-1178.

17. Brown L.M., Stobbs W.M. (1976) 'The work hardening of Cu-SiO$_2$ V Equilibrium plastic relaxation by secondary dislocations', Phil. Mag., 34, 351-372,.

18. Pedersen O.B.(1985) 'Mean field theory and the Bauschinger effect in composites', in Fundamentals of deformation & fracture, Proc IUTAM, Eshelby Mem. Symp.,129-144, April 1984, K.J. Miller & B.Bilby, CUP.

19. Withers P.J. (1991), 'Relaxation', within this proceedings.

APPLICATION OF NEUTRON DIFFRACTION TIME-OF-FLIGHT MEASUREMENTS TO THE STUDY OF STRAIN IN COMPOSITES

D. S. KUPPERMAN, S. MAJUMDAR, and J. P. SINGH
Materials and Components Technology Division
Argonne National Laboratory
Argonne, Illinois 60439

and

A. SAIGAL
Department of Mechanical Engineering
Tufts University
Medford, MA 02155

ABSTRACT. Neutron diffraction time-of flight measurements in the Intense Pulsed Neutron Source at Argonne National Laboratory have been used to study strain (from which stress is calculated) in various metal- and ceramic-matrix composite structures. For example, the effect of fabrication procedures (e.g., thermal cycling) on strain in the constituents of a metal-matrix composite have been studied. In this example, neutron diffraction experiments were used to quantify the process-induced reduction of strain in composite fibers. Measurements carried out to 900°C on this material have been used to validate theoretical assumptions in the prediction of fabrication-induced residual stress. Neutron diffraction studies of a high-temperature ceramic superconducting composite consisting of yttrium-barium-copper-oxide (YBCO) with various volume fractions of silver have also been carried out. The results of these studies have provided information on the effect of Ag content on stoichiometry, strength of interface bonding between Ag and YBCO (indicated by the tensile strains in the Ag), and creep mechanisms during fabrication. Ceramic-matrix composites with randomly dispersed ceramic whiskers have also been studied. Comparison of measured data with analytical-model predictions gave important clues to the micromechanisms involved in the creation of residual stresses. For example, the measured residual strain, as a function of angle, in hexagonal silicon nitride whiskers embedded in an alumina matrix is in excellent agreement with the predictions of a simple linear elastic model. For samples of silicon carbide whiskers embedded in an alumina matrix, temperature-dependent experiments allow determination of the stress-free temperature used in elastic analyses. Problems with determining strain-free lattice parameters and uncertainties in absolute measurements will be discussed.

M. T. Hutchings and A. D. Krawitz (ed.),
Measurement of Residual and Applied Stress Using Neutron Diffraction, 439–450.
© 1992 *Kluwer Academic Publishers.*

1. Introduction

In applying neutron diffraction, the lattice spacings in various crystallographic directions of stress-free powders and/or fibers (which are used to fabricate the composite) are determined first. The shifts in the Bragg peaks of the stressed constituents of the composite are then determined. For any hkl diffraction peak, the lattice strain is given by

$$\varepsilon_{hkl} = (d_{hkl} - d_0)/d_0, \tag{1}$$

where d_{hkl} and d_0 are the average lattice spacings in the stressed and stress-free material, respectively, and hkl represents the Miller indices of the diffracting planes. With a pulsed source, changes in lattice spacing are related to time-of-flight shifts in the Bragg peaks. The relationship of time-of-flight to neutron wavelength is described by equation

$$\lambda = ht/mL, \tag{2}$$

where λ is the neutron wavelength, h is Planck's constant, t is the time of flight for a neutron to reach a detector after leaving its source, L is the flight path of the neutron from its source to the detector, and m is the neutron mass. Bragg's law, $2d\sin\theta = \lambda$, can then be written as

$$t = (2dmL/h)\sin\theta \tag{3}$$

where d is the lattice spacing and θ is the diffraction angle, [fixed for the Intense Pulsed Neutron Source (IPNS), at Argonne National Laboratory], and $\varepsilon_{hkl} = \Delta t/t$.

In general, six independent strain measurements are required to define the stress tensor. However, for composites with fibers, only one measurement with the scattering vector parallel to the fiber axis, and two perpendicular to the fiber axis, are required. If the fibers are single crystals randomly oriented in an isotropic matrix, only one measurement is needed to determine the stress tensor. With the IPNS, the presence of detectors at ± 148, ± 90, ± 60, 30, and $20°$ relative to the neutron beam forward direction permits acquisition of information simultaneously in different directions.

In this paper we apply neutron diffraction time-of flight measurements to study strain (from which stress is calculated) in various metal- and ceramic-matrix composite structures.

2. High-Temperature Ceramic Superconductors

Since the discovery of superconducting materials with relatively high transition temperatures (T_c), considerable effort has been directed toward both understanding the reason for the high T_c and improving the materials mechanical properties, which limit practical applications. The $YBa_2Cu_3O_{7-d}$ (YBCO) compounds have received considerable attention [1-3], because of their high T_c and high upper critical magnetic field. Additions of silver have recently been shown to improve the mechanical properties (toughness and strength) of these compounds [4,5]. During fabrication of YBCO/Ag composites, differential thermal expansion upon cooling can lead to potentially troublesome residual stresses. Because the Ag contracts more than the YBCO, good bonding between the ceramic and Ag could lead [6] to tensile stresses in the Ag and compressive stresses in the YBCO for relatively small percentages of Ag. The mechanical properties and life expectancy of components made from this material may depend on the residual stresses. An understanding of the nature and magnitude of these stresses will help improve the design of these composites.

Time-of-flight neutron diffraction employing the IPNS and the associated general purpose powder diffractometer, was used to measure the residual thermal strains in YBCO composites [6,8].

2.1. SAMPLE PREPARATION

Samples discussed here were fabricated from a mixture of YBCO and Ag powders, with 0, 15, 20, and 30% Ag by volume [8]. Both YBCO and composite powders were formed into rectangular bars ($\approx$5.1 x 0.6 x 0.3 cm) by uniaxial pressing in a steel die at 150 MPa. These specimens were treated in a flowing oxygen atmosphere at 920-980°C. These specimens were subsequently annealed at approximately 500°C. Examination of polished surfaces of sintered YBCO/15% Ag bar specimens by scanning electron microscopy showed that the Ag phase was randomly distributed in discrete globules and no preferred orientation of Ag was observed. Examination of the fracture surface of a sintered specimen of YBCO/20% Ag showed a typical grain size of 15 μm, about twice as large as the grain of YBCO specimens without any Ag addition. This increase in grain size is believed to be due to the presence of a liquid phase formed as a result of Ag addition.

2.2. RESULTS

The IPNS is capable of measuring lattice spacings to an accuracy of ±0.0002 Å. The main advantage of the pulsed source is that many diffraction peaks can be measured at the same time in different spatial directions. Results reported here combine the data from the 90 and 148° detectors.

Figure 1 shows the average bulk strain at room temperature in the Ag as a function of volume percent Ag in the composite and the crystallographic direction. The Ag diffraction peaks for which data are presented are clearly isolated from the YBCO diffraction peaks. The potentially complex loads and the relatively large error in absolute measurement of strain make it difficult to reach any conclusions by comparing strains in various crystallographic directions. However, there is a trend toward decreasing Ag strain as the volume fraction of Ag increases (the relative error in strain between samples of varying Ag content is about half the uncertainty in the absolute values). The strain, however, is expected to be proportional to the difference between the thermal expansion coefficient of Ag and that of the composite (which increases with increasing Ag) [9]. It is unlikely that Y, Ba, Cu or O can change the Ag spacing chemically. The difference between the thermal expansion coefficients of Ag and YBCO is $\approx$2 x 10^{-6}/°C. An increase in Ag content from 15% to 30% should increase the composite coefficient of expansion by $\approx$2%. The observed decrease in strain is much larger than can be explained by the $\approx$2% difference in thermal expansion coefficient. It is possible that the strain is relieved by creep due to increasing amount of Ag in the matrix.

Measuring strains in the $YBa_2Cu_3O_{7-d}$ is difficult because the stoichiometry can change with Ag content. Thus, the shifts in $YBa_2Cu_3O_{7-d}$ diffraction peaks may not be solely the result of strain. Destructive analysis, combined with X-ray diffraction analysis of sections of a 30%-Ag sample [4], shows that the stoichiometry of the 30% sample changes gradually from d = 0.1 at the outer surface (orthorhombic superconducting phase) to d $\approx$0.8 near the center of the specimen (nonsuperconducting tetragonal phase). This finding is consistent with the observation that the 30%-Ag sample shows a dramatic decrease in critical current density, as compared with samples with 0, 15, and 20% Ag.

Because the lattice spacings as a function of stoichiometry for the three principal directions of YBCO are known, it is possible to estimate the diffraction peak shift resulting from strain (for some diffraction lines) by correcting for stoichiometry. For example, because the 111 diffraction peak is a single peak and the shifts are unambiguous, we can predict the change in the spacing of the (111) plane as a function of stoichiometry from the relationship

$$1/d^2_{111} = 1/a^2 + 1/b^2 + 1/c^2, \qquad (4)$$

where the values of a, b, and c as a function of d are determined by neutron diffraction [10]. The value of d for YBCO in the present composites was estimated by comparing relationships that are primarily due to stoichiometry, and not strain, with those of materials with known d. Strains determined from estimates of lattice parameter for strain–free YBCO (111) as a function of Ag content are presented in Fig. 2. We have assumed that the addition of 15%-Ag does not significantly change the stoichiometry of YBCO. The stoichiometry of the 30%-Ag sample is assumed to be nearly tetragonal (bulk average), and the 20%-Ag sample is assumed to be intermediate in stoichiometry between the 15%- and 30%-Ag samples [11]. The lattice parameter is predicted to increase as the stoichiometry approaches the tetragonal phase (with increasing Ag content). However, experimentally, the spacing for the (111) planes decreases as the Ag content increases from 15% to 20%, indicating an average compressive stress in the YBCO. This result qualitatively agrees with expectations, because the Ag strain is tensile. With 30% Ag, the difference between measured and stress–free lattice spacing indicates negligible YBCO strain. This finding is consistent with the low strain measured for Ag in the YBCO/30% Ag composite, and is probably a result of additional composite creep, which can vary with stoichiometry [12].

Strains were also measured at liquid N temperatures for the 15%-Ag sample. Only small changes in strain were detected when the sample was cooled from room to liquid N temperature. The largest increase in strain was observed for the YBCO (111) plane. The compressive strain increased about 0.04%. The absence of any strain relaxation suggests that little or no cracking occurred as the temperature was lowered to the superconducting transition point.

2.3. DISCUSSION

We have shown that neutron diffraction techniques can be used to measure, in the constituent parts of YBCO composites, residual strains caused by differential thermal contraction after fabrication, and to determine the effect of Ag on stoichiometry (and thus on the critical current density). We have observed residual tensile strains in Ag as a function of crystallographic direction; these strains range from as high as 0.085% in 15- and 20%-Ag samples to as low as $\approx$0.02% in a 30%-Ag sample. Compressive strains in the YBCO (111) crystallographic direction were estimated by correcting for the shift in diffraction peak due to changes in stoichiometry with Ag content. The estimated compressive–strain values vary from 0.04% (15% Ag) to 0.09% (20% Ag) to 0.01% (30% Ag), with an uncertainty of $\approx$0.03%. The decrease in strain in YBCO is consistent with the decrease in Ag strain and may be due to stoichiometrically sensitive variations in creep properties of the composite (11). The presence of significant average tensile strain in the Ag, particularly for 15- and 20%-Ag samples, indicates good interface bonding between YBCO and Ag. The hydrostatic stresses in the Ag, estimated from these strain

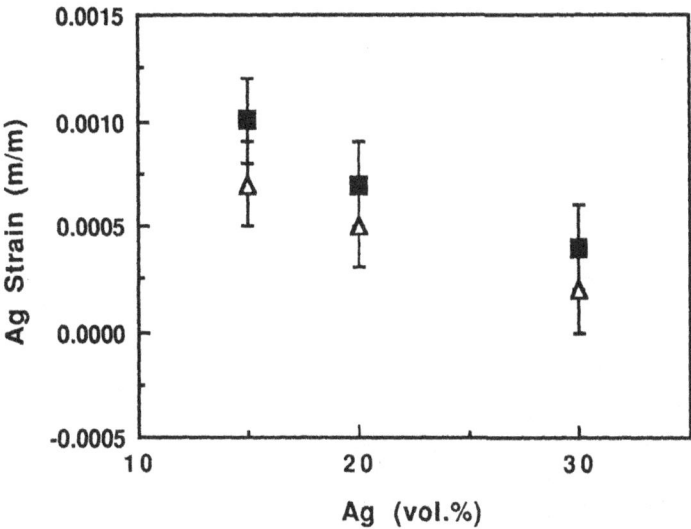

Fig. 1. Room–temperature tensile strain in Ag as a function of Ag content in YBCO/Ag composite and crystallographic direction. The closed symbol is 111 while the open symbol is 220.

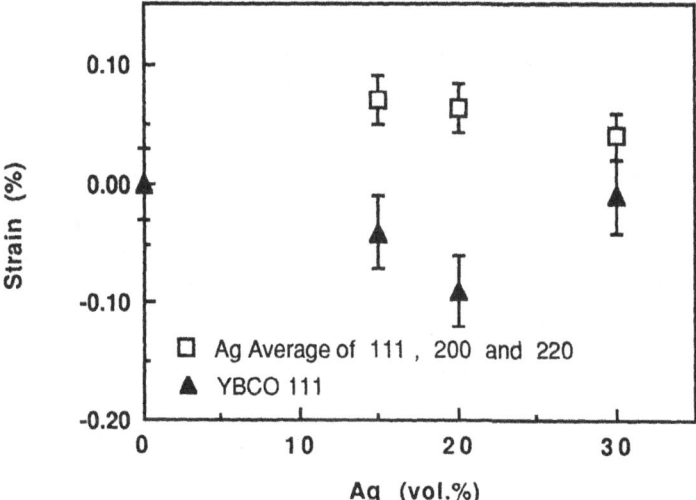

Fig. 2. Ag strain and experimentally measured strains for YBCO (111) plane as a function of stoichiometry.

measurements, vary from 229 Mpa for the 15%-Ag sample to 130 Mpa for the 30%-Ag sample with an uncertainty of ±65 Mpa.

3. Metal-Matrix Composites

Time-of-flight neutron diffraction was used to measure residual strains in a unidirectional fiber-reinforced metal-matrix composite consisting of 35 vol.% SiC fibers embedded in a titanium alloy matrix (Table 1). Columnar grains of the 140-μm-diameter fibers were oriented normal to the fiber axis. As a result of differential thermal contraction, cooling from a fabrication temperature of 950°C resulted in compressive residual stresses in the fiber. Axial and transverse strains in the fibers can be measured simultaneously by aligning the fibers at an angle of 45° to the neutron beam and analyzing neutron diffraction data with detectors ±90° from the neutron beam direction [6,13]. Ignoring the elastic anisotropy of the fibers, and assuming a single infinitely long fiber embedded in an equivalent matrix with the same elastic and thermal properties as those of the composite, Majumdar et al. [14] computed the residual axial and transverse strains in the fibers to be - 0.22 and -0.08%, respectively. The measured -0.21 ± 0.03% axial strain (<220>) and - 0.09 ± 0.02% transverse strain are in excellent agreement with the computed values. The 220 diffraction peak for the transverse strain is significantly broader than that of the axial strain, indicating that the distribution of the transverse strain is much more nonuniform than that of the axial strain.

In another experiment, 35 x 35 x 2 mm samples of a different Ti alloy matrix/SiC-fiber composite were evaluated. Again, the fibers were aligned 45° to the neutron beam and measured with detectors ±90° from the beam direction, providing simultaneous measurements of axial and transverse strains. The beam passed through the 2-mm thickness. We compared strains and calculated stresses for as fabricated, liquid nitrogen dipped (LND), and thermally cycled (100 times to 650°C) specimens. The purpose of these experiments was to determine if processing can be used to reduce the fabrication-induced residual stresses. The general conclusions from the experiments are

(a) Compared to the as-fabricated sample, tensile strains in the matrix and compressive strains in SiC fibers (parallel to fiber) were reduced for the LND and thermally cycled samples; more for the thermally cycled sample.

(b) Matrix and fiber strains perpendicular to the fiber did not change significantly with processing.

(c) Assuming that radial strain (elastic) is equal and opposite to the tangential strain (elastic), force balance calculations are favorable for the as-fabricated, but not for the thermally cycled and LND samples.

The plastic deformation in these specimens may alter the elastic strain distribution in the matrix so that the elastic strain assumption may no longer be valid. There is no doubt that the axial stresses in the fibers are reduced by the processing methods discussed. It is also possible that the matrix is cracked, but this alone can not explain the observed strain variations due to processing. The possibility of matrix cracking is suggested by measuring elastic moduli by ultrasonic velocity of sound propagating perpendicular to the fibers.

Slight decreases were detected in the shear and Young's modulus of LND and thermally cycled samples.

The only diffraction peak that provided useful data for SiC fibers was the 220. Other SiC peaks were mixed with Ti alloy peaks or provided poor statistics because of limited run time and geometry of the samples.

A temperature-dependent experiment was also carried out on the as-fabricated sample. Lattice parameters for various crystallographic directions were measured as a function of temperature from room temperature to 900°C, parallel and perpendicular to the fiber axis. The relative increase in the lattice spacing with increasing temperature, relative to the room-temperature value, for directions parallel and perpendicular to the fiber axis are shown in Figs. 3 and 4. The similarity in expansion curves (parallel to fiber) for the various crystallographic planes of the matrix validates plane strain assumptions in analytical models that predict fabrication-induced residual stresses in these composites. In addition, the anisotropy of the matrix can be seen clearly in Fig. 4 where the expansion curves for different crystallographic directions vary significantly for the matrix. This is the result of small thickness of the sample perpendicular to the fiber axis. This anisotropy can contribute in general to large fabrication-induced matrix stresses.

Table 1. Stress and strain determined by neutron diffraction for as-fabricated, LND, and thermally cycled Ti alloy- matrix/SiC-fiber composites.

Sample	Parallel to fiber		Perpendicular to fiber	
	Strain (%) ±0.03	Stress(MPa)	Strain(%) ±0.03	Stress(MPa)
As-fabricated				
Matrix[a]	0.42	604	-0.29	86
Fiber(220)[b]	-0.19	- 970	-0.05	- 483
LND				
Matrix	0.39	561	-0.29	64
Fiber(220)	-0.12	- 652	-0.05	- 409
Thermally cycled				
Matrix	0.31	446	-0.23	52
Fiber(220)	-0.08	- 470	-0.05	- 366

[a]$E(matrix) = 97$ GPa, $v = 0.33$;

[b]$E(fiber) = 414$ GPa, $v = 0.19$ assume $\varepsilon_r + \varepsilon_\theta = $ constant

4. Ceramic-Matrix Composites

The strength and toughness of fiber-reinforced composites are controlled, to a large extent, by the nature of the bonding between the fibers and the matrix. Ceramic-ceramic composites, in which both the fiber and the matrix are brittle, generally require weak

446

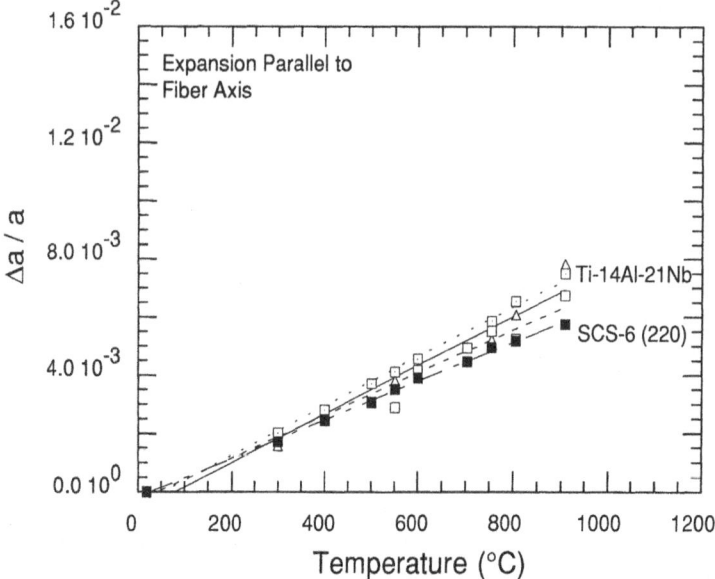

Fig. 3.   Lattice parameter as a function of temperature for various crystallographic
directions measured from room temperature to 900°C. Increases in lattice spacing with
increasing temperature, relative to room-temperature value, parallel to fiber axis, are shown
for SiC fiber {220} plane and several Ti alloy matrix crystallographic directions.

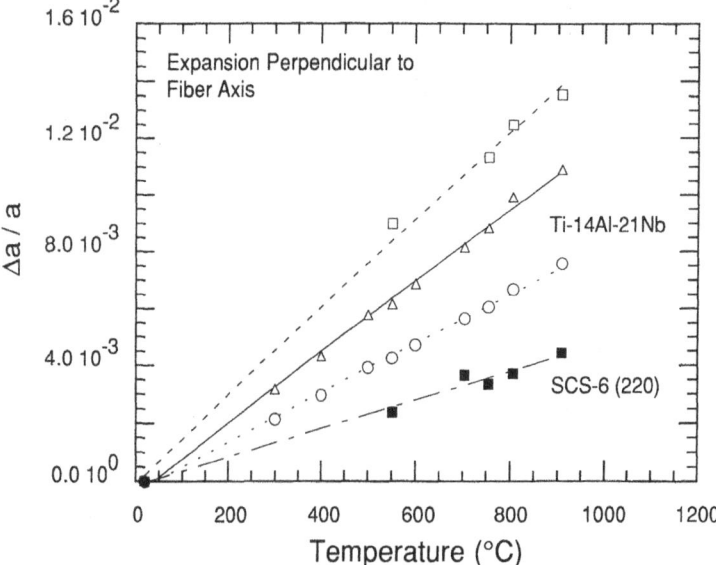

Fig. 4.   Lattice parameter as a function of temperature for various crystallographic
directions measured from room temperature to 900°C. Increases in lattice spacing with
increasing temperature, relative to room-temperature value, perpendicular to fiber axis, are
shown for SiC fiber {220} plane and several Ti alloy matrix crystallographic directions.

bonding to achieve high toughness. In certain systems, chemical bonding is supposed to be nonexistent, and frictional forces at the interfaces provide the necessary link between the fibers and the matrix. Because the frictional forces are dependent on the residual stresses that develop during cool-down after firing, it is important to have an idea of the residual stresses that exist in such composites, particularly at the fiber-matrix interface.

A composite consisting of an alumina (Al_2O_3) matrix with SiC whiskers (15% by weight) was used for this study. The SiC whiskers, approximately 30 µm long and 0.75 µm in diameter, were blended with Al_2O_3 and hot pressed into samples 30 x 25 x 5 mm thick. The whiskers have a high defect density (primarily stacking faults) and their crystallographic structure is not clearly defined. The predominant neutron diffraction peaks for our sample were those associated with the β (cubic) form.

Although the whiskers are randomly distributed in the powder before pressing, the hot-pressing process results in some texturing. This is the result of having few whiskers oriented with the long axis parallel to the pressing direction. (The pressing direction is also parallel to the direction of the neutron beam.) To carry out the neutron diffraction experiment, the sample was cut into two 30 x 10 x 5-mm pieces to partially fill a vanadium tube 10 mm in diameter and 60 mm high. The pressing direction was aligned parallel to the incident neutron beam.

By using two opposing detector banks, we selected out the whiskers with a particular orientation and, at the same time, measured the strains parallel (using the 111 line) and perpendicular (using the 220 line) to the long axis of the whiskers. The test, conducted with the IPNS, showed that, as temperature rises and falls, strain relaxes and intensifies.

Figure 5 shows measured and predicted [6] residual average hydrostatic strain as a function of temperature for whiskers and matrix. The whiskers do not have a well-defined crystal structure, and the handling of this complication is discussed in detail in Ref. 6. The strains in the whiskers are very large, whereas those in the matrix are relatively small. The strains are relieved as the composite is heated toward the fabrication temperature, but the extrapolated point of zero strain is about 1400°C, i.e., 300°C below the fabrication temperature. These observations indicate that, during the temperature drop from 1700°C to about 1400°C, the material is relaxing by creep and is not building up residual stresses. The plot in Fig. 5 shows that satisfactory models to predict hydrostatic strain as a function of temperature can be developed.

As more whiskers are incorporated into our composite, the increased strain in the matrix can cause cracking. It is apparent from experimental and analytical results [15] that as the volume of SiC whiskers increases, the matrix strain (Al_2O_3) increases and the whisker strain (SiC) decreases (becomes less negative). It is desirable to incorporate enough whiskers to toughen the material, but not so much as to cause the matrix to crack.

Analytical estimates of the residual stresses were determined by using two self-consistent models discussed in detail in Ref. 6. In the first model, a plane strain composite cylinder model, the fibers are approximated as infinitely long circular cylinders. In the second model, based on Eshelby, the fibers are approximated as ellipsoids (specialized to prolate spheroids in this case). The two models give almost identical estimates of the stresses and strains in the fibers. The interpretation of the measured strains in the fibers by neutron scattering contains some uncertainty because of the lack of a clearly defined crystal structure of SiC. However, in general, the measured and calculated strains are reasonably close.

In contrast to the poorly defined crystal structure of SiC, Si_3N_4 has a well-defined hexagonal structure. Therefore, various diffraction peaks can be used to determine strain as a function of angle to the long axis of the whiskers (i.e., <0006> uniquely defines the strain along the whisker axis, <1120> uniquely defines the strain perpendicular to the whisker axis, etc.). Calculated strains [14] as a function of angle relative to the c-axis of

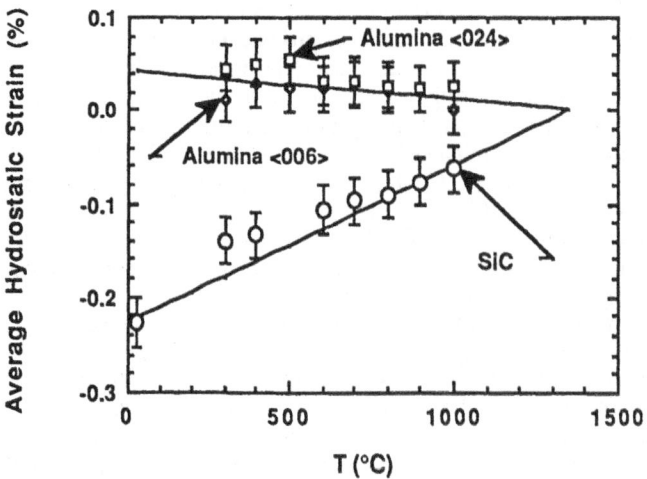

Fig. 5. Measured (symbols) and predicted (solid line) [14] residual average hydrostatic strain as a function of temperature for whiskers and matrix of ceramic composite consisting of 18% volume fraction SiC whiskers embedded in Al_2O_3 matrix.

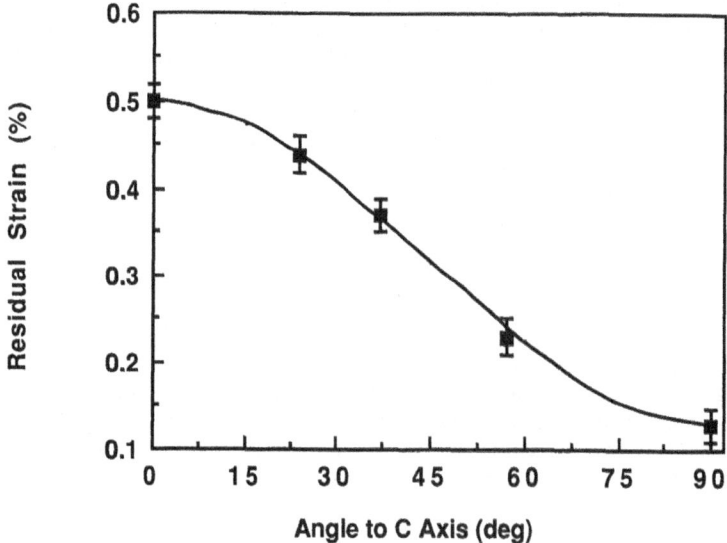

Fig. 6. Calculated (solid line) and measured (symbols) strains [14] as a function of angle relative to c-axis of whiskers compared in a ceramic composite consisting of silicon nitride whiskers embedded in Al_2O_3 matrix.

the whiskers, are compared with measured strains in Fig. 6. The excellent agreement increases our confidence in the measured data and shows that the data are internally consistent, independent of model.

5. References

1. Bednorz, J. G. and Muller, K. A. (1986) 'Possible High-T_c Superconductivity in the Ba-La-Cu-O System', Z. Phys. B-Condensed Matter 64, pp. 189-193.
2. Liang, J. M., Liu, R. S., Chang, L., Wu, P. T. and Chen, L. (1988) 'Structural Characterization of a TlCaBaCu Oxide in T_c Onset=155 K and T_c Zero=123 K Superconducting Specimens', J. Appl. Phys. Lett. 53, pp. 1434-1436.
3. Kohno, O., Ikeno,Y., Sadakata, N. and Goto, K. (1988) 'High Critical Current Density of Y-Ba-Cu-Oxide Wire Without a Metal Sheath', Jpn. J. Appl. Phys. 27, pp. L77-79.
4. Singh, J. P., Leu, H. J., Van Voorhees, E., Goudey, G. T., Winsley, K. and Shi, D. (1989) 'Effect of Silver and Silver Oxide Additions on the Mechanical and Superconducting Properties of $YBa_2Cu_3O_{7-d}$ Superconductors', J. App. Phy., 66 (7), pp.3154-3159.
5. Jin, S., Tiefel, T. H., Sherwood, R. C., Davis, M. E., Van Dover, R. B., Kammlott, G. W., Fastnacht, R. A. and Keith, H. D. (1988) 'High Critical Currents in Y-Ba-Cu-O Superconductors', Appl. Phys. Lett. 52, pp. 2074-2076.
6. Majumdar, S., Kupperman, D. and Singh, J. P. (1988) 'Determination of Residual Thermal Stress in a $SiC-Al_2O_3$ Composite Using Neutron Diffraction', J. Am. Ceram. Soc.71(10), pp. 858-863.
7. Allen, A. J., Hutchings, M. T. and Windsor, C. G., (1985) 'Neutron Diffraction Methods for the Study of Residual Stress Fields', Adv. Phys. 34(4) pp. 445-473.
8. Kupperman, D. and Singh, J. P., Faber, J. Jr. and Hitterman, R. L. (1989) 'Application of Neutron Diffraction to the Characterization of Residual Thermal Strains in $YBa_2Cu_3O_{7-d}$/Ag', J. Appl. Phys. 66(7), pp. 3396-3398.
9. Selsing, J. (1961) 'Internal Stresses in Ceramics', J. Am. Ceram. Soc. 44, pp.419-21.
10. J. D. Jorgensen, Argonne National Laboratory (private communication).
11. H. Shaked, Argonne National Laboratory (private communication).
12. Von Stumberg, A. W., Chen, N., Goretta, K. C. and Routbort, J. (1989) 'High-Temperature Deformation of $YBa_2Cu_3O_{7-d}$', J. Appl. Phys. 66 (5), pp. 2079-2082.
13. Kupperman, D. S., Majumdar, S. and Singh, J. P. (1990) 'Neutron Diffraction NDE of Advanced Composites', Trans. ASME; J. Eng. Mater. Technol. 112 (2), pp.198-201.
14. Majumdar, S., Singh, J. P., Kupperman, D. and Krawitz, A. (1991) 'Application of Neutron Diffraction to Measure Residual Strains in Various Engineering Composite Material', Trans. ASME; J. Eng. Mater. Technol., 113(1), pp. 51-58.
15. Majumdar, S. and Kupperman, D., (1989) 'Effects of Temperature and Whisker Volume Fraction on Average Residual Thermal Strains in a SiC/Al_2O_3 Composite', J. Amer. Cer. Soc., 72(2), pp. 312-313.

6. Acknowledgments

The authors thank Ken Wright for assistance in this project, R. L. Hitterman for assistance in acquiring neutron diffraction data, J. D. Jorgensen, and H. Shaked for helpful discussions an Tani for X-ray diffraction data.This work was supported by NASA, General Electric Aircraft Engines and the U.S. Department of Energy under Contract W-31-109-Eng-38. This work benefited from the use of the Intense Pulsed Neutron Source at Argonne National Laboratory.

DEFORMATION ANALYSIS IN MIXED COMPOSITES

U. SELVADURAI-LASSL
Fachgebiet Qualitätskontrolle
University of Dortmund
Postfach 500 500
W - 4600 Dortmund 50
Germany

H.-A. CROSTACK
Fachgebiet Qualitätskontrolle
University of Dortmund
Postfach 500 500
W - 4600 Dortmund 50
Germany

W. REIMERS
Hahn- Meitner-Institut Berlin GmbH
Dep. N5
Glienicker Str. 100
W - 1000 Berlin 39
Germany

T. VOGT
Institut Laue-Langevin
156X
Centre de Tri
38042 Grenoble Cedex
France

G. ECKOLD
RWTH Aachen and Institut für
Festkörperforschung der KFA Jülich
Postfach 1913
W - 5170 Jülich
Germany

ABSTRACT. The lifetime of plasma sprayed Y_2O_3 partially stabilized zirconia (YPSZ) on metallic substrates under mechanical and thermal cycling is strongly dependent on the conditions of the plasma spraying process. So a detailed characterisation of the coatings is needed.

M. T. Hutchings and A. D. Krawitz (ed.),
Measurement of Residual and Applied Stress Using Neutron Diffraction, 451–459.
© 1992 *Kluwer Academic Publishers.*

Diffraction methods are suited to give information about the phase composition and the residual stress state of the coating. Furthermore, the influence of the crack structure on the deformation behaviour of the composite was investigated by neutron diffraction experiments applying uniaxial external load. The results give evidence for a strong influence of the crack structure not only in ceramic coating but also in the near interface substrate.

1. Introduction

Ceramic coatings are used for protecting metallic substrates against corrsion, abrasion and thermal impact. By the plasma spraying technique also complex geometries as combustion engines and gas turbines can be coated. Since high temperature resistance is needed, often ZrO_2 based ceramics are suited. The pure ZrO_2 exhibits phase transitions from cubic to tetragonal and to monoclinic. The latter one is accompanied by a volume increase of 3 % which would give rise to considerable stresses. For avoiding this effect, ZrO_2 is stabilized by adding Y_2O_3 which suppresses the phase transition under equilibrium conditions /1/. The plasma spraying process, however, is characterized by high cooling rates so that the quenching leads to metastable phases. Hence the phase composition of the coating depends not only from the amount of Y_2O_3 but also from the plasma spraying parameters. So for an understanding of the failure mechanisms of the coating under mechanical and thermal cycling, a detailed phase analysis of the sample material to be investigated is needed which was performed by X-ray diffraction experiments.

The high cooling rate during the plasma spraying process also gives rise to a crack microstructure. So not only the phase transformation and the differences in thermal expansion coefficients between substrate and coating influence the stress state but also the crack formation. Due to these complex interactions an experimental characterization of the stress state in the composites is needed for checking its influence on the failure of the coatings. Herefore X-ray diffraction studies were performed for the near surface state whereas neutron diffraction experiments yielded integral data about the coating and locally resoluted data about the metallic substrates.

The effect of the crack structure on the lifetime of the composites has been demonstrated by thermoshock experiments /2/. By producing an additional macrocrack structure with the help of selected plasma spraying process parameters the mean lifetime could be increased by a factor of 10 compared to macrocrack free coatings. Consequently it was suggested that the vertical segmentation of the coating ameliorates its strain tolerancy. Nevertheless the failure of the composites is known to set in by horizontal spallation in the ceramic coating some ten microns above the metal-ceramic coating interface. A more detailed insight in the deformation behaviour of the composites therefor is needed. Herefor neutron diffraction experiments were performed applying uniaxial external load on the samples. This way the integral deformation of the ceramic coatings and the strain gradient over the thickness of the metallic substrates could be measured.

2. Experimental methods

The investigations performed were done using diffraction methods. Here phase specific results are available. By means of intensity measurements on selected reflections the phase amount in the ceramic coatings of the samples could be determined. The analysis of the reflection position shifts as a function of the orientation of the samples were used for strain stress analysis. Whereas the surface phase composition and stress states were examined by X-ray diffraction, integral results over the ceramic coating and also information about the metallic substrates were obtained by neutron diffraction. Herefore the 90°-scattering technique was applied. By defining the incoming and outcoming beam using diaphragms of 2 mm the near interface state could be investigated. More experimental details are published elsewhere /3/.

3. Sample material and physical properties

For investigating the influence of the material properties on the stress state and the deformation behaviour of the composites samples with different ratios of thermal expansion coefficients and Young's Moduli were used. All samples were coated with partially stabilized $ZrO_2 x7Y_2O_3$ (thickness 450 μm) and NiCrAl as bond coat (thickness 150 μm). As metallic substrates a Ni base alloy (IN 617), cast iron (German grade GG20) and the Al base alloy AlSi12 (thickness 10 mm) were used /4/. The thermal expansion coefficients and the Young's Moduli are summarized in table 1.

The data give evidence for the large variation of the substrate materials as well for the thermal as for the mechanical properties.

Table 1: Physical properties of the composite partners

Material	Youngs-Modulus [GPa]	Poisson-ratio	α [10^{-6}°C^{-1}]	λ [W/mK]
YPSZ	67	0.17	8	1
GG 20	105	0.22	9-10	50
IN617	210	0.31	15.6	91
AlSi12	70	0.34	22.4	220

α : thermal expansion coefficient
λ : thermal conduction coefficient

4. Phase analysis

The monoclinic phase content was determined from the intensity ration $(111)_{monoclinic}$ to $(111)_{tetragonal+cubic}$ /5/. Whereas these reflections are well separated in their diffraction angles, the c/a-ratio of ≈ 1.01 leads to a overlap of the tetragonal and the cubic reflections (fig. 1).

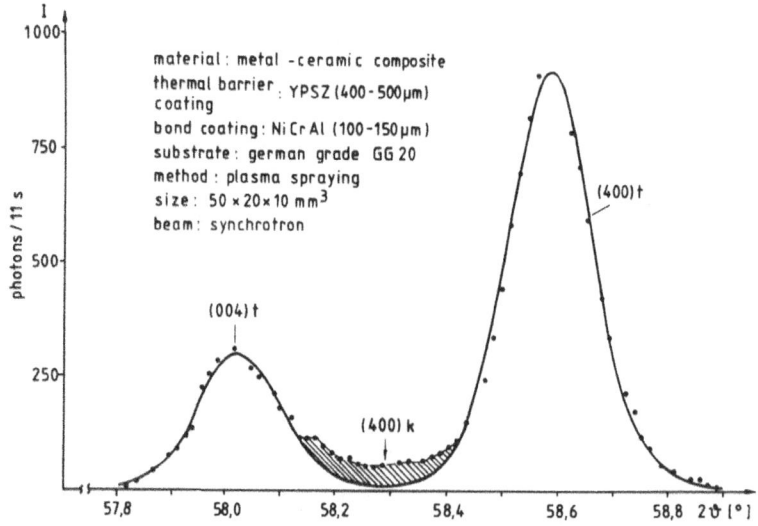

Figure 1. Cubic and tetragonal (400)-reflections

The evaluation of the peak profiles necessary for the analysis of the tetragonal and cubic phase contents can be achieved by Lorentz or Gaussian fits. The difference between the integral intensities of the tetragonal reflections and the measured intensity can be used to determine the cubic phase. The synchrotron diffraction data were taken as standard for this evaluation procedure since they show the best separated profiles and no $K_{\alpha 1}$ - $K_{\alpha 2}$ separation had to be taken into account. Similar phase compositions were obtained also from the X-ray data indicating that the analysis yields reliable results. The phase composition of the ceramic coatings on different substrates are summarized in table 2.

Table 2: Phase compositions of YPSZ coatings

composite	radiation	mkl	tetr.	cub.
YPSZ / GG20	(synchr.)	3	94	3
YPSZ /GG20 (no segment.)	(X-ray) (Cu-Kα) (Cr-Kα)	<1 <1	98 99	2 <1
YPSZ / IN617 (segmentated) (cooled sub.)	(Cr-Kα) (Cr-Kα)	8 <1	89 96	3 4
YPSZ / AlSi12	(Cu-Kα)	<1	99	<1

In all cases the coating exhibits a phase composition of 0 to 8 % monoclinic, 92 to 99 % tetragonal and 0 to 4 % cubic. So the type of the metallic substrates and details of plasma spraying like cooling or heating the substrate do not significantly influence the phase composition.

5. Stress Analysis

For the X-ray and neutron diffraction stress analysis the $\sin^2\psi$-method was applied. As well for the near surface region as for the integral values of the ceramic coating very small stress values were obtained (tab. 3). Since the stress level in the ceramic coating is low, also the substrate materials only exhibit stress values of $\sigma \approx -40 \pm 20$ MPa.

The results show that the stress state is nearly independent of the material properties of the substrate. Especially the comparison of the results obtained on the same substrate GG20, but kept at different temperatures during the plasma spraying process gives evidence that the inhomogeneous shrinking of the composite due to the misfit of the thermal expansion coefficient between ceramic and substrate material does not sinificantly determine the stress state.

Table 3: Residual stresses in YPSZ

composite	radiation	residual stresses
YPSZ on cast ion (substr.-temp.: 90°)	X-ray	+ 2 ± 8
(substr.-temp.: 190°)	X-ray	+ 1 ± 5
(substr.-temp.: 260°)	X-ray	+ 1 ± 5
YPSZ	neutron	- 14 ± 5
Fe-alloy	neutron	- 15 ± 5
YPSZ on Ni-alloy (w/segmentation)	X-ray	- 16 ± 15
YPSZ on Ni-alloy (cooled substr.)	X-ray	+ 4 ± 10
YPSZ	neutron	- 2 ± 8
Ni-alloy	neutron	- 38 ± 5
YPSZ on Al-alloy	X-ray	- 2 ± 8
YPSZ	neutron	+ 17 ± 5
Al-alloy	neutron	- 59 ± 20

So the low stress level found in all samples results from the microcrack formation. Since this effect is always present when plasma spraying the ceramic coating, an influence of the residual macrostress state on the failure of the composites can be excluded. More important, however, is the interaction between crack structure and externally induced strains and stresses, since the spallation of ceramic coatings must be due to shear strains and stresses.

6. Deformation behaviour under external load

The samples of the size 50 x 20 x 10 mm^3 were set under external uniaxial pressure stress up 100 MPa. The load was checked by wire strain gauges on the pressure cylinder. For comparing surface and volume stresses, strain gauges were applied on the ceramic coating and on the metallic substrate and neutron diffraction experiments were performed on these composites partners. These measurements were carried out in reflection and transmission position so yielding the induced strains in the directions of the coating and in the pressure direction. Whereas the deformation values for the ceramics are obtained as integral values due to the small thickness of the coatings the metallic substrate was analysed by two volume elements: one element over an extension of 2 mmm from the coating substrate interface and one element integral over the thickness of the substrate.

The result of the deformation measurements in the pressure directions are summarized in fig. 2.

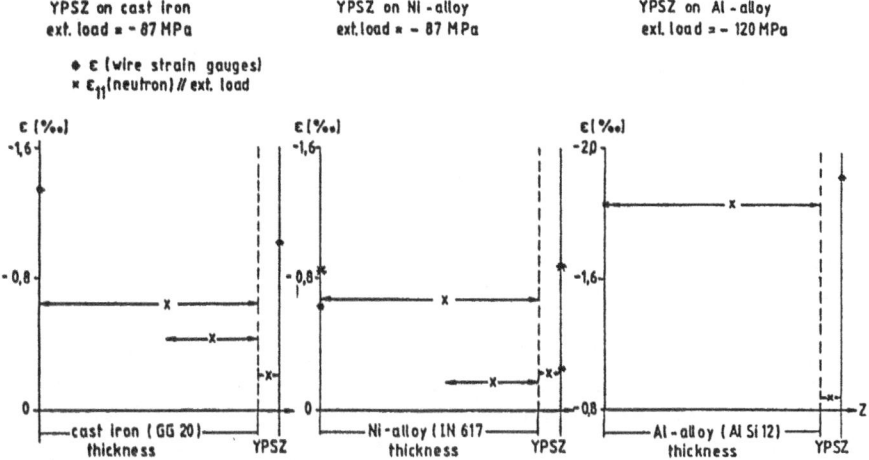

Figure 2. Strain distribution in metal-ceramic composites under external pressure load

The strains measured by strain gauges on the metallic and ceramic side are the same within the experimental accuracy as it is expected under uniaxial deformation. The strains measured by neutrons in the crystalline ceramics, however, are significantly lower. This finding is attributed to the crack closing which is necessary before the deformation in the crystalline material sets in. Since the ceramic "islands" are ridigly attached to the substrate material

458

below, also the substrate deformations in the near
interface zone is hindered. Due to this zone of reduced
deformation in the substrate, the integral deformation
value for the substrate is smaller than the surface
deformation. The small deformations near the interface
leads for material compensating reasons to the formation
of triaxial stresses in this region. This could be
verified by the measurements in the reflecting mode which
give evidence for an increased deformation in the metallic
substrate in the normal to the surface direction (fig. 3).

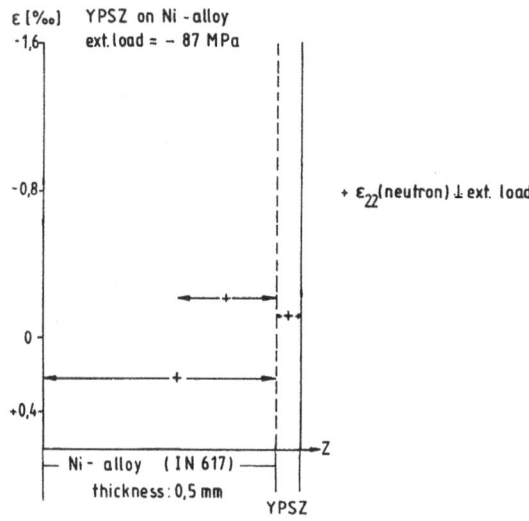

Figure 3. Strain distribution in normal to the surface
direction under external load (IN 617)

7. Discussion

The crack structure of plasma sprayed ceramics seems to be
the decisive factor for its behaviour under external load.
      Under pressure conditions which can occur by
mechanical pressure but also by cooling the composites due
to the small thermal expansion coefficients of the
ceramics, the behaviour in the near interface zone is
characterized by the crack closing. Whereas the ceramic
crystalline material is not yet affected by external
pressure, the metallic substrate is already deformed.
Partly deformation can be compensated by tridimensional
stresses but certainly also compressional strains are
induced over the interface into the ceramic islands. This
effect may lead therefor to the occurence of shear strains
in the near interface ceramic zone. So this finding could
explain the observed horizontal crack formation in cycling
tests. The interpretation also agrees with the observed
increase in lifetime by segmentation of the coatings which

corresponds in this context to an increase of the crack density and hence to a reduction of the ceramic "island" volume. For avoiding the disadvantages of segmentation, which consists e.g. in an important crack opening at high temperature the same positive effects on the lifetime should be achieved by increasing the microstructure crack densities by means of plasma spraying parameters or selected material treatments.

Acknowledgement: The authors are grateful to the German Research Foundation DFG which sponsored the work by the Sonderforschungsbereich 316 and would like to thank Institute Laue-Langevin and KFA Jülich for the allocation of beam time.

8. References

Scott, H. G. (1975) 'Phase relationships in the zirkonia-yttria system', J. Mater. Sci. 10, 1527

Steffens, H.-D., Babiak Z. 'Einfluß der thermischen und Eigenspannungen auf das Thermoschock-Verhalten thermisch gespritzter Schichtverbunde' in: Dortmunder Hochschulkolloqium 1986 (SFB 316), 11.-12.12.1986

Selvadurai, U., Reimers, W. 'Characterisation of phase composition and residual stress state in plasma sprayed ceramic coatings', 7th CiMTEC World Ceramic Congress, Montecatini Terme, Italy, 24.-30.6.1990, in print

Steffens, H.-D., Babiak, Z.,Fischer, U. Arbeitsbericht des SFB 316 (Dortmund 1990), 201-263

Miller, R. A., Garlick, R. G., Smialek, J. L. 'Phase stability in plasma-sprayed partially stabilized zirconia-yttria'; in: Advances in Ceramics, Vol. 3, eds.: A. H. Heuer, L. W. Hobbs, The Amer. Ceramic Soc. (Columbus, OH 1981), 241-253

SURFACE AND NEAR-SURFACE ANALYSIS OF RESIDUAL STRESSES IN ALUMINIUM AND
TITANIUM ALLOYS - EXAMPLES OF THE CASE FOR X-RAY DIFFRACTION

P. HOLDWAY and A. W. BOWEN
Materials and Structures Dept
Royal Aerospace Establishment
The Aerospace Division of the Defence Research Agency
Farnborough, Hants GU14 6TD UK

ABSTRACT. Three examples of the use of X-ray diffraction to measure
surface and near-surface residual stresses in aluminium and titanium
alloys are presented. The first example indicates that surface stresses
in an Al-SiC metal-matrix composite may be significantly different from
those that are measured in the bulk of the material by neutron
diffraction. In cold-expanded holes in 7050 aluminium alloy plate, both
X-ray and neutron diffraction results indicate a compressive stress
field of ~5mm depth, but differences in radial and hoop stresses values
may be due to the different cold-expansion methods employed. The final
example indicates that stress gradients as shallow as $40\mu m$, in this
case in a milled Ti-6Al-4V alloy plate, are readily measurable by the
X-ray diffraction technique. The paper stresses the need for more
combined X-ray and neutron diffraction investigations, to maximise
information retrieval.

1. INTRODUCTION

Surfaces are often the initial source of failure by mechanisms such as
fatigue and corrosion. There is therefore a considerable and continuing
interest in assessing the magnitude and distribution of surface and
near-surface residual stresses in high strength materials in order to
establish their relative contributions to in-service life. X-ray
diffraction has, over the years, proved to be an invaluable non-
destructive tool for measuring residual stresses (see, eg, conference
series ICRS (1987) & (1989)) because of the low penetration of most X-
radiations. For instance, in the materials to be reported upon in this
paper - high strength aluminium and titanium alloys - typical
penetrations of Cu Kα radiation are 28 and $4\mu m$ respectively (Table 1).
The advantages of this surface-specific type of analysis can be
extended further if different radiations are used or material is
selectively removed by electropolishing, since this enables the extent
of stress gradients to be assessed with good depth resolution.
Neutrons, on the other hand, are deeply penetrating and ideally suited
to the bulk analysis of stresses and the determination of the overall
stress state. It is clear, therefore, that X-ray and neutron
diffraction are excellent complementary tools for the overall non-
destructive evaluation of residual stresses; although there are, as
yet, few examples of both techniques being applied to study the same
problems. In this paper, three examples, taken from on-going research,
will be given to illustrate some of the strengths of the X-ray
diffraction technique, where investigations are being carried out using
an omega diffractometer, eleven Ψ tilt angles and the $\sin^2 \Psi$ method to
calculate values of residual stress. In two of these cases,

461

M. T. Hutchings and A. D. Krawitz (ed.),
Measurement of Residual and Applied Stress Using Neutron Diffraction, 461–471.
© 1992 British Crown.

complementary neutron diffraction data from the same types of materials will be included.

2. RESIDUAL STRESSES IN AN Al-SiC METAL-MATRIX COMPOSITE

This work is aimed at establishing the levels of surface stresses produced by different treatments on an Al-SiC metal-matrix composite (MMC). Predictions based on differential thermal mismatch suggest that tensile stresses as high as the yield stress can occur in the aluminium matrix, with corresponding compressive stresses in the SiC.

TABLE 1. Typical penetration depths (from which 63% of the information is received using the equation proposed by Hauk and Macherauch (1983)) at $\sin^2 \Psi = 0.3$ for aluminium and titanium alloys

Radiation	Wavelength Kα (nm)	Diffraction Peak (hkl)	2Θ (deg)	Penetration depth (μm)
Aluminium alloys				
Cr	0.22896	222	156.7	10.0
Fe	0.19359	400	146.0	15.3
Co	0.17888	420	162.2	20.6
Cu	0.15405	422	137.5	27.7
Mo	0.07093	880	164.5	287.6
Titanium alloys				
Cr	0.22896	004	155.8	1.5
Fe	0.19359	203	156.5	2.4
Co	0.17888	114	154.5	2.9
Cu	0.15405	006	161.4	4.4
"	"	213	141.8	4.1

TABLE 2. Residual stress values (in MPa) for an Al-Li alloy 8090-SiC MMC

Radiation	Condition	Analysis No.	L Direction*	ST Direction*
Cu	As-received	1	−43+/−6	−188+/−8
"	Mech polished	2	−129+/−30	−187+/−16
"	100μm removed	3	−141+/−32	−167+/−13
Cr	Repeat	4	−133+/−17	−146+/−8
"	10mm removed	5	−93+/−9	−101+/−5
"	Soln Treat+WQ	6	−127+/−6	−132+/−4
Cu	Repeat	7	−163+/−6	−167+/−8

* Longitudinal (L) and short transverse (ST) directions of the plate

Experimental results, however, suggest that, under conditions such as milling, stresses can be compressive near the surface and that these effects are much more extensive than in un-reinforced alloys (James (1989)). Measurements are being carried out on an Al-Li 8090 MMC containing 20% volume fraction of $3\mu m$ SiC particles. The sample was in the form of a 35mm thick plate that had been forged from a hot isostatically pressed billet. From the analysis of the (422) reflection using Cu Kα radiation, and the (222) reflection using Cr Kα radiation, and using Young's modulus (E) and Poissons's ratio (ν) of 80GPa and 0.33 resp., it has been observed that the as-received condition, which had a machined surface, had a compressive stress (analysis 1, Table 2). This is in agreement with results on fibre-reinforced composites (James (1989)). That these stresses are associated with machining would be in accord with the quite large differences between the stresses in the L and ST directions, corresponding to the directions parallel and perpendicular to that for machining; which is often observed in these types of surfaces (eg Dolle and Cohen (1980); Hauk et al (1982)). Careful grinding and mechanical polishing to remove the machining marks, however, did not significantly change the stress state (analysis 2, Table 2); although the difference in residual stress values in the L and ST directions was reduced. Fig 1a shows the 2θ versus $\sin^2 \psi$ plot for this condition, showing some evidence of the ψ splitting that is indicative of shear stresses. Removal of a further $100\mu m$ from this sample by grinding and mechanical polishing failed to alter the stress state (analysis 3, Table 2). The use of Cr Kα radiation on this surface produced a similar result to that with Cu radiation (analysis 4, Table 2), indicating the presence of a small stress gradient. Removal of a further 10mm from this face resulted in a small reduction in stress (analysis 5, Table 2).

Subsequent solution treatment of the sample at 530^0C for 30m, followed by water quenching (WQ), produced a small increase in residual stress (analysis 6, Table 2); the 2θ versus $\sin^2 \psi$ plot is shown in Fig 1b. Note that the shear stresses had been removed (no ψ splitting) but the stress is clearly compressive. A repeat measurement on this sample using Cu Kα radiation showed that a small stress gradient still existed (analysis 7, Table 2).

These results are clearly in conflict with predictions and the few results available in the literature (eg Ledbetter and Austin (1987)). Two possible explanations can be considered:

1. There is relaxation of the tensile stresses at the surface with the result that X-rays are not sampling the triaxial stress state. However, if the work of Hanabusa et al (1983) is applicable to the matrix as well as the second phase, they showed that the triaxial stress field will exist at depths greater than the inter-particle spacing layer, which is about $5\mu m$ in the present case. Consequently, since the X-rays penetrate about $20-30\mu m$ ie 5-6 layers of particles, they should be sampling the triaxial stress state. However, neutron diffraction results on the same composite, but not the same sample (Root et al (1989), Fig 2), show that at depths greater than 2mm the stresses in the matrix are indeed tensile and between 200 and 300MPa. But by extrapolating these results towards the surface, it could be argued that stress values close to zero, or even compressive, may exist close to the surface. Thus, although X-rays would appear to be sampling the triaxial stress field, there may, in fact, be some relaxation of the stresses close to the surface. Further work needs to be carried out to resolve this point.

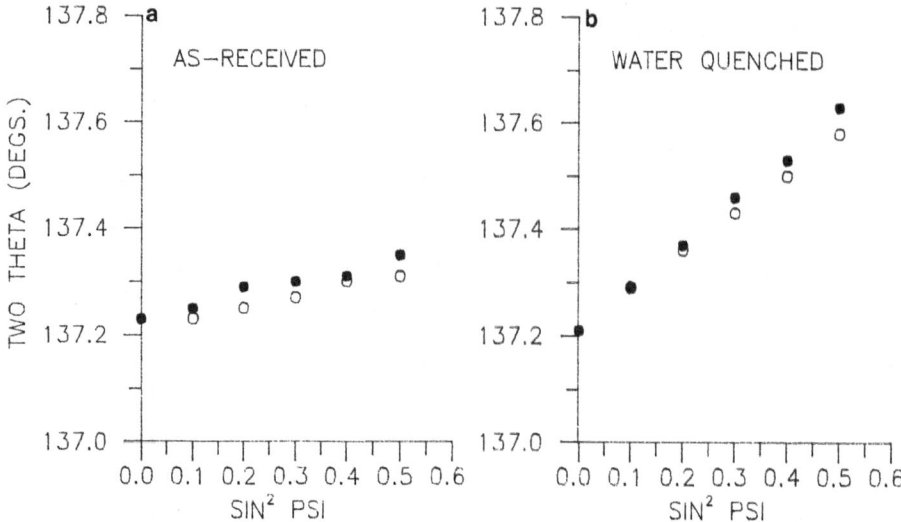

Figure 1. Bragg angle, 2Θ, versus $\sin^2 \Psi$ for (a) as-received and (b) water quenched samples of the 8090-SiC MMC. (open circles = $- \Psi$; closed circles = $+ \Psi$.)

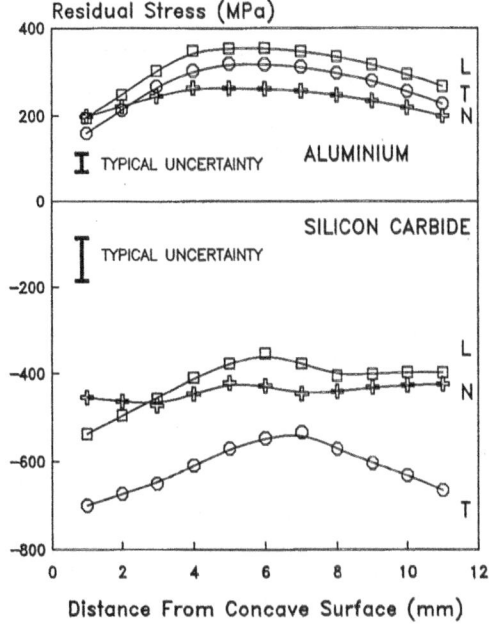

Figure 2. Net residual stresses calculated from measured lattice strains and bulk elastic constants (after Root et al (1989)). Published with the consent of BP Research International and Atomic Energy of Canada Ltd.

2. Surface preparation itself introduces residual stresses. There are instances in the literature of milling introducing a much greater compressive stress field in MMC's compared with the un-reinforced alloys (James (1989)). This poses the question: are the present results merely a reflection of the way in which the surface was prepared? If so, this will have a significant impact on the techniques for preparing MMC surfaces. Further work to resolve this point, and hopefully throw some light on point 1 above, is currently underway on the 8090 MMC sample that was used to produce the results in Figure 2, using both X-ray and neutron diffraction.

3. RESIDUAL STRESSES AROUND COLD-EXPANDED FASTENER HOLES IN 7050 ALUMINIUM ALLOY PLATE

Since compressive stresses are known to increase fatigue life, the cold expansion of fastener holes in aircraft structures will induce compressive stresses and hence enhance fatigue life (eg Buxbaum and Huth (1987); Ozelton and Coyle (1986); Link and Sandford (1990)). The objective of this study is to determine the form of the stress gradients introduced into fastener holes in aluminium alloy plate as a consequence of cold expansion, particularly with respect to its distribution along the radial and hoop directions (Holdway and Bowen

TABLE 3. Residual stress values around 9mm cold-expanded hole in Al7050

Direction/Position	Residual Stress (MPa)
Radial (at hole)	-301+/-20
" (18mm from hole)	-25+/-16
Hoop (at hole)	-135+/-16
Hoop (opposite side)	-131+/-19

(1991)). Measurements are therefore being carried out on a series of plates of the aluminium alloy 7050 containing holes that have been produced by two different methods. Cu Kα radiation is being used to measure the shift in the (511)/(333) reflection using a reduced beam size of 1.5x1 mm^2 and E=72GPa and ν=0.34. This small beam size permitted the shape of the stress gradient away from the holes to be established with an acceptable degree of positional resolution in this initial phase of the work. Two sets of holes were examined:

(a) those 6mm in diameter, which were expanded by the method developed by Fatigue Technology Inc. of the USA (and referred to as the FTI method), where a bulbed mandrell is drawn through a lubricated sleeve inserted into the hole. Cold expansion is achieved by making the bulb diameter plus twice the sleeve thickness to be greater than the hole diameter by ~4%. This hole was de-burred by mechanical polishing before measurement to provide a flat surface near the hole edge. Fig 3 shows a plot of the variation in radial and hoop stresses with distance away from the hole. The highest stresses are those in the hoop direction, and the values in both directions show a significant reduction within 5mm of the hole edge. Similar results have been reported for steels by Dietrich and Potter (1976).

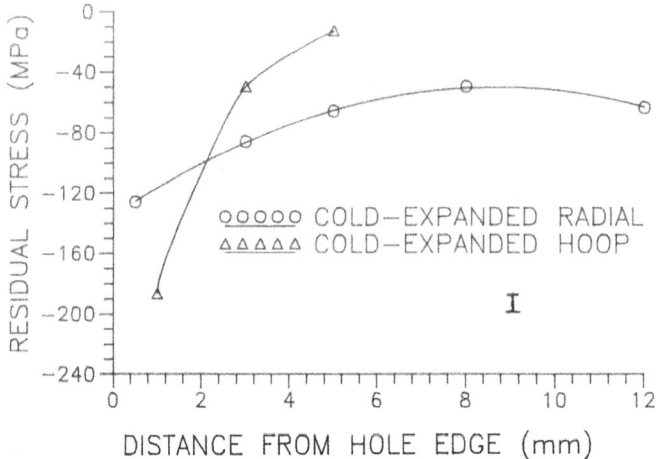

Figure 3. Residual stress values as a function of distance from the edge of a cold-expanded hole in an Al 7050 alloy plate. The typical uncertainty is indicated by the error bar.

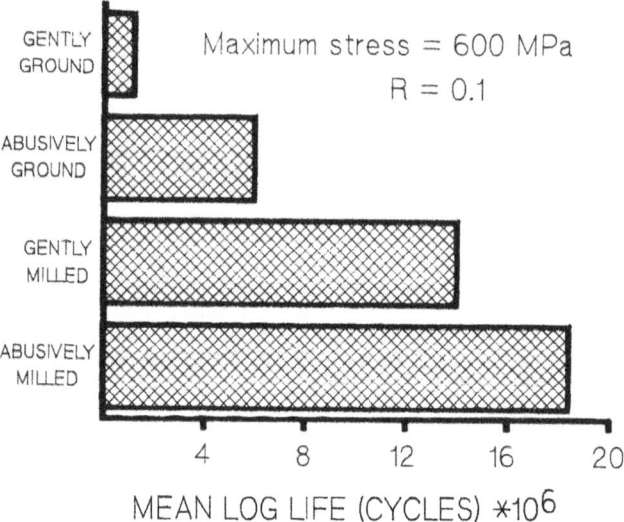

Figure 4. Mean fatigue life dependence on machined surface condition for single edge notched test pieces cut from a Ti-6Al-4V alloy bar.

467

(b) those 9mm in diameter, which were expanded by a tapered roller/mandrell system at the Open University (Edwards (1991)). The degree of expansion is related to the degree to which the tapered mandrell expands the tapered rollers (which are held together in a cage). Table 3 lists the measured compressive stresses within 1mm of the hole edge. These were compressive in both the radial and hoop directions ie 301 and 135 MPa respectively. At a distance of ~18mm from the hole the stress was reduced to -25MPa in the radial direction. These results are in reasonable accord with neutron diffraction data, on the same sample, which showed that the radial and hoop stresses at approximately equivalent positions, but 0.5mm below the surface, were -200 and 0 MPa resp. (Edwards (1991)). The extent of the compressive stress field, as measured by neutron diffraction, was approximately the same as that measured by X-ray diffraction for the 6mm hole ie ~5mm in depth (Edwards (1991)).

TABLE 4. Grinding and milling conditions for the Ti-6Al-4V alloy

GRINDING

	Gentle grinding	Abusive grinding
Wheel type	CG60-H5-VG	C80-J11-VG
Wheel speed (m/min)	840	840
Downfeed	(a)	(a)
Total depth (mm)	0.25	0.25
Crossfeed (mm/pass)	0.1	0.1
Table speed (m/min)	12	12
Grinding fluid	Ultagrind S (80:1)	Ultragrind S (80:1)
Diamond traverse (s/mm)	0.28	0.55

(a) First 0.2mm at 0.0125 mm/pass, next 0.02mm at 0.01 mm/pass and final 0.03mm at 0.005mm/pass. Wheel re-dressed before final removal of 0.03mm.

MILLING (End mill-peripheral cut)

	Gentle milling	Abusive milling
Tool material	HSS	HSS
Cutter dia (mm)	51	51
Number of teeth	8	8
Cutting speed (mm/min)	12	12
Depth of cut (mm)	0.76	0.76
Feed (mm/tooth)	0.025	0.14
Tool wear (initial) (mm)	0	0
Cutting fluid	Tellus	Tellus

There are two points to stress from the present work:

1. It is possible to map the stress field around cold-expanded fastener holes in aluminium alloy plate with adequate positional resolution using both X-rays and neutrons. Additionally, X-ray diffraction line profile analysis can be used, destructively, to measure plastic strains with a high degree of depth resolution (Holdway and Bowen (1991)).

2. The maximum values of radial and hoop stresses are dependent on the method of cold expansion (cf data in Figure 3 and Table 3). Further work is in hand to study these differences in more detail.

4. STRESS GRADIENTS IN MILLED Ti-6Al-4V ALLOY SURFACES

This programme is a continuation of earlier work (Holdway and Bowen (1989)) that is aimed at quantifying the effects of machining practice on residual stress, and hence fatigue life (Gardiner et al (1986); Wilson et al (1986)), of the Ti-6Al-4V alloy. These machining conditions were chosen to simulate good and bad workshop practice (slow and fast tool traverse times) on a normal day and not the extremes of damage that are possible. Details of these conditions are given in Table 4. In the initial phase (Holdway and Bowen (1989)) it was found that grinding did not significantly change the stress from that in the as-received condition but that milling (either gently or abusively) produced compressive stresses of the order of 300-500MPa (Table 5). This was in general agreement with the trend in fatigue life results (Gardiner et al (1986); Wilson et al (1986), Figure 4) but it was not possible to distinguish between gentle and abusive machining. Plastic strain values calculated from the line broadening (typical examples are shown in Figure 5) are given in Table 5. They also show differences between the different machining conditions but also do not distinguish between the fatigue results for gentle and abusive machining (Table 5).

In this second phase, the extent of stress gradients is being established to ascertain if gentle and abusive milling produce sufficiently different gradients to explain the finer details of Figure 4. Triaxial stress state measurements have been made on the abusively milled surface at 0, 45 and 90^0 to the milling direction, and the measurements have then been repeated after periodic selective layer removal by electropolishing. Residual stress values were calculated from the shift in the (213) reflection, using Cu Kα radiation and the elastic constants E=115GPa and ν=0.32. Figure 6 shows representative sin^2 Ψ plots from measurements parallel, and perpendicular, to the machining direction at the surface, and after ~20μm removal by electropolishing. Shear stresses are evident in the Ψ splitting. The alloy had a strong texture but this did not produce severe oscillations in the sin^2 Ψ plots. Table 6 shows the triaxial stress tensors determined at different depths and confirms that a steep stress gradient exists near the machined surface; the compressive stresses being reduced from 300-500MPa to zero within ~40μm. At the same time, significant reductions in peak width were observed. The large reduction between 10 and 20 μm is in good agreement with the damage layer depths of between 6 and 20 μm estimated from metallographic examination (Gardiner et al (1986); Wilson et al (1986)).

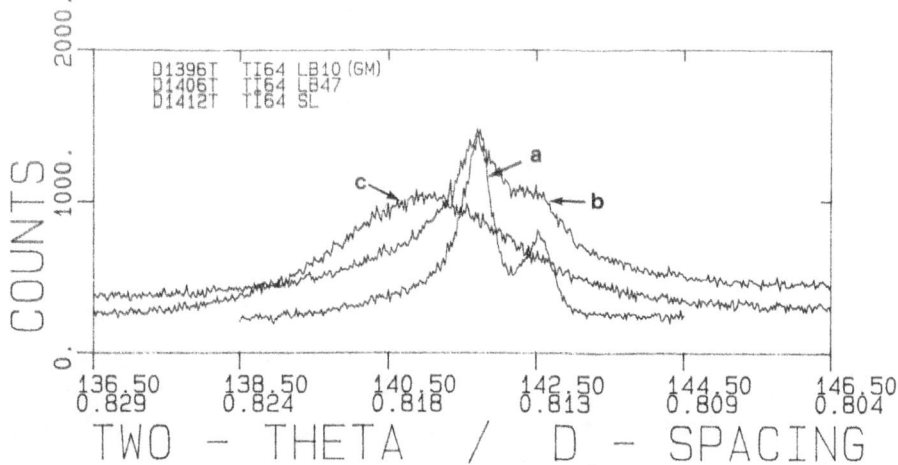

Figure 5. Typical X-ray diffraction line profiles for (a) as-received, (b) gently ground and (c) gently milled surfaces of the Ti-6Al-4V alloy.

TABLE 5. Values of residual stress and plastic strain for Ti-6Al-4V alloy test pieces in the as-received condition, and after grinding and milling.

Surface finish and directions	Residual stress (MPa)	FWHM+ (deg)	Plastic strain %
AR# L*	-29.8+/-6.6	0.48	0.04
AR ST*	-22.6+/-10.8	0.48	0.04
GG# L	-16.6+/-20.6	0.93	0.04
GG ST	-42.7+/-8.9	0.93	0.04
AG# L	-27.9+/-34.5	1.46	0.18
AG ST	-110+/-25.9	1.46	0.18
GM# L	-364+/-19.5	2.56	0.34
GM ST	-546+/-16.8	2.56	0.34
AM# L	-372+/-51.8	2.53	0.34
AM ST	-494+/-11.8	2.53	0.34

+ FWHM - Full width at half maximum of the (213) reflection

AR - as-received condition
  GG - gently ground
  AG - abusively ground
  GM - gently milled
  AM - abusively milled

* L and ST correspond to the longitudinal (L) and short transverse (ST) directions in the bar.

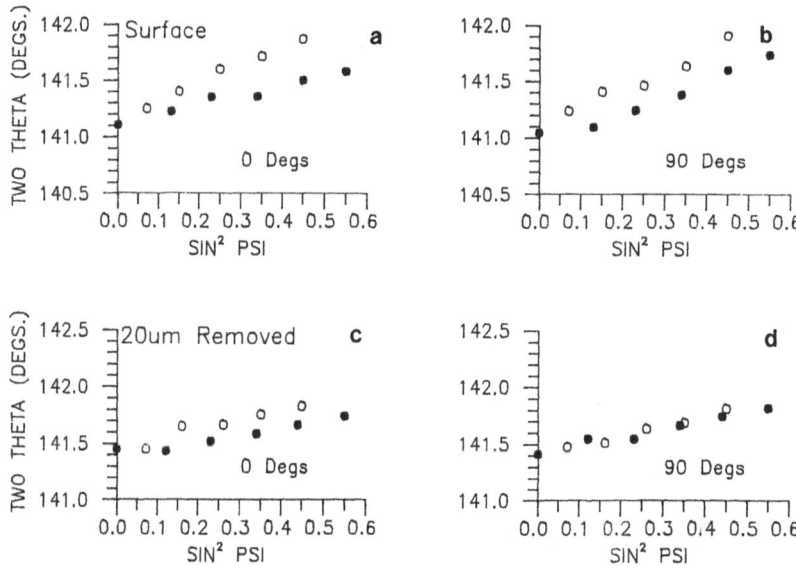

Figure 6. Bragg angle, 2Θ, versus sin² Ψ for (a)&(b) abusively milled surface of Ti-6Al-4V alloy and after (c)&(d) a 20μm layer removal by electro-polishing. (open circles = - Ψ ; closed circles = + Ψ .)

TABLE 6. Triaxial stress state (in MPa) for the surface of an abusively milled test piece of Ti-6Al-4V, and at different depths below the surface.

Surface

-374+/-109	-57+/-53	42+/-15
-57+/-53	-384+/-109	33+/-15
42+/-15	33+/-15	0+/-58

10μm below the surface

-309+/-94	-32+/-45	38+/-12
-32+/-45	-357+/-94	22+/-12
38+/-12	22+/-12	0+/-50

20μm below the surface

-187+/-43	-15+/-21	20+/-6
-15+/-21	-197+/-43	1+/-6
20+/-6	1+/-6	0+/-23

40μm below the surface

-12+/-40	-1+/-19	-4+/-5
-1+/-19	-22+/-40	-7+/-5
-4+/-5	-7+/-5	0+/-21

REFERENCES

Buxbaum, O. and Huth, H. (1987) 'Expansion of cracked fastener holes as a measure for extension of lifetime to repair', Eng Fract Mech 28, 689-698.

Dietrich, G. and Potter, J. H. (1976) 'Stress measurements on cold-worked fastener holes', Adv in X-ray Anal 20, 321-328.

Dolle, H. and Cohen, J. B. (1980) 'Residual stresses in ground steels', Met Trans 11A, 159-164.

James, M. R. (1989) 'Residual Stresses in Metal Matrix Composites', in G. Beck et al (eds.), Proc 2nd Int Conf on Residual Stresses ICRS2, Elsevier Applied Science, London, pp429-435.

Edwards, L (1991) Open University UK, unpublished work.

Gardiner, R. W. et al, (1986) 'Influence of machining parameters on fatigue properties of textured T-6Al-4V alloy plate', in Designing with Titanium, Inst of Metals, London, pp277-286.

Hanabusa, T. et al (1983) 'Criterion for triaxial X-ray stress analysis', Zeit Metallkde 74, 307-313.

Hauk, V. and Macherauch, E. (1983) 'A useful guide for X-ray stress evaluation (XSE)', Adv in X-ray Anal, 27, 81-99.

Hauk, V. et al (1982) 'The state of residual stress in near surface region of homogeneous and heterogeneous materials', Met Trans 14A, 1239-1244.

Holdway, P. and Bowen, A. W. (1989) 'Internal Strains in Machined Surfaces of Strongly Textured Ti-6Al-4V', in G. Beck et al (eds.), Proc 2nd Int Conf on Residual Stresses ICRS2, Elsevier Applied Science, London, pp429-435.

Holdway, P. and Bowen, A. W. (1991) 'Internal strains near cold-worked fastener holes in high strength aluminium alloy sheet', to be presented at 3rd Int Conf on Residual Stresses, 24-26 July, Tokushima, Japan.

ICRS (1987) & (1989), Proc of 1st Int Conf on Residual Stresses, Ed V Hauk & E Macherauch, DGM Oberursel (1987); Proc of 2nd Int Conf on Residual Stresses, Ed G Beck et al, Elsevier, London (1989).

Ledbetter, H. M. and Austin, M. W. (1987) 'Internal strain (stress) in a SiC-Al particle-reinforced composite: an X-ray diffraction study', Mat Sci & Eng 89, 53-61

Link, R. E. and Sandford, R. J. (1990) 'Residual strains surrounding split-sleeve cold expanded holes in 7075-T651 aluminium', J of Aircraft 27, 599-604

Ozelton, M. W. and Coyle, T. G. (1987) in 'Fatigue in mechanically fastened composite and metallic joints', ASTM STP 927 pp53-69.

Root, J. H. et al (1989) Atomic Energy of Canada, Canada, unpublished work.

Wilson, R. N. et al, (1986) 'Effects of severity of machining and crystallographic texture upon the fatigue strength of Ti-6Al-4V alloy plate', Paper C256/86, Fatigue of Eng Mat and Struct, Inst of Mech Eng, London, pp407-416.

ACKNOWLEDGEMENTS

The authors wish to thank BP Research International for the supply of the 8090 MMC, and Dr L Edwards (Open University) and Mr R Cook (RAE) for the supply of samples containing cold-expanded holes. BP Research International and Dr T M Holden (AECL, Canada) are thanked for permission to include Figure 2. Dr L Edwards is thanked for access to unpublished results. Dr Edwards, Dr Holden and Mr Cook are also thanked for valuable discussions. © Controller HMSO London 1991.

RESIDUAL STRESSES IN BRAZED CERAMIC-METAL COMPOUNDS*

L. PINTSCHOVIUS, N. PYKA
Kernforschungszentrum Karlsruhe, INFP
Postfach 3640, W-7500 Karlsruhe, Germany

R. KUSSMAUL, D. MUNZ
Kernforschungszentrum Karlsruhe, IMF II
Postfach 3640, W-7500 Karlsruhe, Germany

B. EIGENMANN, B. SCHOLTES
Institut für Werkstoffkunde I, Universität Karlsruhe
Kaiserstraße 12, W-7500 Karlsruhe, Germany

ABSTRACT. The residual stresses in metal-ceramic compounds were investigated by X-ray and neutron diffraction. The neutron data show clear evidence for plastic flow in the steel so that elastic models cannot be expected to give a satisfactory explanation of the residual stress state. Finite- element calculations taking into account the temperature dependent elastic-plastic behavior of the materials reproduce the experimental results very well.

1. Introduction

The use of metal-ceramic compounds allows to benefit from both the high ductility of metals and the high resistance of ceramics against corrosive media and / or abrasive wear. Joining is a key technology and many different techniques have been employed to join ceramics on a metallic substrate. Brazing seems to be particularly promising for joining, but the residual stresses originating from the brazing procedure may be very detrimental and even cause cracks without external load. The residual stresses can be calculated from the material properties of the partners, their dimensions and the thermal history. The results can be checked by residual stress measurements at the surface of the specimen by X-ray diffraction. However, a much more stringent test is a comparison of the calculated residual stress distribution in the interior of the specimen with experimental values, which can be obtained by neutron diffraction only. In the following we report on

---

* the paper was presented by W. Reimers

M. T. Hutchings and A. D. Krawitz (ed.),
Measurement of Residual and Applied Stress Using Neutron Diffraction, 473–477.
© 1992 *Kluwer Academic Publishers.*

results of a neutron diffraction-study of the residual stresses in Si_3N_4- and ZrO_2-plates brazed onto steel.

2. Samples

ZrO_2 or hot pressed silicon nitride (HPSN) was brazed onto a plain carbon steel, mostly of type Ck45. The samples consisted of ceramic platelets of dimensions $20 \times 20 \times 4$ mm^3 brazed onto steel platelets of dimensions $20 \times 20 \times X$ mm^3 with $3.5 \le x \le 10$. The difference in thermal expansion between steel and the ceramic material gives rise to large residual stresses after cooling down from the brazing temperature and thereby to a deformation of the compound as illustrated in Fig. 1.

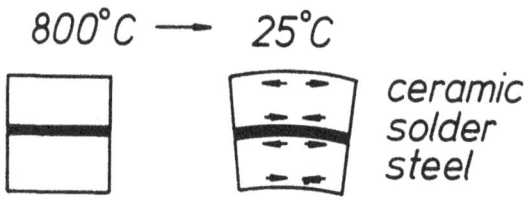

Fig.1 Schematic representation of the specimen and its deformation after cooling down from the brazing temperature.

3. Neutron measurements

The neutron measurements were done on the G4.3 triple-axis spectrometer at the ORPHEE-reactor, Saclay. The instrument allowed to vary the wavelength of the neutrons and thereby to operate always at $2\Theta \approx 90^0$. Slits of dimensions 0.5×10 mm^2 in the incoming and the diffracted beam gave a gage volume of $0.5 \times 0.5 \times 10$ mm^3. Thus, the spatial resolution in direction of the large stress gradients perpendicular to the interface was about 0.5 mm. Strains were measured in two directions parallel and perpendiccular to the interface along a central line from surface to surface. Additional measurements in a third direction confirmed that the strain distribution was to a good approximation two-axial with $\varepsilon_y = \varepsilon_x$.

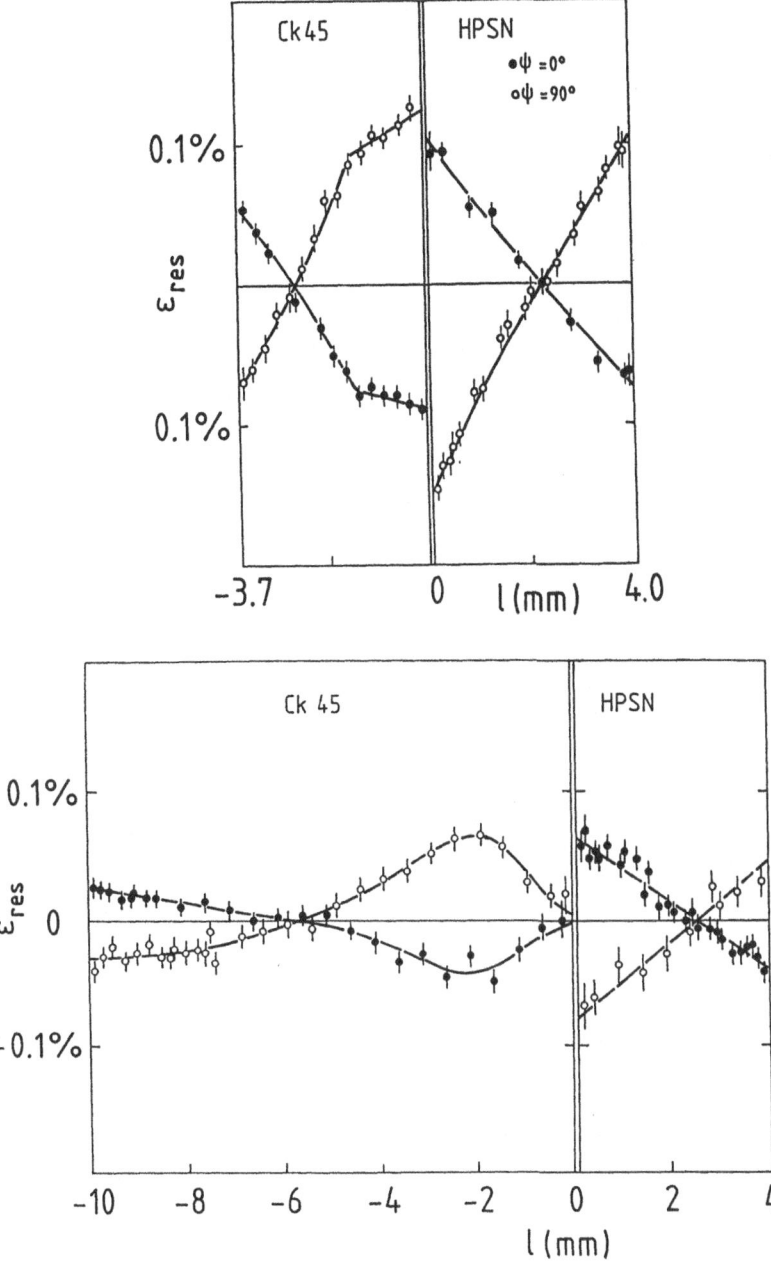

Fig. 2 Examples of residual strain distributions in metal-ceramic compounds measured by neutron diffraction.

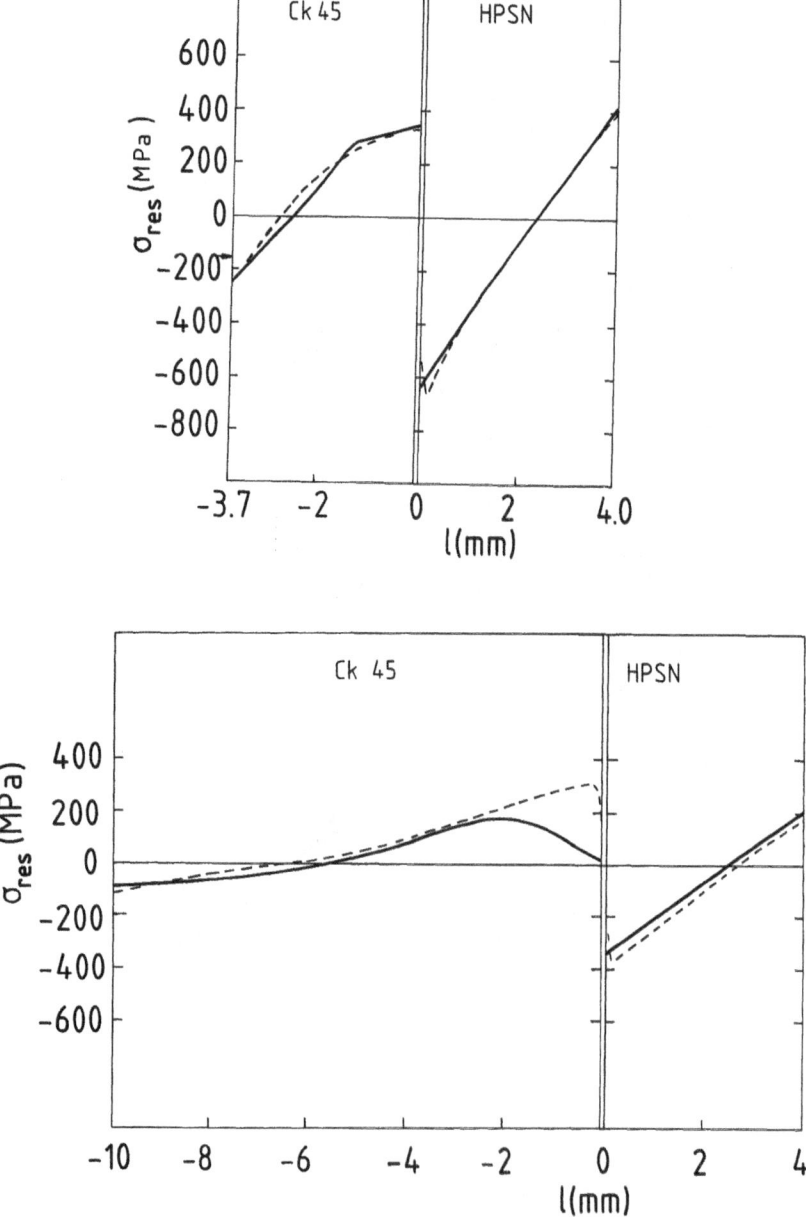

Fig. 3 Comparison of calculated and observed residual stress distributions in two metal-ceramic compounds

4. Experimental results

Examples of residual strain distributions are shown in Fig. 2. The labelling of the two components by $\psi = 0^0$ (ε_z) and $\psi = 90^0$ ($\varepsilon_x = \varepsilon_y$) corresponds to the X-ray notation. The linear variation of the strains in the ceramic part versus distance from the interface indicates that the deformations were entirely elastic. On the other hand, the strain distribution in the steel part shows clear evidence for plastic flow. This behavior was not anticipated and showed the need to go beyond an earlier attempt to calculate the residual stresses with an elastic model [1].

5. Finite-element calculations

Calculations of thermal residual stresses in the elastic regime can be performed analytically [1]. However, the elastic-plastic behavior of the materials can be modelled only by finite-element calculations requiring considerable computer time. Our calculations were based on published data for the material properties of the ceramic, the steel and the solder. There are no adjustable parameters, but the scarcity of the data for high temperatures leave some margins. The results come surprisingly close to the observation, especially for those cases were the steel plates are not very thick (see Fig. 3.) For the sample with a 10 mm thick steel plate, the calculation somewhat underestimates the plastic flow in the region close to the interface, but nevertheless gives a good overall description of the observed residual stress distribution.

6. Conclusions

An earlier study using only X-ray stress analysis at the surface concluded that the residual stresses in brazed metal-ceramic compounds can be largely understood by simple elastic models. Neutron measurements of the residual stresses in the interior of the specimen showed that this is not the case. Only finite-element calculations taking into account the full temperature dependent elastic-plastic behavior of the materials can reproduce the observed residual stress distributions. Metal-ceramic compounds are an example for the case that neutron data allow a much more stringent validation of the theory than X-ray data.

7. Reference

1. Jancu, O., Munz, D., Eigenmann, B., Scholtes, B. and Macherauch, E. (1990) "Residual stress state of brazed ceramic/metal compounds, determined by analytical methods and X-ray residual stress measurements", J. Am. Ceram. Soc. 353, 1144

7. MEASUREMENTS ON BULK COMPONENTS

RESIDUAL STRESS DISTRIBUTION IN CRACKED AUTOFRETTAGED TUBING

M.A.M. BOURKE, H.J. MacGILLIVRAY*, G.A. WEBSTER* and
P. J. WEBSTER**
LANSCE, Los Alamos National Laboratory
Los Alamos, NM, 87545, USA
*Department of Mechanical Engineering
Imperial College, London, SW7 2BX, UK
**Department of Civil Engineering
University of Salford, Salford, M5 4WT, UK

ABSTRACT. Steel tubing used for a variety of critical high pressure applications is often autofrettaged to inhibit crack initiation and fatigue propagation from the bore. The autofrettage process introduces compressive residual hoop stresses at the bore balanced by tensile stresses at larger radii. Neutron strain scanning has been used to determine the residual stress distribution in a ring sample sliced from a high strength, low-alloy steel autofrettaged tube, with a bore of diameter 60 mm and a wall of thickness 32 mm, fatigue cracked through half the wall thickness into a region which originally contained tensile residual stress. A substantial increase in residual hoop tension, relative to that measured on the uncracked side, is observed ahead of the crack tip. Qualitative agreement is achieved between the measured hoop stress distribution and theoretical values predicted using a boundary element method.

1. Introduction

Thick-walled autofrettaged steel tubing is used in a variety of critical high pressure military and chemical engineering applications. When tubes are subject to cyclic internal pressure fatigue cracks can propagate outwards from the bore and subsequently cause failure. The autofrettage process, which is applied prior to service, provides some protection against fatigue cracking from the bore by applying an internal over-pressure of sufficient magnitude to cause tensile yielding partially through the wall thickness. This introduces a compressive residual hoop stress at the bore which is balanced by a tensile stress at larger radii. The magnitude of the over-pressure determines the limit of the plastically deformed zone and the shape of the stress distribution through the wall thickness. Subsequent applied loading must first overcome the compressive residual hoop stress before cracks can propagate from the bore. The benefits of autofrettage in improving the fatigue life of tubing are documented in detail elsewhere [1].

The residual stress produced by autofrettage in uncracked specimens has already been measured both by neutron diffraction and conventional Sachs Boring methods [2]. In this study we have examined the redistribution of residual stress resulting from the growth of a fatigue crack from the bore through the compressive region to a position half-way through the tube wall at which the autofrettage induced stresses were known to be tensile. While a

M. T. Hutchings and A. D. Krawitz (ed.),
Measurement of Residual and Applied Stress Using Neutron Diffraction, 481–492.

crack remains in the compressive region little redistribution is expected since the crack plane can still transmit a compressive force. However, when the crack traverses a region which originally contained a tensile residual stress, redistribution must occur since the cracked region cannot transmit a tensile load across it. This paper describes how neutron strain scanning was used to determine the residual stress pattern on the cracked and uncracked sides of an autofrettaged steel tube.

2. Specimen Details

Thick-walled tube of internal diameter 60 mm and external diameter 124 mm (diameter ratio 2.07) was subjected to an internal autofrettaging pressure of 662 MPa. The bore is defined as R = 0 and the outer edge as R = 32 mm. At this pressure yielding through 40% of the wall thickness, to R = 13 mm, was expected. The tube material was a high strength, low-alloy steel, AISI 4333 M4, containing principal alloying additions of 0.34% C, 1.67% Ni and 0.8% Cr. The yield stress was 1070 MPa and the ultimate tensile stress was 1150 MPa. The autofrettage was applied without producing end loading on the tube and corresponded to plane stress.

A ring of thickness 5 mm was removed from the tube, after autofrettage, as shown in figure 1. Previous studies suggest that the removal of thin rings from autofrettaged tubing does not substantially alter the residual hoop stresses [3], although it is worth noting that finite element analyses of *thin*-walled tubing do indicate a significant reduction in the hoop

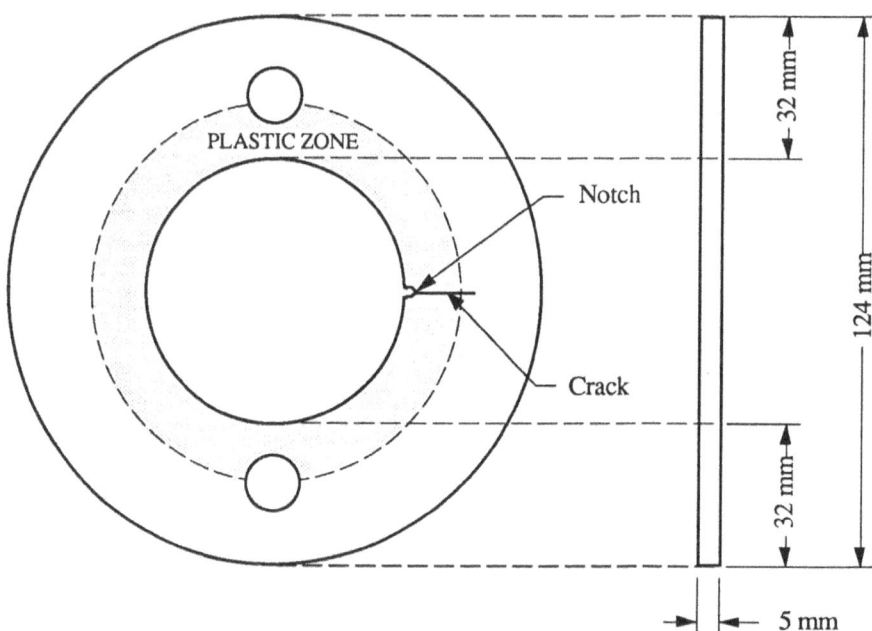

Figure 1.   Dimensions of the autofrettaged, fatigued, ring showing the notch, crack position and regions of plastic and elastic strain.

stress on removal of thin slices [4]. In practice the 5 mm thickness of the specimens was selected to give reasonably fast count times thus making efficient use of the neutron beam time available. A 2.5 mm deep notch was introduced at the bore of the ring from which the fatigue crack 16 mm in length (including the notch) was propagated. The crack length, to R = 16 mm, corresponded to failure through 50% of the wall thickness. The previous measurements cited above indicated that the change from compression to tension occurs approximately 8 mm from the bore. Thus, relative to the original hoop stress distribution, the crack tip lies 8 mm into the tensile region, as indicated in figure 1.

3. Measurements

The neutron diffraction technique uses the interplanar lattice spacings of selected Bragg reflections to act as internal strain gauges [5]. The strain measurements were made using the High Flux Beam Reactor at the Institut Laue Langevin in Grenoble, France. Strains were determined by measuring the ferritic (211) reflection, using the high resolution powder diffractometer, D1A, which fell at a detector angle 2θ of about 109.5°, for a wavelength of 0.19 nm. The (211) reflection was selected because it has a compliance close to the bulk macroscopic response and is at a convenient Bragg angle. Count times were typically 20 minutes per measurement giving an accuracy of ±40 microstrain.

In principle the presence of a crack disrupts the circular symmetry of the stress distribution produced by the autofrettage process. Consequently a compact sampling volume was required to resolve changes along the crack plane and a gauge volume of 2 × 2 × 2 mm was used for all measurements. Hoop, radial and axial strains were measured along a radius from the notch to the outside edge, and along the opposite uncracked radius, on the centre plane at mid-thickness (see Figures 2 and 3).

Strains are determined from the change in angular position of the measured Bragg reflection from the position of a reflection from similar unstrained material. The unstrained position is best measured on material which is identical in composition to the specimen but which is also small enough to be assumed to be free of significant macroscopic stress or is known to be stress free. Unfortunately such a reference was not available in this case and so the unstrained value was calculated by assuming that the forces acting across the plane of the uncracked side must balance. A moment balance was also performed and the stresses calculated from the two methods of obtaining the unstrained value differed by less than 10 MPa.

4. Results

4.1 UNCRACKED SIDE

The strains for the uncracked side, figure 2, show that even with the fatigue crack half-way through the bore on the other side of the ring all three curves, for hoop, radial and axial directions, are smooth within the experimental error and there is no detectable discontinuity or apparent redistribution of strain on the uncracked side. In fact, within experimental accuracy the stresses on the uncracked side were the same as previously measured in a completely uncracked ring [2]. The unmodified autofrettage hoop stress, shown dashed in figure 4, shows a maximum compressive value at the bore of -250 MPa becoming zero at R = 7 mm then attaining a broad maximum tensile value of about 80 MPa from R = 10 mm. The radial and axial stresses, figures 5 and 6, remain mostly well within ± 50 MPa throughout the wall thickness.

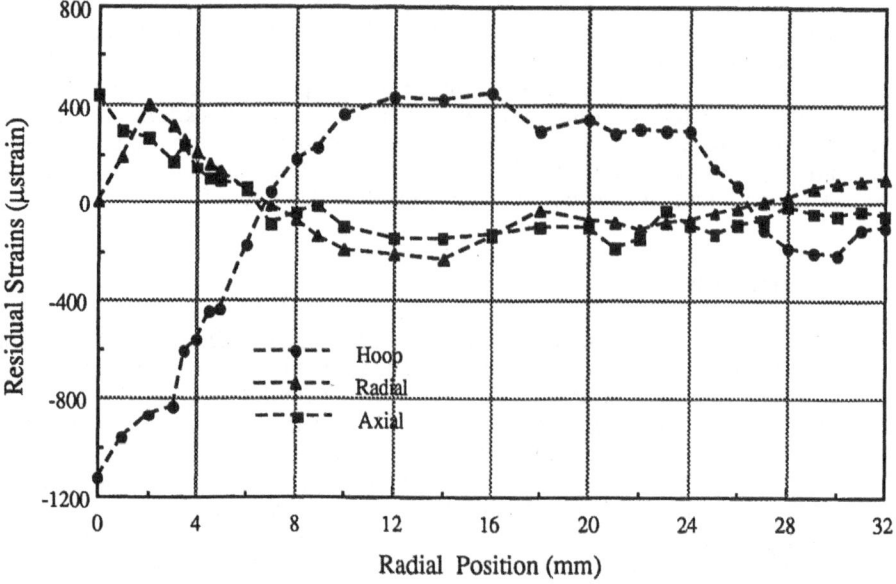

Figure 2.   Hoop, radial and axial residual strains in the uncracked side of the ring.

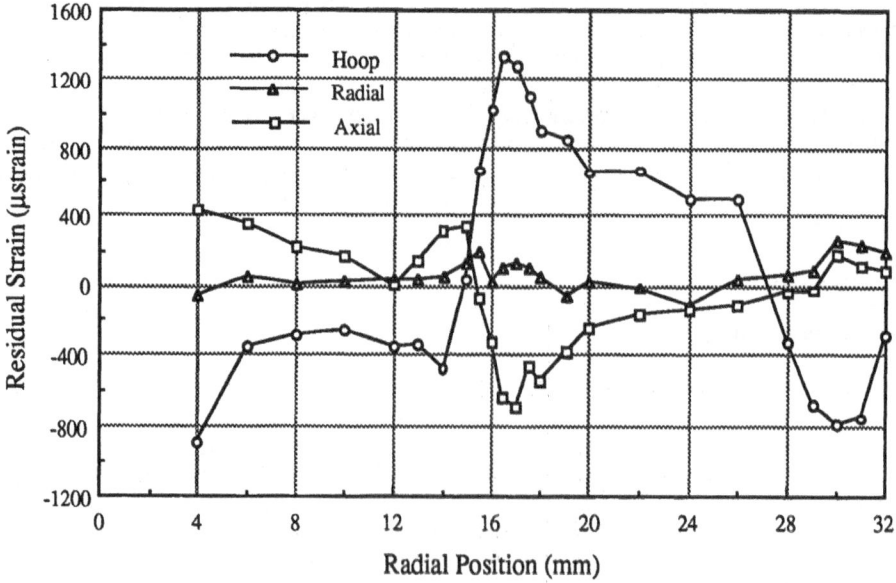

Figure 3.   Hoop, radial and axial residual strains in the cracked side of the ring.

4.2 CRACKED SIDE

The hoop strains, figure 3, along the crack demonstrate a characteristic shoulder between R = 6 and R = 14 mm along which the strain is approximately constant and compressive. Beyond R = 14 mm the hoop strain rapidly becomes increasingly tensile to a maximum of 1300 microstrain between R = 16 and R = 17 mm. The radial strains are low through the entire wall thickness and the axial strains are approximately an inverted mirror of the hoop strains but of about half the magnitude.

The corresponding stresses are shown as continuous lines on figures 4, 5 and 6. The compressive shoulder for the hoop stresses is at about -70 MPa and the maximum tensile stress of 300 MPa is reached at R = 17 mm before reducing to 120 MPa at R = 26 mm prior to falling rapidly to -170 MPa at R = 29 mm. The radial stresses are generally low except in the vicinity of the crack tip where they rise to a maximum of about 100 MPa. The axial stresses show broadly similar behaviour.

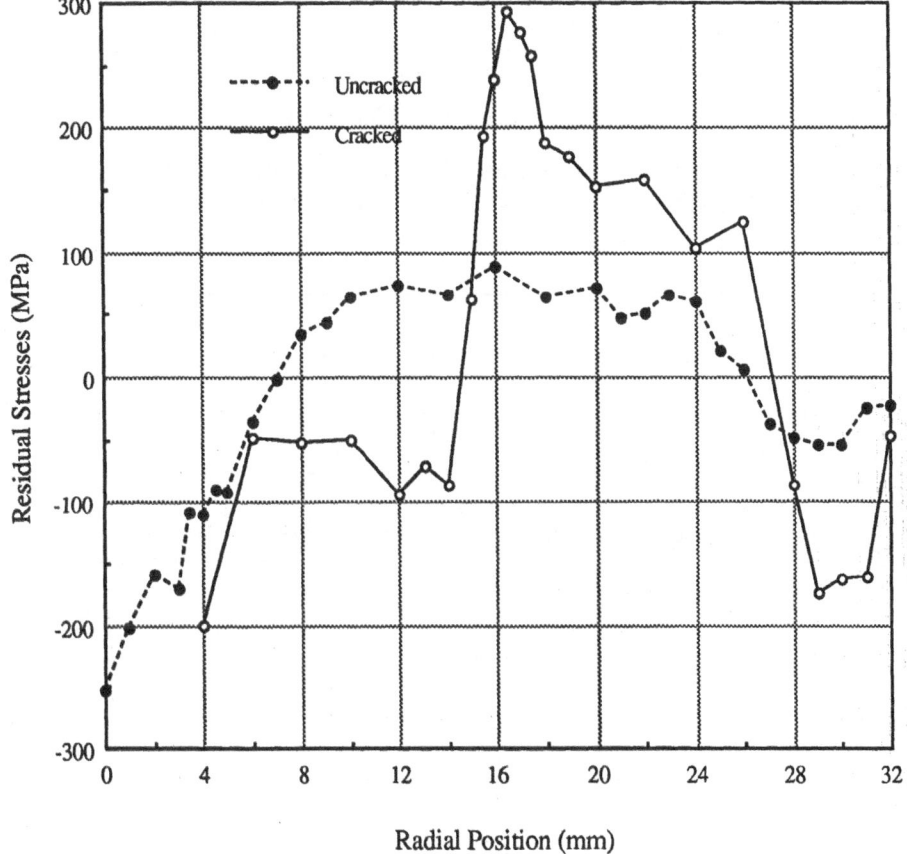

Figure 4.   Comparison of the residual hoop stresses for the cracked and uncracked (dashed line) sides of the autofrettaged, fatigued ring.

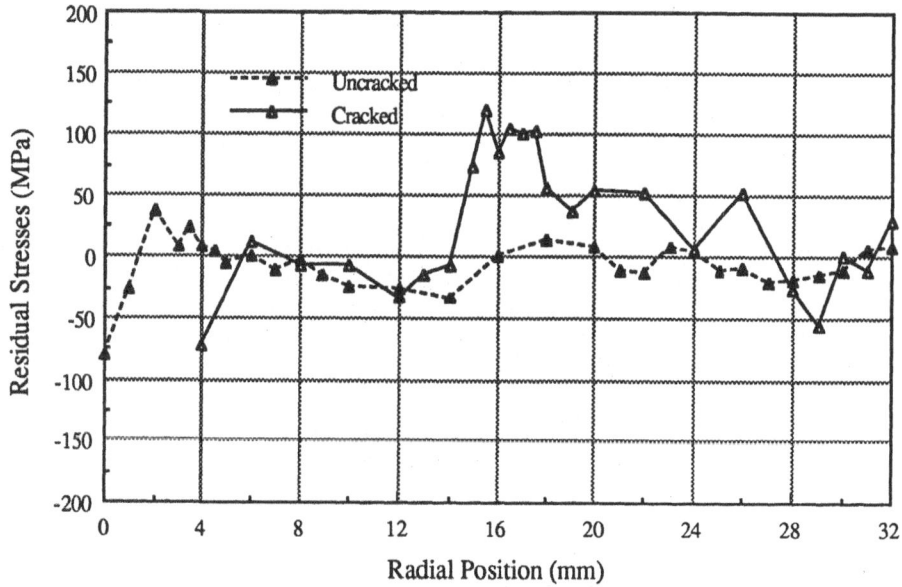

Figure 5.   Comparison of the residual radial stresses for the cracked and uncracked
              (dashed line) sides of the autofrettaged, fatigued ring.

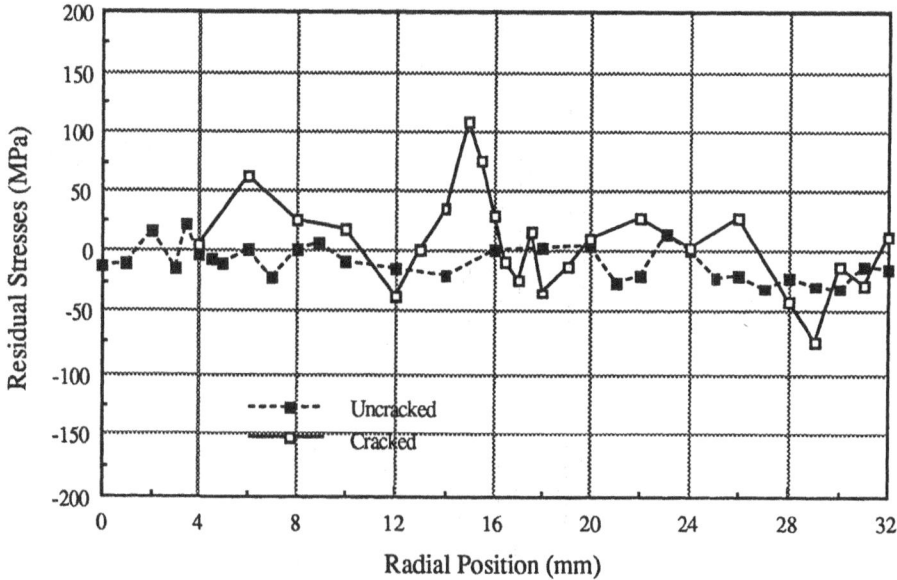

Figure 6.   Comparison of the residual axial stresses for the cracked and uncracked
              (dashed line) sides of the autofrettaged, fatigued ring.

4.3 COMPARISON OF CRACKED AND UNCRACKED SIDES DATA

On the uncracked sides the radial and axial residual stresses are small and there is little change on the cracked side except, in both cases, in the vicinity of the crack tip where the stresses rise to and by about 100 MPa. The most significant stresses, and the largest changes, are observed in the hoop direction. From R = 6 mm to 15 mm and from 28 mm to 32 mm the redistribution is towards compression whereas between R = 15 mm and 27 mm the shift is towards tension.

The crack tip was determined from visual observations to be at R = 16 mm but the change to tensile residual hoop stress is apparently at 15 mm which is not possible because the crack would open. However, it should be remembered that the minimum dimensions of the gauge volume are 2 mm plus penumbra effects due to the horizontal divergence and vertical focussing of the neutron beam. The data around the crack tip are thus averaged over a significant distance over which the stresses are rapidly changing in size and in sign. Consequently, in this region, apparent peak shifts, rounding, and gradient decreases should be expected. Additionally, at the time of the measurements, the positioning techniques were accurate only to within 1 mm so absolute errors of this magnitude are possible in the position data. Subsequent modifications to the equipment, and techniques, have reduced positioning errors to 0.1 mm and the gauge volume is now better defined.

4.4 PEAK WIDTHS

Another feature of the measurements concerns the widths of the Bragg reflections. The shapes and widths of the diffracted peaks can, in principle, provide information about the plastic deformation and microstrain present in the material although care must be taken to separate broadening effects due to the finite sampling volume in regions of steep stress gradients. Comparison of the data from the cracked and uncracked sides showed almost the same effects in all three principal directions. The hoop peak width data is shown in figure 7 as an example.

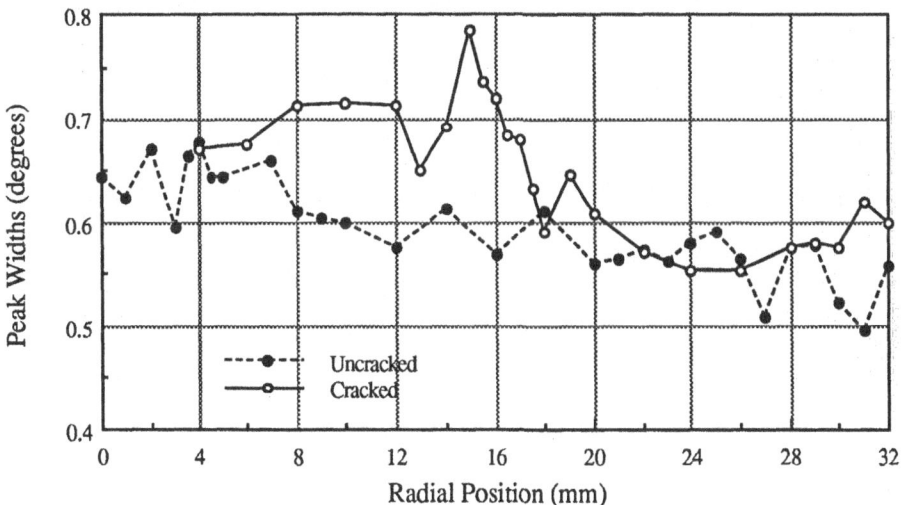

Figure 7. Comparison of the peak widths for the hoop directions for the cracked and uncracked (dashed line) sides of the autofrettaged, fatigued ring.

It was expected from the autofrettaging parameters that plastic deformation would be produced to a distance of R = 13 mm, 40% of the way through the tube wall. On the uncracked side the peak width is steady from the bore to R = 5 mm at about 0.65° after which it declines steadily before levelling out for several millimetres at about the limit of the plastic zone before slowly declining again to around 0.55°. On the cracked side there is a substantial increase to about 0.75° near the crack tip and to near 0.7° below. Some of the additional broadening at the crack tip may be related to the macrostress gradient but the majority is related to fatigue enhanced microstrain.

5. Theoretical calculations

5.1 THE BOUNDARY INTEGRAL METHOD

The boundary integral (BIE) or boundary element method is a stress analysis tool which provides solutions to engineering problems for which no analytical solution is available. It is a numerical technique which converts the partial differential equations which describe the elastic behaviour of a material to a surface integral applicable over a boundary. The conversion is exact and approximation is only introduced by the subsequent numerical solution of the integral. By applying tractions and displacements to points on a boundary the distribution of tractions and displacements at other points on the boundary can be predicted subject to any enforced conditions.

Boundary element methods combine known solutions of simple problems to give solutions to more complicated problems. If a force of known direction and magnitude acts at a point on the boundary of a material then subject to the laws of elasticity an analytical solution exists for the forces and displacements which act on any other point. By considering the application of multiple individual forces the fundamental solution associated with each can be superposed and by applying boundary conditions the effective traction and displacement over the remainder of the boundary can be predicted. Application of boundary integral methods to stress analysis is discussed by Fenner [6] and has been applied for calculating stress intensity factors in autofrettaged rings [7].

Unlike finite element methods, in which the whole of the relevant domain must be discretised, only the boundary is considered. This simplifies the model and reduces the dimensionality of the problem by one. Fewer elements are needed than equivalent finite element models which reduces the preparation and computing time. No approximation of the internal region is necessary which improves accuracy and produces less unwanted information. Selection of an appropriate mesh makes it easier to focus on a region of interest. Cracks can be introduced by creating new boundaries which can be external or internal.

Within each element on the boundary the displacement and tractions are described by nodal values together with a shape function for interpolation. Integrations along the boundary can be complicated and the solution is usually effected using Gaussian quadrature applied to the boundary elements. Small strains are implicit otherwise the shape of the domain will be altered.

5.2 APPLICATION OF BOUNDARY INTEGRATION

The features in the cracked rings which are open to modeling using the boundary element method are the notch, which initiates the fatigue crack, and the presence of a crack in the tensile region. Removal of material to form a notch permits the faces to compress in response to the residual stress at the bore and thus no compressive load is transmitted between R = 0 and R = 2.5 mm. Similarly, when the crack has propagated beyond 25% of the wall thickness it relieves the tensile hoop stress because the cracked region cannot transmit a tensile force. In the tensile region, for R > 7 mm the crack is drawn open and the stress must redistribute.

A boundary element program [8] was used for the calculations. The program assumes plane strain conditions and uses Hermitian cubic elements. Cubic variations of displacements and tractions are permitted along each element. In figure 8 the original autofrettaged hoop stress distribution is illustrated across the wall of a ring. In figure 9 the position of the crack tip is indicated and the shaded regions AB and CD denote the relieved regions corresponding to the notch and to the region of the ring where the crack lies in an originally tensile stress region.

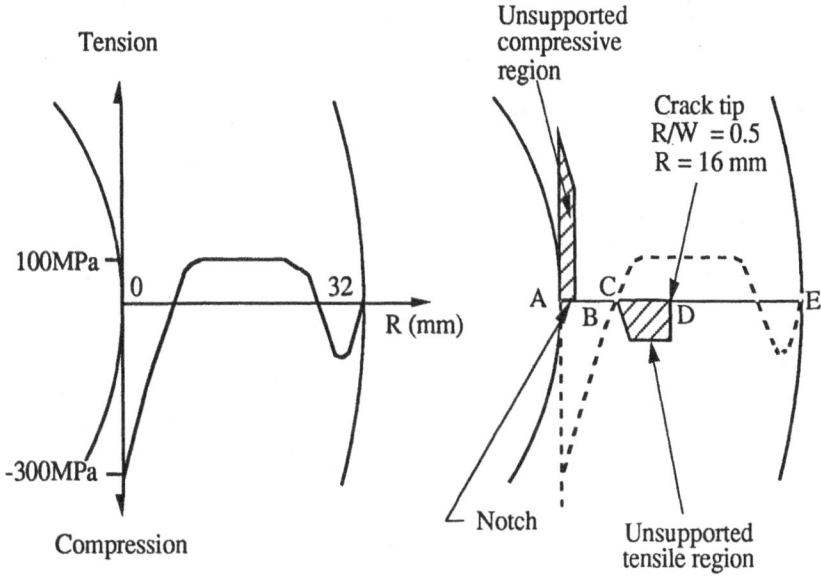

Figure 8.   Original autofrettaged hoop   Figure 9.   Applied stress for BIE
             stress   calculations

The BIE program was used to predict the stresses induced in the regions BC and DE for an applied stress distribution corresponding to the shaded regions in figure 9. Assuming superposition and adding the original hoop stress distribution, figure 8, to the predicted stress distribution due to the applied stresses in figure 9 the stresses in the region AB and CD are negated. The modified stresses in regions BC and DE are then the residual stresses of interest.

Symmetry requires that only half of the ring needs to be modeled. The mesh used for the calculations is shown in figure 10. Fifty eight nodes were used around the mesh with 20 on each flat section. The boundary conditions are shown in figure 11. In regions AB and CD tractions are applied of equal magnitude but opposite sign to those existing in the uncracked ring and the elements are permitted free displacement in the hoop direction. In regions BC and DE no tractions were applied and the elements were permitted no displacement normal to the crack plane, the hoop direction. On the uncracked side no

490

tractions were applied and the nodes were held at zero displacement in the hoop direction. All other nodes were completely free.

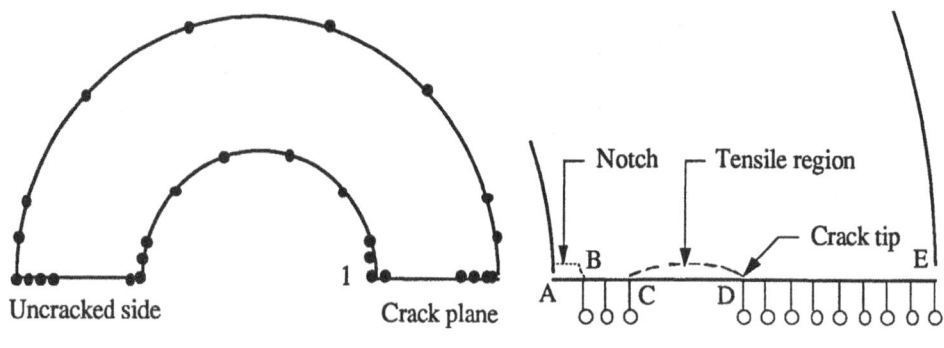

Figure 10.   Boundary element mesh   Figure 11.   Boundary conditions

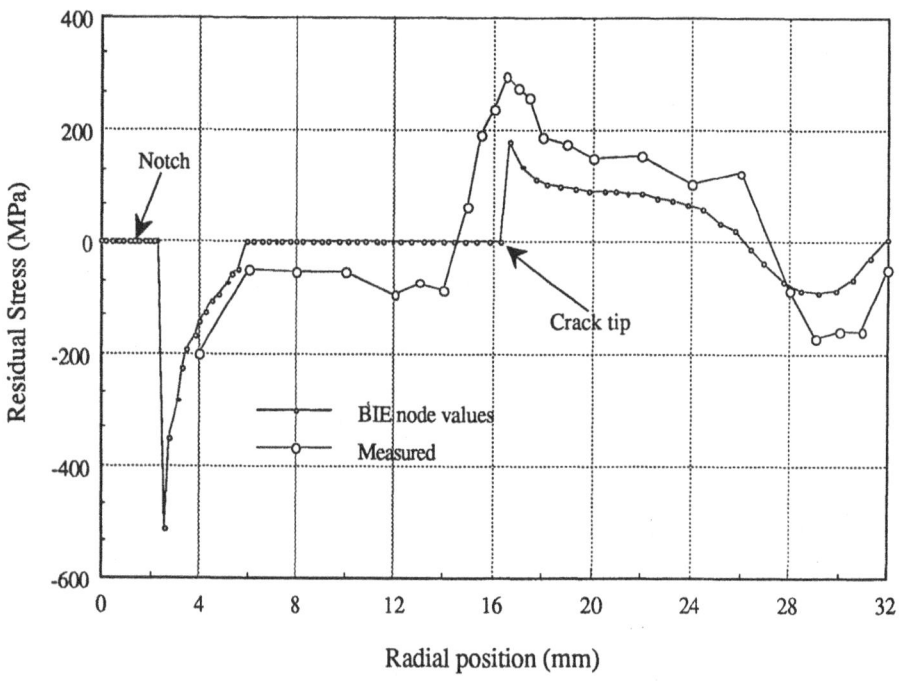

Figure 12.   Crack side residual hoop stress and BIE prediction.

In the first analysis the BIE program predicted some overlap of the crack plane to the left of point C, corresponding to R = 7 mm at the limit of the compressive hoop stress. When the crack face is shut the solution must not predict overlap of the crack face. Since the contact area was not known in advance, an iterative approach was adopted [9] and the reversed stress distribution was extended backwards to R = 5.9 mm. Physically this corresponds to the tensile stress holding the crack open at the end of the compressive zone. The crack tip, point D in figure 11, corresponded to R = 16.3 mm. The BIE prediction of the redistributed stress is plotted with the measured values in figure 12.

On the uncracked side even when the effect of reversing the stress over the whole length of the crack is considered the predicted change to the stresses are less than 15 MPa. This is in accordance with the experimental measurements in which no change was observed on the uncracked side, even for the longest crack length.

5.3 DISCUSSION

The BIE analysis presented here is limited by the coarse mesh size and the limited spatial resolution of the measurements. However, when the crack tip is assumed to be at R = 16.3 mm and the crack face is drawn open to R = 5.9 mm qualitative agreement between the measured results and the BIE predictions is achieved. In figure 12 the first node beyond point D is 0.5 mm away. Refinement of the mesh to give the values closer to point D would predict larger tensile stresses which would improve the agreement with the observed data. However, there is considerable discrepancy in the position of the crack tip. The hoop stress results imply that the crack tip lies at or near R = 14.5 mm since the crack cannot support a tensile stress. However this is shorter than the optical assessment of 16 mm. The BIE prediction gives nodal values and strictly some account of the finite size of the sampling volume should be made. Taking the finite sampling volume of the neutron technique into account would improve the agreement between the BIE prediction and the measured values.

In the BIE prediction the stress in the originally tensile region along the crack plane is negated to leave zero stress. This does not account for the constant compressive hoop stress which exists along the crack plane behind the crack tip. Its origin is unclear but may be associated with the plastic wake produced by the passage of the crack.

6. Conclusions

Propagation of fatigue cracks through half the wall thickness of autofrettaged specimens has negligible effect on the stress distribution on the opposite uncracked side. Cracks which remain in the originally compressive region of an autofrettaged ring ring do not modify the residual stresses except due to the artificially introduced notch. However, once a crack is propagated into the tensile region distinctive changes occur as the stress redistributes. When a crack had propagated through 8 mm of the tensile region the tensile stress ahead of the crack tip increased and reached a maximum of 300 MPa which is three times the value in an uncracked ring. A constant compressive hoop stress of -80 MPa was observed over much of the crack plane except near the bore where the compressive stress was increased due to the removal of the notch material.

The redistribution caused by the crack has been addressed using a BIE and superposition approach. The effects of the stress relieved by the crack were calculated and combined with the uncracked stress distribution using the principle of superposition. Qualitative agreement was achieved between the predicted and measured values. The simple analysis fails to account for the compressive shoulder along the crack plane and uncertainty in the spatial resolution of the experimental measurements precluded more detailed attempts to model the stress distributions.

492

7. Acknowledgements

The neutron work described here involved the use of facilities at the ILL with the support of the Directors and staff. Financial resources were provided by the UK Science and Engineering Research Council via grants GR/D63196 and GR/E55556.

8. References

1. Stacey, A. and Webster, G.A. (1984) 'Fatigue crack growth in autofrettaged thick walled high pressure tube material', High pressure in science and technology, Mat. Res. Soc. Symp. Proc., Part III, 22, 215-219.
2. Stacey, A., MacGillivray, H. J., Webster, G. A., Webster, P. J. and Ziebeck, K. R. A. (1985) 'Measurement of residual stresses by neutron diffraction', J. Strain Analysis 20, 93-100.
3. Pintschovius, L., Macherauch, E.and Scholtes, B. (1986) 'Determination of residual stresses in autofrettaged steel tubes by neutron and X-ray diffraction' Proc. Int. Conf. on Residual Stresses, Garmisch-Partenkirchen, 1986, 159-165.
4. Sorem, Jr.J.R., Shadley, J.R. and Rybicki, E.F. (1990) 'Experimental method for determining through thickness residual stresses in thin walled pipes and tubes without inside access' Strain, February 1990, 7-14.
5. Allen, A., Hutchings, M. T. and Windsor, C. G. (1985) 'Neutron diffraction methods for the study of residual stress fields' Adv. Phys. 34, 445-473.
6. Fenner, R.T. (1983), 'The boundary integral equation (boundary element) method in engineering stress analysis' J. Strain Analysis 18, 199-205.
7. Webster, G.A., Klintworth, G.C. and Stacey, A. (1983) 'Stress intensity factors for cracked C-shaped and ring type test-pieces' J. Strain Analysis 18, 225-230.
8. Watson, J.O. (1982) 'Hermitian cubic boundary elements for plane problems of fracture mechanics' Res. Mechanica. 4, 23-42.
9. Becker, A.A. and Plant, R.C.A. (1987) 'Contact mechanics using the boundary element method' Tribology conference, Institute of Mechanical Engineers, C227/87, 975-980.

MEASUREMENT OF STRESSES IN METAL ADHESIVE JOINTS

A. LODINI* - J. LI* - M. PERRIN** - F. DUNSTETTER** -
L. RIMLINGER*

* Laboratoire Science des Matériaux
Université de Reims Champagne Ardenne
51100 Reims
France

** Laboratoire Léon Brillouin (CEA-CNRS)
CE Saclay
91191 Gif-sur-Yvette Cedex
France

ABSTRACT. Among the different techniques, neutron scattering is the only one which can investigate directly inside the metal-adhesive interface because the linear absorption coefficients are much lower than those observed with the X-Rays. The specimen was assembled by a single component epoxy adhesive. The quality of the bonded joint was controlled by neutron radiography. The assembly was subjected to a macroscopic strain by means of a special set-up. The measurement of stresses near the interface made under different experimental conditions. The results obtained are compared with those from finite element calculations.

1. Introduction

For two years now, we have used the G4-2 spectrometer of the laboratory Léon Brillouin of Saclay for the evaluation of stresses in metallic materials. The G4-2 spectrometer is a two axis spectrometer with Euler equipment using neutrons coming from one of the cold guides of the Orphee reactor. During this period, we have studied several different subjects with this spectrometer : welding, adhesive bonds, composite materials, enamel of tooth. In this work, we present the measurement of stresses in the adhesive bond of an aluminium alloy. (Figure 1). We work with the classic condition of constant wavelength. The method consists of the precise evaluation of the interplanar distance to give the strain.

$$\varepsilon \phi \psi = \frac{d - do}{do}$$

When we differentiate the Bragg relation with respect to the diffraction angle θ, we can obtain the usual formula, for strain :

M. T. Hutchings and A. D. Krawitz (ed.),
Measurement of Residual and Applied Stress Using Neutron Diffraction, 493–501.
© 1992 Kluwer Academic Publishers.

$$\varepsilon\phi\psi = -\Delta\theta \cot g\ \theta$$

The $\Delta\theta$ value becomes more important as the $\cot g\ \theta$ value is smaller. However, if we choose a large θ angle, the peak and the volume analysed are large. So, we used a 2θ angle of $90°$.

2. Experimental conditions

We have employed three complementary experiments for the stress measurement in the adhesive bonds.

2.1. 1D METHOD :

We directly measure the strain $\varepsilon\phi\psi$ with the equation :

$$\varepsilon\phi\psi = \frac{d-do}{do}$$

It is only necessary to know the value of do, which is obtained before the loading of the adhesive specimen.

Figure 2 indicates the principle of the method. At a particular location in the sample we can obtain this main strain in particular orientation with the cradle Eulerian equipment.

$$\varepsilon_{xx} \text{ is determined for } \psi = 90° \qquad \phi = 0°$$
$$\varepsilon_{yy} \text{ is determined for } \psi = 90° \qquad \phi = 90°$$
$$\varepsilon_{zz} \text{ is determined for } \psi = 0° \qquad \phi = 0°$$

This method is very interesting when we want to realize a cartography of the distribution of strains in a particular direction and plane.

2.2. 2D MEHOD : $\sin^2\psi$ METHOD

It is important to improve the precision of the measurement using a statistical analysis. Therefore, we use a large number of experimental values, typically 15 points.

If we consider in the isotropic approximation an element with an ellispsoid shape under the action of an homogeneous elastic stress, $\varepsilon\phi\psi$ can be calculated with the formula :

$$\varepsilon\phi\psi = \sin^2\psi\ (\varepsilon_{xx}\cos^2\phi + \varepsilon_{yy}\sin^2\phi - \varepsilon_{zz} + \varepsilon_{xy}\sin2\phi) + \sin2\psi$$
$$(\varepsilon_{yz}\sin\phi + \varepsilon_{xz}\cos\phi) + \varepsilon_{zz}$$

If we assume the main axes of strains coincide with those of the specimen, we can use the formula :

Figure 1 : Aluminium alloy specimen assembled by a single component epoxy adhesive.

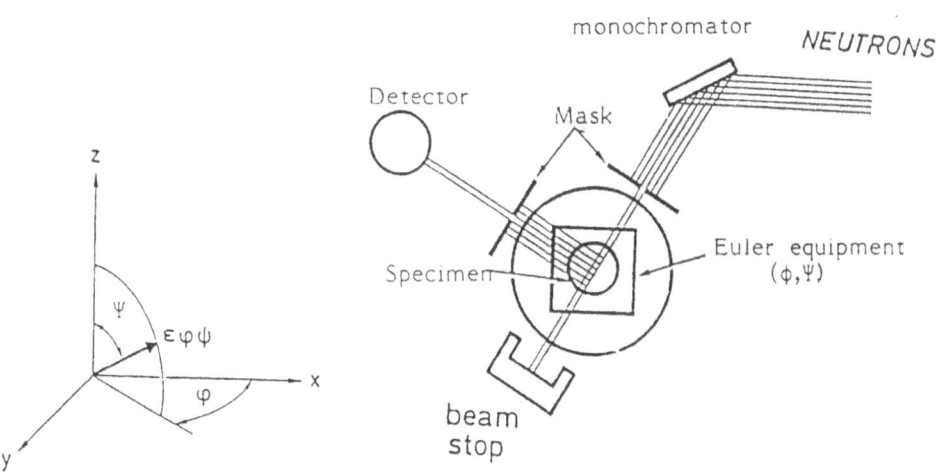

Figure 2 : Principle of the method with the 2 axis spectrometer.

$$\epsilon\phi\psi = \sin^2\psi \, (\epsilon_{xx} \cos^2\phi + \epsilon_{yy} \sin^2\phi - \epsilon_{zz}) + \epsilon_{zz}$$

and with linear regression we can determine ϵxx, ϵyy and ϵzz.

This method can be used with a defect free adhesive specimen under on applied load.

2.3.   3D METHOD : GENERALIZED METHOD

In a case of an adhesive specimen with a defect, the orientation of the main stresses is no longer known around this defect. It is then necessary to use a generalized method. An homogeneous and isotropic solid with an eleastic deformation is characterized by a tensor :

$$\varepsilon = \begin{vmatrix} \varepsilon_{xx} & \varepsilon_{xy} & \varepsilon_{xz} \\ \varepsilon_{yx} & \varepsilon_{yy} & \varepsilon_{yz} \\ \varepsilon_{zx} & \varepsilon_{zy} & \varepsilon_{zz} \end{vmatrix}$$

If we suppose, once again, a volume element with an ellipsoid shape :
ε is determined with the formula

$$\epsilon\phi\psi = a_1{}^2\, \varepsilon_{xx} + a_2{}^2\, \varepsilon_{yy} + a_3{}^2\, \varepsilon_{zz} + 2\, a_1 a_2\, \varepsilon_{xy} + 2\, a_2 a_3\, \varepsilon_{xy} + 2\, a_1 a_3\, \varepsilon_{xz}$$

It is again necessary to know the measurement of do ; the value of which is obtained before the loading of the adhesive specimen. An ellipsoïd regression is determined with an important number of experimental values, typically 20 points. The principal strains and their directions are determined by diagonalization of this matrix. It is then possible to determine the main stresses for an elastic isotropic material using the generalized Hooke Law :

$$\varepsilon_{xx} = \frac{1}{E}\, (\sigma_{xx} - \upsilon\,(\sigma_{yy} + \sigma_{zz}))$$

$$\varepsilon_{yy} = \frac{1}{E}\, (\sigma_{yy} - \upsilon\,(\sigma_{xx} + \sigma_{zz}))$$

$$\varepsilon_{zz} = \frac{1}{E}\, (\sigma_{zz} - \upsilon\,(\sigma_{xx} + \sigma_{yy}))$$

Young Modulus E and Poisson's ration υ are the elastic constants of the plane (h, k, l) which is used. A preliminary calibration is necessary to determine these constants.

3.   Calibration of the different experimental methods

In a preliminary study, it was interesting to measure the distribution of the stresses in the aluminium alloy material before sticking, using the technique of neutron diffraction, and then to compare this method to that of finite element calculation.

For this purpose we have used the experimental set-up of Schmank and Krawitz (1982). The specimen was an aluminium alloy with U shape and a square section of 25.4 x 25.4 mm2. If we compress the ends of the U shape specimen we obtain a bi-axial stress. (The displacement was $\Delta_x = 15,5$ mm); If one end of the U shape is compressed in two directions, we obtain a triaxial stress. (The displacement was $\Delta_x = \Delta_y = 12$ mm).

In the 1D method, the volume sampled by a neutron beam was 3 x 25.4 x 3 mm3 and we used a wavelength of 3.29 Å and the Bragg plane (111).

We determined the diffraction peak using 40 experimental settings of angle every 0.05 degree during about 4 hours. We calculated the position of the peak with a gaussian profile.

The difference between the results of the neutron and the finite elements methods is small. Figure 3 shows the different values of the principal stress $\sigma_{xx} = f(z)$. For stress values between 0 and 400 MPa we obtain a difference less than 40 MPa.

In 2D method, the set up is similar, the gauge volume is also 3 x 25.4 x 3 mm^3. We now make a ψ angle rotation. The large number of measurements makes it possible to make a good statisitical analyis. Figure 4 shows that for the principal stress $\sigma_{xx} = f(z)$, there is a small difference between the neutron and the finite elements. For stresses between 0 and 400 MPa we obtain a difference less than 30 MPa.

In the 3D method, the simultaneous rotation of the ϕ and ψ angles involves a much smaller volume : 3 x 3 x 3 mm^3. Because of the large number of experimental points necessary for the calculation of the regression, the measuring time is now about 2 weeks. However, in this case the difference between the neutrons and the finite elements is much higher.

Figure 5 shows the variation of the principal stress $\sigma_{xx} = f(z)$. For stress values between 0 and 320 MPa we obtain a difference less than 80 MPa.

4.   Determination of stresses in adhesive metal interface

We have also used these methods for the bonded structure. The assembly was in the form of two beveled half-test bar of Aluminium alloy. It was assembled by a single component epoxy adhesive of 0.4 mm thick.

The quality of the bonded joint was controlled by a neutron radiograph. This method is quite appropriate to a bonded assembly.

The assembly is subjected to a load by means of a specific set up which enables one to measure the strain by neutron diffraction. The loading of specimen is 11000 N. (figure 6).

In the 2D method, the measurement is made in the metal at several points distributed in a parallel direction to the bonded joint, 2 mm away from the interface. Figure 7 shows the results obtained with finite element caculations and neutron difffraction method for the σ_{xx} direction. We observe similar results between the two methods for most of the points, only one point is markedly different.

498

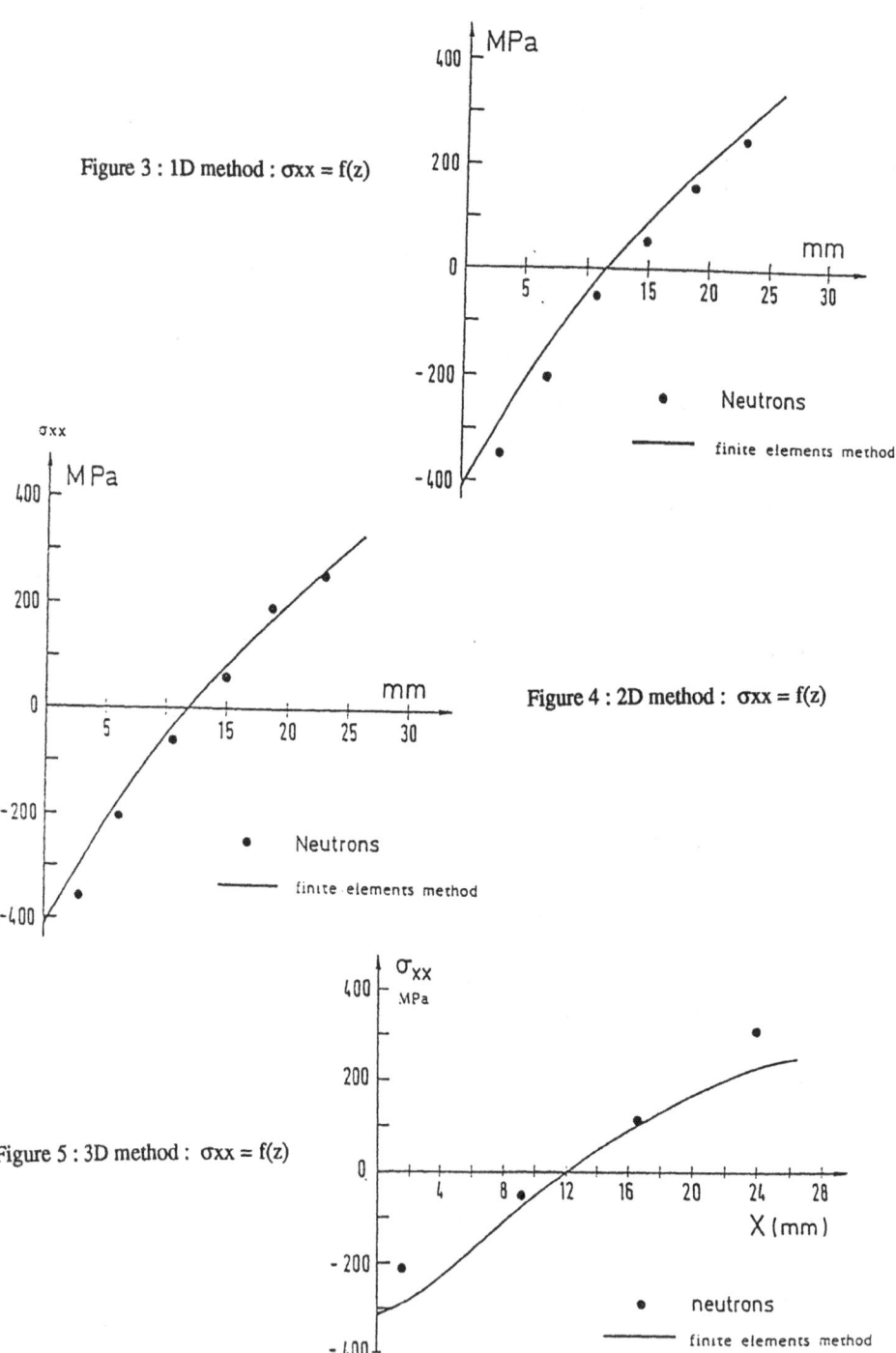

Figure 3 : 1D method : σxx = f(z)

Figure 4 : 2D method : σxx = f(z)

Figure 5 : 3D method : σxx = f(z)

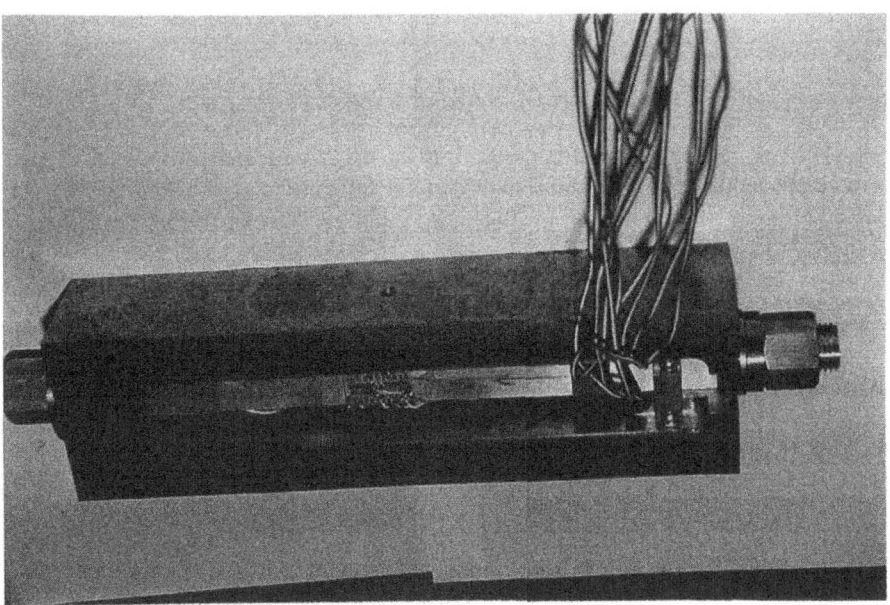

Figure 6 : specific set up for loading the assembly.

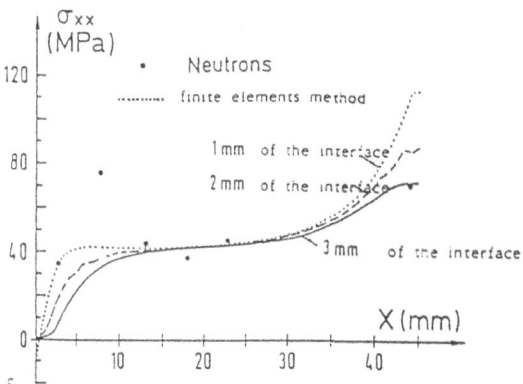

Figure 7 : 2D method in metal - adhesive joints.

With the 1D method, the next figure represents a mapping of the isodeformations for a section plane perpendicular to the adhesive surface and for a direction parallel to the direction of the load. The maps (Figure 8) realized before and after loading clearly show the influence of the 11000 N load on the redistribution of the strains.

With the 3D method, the maximum imprecision is 80 MPa, and difficults make a comparion of the measurement on the adhesive joints. However we have made measurements with the 3D method. In particular, we observe a compressive region, near the defect :

$$\sigma_{xx} = - 106 \, \text{MPa}$$
$$\sigma_{yy} = - 16 \, \text{MPa}$$
$$\sigma_{zz} = - 90 \, \text{MPa}$$

We now find important differences with the method of the finite elements

$$\sigma_{xx} = + 52,6 \, \text{MPa}$$
$$\sigma_{yy} = - 0,8 \, \text{MPa}$$
$$\sigma_{zz} = - 0,1 \, \text{MPa}$$

5. Conclusion

As a conclusion, we can say that the measurement of stress by the neutron technique can be made in adhesive joints between highly thick metal.

The modeling of the stresses with an ellipsoidale deformation is adapted to the neutron technique. We have to assure a homogeneous material, with low texture and elastic deformation.

Yet, it is necessary to work with a nuclear reactor with an adequate flux. Even in this case the measurement time is of one or two weeks.

The size of the gauge volume sampled gives the average of the stress over a large distance of interface.

The 2D method clearly shows the evolution of the stresses near the interface.

The 1D method shows the evolution of the isodeformations under the influence of the load.

The 3D method have more particularly used for the triaxial stresses developed in an adhesive specimen with a defect. We have observed the influence of the defect on the redistribution of the stress.

References

Allen, J. ; Hutchings, M. and Windsor, C. (1985) 'Advances in physics' 4, 445-473.

Holden, T.M. ; Root, J.H. ; Fidleris, V., Hoct, R.A. and Roy, G. (1988) 'Materials Science Forum ' 27, 359-370.

Pintschovius, L. ; Hauk, V. and Krug, W.K. (1987) 'Materials science and engineering' 92, 1-12.

Lodini, A. and Perrin, M. (1989) 'Mémoires et études scientifiques', Revue de métallurgie 11, 715-722.

Schmank, M.J. and Krawitz, A.D. (1982) 'Metallurgical transactions' A, 13A, 1069-1976.

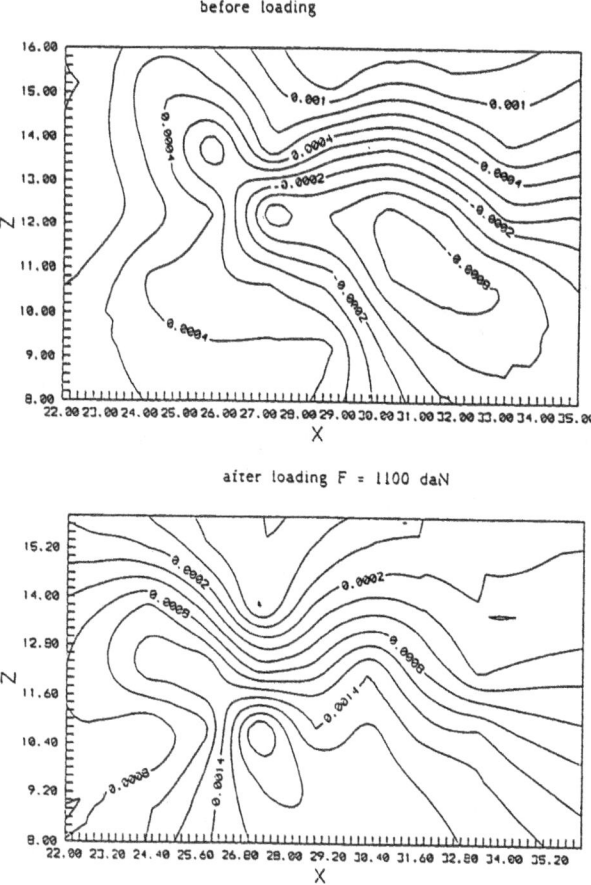

before loading

after loading F = 1100 daN

Figure 8 : 1D method : isodeformations.

RESIDUAL STRESS MEASUREMENTS IN ARMAMENT-RELATED COMPONENTS

H. J. PRASK and C.S. CHOI[1]
Reactor Radiation Division, MSEL
National Institute of Standards and Technology
Gaithersburg, MD 20899
USA

ABSTRACT. Energy-dispersive neutron diffraction has been developed at the NIST reactor as a probe of sub- and near-surface residual stresses in engineering samples. Application of the technique has been made to a variety of metallurgical specimens which includes the determination of tri-axial stresses as a function of depth in a number of uranium-3/4wt%Ti samples with different thermo-mechanical histories, and in two types of 7075-T6 aluminum "ogives"- of interest to the Army. Most recently, results have been obtained for two induction-hardened steel shafts, fatigue lifetime test specimens for the Society of Automotive Engineers.

1. INTRODUCTION

Among the first references to the use of neutron diffraction "depth-profiling" to solve engineering problems was in 1979, and was in the area of crystallographic texture characterization [1]. The first descriptions of the application of neutron diffraction to the measurement of residual stress gradients for engineering problems was in 1981 [2]. Since then the field has expanded enormously, as evidenced by the scope of contributions to this workshop. In the following we report on a portion of the neutron diffraction residual stress work that has taken place at the NIST Research Reactor, chosen to illustrate the variety of applications. Some of the results have been presented elsewhere [3,4].

2. METHODOLOGY

2.1 Instrumental Details

A unique advantage of neutron diffraction arises from the different relative scattering cross-sections and penetration of neutrons relative

---

[1]Guest researcher from ARDEC, Picatinny Arsenal, NJ, USA

M. T. Hutchings and A. D. Krawitz (ed.),
Measurement of Residual and Applied Stress Using Neutron Diffraction, 503–516.
© 1992 *Kluwer Academic Publishers.*

to x-rays. This is illustrated in Table 1 in which $t_{1/2}$, the thickness at which half the beam intensity is lost through scattering and absorption processes, is listed for selected metals. The values are based on cross-sections from standard references and the difference in wavelengths used for neutrons (0.108 nm) and x-rays (0.154 nm) is not significant.

Table 1.   X-Ray/Neutron Comparison

Element(At. No.)	$t_{1/2}$(X-rays)	$t_{1/2}$(Neutrons)
Al (13)	0.0530 mm	71.0 mm
Ti (22)	0.0076	15.9
Fe (26)	0.0027	6.1
Cd (48)	0.0035	0.057
W (74)	0.0021	6.5
U^{238} (92)	0.0015	13.6

The $t_{1/2}$ values clearly show that neutrons in the normal diffraction wavelength range are several orders-of-magnitude more penetrating than x-rays. Also, the penetration does not decrease monotonically with atomic number as with x-rays but is, essentially, a random function of atomic number. It is clear from the Table that both depleted uranium and aluminum are very good materials for neutron examination, while iron/steel is less so.

The properties of neutrons presented the possibility of measuring sub-surface residual stress gradients by employing tight collimation and ~90° scattering geometry (i.e. "depth-profiling"). In our measurements, we have made use of the fact that Bragg-condition resonances can be observed at fixed scattering angle, with varying wavelength. With the scattering angle fixed the gauge volume, ΔV, remains the same throughout each scan. Also in this mode, the intensity profile distortion which occurs when part of the ΔV is outside the sample may be lessened compared to angle-dispersive scans [5].

The instrument that we use for energy-dispersive neutron diffraction (EDND) is a triple-axis spectrometer. Crystals of known d-spacing are placed before and after the sample (monochromator and analyzer, respectively); the Bragg relation is then used to select and step the wavelength incident on the sample. In principle, the analyzer crystal - which we step at the identical wavelength as the monochromator - is not needed. However, for a given scattering configuration, utilization of the analyzer significantly enhances instrumental resolution. Other instrumental details are given elsewhere [3].

2.2 ANALYTICAL FORMALISM

The relation between stress and strain applicable to diffraction measurements has been presented, for example, by Evenschor and Hauk [6]. With reference to Figure 1, r, θ, and z are specimen-fixed axes, and the strain $\epsilon'_{\phi\psi}$ is measured along $\vec{L}'_3$; then

$$\epsilon'_{\phi\psi} = (d_{\phi\psi} - d_0)/d_0 \tag{1}$$

where $d_{\phi\psi}$ is the lattice spacing along $\vec{L}_3'$ and d_0 is the unstressed lattice spacing. The stresses are related to the measured strains through

$$\epsilon_{\phi\psi}' = \tfrac{1}{2}S_2(hkl)[\sigma_{rr}\cos^2\phi\sin^2\psi + \sigma_{\theta\theta}\sin^2\phi\sin^2\psi + \sigma_{zz}\cos^2\psi$$

$$+ \sigma_{r\theta}\sin2\phi\sin^2\psi + \sigma_{rz}\cos\phi\sin2\psi + \sigma_{\theta z}\sin\phi\sin2\psi] \qquad (2)$$

$$+ S_1(hkl)[\sigma_{rr} + \sigma_{\theta\theta} + \sigma_{zz}].$$

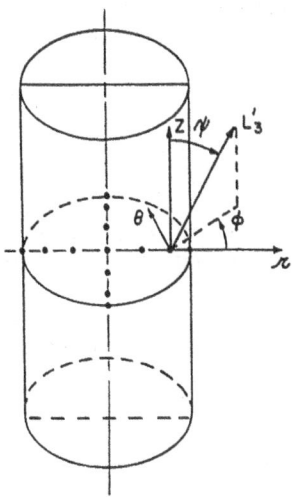

Figure 1. Coordinate system and possible measurement mesh for the neutron diffraction measurements. The solid circles represent the centers of the ΔVs examined.

The $S_i(hkl)$ are diffraction elastic constants ("XEC") for the (hkl) reflection which, in general, depend on the material and the reflection examined. For an elastically isotropic solid the XEC are given by

$$\tfrac{1}{2}S_2(hkl) = (1+\mu)/E \quad \text{and} \quad S_1(hkl) = -\mu/E \qquad (3)$$

where μ, E are Poisson's ratio and Young's modulus, respectively.
Since the determination of residual stress in technological samples by means of Eq. 2 depends directly on measurement of strain values, a precise value for the unstressed d-spacing, d_0, is essential. Often d_0 can be obtained from powders or stress-free samples, but this is not always possible. Alternatively, we utilize the overall equilibrium conditions required by elasticity theory to determine d_0. That is, since the body is static with no external force applied, residual stresses normal to any plane must balance such that in cylindrical geometry:

$$\int \sigma_{zz} r\,dr\,d\theta = 0 \quad \text{and} \quad \int \sigma_{\theta\theta}\,dr\,dz = 0 \qquad (4)$$

and at any surface the stress orthogonal to that surface must vanish. Stresses inferred from measured strains can be adjusted, by varying d_0, to fulfill the equilibrium conditions.

3. RESULTS

3.1 U-0.75wt%Ti ("U-3/4Ti")

U-3/4Ti is a high-density, high-strength alloy of considerable importance in several military applications. In the present work, two different types of U-3/4Ti samples were studied. Starting with 3.3 cm diameter stock, all material was γ-phase solutionized at 800°C (by induction or oven heating), (spray or bath) water quenched, and rotary straightened. Of these, two were machined to the final 2.4 cm diameter. Additional cold working was performed on two of the samples before final machining: after turning to ~2.9 cm diameter, both were swaged in a single pass (7% R.A.); one of the two was then swaged in three additional passes to a final 31% R.A. Both were then turned to the final diameter. In all cases aging after quenching was 1-2 hours at 400°C or less. For convenience, the residual stress determinations were made on 10 cm long pieces cut from the mid-point of 46 cm long rods.

One difficulty with U-3/4Ti material prepared as described above is that it is not single-phase. Very rapidly cooled material consists of a single martensitic phase, α'-U, whereas slowly cooled material consists of α-U (containing some dissolved Ti) and U_2Ti. α- and α'-U are orthorhombic and almost identical in structure [7], and 3.3 cm U-3/4Ti rods of the above type are ~88 vol% α'-U at the surface, ~67% at the centerline after quenching and aging [8]. The machined and cold-worked samples examined with neutrons were estimated to be 85 or (for the 31% swaged sample) 87 vol% α'-U at the surface.

Utilizing the (112) reflection and a 4x4x4 mm³ gauge volume we have determined d-spacings in the midpoint r-θ plane of each sample for the measurement grid shown in Fig. 1. Determination of stresses was made using eqns. 2-4 and isotropic elastic constants [9], simply averaged according to composition at each point. Two d_0's were determined for each sample from eqn. (4), with the effect on d of compositional gradients taken into account. Stress distributions for the two unswaged samples and the 7% swaged sample have been presented elsewhere [3]. The results for the unswaged samples were in excellent quantitative agreement with measurements made on similar samples by the Sach's boring-out technique (references cited in [3]).

An internal check of the utilization of eqn. 4 to determine d_0 is possible as follows. The d_0's obtained from the σ_{zz}- and the $\sigma_{\theta\theta}$- balance conditions, if the procedure is correct, should be approximately equal (assuming that $\sigma_{\theta\theta}$ at any r is essentially constant along the length of the cylinder). In Table 2 a summary of results for four U-3/4Ti samples is presented. The d_0's obtained for the swaged and unswaged pairs differ because scattering angles, Ω, in the two cases were not determined absolutely. The agreement between d_{0z} and $d_{0\theta}$ is excellent for each sample except the 31% swaged material. The $d_0\theta$'s for each unswaged sample are also in excellent agreement, whereas the d_0's of the two swaged materials are significantly different from each other.

Table 2. U-3/4Ti Summary

Sample	Ω(nom.)	$\underline{d}_{0_z}$	$\underline{d}_{0_\theta}$	$\sigma_{\theta\theta}$(r=0)	$\overline{\sigma_{\theta\theta}}$(r=R)
Ind. heat/spray	100°	0.179094nm	0.179104nm	297±15MPa	-314±12MPa
Vac. heat/bath	100°	0.179095	0.179091	250±12	-327±14
7% RA swaged	90°	0.178551	0.178551	-113±12	84±13
31% RA swaged	90°	0.178676	0.178600	(-71±13)	(-16±14)

The measured textures of these several samples show that both of the swaged samples possess fiber texture ([072] fiber axis), with the (112) poles aligned perpendicular to the cylinder axis; the unswaged samples are essentially untextured. The degree of orientation is ~2.5 times as great in the 31% swaged as the 7% swaged sample. Since the U-3/4Ti specimens are in fact three-phase materials, we speculate in the following way. We assume that microstrains have been produced in the cylinder-axis direction in either the α-U or U_2Ti phases by severe plastic deformation, leading to $d_{0_z} \neq d_{0_\theta}$. To correct for this we add a uniform Δd_z to all measured d_z's until eqns. (4) are satisfied. Two interesting results come out of this: first, the d_0 which balances stresses in both directions is 0.178560nm, in quite good agreement with the 7% swaged sample; secondly, the Δd_z (=0.00022nm) is comparable to $d_\parallel$(112)-$d_\perp$(112) (=0.00031nm) measured for a 1cm long x 1cm diam. texture specimen taken from the center of the 31% swaged sample, and totally immersed in the neutron beam. Some representative residual stress values are shown in Table 2; those for the 31% swaged sample are after the empirical correction.

3.2 7075-T6 OGIVES

In the past several years ogive failures have occurred during ballistic acceptance testing of the 155 mm M483A1 projectile. The ogive, which attaches to the front end of the shell body and contains an explosive charge to expel the payload through the rear of the shell, is manufactured from 7075-T6 aluminum and is pictured in Figure 2. Production ogives are manufactured by two different suppliers who use somewhat different manufacturing methods. Manufacturer B produces the ogives by cold forging the cavity to finished dimensions, heat treating, and machining the outside dimensions. Manufacturer A produces a preform by forging at 332°-382°C, machines the cavity, heat treats, then finish machines the outside dimensions. Both suppliers start with identical aluminum. In full-scale tests the primary failure mode, exclusive to the B-type ogive, is a circumferential fracture at the first loaded thread.

508

Several material characterization studies, including X-ray and hole-drilling residual stress measurements with layer removal, and simulation tests have been conducted (summarized in ref. [3]).

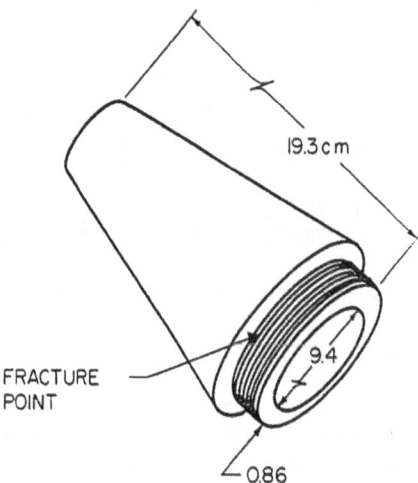

Figure 2. M483A1 ogive with forward end at upper left. The region of interest is under the first thread, as indicated.

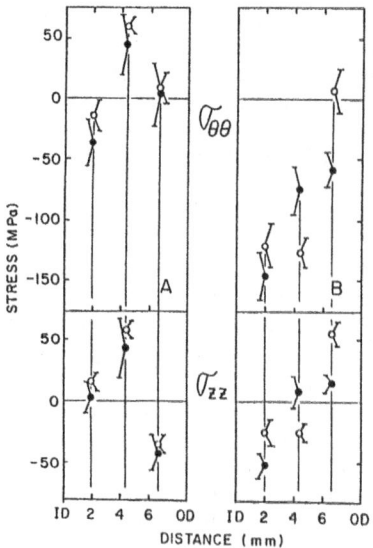

Figure 3. Stress distributions of the A and B type ogives. "OD" is the thread root position. The open and closed circles correspond to positions 180° apart on the same diameter.

No substantial material property differences, including residual stresses, were observed. The results of a simulation study, in which ogives were sealed and a charge exploded within, showed substantial differences in behavior. The B-ogives failed at the thread at containment pressures substantially below that which is required to fail A-ogives. Furthermore, these tests showed that the A-ogives failed by ductile rupture with the crack running in the longitudinal direction.

In our work one ogive of type B and one of type A were studied by EDND. Six 4x4x4mm³ beam-spot positions along a diameter in the plane of the potential failure site were used. Axial-direction equilibrium (eqn. 4) was used to determine $d_0(200)$, assuming that the average of the 0° and 180° data was representative of the full circumference. The d_0's arrived at to balance σ_{zz} stresses were in reasonable agreement for both types of ogives. The diffraction elastic constants used in the stress/strain calculations were the theoretical values of Bollenrath et al [10].

In Figure 3, final absolute residual stress values are shown for both ogive types for 0° and 180° positions at the first thread position. In contrast to previous measurements, the nondestructive EDND stress determination indicates very significant differences in the two ogive types. In the A-type ogive it is found that near the thread root ("OD") the σ_{zz} stresses are about -40 MPa (compressive) whereas the B-type show +10 to +55 MPa (tensile) stresses. In contrast, σ_{zz} and $\sigma_{\theta\theta}$ near the ID of the A-type ogive are zero or somewhat tensile, but in the B-type are clearly compressive at this position. These results are consistent with the observed failure modes for the two ogive types.

3.3 1045 STEEL TEST AXLES

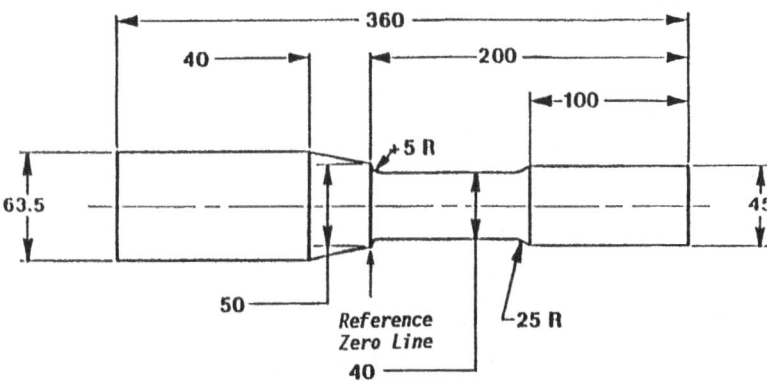

Figure 4. SAE biaxial fatigue test "axle". The 40 mm length on the left is the stationary portion in fatigue cycling.

The Society of Automotive Engineers Fatigue Design and Evaluation Committee has been conducting a long-term program aimed at the development of a predictive capability for fatigue life. The first phase of this program examined unhardened SAE 1045 steel shafts subjected to a variety of torsion and/or bending strains [11]. In Phase

510

II of this program the multiaxial fatigue behavior of induction-hardened and tempered SAE 1045 notched shafts is being examined. In this phase a more comprehensive, integrated engineering approach is being applied to the problem than in Phase I. It includes materials properties characterization, finite element analysis prediction of stress/strain, residual stress characterization, fatigue-life measurement, and fatigue lifetime prediction. Some depth-profiling of residual stresses has been done by x-ray diffraction with layer removal [12]; however, the nondestructive character of neutron diffraction offers an excellent means for both characterizing the residual stress distributions and in testing the x-ray layer-removal approach.

The geometry of the component-like test "axles" is shown in Figure 4. Two samples were studied with neutron diffraction: an axle not subjected to any strains, and an axle which had been subjected to bending and torsion fatigue cycling to one-half the fatigue life (based on fatigue cycling to failure of essentially identical specimens). Induction hardening of the surface region in these samples introduces complications to what would have been - except for the relatively high attenuation of thermal neutrons by iron - a straightforward stress determination. In Figure 5, measured hardness versus depth is shown for several positions on the unfatigued shaft [13]. The near-surface region (0 - 3 mm depth) is predominantly tempered martensite, changing to a mixture of pearlite/ferrite at ~10 mm depth. To determine the reference d_0's for the different regions, a third axle was cut by EDM into a 3 mm thick "martensite" ring which included the surface, and a shaft-centered 20 mm diameter pearlite/ferrite plug, each 1 cm wide. It should be mentioned that in the utilization of Eqn. 4 to determine d_0, absolute values of d_0 are not required, only an internally consistent set for the conditions of the experiment.

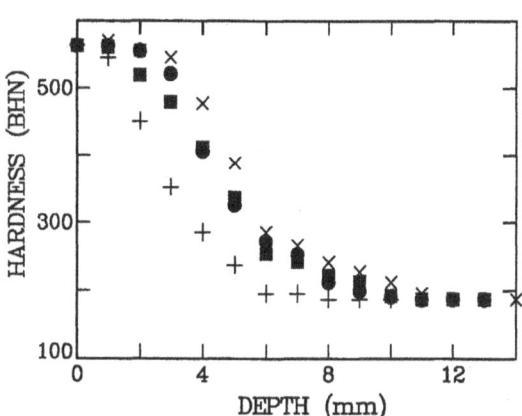

Figure 5. Hardness vs. depth for the induction hardened shafts [13] for various distances for the reference zero line: ● = 10 mm ("B"), X = 25 mm, ■ = 50 mm (interpolated, "D"), + = 110 mm.

For all of the measurements the (110) reflection of steel was employed at a scattering angle of 80⁰ from the sample. The d_0's for the martensite and pearlite/ferrite structures were obtained from the ring and plug pieces, respectively, with 2x2 mm² apertures and the macroscopic equilibrium conditions for axial stresses. Strain measurements were made at a total of seventeen points along two mutually perpendicular diameters - eight points in the ring - from which d_0(ring)= 0.20323 nm and d_0(plug)=0.20303 nm were obtained by requiring balance of axial stresses in the central r-θ plane of each piece. XEC determined by Prevey [12] for the (211) reflection of SAE 1050 steel were used. Current theory indicates that the XEC for the (211) and (110) reflections are identical [10].

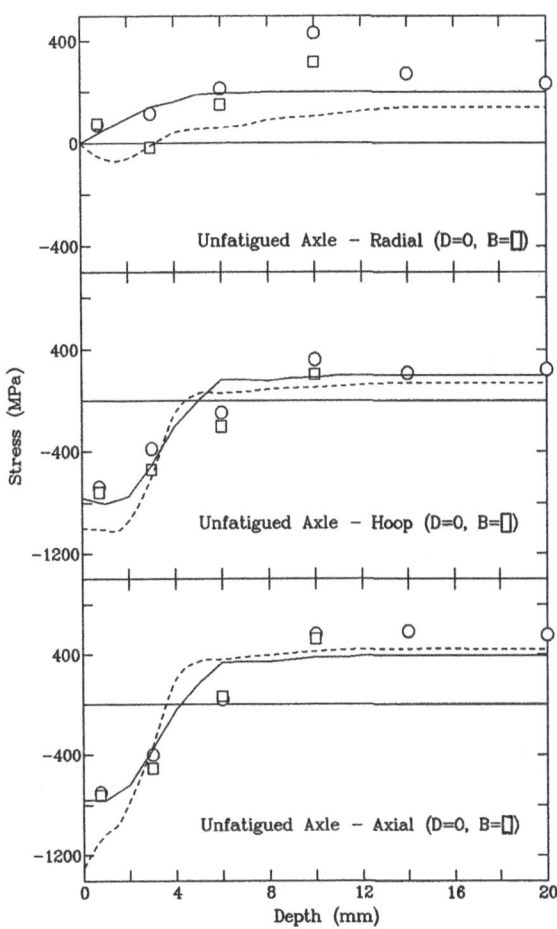

Figure 6. Neutron diffraction determined residual stresses for the unfatigued axle compared to the finite element analysis [14]. The dashed line is the FEA result at 5 mm from the *reference zero line*; the neutron "B" values are at 10 mm (see Figure 4).

512

Because of the high neutron attenuation of steel, 6x6 mm² apertures were used to determine d-spacings at r=0 and r=6 mm in the unfatigued axle at position "D" (50 mm from the *reference zero line*[1] indicated in Figure 4). Other measurements on the two 20 mm radius shafts were made with 2x5 mm² apertures oriented either horizontally or vertically for measurements at r= 10, 14, and 17 mm, and with 1x5 mm² apertures at r=19.2 mm; the 5 mm dimension was parallel to the cylindrical axis of the shaft in either orientation. Data were obtained along single radii at positions "B" (= 10 mm from the *reference zero line*) and "D" in the

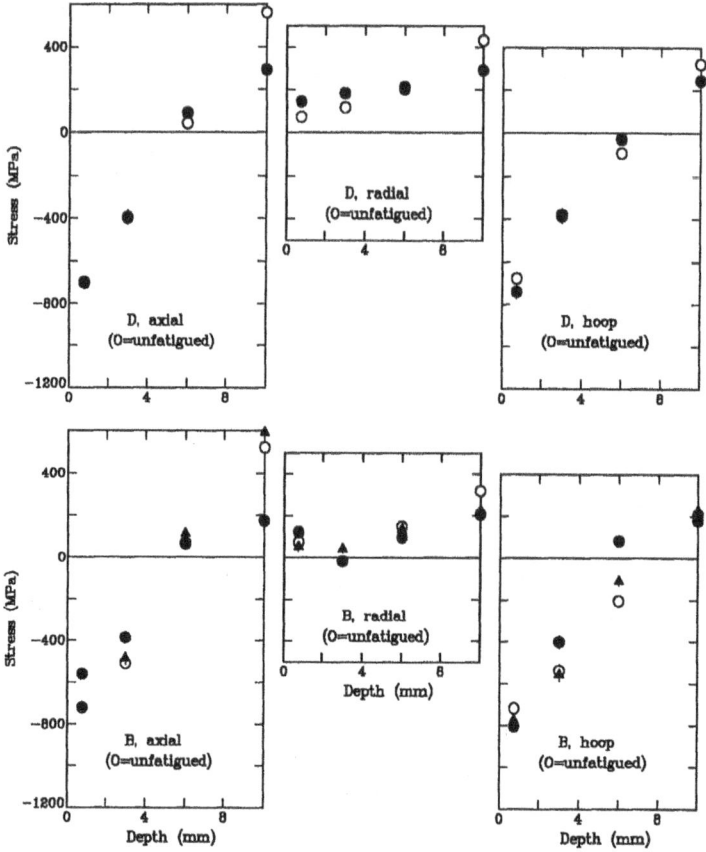

Figure 7. Neutron diffraction determined residual stresses at the 10 mm ("B") and 50 mm ("D") positions for the fatigued (●) and unfatigued (○) samples. The triangles correspond to stresses at a radius 90⁰ to the bending axis.

---

[1]It should be noted that the "B" and "D" positions used here are close to but not exactly those used in the originally reported hardness measurements. In this paper, we define "B" and "D" with respect to the *reference zero line* of Figure 4.

unfatigued shaft; at "B" and "D" on the radii parallel to the bending
direction, and at the "B" position at 90⁰ from the bending direction in
the fatigued axle. All d-spacing data were normalized to Fe(110) of an
iron powder which was scanned for each change of apertures.

A test of the ring-plug d_0's was made using the unfatigued axle data
at the "D" position. Using the hardness vs. depth results of Langner
[13], d_0's for each depth examined by neutrons can be inferred, with
the ring-plug d_0's at the extremes. The hardness at the "D" (= 50 mm)
position was obtained by a linear interpolation of the 25 and 95 mm
measured hardness values. If the approach is correct, axial stresses
measured in the r-θ plane at "D" should balance with no parameter
adjustment for appropriately chosen area segments for each point of
measurement (e.g. the 2x2 mm² beamspot cross-section centered at a
depth of 3 mm is assumed to represent the full 2 mm wide ring of area

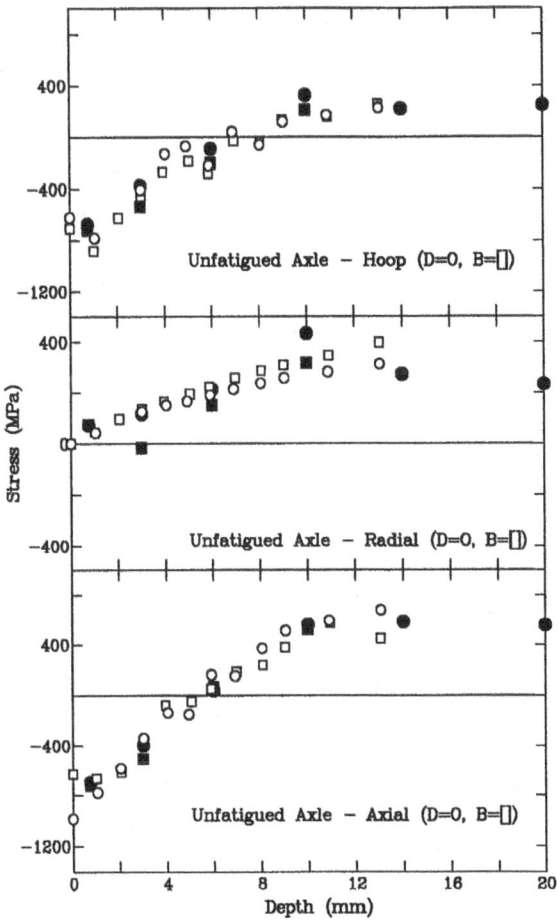

Figure 8. Comparison of the nondestructive, neutron diffraction
determined residual stresses (solid points) and x-ray diffraction
determined stresses using layer removal [12] for the unfatigued axle.

from r=16 to 18 mm of which it is a small part, etc). The axial stresses balance to 10 MPa, which is to say that perfect balance would be achieved if the axial stress determined at each of the six measurement points was shifted by 10 MPA. Since the standard deviation on each measured axial stress value is typically ~20 MPa, this balance-of-stresses represents an excellent confirmation of the approach.

Ochsner [14] has made a finite element analysis (FEA) of an induction hardened steel shaft to obtain the thermally-produced residual stress distribution. The results are shown in Figure 6 along with the neutron-diffraction-determined stresses at positions "B" and "D" in the unfatigued shaft. The effect of fatigue on the residual stress distributions is illustrated in Figure 7 where results for the two shafts are shown at corresponding positions. In Figures 8 and 9 a comparison is made of the results of the nondestructive neutron determination and the x-ray diffraction measurement of residual stress employing layer removal [12]. It should be pointed out that x-ray and neutron measurements were at the same positions on the unfatigued axle, whereas two x-ray measurements on the fatigued axle bracketed the neutron "B" position.

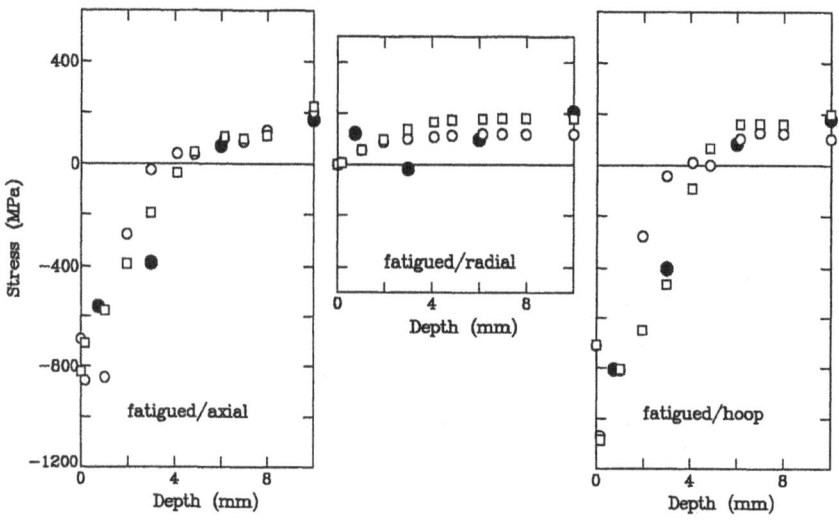

Figure 9. Comparison of the nondestructive, neutron diffraction determined residual stresses (solid points) and x-ray diffraction determined stresses using layer removal [12] for the fatigued axle. The x-ray stress values were obtained at 5 mm on either side of the center of the neutron measurements.

4. Discussion

The neutron diffraction determined residual stresses for the the uranium samples confirm that for heavy metals this technique is the only one that can provide results nondestructively in some cases. The empirical correction utilized with apparent success for the highly swaged sample suggests an avenue to be pursued for the elucidation of pseudomacrostresses in a simple geometry.

The ogive study is a classic case of the need for a nondestructive probe which would not be affected by surface conditions (i.e. threads). Although the correct qualitative results were obtained by one hole-drilling determination close to the threaded region, other contradictory results placed doubts on the determination.

In the case of the steel axles, several aspects are of interest. Overall, the agreement between the nondestructive neutron results and the x-ray diffraction results, obtained employing layer removal, is excellent. Although layer removal for symmetric specimens has been known to be a reliable technique for subsurface stress determination with x-rays, the agreement obtained in this case to a depth of 13 mm seems remarkable. The only point at which there is a significant difference occurs at a depth of 3 mm for the fatigued axle at the "B" position for all stress components (Figure 9), and at the 3 mm depth for the radial stress component at the "B" position in the unfatigued axle. The fact that the neutron results for σ_{rr} at 3 mm depth for both the fatigued and unfatigued axles are slightly compressive at the "B" position, and that these results are in qualitative agreement with the finite element calculation, supports the correctness of the neutron determination at this point.

The FEA results are in reasonably good agreement with the stress measurements for the unfatigued axle. However, except for the radial stress at 3 mm depth, the measured stresses seem to show much less "B"- vs. "D"-position difference than the FEA predicts. No FEA results are yet available for the fatigued-axle case. Experimentally, the measurements at the "D" position show very little stress relaxation with fatigue except at 10 mm depth for σ_{zz}. In contrast, except at this depth, there seems to be a tendency for the axial and hoop stresses to become more tensile with fatigue at the "B" position. It is worth noting that at the "B" position, the stresses measured with neutrons at 90° to the bending direction are very close in magnitude to the unfatigued stresses, as expected.

5. References

1. Choi, C.S., Prask, H.J. and Trevino, S.F. (1979) 'Non-destructive investigation of texture by neutron diffraction', J. Appl. Cryst. 12, 327-331.
2. Allen, A., Andreani, C., Hutchings, M.T. and Windsor, C.G. (1981) 'Measurement of internal stress within bulk materials using neutron diffraction', NDT International, 249-254; Krawitz, A.D., Brune, J.E. and Schmank, M.J. (1982) 'Measurement of stress in the interior of solids with neutrons', in E. Kula and V. Weiss (eds.), Residual Stress and Stress Relaxation, Plenum Press, New York and London, pp. 139-155; Pintschovius, L., Jung, V., Maucherauch, L., Schäfer, R. and Vöhringer, O. 'Determination of

residual stress distributions in the interior of technical parts by means of neutron diffraction', ibid., pp. 467-482.

3. Prask, H.J. and Choi, C.S. (1987) 'Residual stress measurements in armament-system components by means of neutron diffraction', in W.B Young (ed.), Residual Stress in Design, Process and Materials Selection, ASM International, pp. 21-26.

4. Prask, H.J. and Choi, C.S. (1990) 'Residual stress measurements by means of neutron diffraction', Mat. Res. Soc. Symp. Proc. 166, 293-298.

5. Prask, H.J. (1984) unpublished.

6. Evenshor, P.D. and Hauk, V., Z. Metallkde. 66, 167-8 (1975).

7. Choi, C.S., Prask, H.J. and Ludtka, G.M. (1987) 'Neutron diffraction study of age-hardened U-0.75 wt% Ti alloy', J. Nucl. Matls. 150, 85-92.

8. Eckelmeyer, K.H. (1986), Sandia National Laboratory letter report, unpublished.

9. Llewellyn, G.H., et al. (1989) 'Experimental and analytical studies in quenching uranium-0.75% titanium alloy cylinders', Oak Ridge Y-12 Plant Report No. Y-2397.

10. Bollenrath, F., Hauk, V. and Muller, E. (1967), Z. Metallkde. 58, 76-82.

11. Leese, G. and Socie, D. (eds.), (1989) Multiaxial Fatigue: Analysis and Experiment (SAE AE-14), SAE Inc.

12. Prevey, P. (1991) 'X-ray diffraction determination of sub-surface residual stresses in a biaxial fatigue test specimen', April Meeting SAE Fatigue Design & Evaluation Committee Minutes, unpublished.

13. Langner, M. (1988) 'Hardness profiles in a biaxial fatigue test specimen', October Meeting SAE Fatigue Design & Evaluation Committee Minutes, unpublished.

14. Ochsner, J.K. (1988) 'Residual stresses in SAE biaxial fatigue test specimen', April Meeting SAE Fatigue Design & Evaluation Committee Minutes, unpublished.

PROBLEMS WITH RAILWAY RAILS

P. J. WEBSTER, X. WANG and G. MILLS
Department of Civil Engineering
University of Salford
Salford M5 4WT
England

ABSTRACT. Neutron diffraction is now being applied to measure macrostrain distributions inside components of engineering interest. The technique has several unique advantages but there are practical constraints which limit its widespread adoption. Compromises are often necessary to bridge the gap between what is required by engineers for use in stress analysis calculations and what it is practical to measure using the technique, especially when large sized components are involved.

An example is given of how neutron strain scanning may be applied, in a limited but cost effective manner, to the problem of the "non-destructive" measurement of residual stresses in complex shaped large components such as railway rails.

1. Introduction

Strain measuring techniques of most interest and of most use to engineers are those that can be used on-site by technical staff to obtain accurate strain data, non-invasively, cheaply and reliably, preferably with on-line digital and graphical output which may be expressed as either strains or stresses in appropriate units. Not surprisingly it is rare to find a technique which meets all these desirable criteria for solving a particular problem.

The ubiquitous strain gauge has many desirable features when used to measure applied stresses at a surface but has significant limitations when internal or residual stresses are to be measured. For internal stresses theoretical extrapolations of sometimes doubtful validity are required if surface strain gauge data is used. Alternatively if gauges are inserted internally, or if residual stresses are to be measured, stresses in the component must be to some extent unavoidably or deliberately relaxed as the component is partially or substantially destroyed.

The X-ray diffraction technique has the attraction of portability whilst being non-destructive. It is an absolute method used to measure the lattice strain and consequently residual stresses can in principle be determined without relaxation if the stress free lattice spacing is known. It is very effective for surface measurements when used in the backscattering mode but is not suitable for internal measurements due to strong absorption.

Neutron strain scanning is now being used to determine residual stress distributions inside components of engineering interest. The neutron technique has several unique advantages but there are practical constraints which limit its range and rate of application. It is, like the X-ray method, an absolute technique which can be used non-destructively to measure residual stresses, but in contrast to X-rays neutrons can penetrate several centimetres into many engineering materials and can be used directly to measure internal

M. T. Hutchings and A. D. Krawitz (ed.),
Measurement of Residual and Applied Stress Using Neutron Diffraction, 517–524.
© 1992 *Kluwer Academic Publishers.*

518

strains. Its principal restriction is that an intense neutron source is required and these are not portable. All neutron strain measurements must be made at central neutron facilities to which the component must be transported. Additionally, handling and absorption constraints restrict the size and shape of the component that can be economically measured.

Compromises are often necessary to bridge the gap between what is desired by engineers for use in stress analysis calculations and what it is practical to measure using any technique, and this also applies to neutron scanning especially when large or awkward shaped components are involved.

The measurement of residual stresses in railway rails provides a typical example of an engineering problem where an indirect approach is required. Railway rails in use are very long, their cross-section is a complex shape and in the thickest part, the head, attenuation of the neutron beam is severe. However, it is possible to make extensive but limited neutron measurements that are of value to the engineer.

2. The Neutron Strain Scanner

The neutron technique and strain scanning have been described in detail elsewhere [1-8]. A diagram showing the principal features of a neutron strain scanner as used on a steady state source is shown in figure 1.

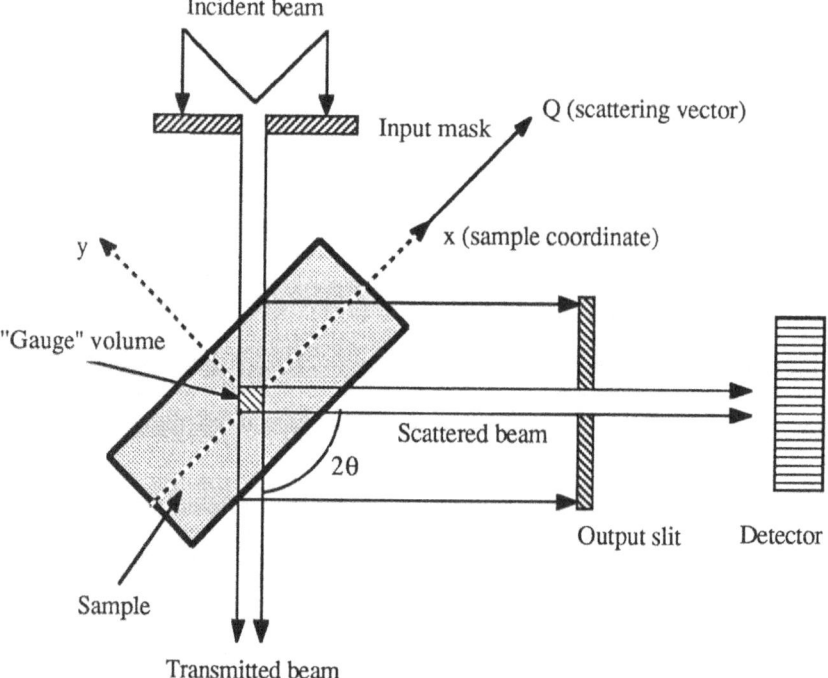

Figure 1.   Principles and outline of the neutron strain scanner.

In a neutron strain scanner the "gauge volume" is defined by absorbing masks in the incident and output beams placed as close as possible to the sample to minimise the defocussing effects of beam divergence. The direction of the strain measurement is along Q which bisects the angle between the incident and diffracted beams. Different strain directions are selected by rotating the sample about appropriate axes. Different locations in the sample are investigated by translating the sample through the gauge volume using a combination of horizontal and vertical orthogonal translations.

2.1. PROBLEMS AND CONSTRAINTS

If the sample is reasonably small and symmetrical in shape series of scans along principal directions generally pose few problems and comprehensive and precise data may be efficiently collected. On the other hand, if the component is large, heavy, and an awkward shape the problems may be severe. The masks may have to be withdrawn radially to enable the component to be mounted on the scanner, reducing the spatial resolution. The weight may exceed tolerance limits and make translation and rotation, particularly about an axis that is not vertical, difficult. In addition, if size, shape and weight are a manipulation problem there is often severe neutron beam attenuation. It will require a considerable time to collect data with even poor statistical accuracy and subsequent calculations will be subject to large errors. Enlarging the gauge volume can reduce counting times but only at the cost of reduced spatial resolution which may involve unacceptable loss of detail.

2.2. THE PRACTICAL ECONOMIC APPROACH

When sample size and shape make it impractical economically to collect strain data on the complete component it may be better to make series of measurements on a number of sections cut from the component, to which strain gauges have been attached before cutting, and then to deduce the stresses in the original component by calculation. Some of the original stresses will have been partially relaxed, but some of the relaxation will have been recorded by the strain gauges and if a careful choice is made when sectioning it is often possible substantially to reconstruct the original stress distribution. If necessary, confirmatory measurements may subsequently made at chosen points on a full-sized uncut component.

The measurements may not be exactly what the engineer would ideally specify but they are probably good enough for most practical purposes and far better than what is possible to achieve by most other techniques.

3. Measurements on Railway Rails

Residual stresses are generated in railway rails during the manufacturing process as a result of plastic deformation caused by differential cooling rates and phase changes that occur after hot rolling and heat-treatments and then by any subsequent roller straightening. In service the initial residual stresses are modified by further repeated plastic deformations caused by the weight of passing trains. After an initial "shakedown" period characteristic residual stress patterns evolve. Eventually these may lead to fatigue failures such as squats, taches ovales, shelling or perhaps in extreme cases fracture and possible train derailment.

Railway engineers, if they are to predict and prevent such failures, need to know the condition of rails and the residual stresses within them. Rails are regularly inspected on site visually and by acoustic or electromagnetic sensors which are designed to detect cracks at an early stage. However, the measurement of internal residual stresses invariably involves the removal of a section of rail for laboratory measurement. This creates the following problems which are typical of most measurements on large components.

    a   It is expensive and inconvenient to have to remove a sample section of rail from a length of track.

    b.   The rail sample should be at least twice as long as its height or width so that, although longitudinal stresses at the ends will be zero, the stresses at the centre will be substantially the same as in a long rail in situ. This usually implies a minimum length of about 350 mm.

    c.   The weight and shape of a 350 mm long steel rail sample make it difficult to manipulate in a neutron strain scanner and to define the "gauge volume" precisely.

    d.   Neutron beam attenuation through the material thickness on a long rail precludes economically viable measurements in the centre of the rail head and in longitudinal directions in some other regions.

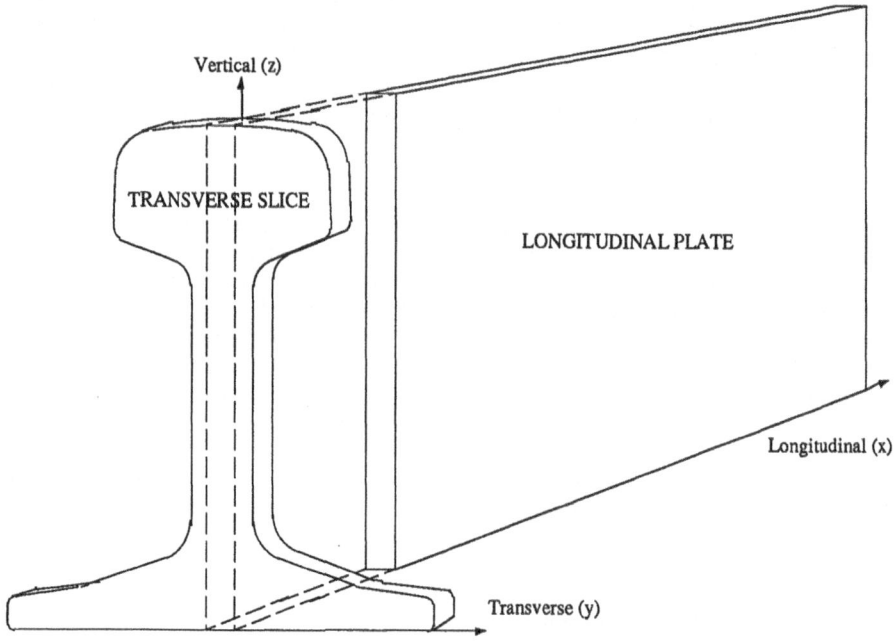

Figure 2.   Transverse slice and longitudinal plate samples cut from a railway rail.

The alternative, economic approach is to cut transverse slice and longitudinal plate sections from the rail sample as indicated in figure 2. If the thicknesses are not more than about 10 mm then attenuation of the neutron beam will be relatively low and it should be possible to obtain economically a large number of accurate strain measurements in the three principal orientations and then to derive the residual stresses throughout the sections. In the transverse slice the longitudinal residual stresses will have been substantially relaxed but the transverse and vertical stresses, although modified, should not differ to any great extent from those in the uncut rail. In the longitudinal plate the longitudinal and vertical stresses will have suffered the least relaxation but the the transverse stresses will have been substantially relaxed. In both sets of measurements the vertical stresses measured down the centre lines should be similar to those in the uncut rail because in neither case has the vertical dimension been reduced by cutting. On the other hand the longitudinal and transverse pairs of measurements should illustrate the extent of the relaxation caused by the sectioning [9-10].

Figures 3, 4 and 5 show respectively the longitudinal, transverse and vertical stresses, in transverse and longitudinal sections cut from a used railway rail, derived from the corresponding strains measured by neutron scanning down the centre lines of the two 10 mm thick slice and plate sections.

In figure 3 the longitudinal residual stresses measured in the plate show a characteristic and pronounced variation. There is tension in the foot and in the head with balancing compression in the web. Near the running surface at the top of the head there is a thin layer of traffic induced compression which rapidly reduces to become tension in the centre of the head. The corresponding longitudinal stresses measured in the transverse slice show minimal variation. They have been almost completely relaxed with only a small amount of compression at the extremities and a low and essentially constant tension in between. There is no apparent similarity between the two patterns.

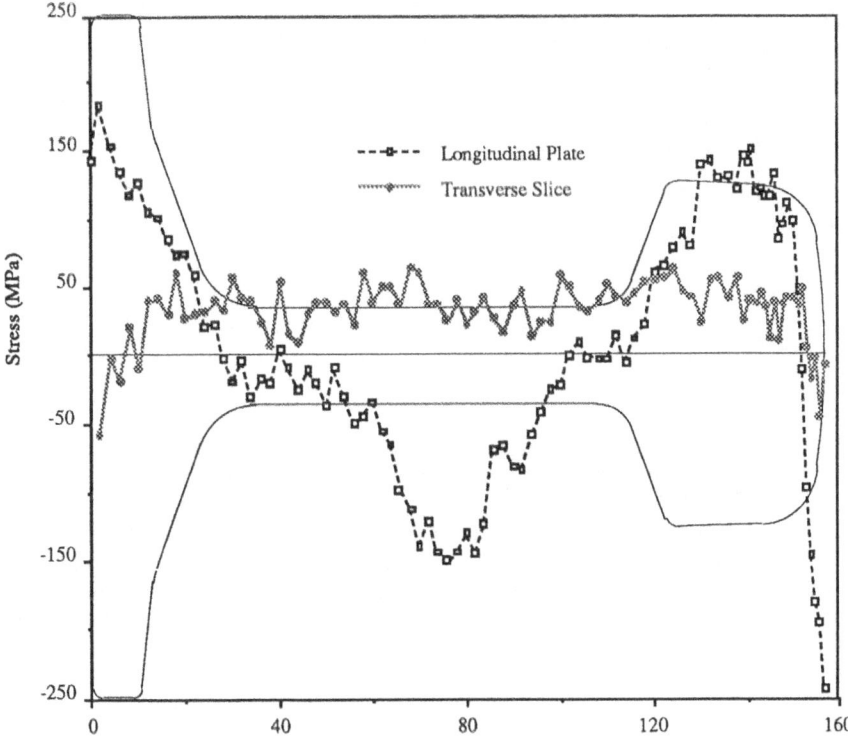

Figure 3.   Longitudinal stresses measured down the centre-lines of transverse slice and longitudinal plate samples cut from a used railway rail.

522

The transverse stresses shown in figure 4 have a less pronounced variation than the longitudinal stresses. In the slice there is some compression in the foot and low stresses in the web. In the head there is once again tension in the centre region and compression just below the running surface. In the plate the low stresses in the foot and web are virtually unchanged but there is evidence of substantial relaxation in the head where much of the thickest section has been cut away.

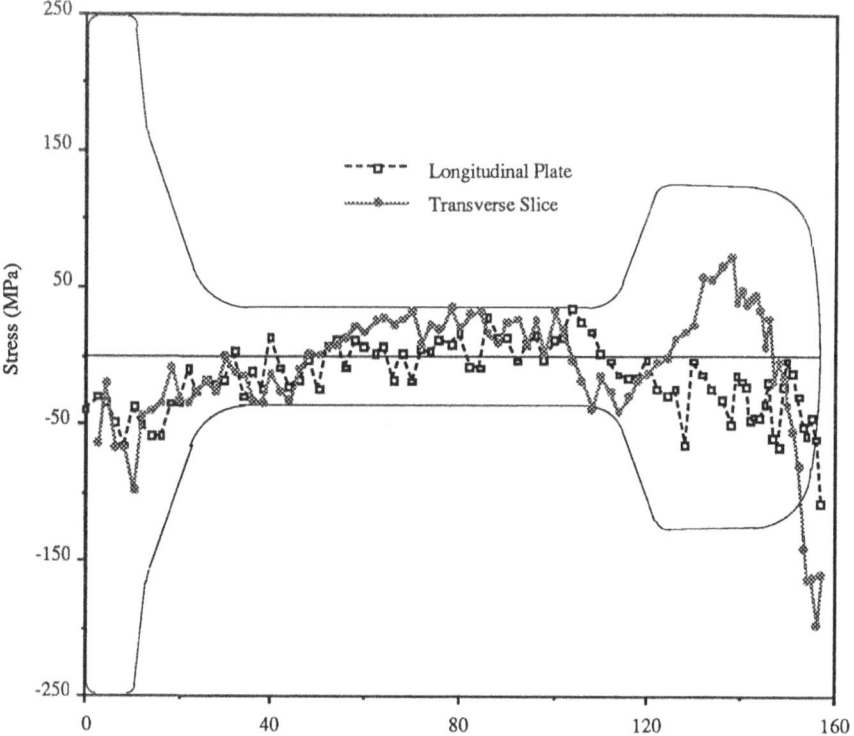

Figure 4.   Transverse stresses measured down the centre-lines of transverse slice and longitudinal plate samples cut from a used railway rail.

The vertical stresses are shown in figure 5. As expected they show the least evidence of stress relaxation although the two patterns do differ due to the rebalancing of stresses that must occur on sectioning. In the foot the two patterns are very similar but in the web and head they diverge to give a significant shift between the two patterns whilst retaining a recognisably related profile.

4.   Discussion

It is not often practical to make extensive neutron scanning measurements on most large thick samples because of the difficulties associated with precise manipulation and the errors introduced by severe neutron beam attenuation. However, cost effective detailed and

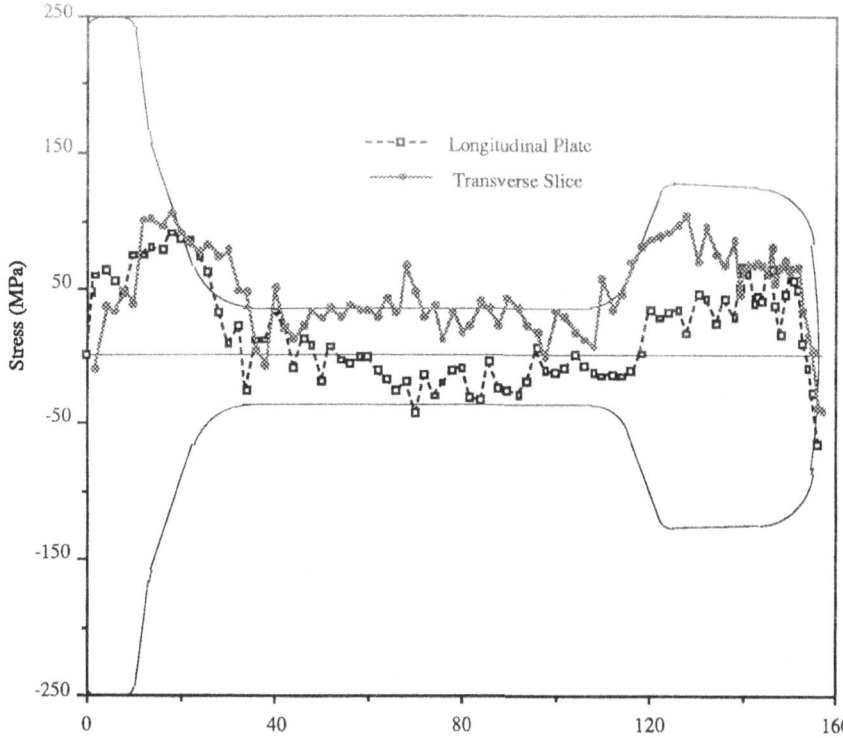

Figure 5.   Vertical stresses measured down the centre-lines of transverse slice and longitudinal plate samples cut from a used railway rail.

precise measurements can be made on relatively thin sections. The results are not exactly what are required by the engineer, as important stresses will have been relaxed and the component has to be partially destroyed, but the combined data can provide unique and useful information for comparison with computer models. If necessary validation measurements can then be made at specific locations on an uncut component to quantify the extent of any relaxation effects. In the case of railway rails a combination of measurements on transverse and longitudinal sections enables a detailed, quantitative but somewhat distorted, pattern of the residual stresses in the principal directions to be obtained over the entire rail profile.

5. Acknowledgements

Much of the work discussed here has involved the use of facilities at the ILL with the involvement and support of the Directors and staff. The railway rails were supplied by British Rail as part of a Co-operative project with Imperial College. Financial support was provided by the UK Science and Engineering Research Council via grants GR/D63196, GR/E55556 and GR/F02427.

6. References

1. Allen, A., Andreani, C., Hutchings, M. T. and Windsor, C. G. (1981) 'Measurement of internal stress within bulk materials using neutron diffraction' NDT International 14, 249-254.
2. Pintschovius, L., Jung, V., Macherauch E. and Vohringer, O. (1983) 'Residual stress measurements by means of neutron diffraction' Mater. Sci. Eng. 61, 43-50.
3. Allen, A., Hutchings, M. T. and Windsor, C. G. (1985) 'Neutron diffraction methods for the study of residual stress fields' Adv. Phys. 34, 445-473.
4. Stacey, A., MacGillivray, H. J., Webster, G. A., Webster, P. J. and Ziebeck, K. R. A. (1985) 'Measurement of residual stresses by neutron diffraction' J. Strain Analysis 20, 93-100.
5. Krawitz, A. D. and Holden, T. M. (1990) 'The measurement of residual stresses using neutron diffraction' MRS Bull. XV, 57-64.
6. Lorentzen, C. (1988) 'Non-destructive evaluation of residual stresses by neutron diffraction' NDT International 21, 385-388.
7. Webster, P. J. (1990) 'The neutron strain scanner: a new analytical tool for engineers' Steel Times 218 No. 6, 321-323.
8. Webster, P. J. (1991) 'Neutron strain scanning' Neutron News (In Press).
9. Webster, G. A., Webster, P. J., Bourke, M. A. M., Low, K. S., Mills, G., MacGillivray, H. J., Cannon, D. F. and Allen, R. J. (1990) 'Neutron diffraction determinations of residual stress patterns in railway rails' Int. Conf. on Residual Stress in Rails: Effects on Rail Integrity and Railroad Economics, Krakow, April 1990 (In Press).
10. Webster, P. J., Low, K. S., Mills, G. and Webster, G. A. (1990) 'Neutron measurement of residual stresses in a used railway rail' Mat. Res. Soc. Symp. Proc. 166, 311-316.

NEUTRON MEASUREMENTS OF RESIDUAL STRAIN IN SOME
TECHNOLOGICAL MATERIALS AND COMPONENTS.

G.ALBERTINI[1], M.CERETTI[2], R.COPPOLA[3], A.LODINI[4],
M.PERRIN[5], F.RUSTICHELLI[2]

1)   Dipartimento di Scienze dei Materiali e della
     Terra, Università, 60131 Ancona (Italy)
2)   Istituto di Fisica Medica, Università, 60131
     Ancona (Italy)
3)   ENEA Casaccia -Roma (Italy)
4)   Lab.Sciences des Materiaux, Ecole Superieure
     d'Ingenieurs en Packaging, Universite', Reims
     (France)
5)   Laboratoire Leon Brillouin, Saclay
     (France)-(Lab.commun. CEA-CERN)

ABSTRACT.   Strain in materials for new applications is
investigated by using neutron diffraction. In particular the
strain induced by cyclic thermal treatment is studied in two
components of AISI 316L steel, considered as mock-ups of
first wall for nuclear fusion reactors. Results concerning a
TIG welded AISI 304 steel are also reported: the strain in a
fixed direction is studied inside the weld and in the
thermally affected region around it, by considering points
along two lines in a given plane. These results are a part
of a wider study aiming to obtain an iso-strain map.

1. INTRODUCTION

Neutron diffraction technique for the study of residual
strain is assuming an increasing importance in the last
years, in particular for what concerns the technological
materials.
     In this paper we present the status of some experiments
on that subject, which are performed in the frame of
research activities developed at the University of Ancona
(Italy), in collaboration with other institutions.
     Some of the investigations are connected with
structural materials to be used in fusion reactors in
general and in the "first wall" in particular.

M. T. Hutchings and A. D. Krawitz (ed.),
Measurement of Residual and Applied Stress Using Neutron Diffraction, 525–534.
© 1992 Kluwer Academic Publishers.

2. STEEL COMPONENTS FOR FUSION REACTORS

In Europe the main fusion project is NET (Next European
Torus), a controlled fusion reactor of the TOKAMAK type,
using magnetic confinement. One of the most critical
components of the reactor is the so called "first wall",
that is the first solid structure, directly exposed to flux
of energy (high energy neutrons and ions, electromagnetic
radiation) produced by the fusion reaction.
    Among the stresses to which the first wall is
submitted, the cyclic thermo-mechanical ones, connected to
the pulsed nature of the machine, are to be considered. They
are induced in the component by two opposite effects:
heating by high energy particles and cooling by coolants.
    In the framework of the European program for fusion
technology, a variety of mock-ups differing in size,
geometry, and manufacture technology are at present being
studied under thermal fatigue test, in the aim of providing
the NET design team with experimental evidence of the first
wall behaviour under the effect of this critical failure
mechanism [1-3].
    In our case the behaviour under thermal fatigue of
first wall mock-ups of AISI 316L steel, the reference
material for the first wall of the experimental reactor, was
studied.

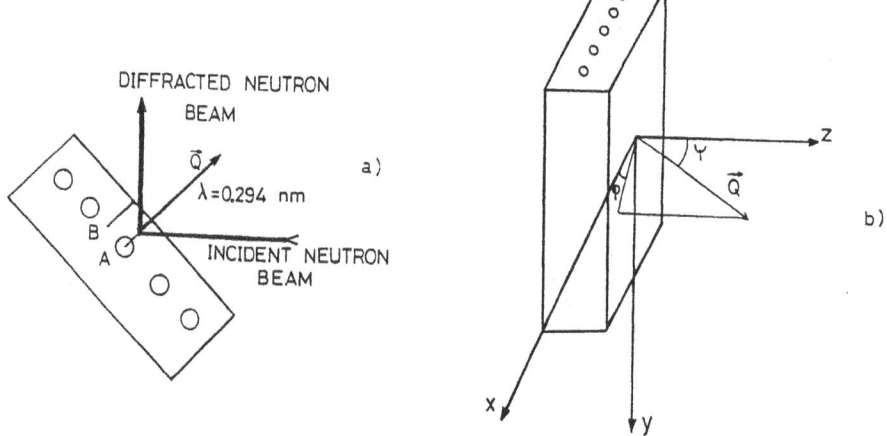

Figure 1.   Sketch of the first wall mock up with simplified
geometry (a) and geometry of the experiment (b).

    A component of simplified geometry was first
considered [4]: a parallelepiped with dimensions
300 x 78 x 44 mm^3. Five parallel channels having internal
diameter of 8mm were drilled along the longest side (fig.1).
It was submitted to cyclic surface heating by a device,
which simulates the flux from plasma, and cooling by a
coolant flowing through the parallel channels. The coolant

was low pressure water at an inlet temperature of 10°C. The sample was submitted to 27000 thermal cycles in an experimental station which was built up to this purpose at the JRC Ispra (Italy)[5]. The thermal flux varied from 0 to 500 KW/m² at a rate of 5mHz.

Neutron measurements were performed at the two axes diffractometer G4.2 of the Laboratoire Leon Brillouin in Saclay (France). A 0.294 nm neutron wavelength was chosen in order to obtain a Bragg diffraction angle $\Theta=45°$ from the {111} planes.

The incident neutron beam, having a 30 mm high and 1.3 mm wide cross section, and the beam diffracted at $2\Theta\approx 90°$, having a 30 mm high and 1.3 mm wide cross section, determined the volume under investigation. The dimensions of that volume were considered as optimal in order to achieve a good spatial resolution and a high diffracted neutron flux.

The geometry of the experiment is shown in fig. 1. The direction of the exchanged momentum Q with respect to an orthogonal reference system (having y axis parallel to the cooling channels and z axis perpendicular to the sample surface) is given by the Ψ and Φ angles defined in fig. 1.

The reference interplanar distance d_{111} was first determined by considering an AISI 316L piece not strained and belonging to the same bar of the stressed material.

Due to the geometry of the component, a two dimensional strain distribution was assumed and thus neutron measurements were performed at $\Phi = 0$.

Two sets of measurements were performed: one along the z direction from the surface to the axis of the central hole (line A) and the other along a line parallel to the former and crossing the central point between two axes: the axis of the central hole and the axis of the next hole (line B). Five points at different depth were considered for each set. At each point different orientations Ψ were considered, in order to evaluate the strain components by:

$$\epsilon(\Phi=0,\Psi)= \epsilon(\Psi) = \sin^2\Psi \cdot (\epsilon_{xx} - \epsilon_{zz}) + \epsilon_{xz} \cdot \sin(2\cdot\Psi) + \epsilon_{zz}$$

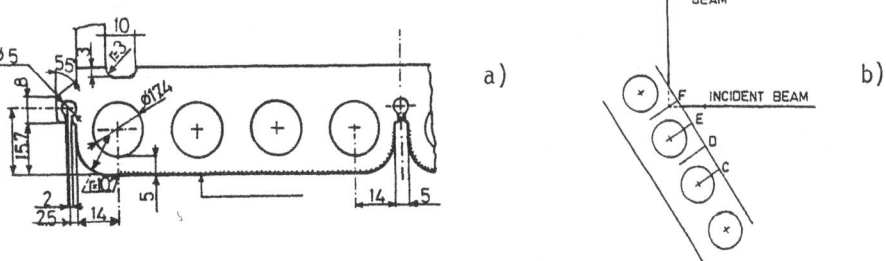

Figure 2. Sketch of the first wall brazed mock up (a) and investigated regions C, D, E and F (b).

528

As a second step in the study of effects induced by cyclic variation of heat flux on AISI 316L components, a more realistic first wall mock-up was considered [6]. The channels for cooling circuit were obtained by brazing two parts. Fig. 2a shows the region studied by neutron diffraction, the overall dimensions of the component being about 30 cm, 60 cm, 90 cm and its weigth being about 60 kg.

Also in this case the strain map was supposed to be two dimensional and thus the Q vector was kept perpendicular to the channel axes. Two directions of Q were considered in this plane: the parallel and the perpendicular to the surface.

Measurements were performed along lines from the surface to the axis of a channel (line C and E in fig.2b) and from the surface to an interassial region (line D and F in fig.2b).

The two components ϵxx and ϵzz were thus directly measured.

The obtained $\epsilon(\psi)$ along the line B are reported vs. $\sin^2\psi$ in fig. 3a-e. In all cases $\epsilon_{xz}=0$: that means that x and z are axes of principal deformation and no shear deformation is present.

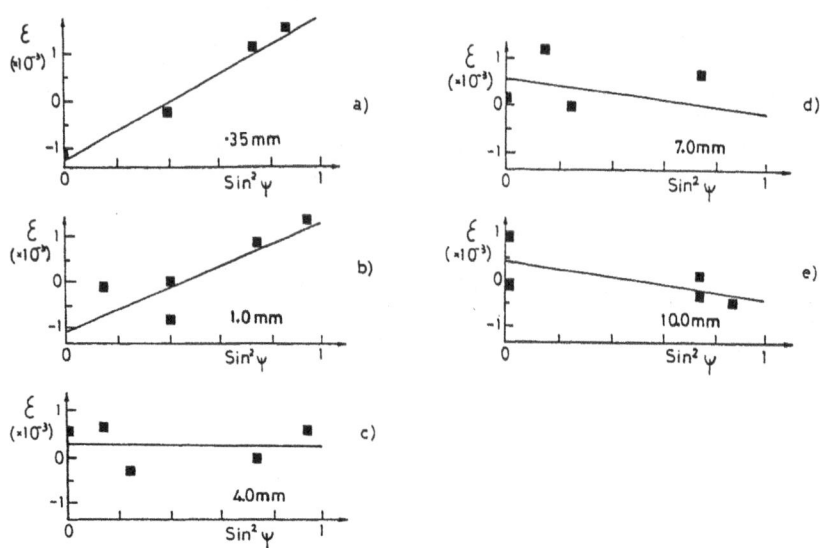

Figure 3. Measured values of ϵ vs. $\sin^2\psi$ ($\Phi=0$) along the line B at depths $-z=$ .35 mm (a), 1.0 mm (b), 4.0 mm (c), 7.0 mm (d) and 10.0 mm (e).

The ϵxx and ϵzz in each point were obtained by considering that $\epsilon(\Phi=0,\sin^2\psi=0)=\epsilon zz$ and $\epsilon(\Phi=0,\sin^2\psi=1)=\epsilon xx$. The so obtained values of the two components are reported in fig.4 as functions of the depth along the B line.

In order to compare the results with those expected from theoretical calculation, a mapping of the x and z elastic residual strain of the component was performed by means of the finite element technique. The numerical analysis was performed using a 3D analytical model formed of 2667 nodes and 448 twenty-nodes elements. The elastic strain was then deduced from the isostress contour plot [4].

The behaviour of ϵxx and ϵzz along the B line expected from numerical calculation is also indicated in fig.4.

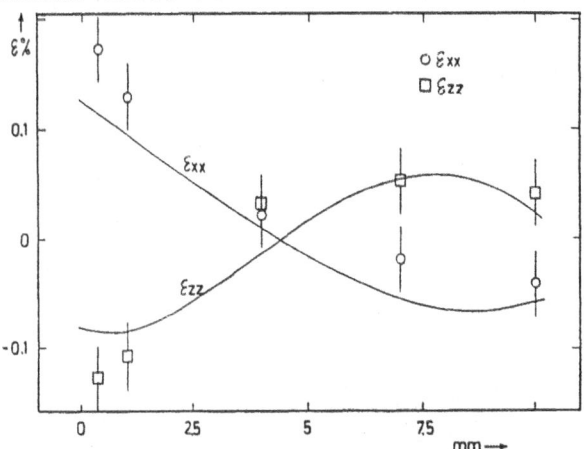

Figure 4. Values of ϵ_{xx} (circle) and ϵ_{zz} (square) as functions of depth along the line B (from surface to the central point between two succesive holes). The lines represent the expected behaviour according to numerical calculation.

The agreement between theory and measurements is satisfactory, also considering that the theoretical values are calculated at the end of the cooling phase and not for a sample in isothermal equilibrium at room temperature and that the component was also subjected to thermal variations, not considered in the numerical calculation, connected to preparation and handling.

We remark that neutron diffraction gives information on the actual state of the component, which includes also the effects of undesirable thermal variations, not easy to be considered in the theoretical computation. That capability of the neutron method, together with the high penetration power of the neutron beam make the neutron diffraction a refined tool for studying residual strains, expecially in the case of welding or soldering.

3. NEUTRON MEASUREMENTS IN TIG WELDED AISI 304 STEEL

The knowledge of residual stresses in welding zones is of

530

great practical importance. In fact, internal strains can arise in welding zones during the solidification, owing to the volume reduction, which is resisted by the surrounding cooler metal.

Generally strong stress gradients can be present in the vicinity of the welding, even if thermal treatments are performed in order to reduce these tensions. From this point of view the neutron diffraction technique provides a powerful tool for accurate and non destructive measurements of the internal stress field.

A particularly interesting method for the welding control is obtaining a map of the "iso-strains" through a given plane [7]; it requires a two-dimensional movement of the tested point with respect to the sample and only one strain direction is considered.

Recently, we have evaluated the residual strain in a TIG weld in order to obtain the map of the iso-deformations [8].

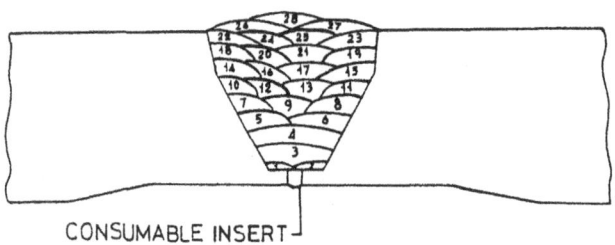

CONSUMABLE INSERT

Figure 5. Preparation of the TIG weld and pass sequence of the weld in AISI 304 steel.

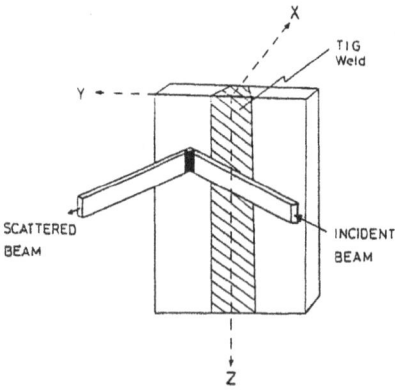

Figure 6. Geometry of the experiment in the welded component.

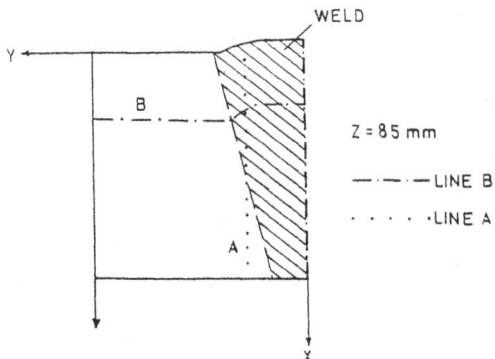

Figure 7. Lines of investigation A and B in the TIG weld.

The analysed sample was a section cut from an head-to-head weld between two part of an AISI 304 steel tube, as shown in fig. 5. Dimensions are 135 mm high, 175 mm long and 35 mm thick, were 135 mm and 35 mm are the dimension of the welding cross section.

The chemical composition of AISI 304 steel is reported in table 1. The low carbon content of this austenitic steel assures of a good corrosion resistance in welded structures.

The preparation of the weld and the typical weld pass sequence are shown in figure 5: fusion of the consumable insert occured first; then of two TIG passes, and, at last, of 26 welding electrodes passes.

The coordinate system in the sample is shown in fig. 6. The volume under investigation ($2 \times 2 \times 20$ mm^3) is determined by the intersection of the incident and diffracted beam.

TABLE 1. Weigth composition of AISI 304 steel

Element	wt%
Cr	18.12
Ni	10.56
Mn	1.37
C	0.069
Mo	0.37
S	0.014
P	0.027
Si	0.34
Fe	bal

Neutron diffraction experiments were carried out at the LLB (Laboratoire Commun. CEA-CNRS, Saclay France) with

the two axes spectrometer G4.2 situated on a cold neutron
guide of the reactor Orphée.

A neutron wavelength of 0.33 nm was use. The
reflection (111) was considered in order to evaluate the
lattice parameter.

The Bragg diffraction peaks consisted of about 30
experimental points, spaced 0.1 degree each other. In order
to have statistical errors of the same order of magnitude,
different measurement times were used at different depths.
The interplanar distance d was calculated by fitting the
experimental data with a gaussian curve. The error on the
peak centre position is around 1.7×10^{-5} nm.

The lattice deformation ϵ were calculated, by using
the equation:

$$\epsilon = (d-d_0)/d_0$$

where d_0 is the reference unstrained parameter. This latter
was measured from a powder of AISI 304 stainless steel,
obtained from a piece belonging to the same bar of the
sample.

The direction of the measured deformations is that of
the exchanged wave vector Q: the direction perpendicular to
the lateral surface of the sample, i.e. parallel to the y
axis, was selected.

Two sets of data are reported, which concern
measurements in a plane having the Z value constant and
lying in the central part of the component, in order to
avoid boundary effects. One set corresponds to points along
a line parallel to the weld section axis at about 1 cm from
it (line A of fig.7) and the other along a line parallel to
the back surface contour at 1 cm from it (line B of fig.7).

In the latter case (line B), seven significant
positions at different distance from the weld centre were
analyzed. In the former case (line A), different points were
considered, each other spaced by 2 mm, the first analysed
positions being inside the weld zone, while the others are
in the thermally altered zone of the sample.

The weld was considered symmetrical with respect to
the XZ plane. The central part was not explored more in
detail, in order to avoid the problems connected with its
high absorption.

The obtained deformation values along the lines A are
reported in fig. 8 as a function of the thickness; those
along the line B are reported in fig. 9 as a function of the
distance from the weld center.

It is evident, according to fig. 9, that the tension
strains increase approaching the weld centre. The strain
inside the weld increases from the surface to the inner
part, as shown in fig. 8, and then decreases passing from
the weld to the thermally affected zone.

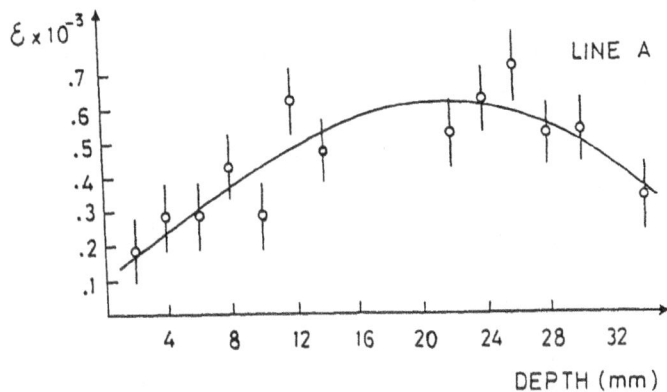

Figure 8. Єz deformation as a function of the distance from the front surface, along the line A. (The line is a guide to eye).

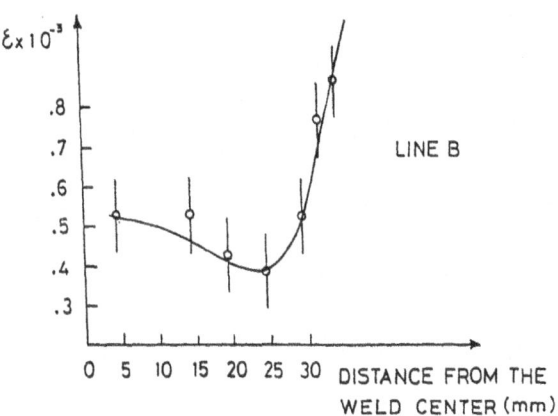

Figure 9. Єz deformation as a function of the distance from the weld center, along the line B. (The line is a guide to eye).

REFERENCES

[1]   Guerreschi, U., Cardella, A. and Matera, R. (1985) "Fabrication, Quality Control and Thermal Fatigue Testing of a Panel Type First Wall", Fusion Technology 1, 285, Pergamon Press, Oxford.

[2]   Avanzini, P.G., Brossa, F., Guerreschi, U., Lazzaretti, M., Persano Adorno, G., Simbolotti, G.,

534

Zampaglione, V., Vieider, G. and Matera, R. (1990) "Technological aspects of the fabrication, control and testing of a representative first wall panel for NET", 16th Symposium on Fusion Technology, London, September 3-7.

[3]   Besson, D., Michel, B., Archer, J., Libin, B., Matera, R. and Vieider, G. (1990) "NET First Wall Test Sections", 16th Symposium on Fusion Technology, London, September 3-7.

[4]   Albertini, G., Andre', G., Matera, R., Miele, P., Merola, M., Perrin, M. and Rustichelli, F. (1991) "Residual strain in fusion reactor first wall component subjected to thermal fatigue", Materials Science & Engineering - Submitted.

[5]   Matera, R., Merola, M., Biggio, M., Cicchetti, E., Renda, V. and Eto, M. (1990) "Behaviour of First Wall components under Thermal Fatigue", J. of Nuclear Materials - In press.

[6]   Albertini, G., Ceretti, M., Coppola, R., Lodini, A., Mariani, P., Perrin, M., Rustichelli, F., Vieider, G. and Jakeman, R.R., "Residual strain in AISI 316L first wall mock-up" - To be published.

[7]   Webster, Peter J. (1990) "The neutron strain scanner: a new analytical tool for engineers", Steel Times, June.

[8]   Albertini, G., Ceretti, M., Ghia, S., Lodini, A., Mariani, P., Perrin, M. and Rustichelli, F., "Residual strain map of a TIG welded component in AISI 304 steel, obtained by neutron diffraction" - To be published.

DEVELOPMENT OF THE NEUTRON DIFFRACTION TECHNIQUE FOR THE DETERMINATION OF NEAR SURFACE RESIDUAL STRESSES IN CRITICAL GAS TURBINE COMPONENTS

A.N. EZEILO[+], P.S. WEBSTER [++], G.A. WEBSTER[+] and P.J. WEBSTER[+++]

+   *Department of Mechanical Engineering,*
        Imperial College, London, SW7 2BX, UK

++   *Rolls-Royce plc, P. O. Box 31, Derby, DE2 8BJ, UK*

+++   *Department of Civil Engineering,*
        University of Salford, Salford, M5 4WT, UK

ABSTRACT. Near surface residual stresses contribute significantly to the life of structural engineering components. A method of producing compressive residual stresses in the surface region of components to give improved resistance to fatigue fracture is shot peening. Neutron diffraction methods are successfully being used to determine through thickness engineering strains in polycrystalline materials. In this paper the results are reported of measurements that have been made on shot peened samples of a nickel superalloy. Good instrumental characteristics, sample positioning in the beam and the correction of near surface data have resulted in the accurate determinations of the residual stress fields. It is observed that the stress profile has a characteristic shape irrespective of material or of the peening intensity employed, and also that the distribution of stresses balancing the surface compressive field results in a peak tension of approximately one third of the maximum surface compression.

1. Introduction

In the quest for improved performance from aeroengines (without compromising safety) it is imperative to utilise the maximum capability of a given material. This requires a full understanding of the behaviour of the material and a key factor in achieving this objective is the ability to characterise the residual stress state and surface condition of each component. This is illustrated in the following example. Figure 1 shows a typical Gas Turbine High Pressure Compressor. The lowest weight design solution is one in which individual compressor discs are welded together to form an assembly . Undesirable residual stresses are produced as a result of the welding operation and so a post weld heat treatment is prescribed to reduce their effect. Table 1 demonstrates the influence of two different post weld heat treatment conditions (time, temperature) on fatigue performance. By increasing the heat treatment temperature (and thereby reducing harmful tensile residual stresses) a sixfold increase in fatigue life was obtained.

Another source of residual stresses results from the machining operation. In this instance machining has produced beneficial compressive stresses at the surface and the

M. T. Hutchings and A. D. Krawitz (ed.),
Measurement of Residual and Applied Stress Using Neutron Diffraction, 535–543.
© 1992 *Kluwer Academic Publishers.*

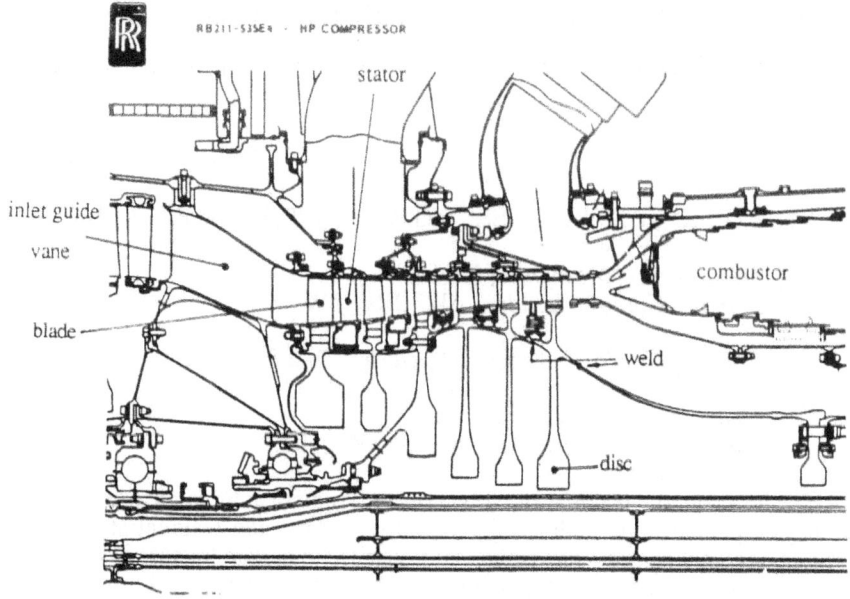

Figure 1 RB211-535E4 - High Pressure Compressor

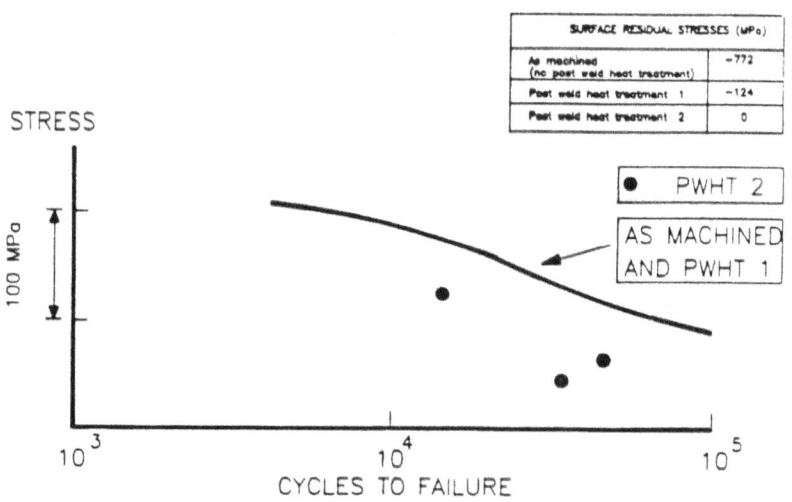

Figure 2 Reduction in surface fatigue strength resulting from Post Weld Heat Treatment

Table 1 Welding residual stress Effects (titanium alloy welded compressor drum)

Titanium alloy welded compressor drum

Post Weld Heat Treatment	1	2
Residual Stress (MPa)	+463	+232
Rig Test Cyclic Stress Range (MPa)	+324	+324
Rig Test Fatigue Life	X Cycles	6X Cycles

effect of the post weld heat treatment is to diminish the benefit. It was, however, necessary to choose the second post weld heat treatment to obtain adequate fatigue performance from the weld region, Figure 2 indicates that a reduction in surface fatigue strength results from this decision. This example serves to illustrate the necessity of making a full evaluation of residual stresses from all sources and of having a thorough understanding of their influence.

Other steps can be taken to modify residual stresses. Shot-peening for example, is one well attested method of introducing compressive stresses into the surface of components. For instance, in the previous example if the resultant fatigue life had been inadequate, then this would have been a worthwhile consideration. Figure 3 indicates, for a nickel base superalloy, the influence that shot-peening can have on fatigue performance. What this highlights is the importance of having a capability to investigate non-destructively the surface condition to;

a) ascertain the stress conditions which are detrimental to fatigue performance.
b) determine optimum surface treatment parameters
c) understand how residual stresses are affected by service environments.
d) incorporate this information explicitly into assessments of component fatigue lives.

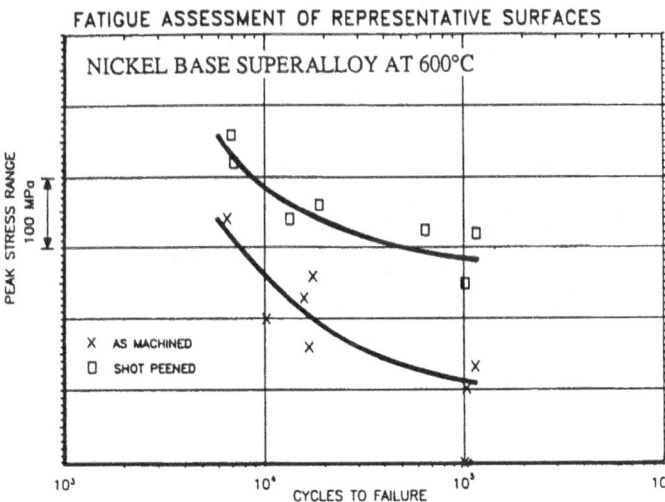

Figure 3 The influence of shot peening on fatigue life

538

Neutron diffraction is emerging as a valuable engineering tool for determining the through thickness state of strain in crystalline materials non-destructively (1-7). High spatial resolution is required in the case of shot peened parts because of the very steep stress gradients generated. For near surface measurements it is necessary to develop procedures for interpreting data when the sampling volume is only partially immersed. In addition, for these near surface strain measurements particularly good instrumental characteristics are required.

To perform satisfactory engineering strain measurements with neutrons, a high flux neutron source is necessary if measurements are to be made at any appreciable depth.

2. Neutron Diffraction Strain Scanning

All the neutron diffraction measurements described in this paper were made on instrument D1A at the high flux reactor at the ILL in Grenoble. D1A, with its high spatial and angular resolutions, and because only one well collimated detector is employed during strain scanning, is not susceptible to the severe instrumental peak shifts that occur when using most multi-detector instruments if only part of the gauge volume is within the sample. This makes it particularly suitable for surface measurements as the only corrections required are simply geometrical to the co-ordinate position of the centre of the sampling volume which shifts from being coincident with the centre of the gauge volume as it passes through the surface. A collimated beam was masked by precisely made rectangular slits to define the small sampling volume shown in fig 4. To obtain the through thickness residual stress profile the sample was moved through the beam in steps of 0.1mm in the surface and in greater steps elsewhere. Measurements were made in the three principal x, y and z directions.

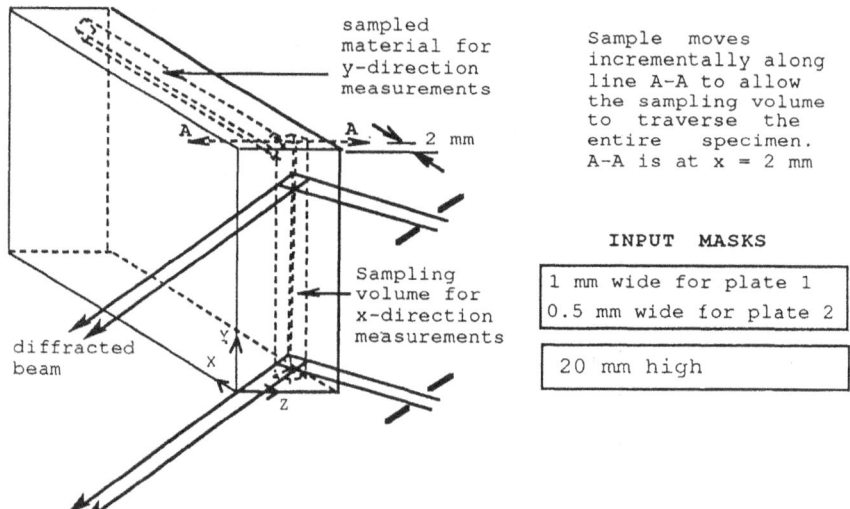

sampled
material for
y-direction
measurements

2 mm

Sampling
volume for
x-direction
measurements

diffracted
beam

Sample moves
incrementally along
line A-A to allow
the sampling volume
to traverse the
entire specimen.
A-A is at x = 2 mm

INPUT MASKS

| 1 mm wide for plate 1 |
| 0.5 mm wide for plate 2 |

| 20 mm high |

Figure 4 Shot peened plate sampling volume geometry. Measurements of strain in a plate of 38 x 38 x x 12.75 mm dimensions.

The strain tensor ε_i for a neutron beam diffracted through an angle $2\theta_i$ can be obtained from

$$\varepsilon_i = -(\theta_i - \theta_0) \cot\theta_0 \tag{1}$$

where $2\theta_0$ is the diffraction angle for unstrained material.

Assuming crystallographic isotropy, an elastic modulus E and Poissons ratio ν, the principal stresses σ_i are given in terms of the strains by;

$$\sigma_i = \frac{E}{(1+\nu)(1-2\nu)}\left[(1-\nu)\varepsilon_i + \nu(\varepsilon_j + \varepsilon_k)\right] \tag{2}$$

or from equation (1),

$$\sigma_i = \Gamma\left[\theta_i(1-\nu) + \nu\theta_j + \nu\theta_k - \theta_0(1+\nu)\right] \tag{3}$$

where $\qquad \Gamma = \{-E\cot\theta_0/(1+\nu)(1-2\nu)\}\{\pi/180\} \tag{4}$

where the diffraction angle is measured in degrees.
The magnitude of σ_i is affected by the value of θ_0 and the accuracy of resolution of $\theta_{i,j,k}$.

3. Experiments

Neutron diffraction residual stress measurements were made on two separate plates of a nickel base superalloy with a bulk elastic modulus of 220 GPa and yield strength of 1200 MPa. Table 2 shows the shot peening parameters employed. Plate 1 had different peening operations applied to each face while plate 2 was peened identically on both faces. Matchstick shaped sampling volumes were employed as shown in Fig 4. Cadmium masks were inserted in both the incoming beam and the diffracted beam to define sampling volumes of 1x1x20 mm for plate 1 and 0.5x0.5x20 mm for plate 2.

Table 2 Shot Peening Conditions

Peening Parameters	Plate 1		Plate 2
	face 1	face 2	face 1 and 2
Shot size	MI. 330	MI. 330	MI. 110 H
Almen intensity	0.018 to 0.02 A	0.018 to 0.02 A	0.006 to 0.008
Coverage	200%	200%	125%
		+	
		MI. 070 H	
		0.015 to 0.020 N	
		200%	

The neutron diffraction data were recorded for the (311) plane. This plane was chosen because it produced a satisfactory neutron count rate, gave good spatial resolution and had an elastic modulus close to that of the bulk material. A 15 minutes count time was employed for the 20 mm^3 sampling volume and 35 minutes for the 5 mm^3 volume to give profiles with a peak height between about 100 and 200 neutrons. These times were doubled near surfaces where the irradiated volume was reduced. For plate 1 a value for the stress free diffraction angle $2\theta_0$ was obtained by averaging six independent measurements made on a small block of the same material expected to have low residual stresses. A force equilibrium balance was also made as a check. In the case of plate 2, $2\theta_0$ was obtained by taking measurements in the interior of the block as well as by performing a force equilibrium check. Agreement to within 12 MPa was achieved between the equilibrium and stress free zero determinations.

4. Results

The position of the sampling volume within each plate was determined optically and by observing the change in neutron count rate as the sampling volume penetrated a surface. For any measurements where the sampling volume was not fully immersed in the sample the location of the average strain within the volume was taken to be at the centroid of the immersed cross-section. An example of the raw data peak positions obtained for plate 1 in the x, y and z directions is shown in Fig 5. Fig 6 shows the peak width variation recorded in the y direction with depth for each plate. The peak broadening shown in Fig 6 extends to a depth of about 0.5 mm for plate 1 and 0.25 mm for plate 2 . Peak broadening results from either strain gradients within the sampling volume or microstress effects typically arising from plastic deformation. The data suggest that the depth of plastic deformation caused by the peening was no greater than about 0.5 mm.

The complete through thickness residual stress profiles for plates 1 and 2 are shown in Figs 7 and 8 respectively. The results show that an increase in peening intensity causes an increase in the maximum compressive and tensile residual stresses measured but that all the stress distributions have a similar characteristic shape. At the highest intensity the maximum surface compression is in the region of the material yield strength. Also it would seem that the depth of peak tension, which is in magnitude about a third of the corresponding maximum compression, increases with peening intensity: Little residual stress is observed in the centre of the plate.

5. Conclusions

The role of residual stress in improving the fatigue performance of a compressor disc assembly has been outlined. Neutron diffraction data of residual strains obtained from the (311) lattice spacings through the surfaces of two shot peened nickel base superalloy plates have been reported. It has been found that a characteristic shape of the residual stress profiles is obtained irrespective of the peening intensity employed. With the highest peening intensities compressive surface residual stresses in the region of the material yield strength have been observed. A peak subsurface residual tension of approximately a third of the corresponding maximum compression has been identified.

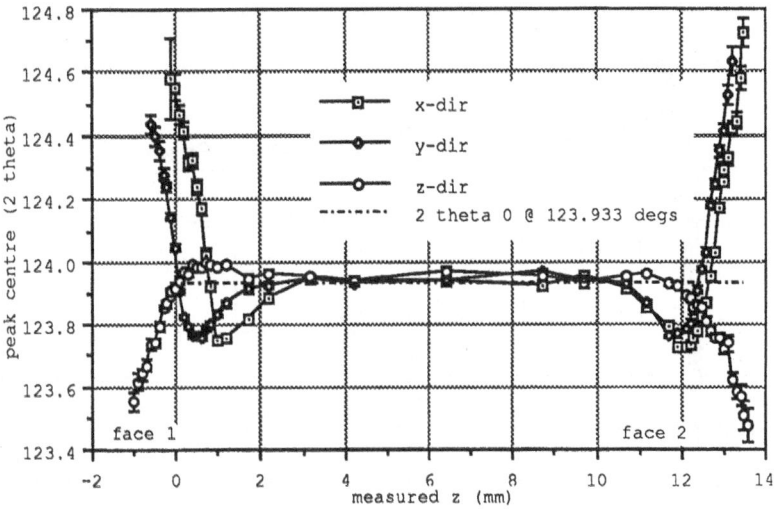

Figure 5 Neutron diffraction peak positions through thickness of plate 1

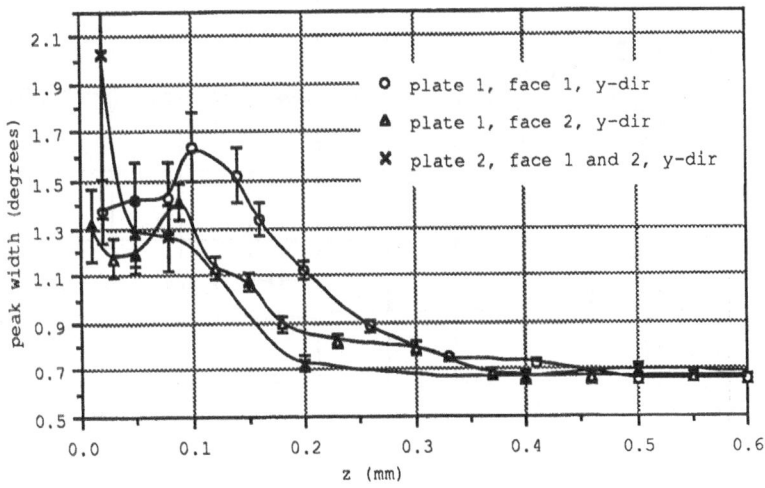

Figure 6 Neutron diffraction peak widths for plates 1 and 2.

542

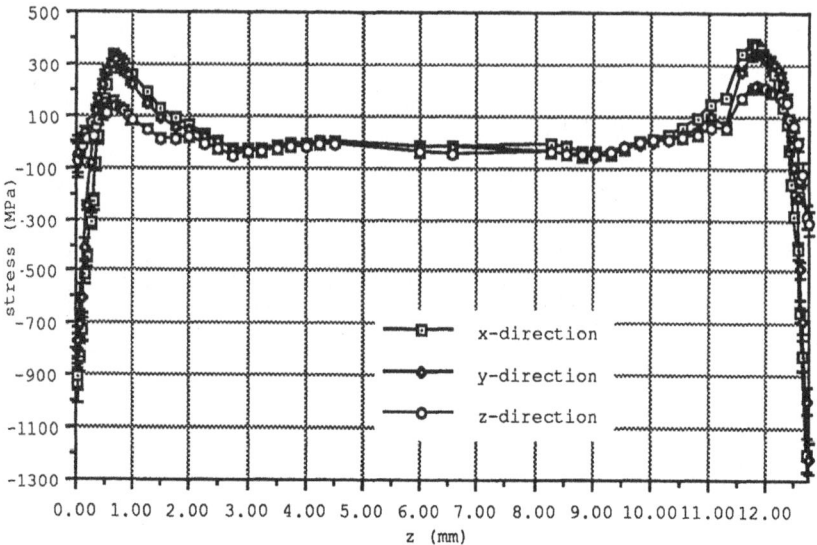

Figure 7 Residual stress field through thickness of plate 1

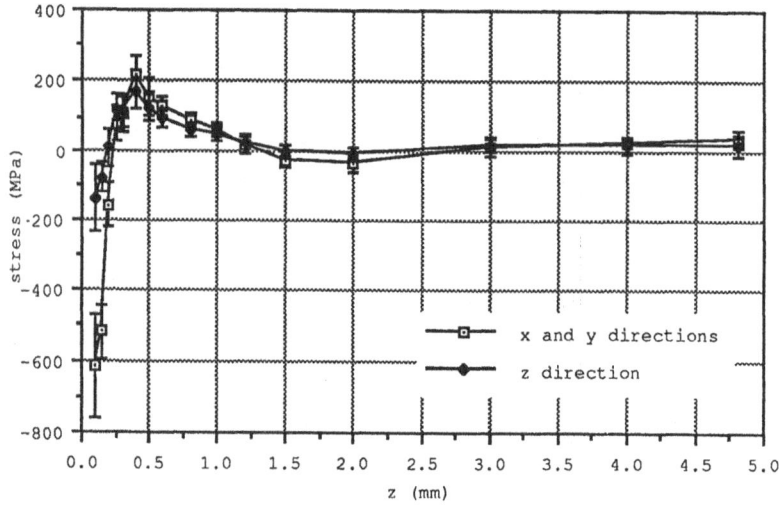

Figure 8 Residual stress field through half thickness of plate 2

6. References

1. Pintschovius, L., Jung, V., Macherauch, E and Vohringer, O. (1983) 'Residual stress measurements by means of neutron difrraction' Mater. Sci. Eng. 61, 43-50.

2. Stacey A., MacGillivray H. J., Webster G. A., Webster P. J., Ziebeck K. R. A. (1985) 'Measurement of residual stresses by neutron diffraction' J. Strain Analysis 20, 93 - 100

3. Allen, A. J., Hutchings M. T., Windsor C. G., Andreani C. (1985) 'Neutron diffraction methods for the study of residual stress fields' Adv. Phys, 34, 445 - 473.

4. Bourke M.A.M., MacGillivray H. J., Webster G. A., Low K. S., Webster P. J. (1987) 'Improving the Resolution of Neutron Diffraction Residual Stress Measurements in Engineering Components' Paper presented at I.I.T.T. International Conference on Fatigue and Stress, Paris.

5. Smith D. J., Leggatt R. H., Webster G. A.,MacGillivray H. J.,Webster P. J., Mills G. (1988) 'Neutron Diffraction Measurements of Residual Stress and Plastic Deformation in an Aluminium Alloy Weld' J. Strain Analysis 23, 201 - 211

6. Ezeilo A. N. , Webster G. A., Webster P. J. , Roth M., Muster W.J. (1991) 'Comparison of X-ray and neutron diffraction determinations of residual stresses in a laser treated martensitic steel' Euromat 91, Institute of Metals, Cambridge July 1991, to be published.

7. Krawitz, A. D. and Holden, T. M. (1990) 'The measurement of residual stresses using neutron diffraction' MRS Bull. XV, 57-64

RESIDUAL STRESSES AT COLD EXPANDED FASTENER HOLES

L. EDWARDS AND A. T. OZDEMIR
Fracture Research Group
Materials Discipline
Faculty of Technology
The Open University
Milton Keynes
MK7 6AA
U.K.

Abstract   Cold expansion of holes to combat fatigue is now common practice. Expansion is usually achieved commercially by the split sleeve process using prescribed levels of mandrel interference. A novel sleeveless method which allows the degree of cold expansion to be infinitely varied in a simple manner has been developed and applied to 5mm thick 7050 aluminium alloy plate. It is shown that this technique can produce significant increases in fatigue life. The 3D residual stress distribution produced by this method has been studied using neutron diffraction and compared to that estimated by a modified Sachs technique. The relative spacial accuracies of the two methods make direct comparison of the estimated residual stress fields difficult but good qualitative agreement is achieved. Furthermore, the neutron method provides information about the three dimensional residual stress distribution that cannot be obtained from mechanical methods of stress analysis.

1. Introduction

The use of cold expansion to combat fatigue at holes in aircraft structures is well known. It is used primarily for repair or retrofit in military aircraft but it is beginning to be used at the construction stage of new civil aircraft. The technique basically involves plastically deforming a hole so as to permanently increase its diameter. Cold expansion techniques have been used for over thirty years to produce fatigue life enhancement and have been recently reviewed by Champoux, (1).

All involve expansion of the hole by means of the insertion of a hard tool that is bigger than the initial hole diameter. When the tool is removed, the elastic bulk of the workpiece surrounding the now plastically deformed hole forces it to spring back so that the vicinity of the hole experiences compressive residual stresses. Superposition of these residual stresses with service loads then leads to improvement in fatigue life either by reducing or suppressing crack initiation or more often by reducing fatigue crack growth rates.

545

M. T. Hutchings and A. D. Krawitz (ed.),
Measurement of Residual and Applied Stress Using Neutron Diffraction, 545–553.
© 1992 *Kluwer Academic Publishers.*

2. Methods of Cold Expansion

The most common method presently used in the aircraft industry involves expansion of a lubricated split sleeve process by an oversize mandrel using commercially available equipment and prescribed levels of mandrel interference, (2). The optimum degree of mandrel interference, and hence expansion for a particular application, will however depend on the local geometry of the component. Furthermore, fatigue life predictions of structures containing such expanded holes rely critically on estimates of the residual stress distribution surrounding the hole.

A sleeve-less method for cold expanding circular holes has been developed at the Open University which allows the expansion to be infinitely varied in a simple manner. This system greatly facilitates the study of the effect of degree of cold expansion on both the residual stress distribution and its subsequent effect on fatigue life. It basically consists of a tapered mandrel of circular section around which are located a set of three caged rollers. When the mandrel is rotated and advanced into the cage, the rollers rotate in a planetary motion around the circumference of the hole, increasing its diameter.

By arranging for the taper on the rollers to "cancel" out that on the mandrel the opposite sides of the hole remain parallel after treatment. The tooling can be used on a pillar drill or standard milling machine. The specimen is first clamped in the vice and the hole position is located under the milling head. The initial hole is then drilled using a suitable lubricant. The expander is manually positioned in the hole and the stationary centre mandrel is lowered until the rollers lightly grip the hole.

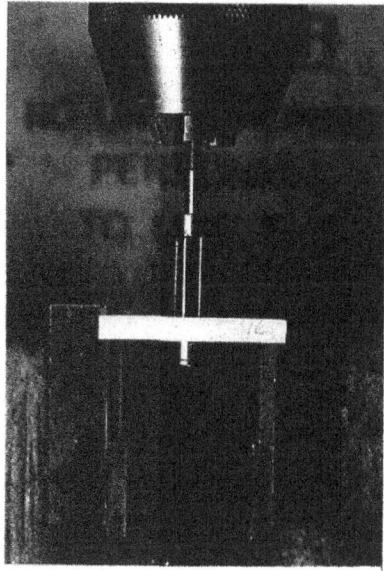 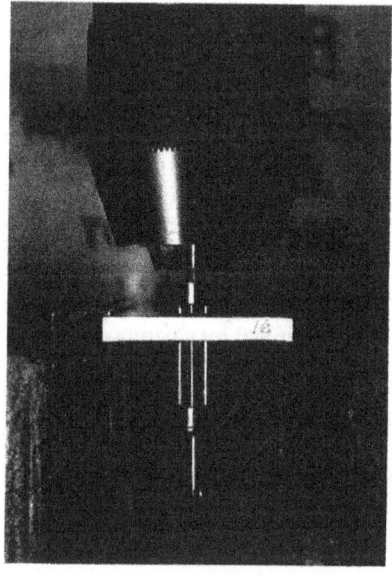

Figure1 Before Expansion   Figure 2 After Expansion

The mill is then started, lubricant is applied and the centre roller is fed into the cage so expanding the hole. This sequence of events can be visualized by comparing figures 1 and 2 which show the tooling at the beginning and end of expansion respectively. Note that the mandrel travels through the cage as both slowly precess down through the hole. Finally the hole is deburred and reamed to the appropriate size.

This cold expansion technique is somewhat similar to roller burnishing, (3), which also typically uses tooling consisting of caged tapered rollers revolving around a tapered mandrel. However, conventional roller burnishing involves the use of fixed working diameter tooling where the rollers are set to a given size before being fed sequentially through the hole in the workpiece. If the rollers are inclined at a slight angle to the hole axis then the tooling is self feeding Either mechanism of operation can only be used for relatively small expansions and conventional roller burnishing is mainly employed for either improving surface texture or finishing holes to tight tolerances.

Although higher expansions have been achieved using multiple passes, (4), and variable diameter tooling has been mentioned in the technical literature, (3), the authors are unaware of the use of either tooling system to produce high levels of cold expansion in one operation and thus we believe the technique developed at the Open University to be novel. It has been shown to produce increases in fatigue performance in thin, (<2mm), sheet, (5,6), whilst the present work presents results on thicker, (5mm), material.

3. Effect of Cold Expansion on Fatigue Life

In order to assess the effect of of expansion on fatigue life, specimens of 7050 alloy were prepared with both expanded and plain holes. Specimens containing a common fastener hole size, 9.52mm,(3/8"), were prepared. To produce the cold expanded specimens holes 8.9 mm in diameter were drilled centrally in 5 x 40 x 300 mm blanks and then reamed out to 9.12 mm. The resulting holes were expanded to 9.3 mm, (a 2% expansion in diameter), and then deburred before being reamed to 9.52mm, (3/8").

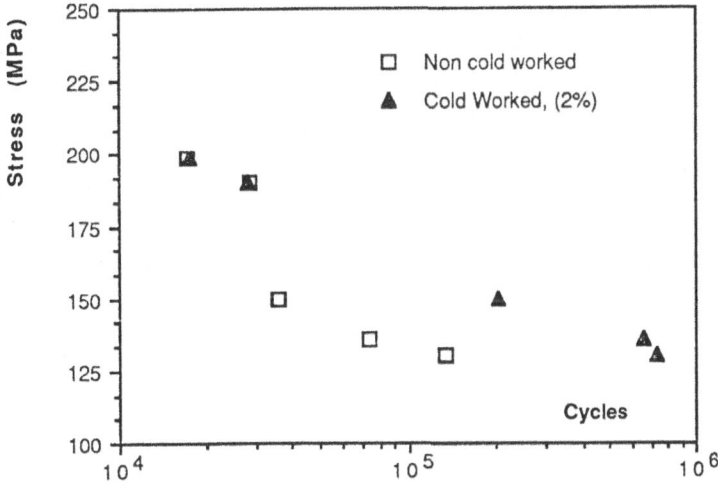

Figure 3 Effect of Cold Expansion on Fatigue Life.

Specimens containing plain holes were prepared by drilling and subsequently reaming holes of final size 9.52 mm in similar blanks. Figure 3 compares the fatigue life of these two specimen types. It can be seen that cold expansion produces almost an order of magnitude improvement in fatigue life at intermediate stresses. As may be expected, no improvement is seen at high stresses, where yielding at the hole effectively washes out the beneficial effects of the induced compressive residual stresses.

4. Neutron Diffraction

Residual stresses were measured in a specimen that had been expanded 2% by the technique described above using the high resolution powder diffractometer, D1A, at the Institut Laue Langevin (ILL) in Grenoble, France. The theory of the technique is now well known and we shall not repeat it here. Furthermore, the use of this particular diffractometer for the measurement of residual stresses in a bulk body has previously been described in the literature, (7,8), and is covered in detail by Webster, (9), in this volume.

The main criterion affecting our particular experiment was the high stress gradients expected near the hole which demanded measurements of a high spatial accuracy and, thus, a small sampling volume was used. Specifically, cadmium masks containing thin slits were aligned so that a sampling volume of the order of $1mm^3$ was obtained. Additionally, as similarly described by Webster, (9), the specimen was mounted on three orthogonal micro-manipulators and accurately aligned with respect to both the incoming beam and the detector so that position of the diffracting volume within the specimen could be accurately measured and controlled to an accuracy of better than $100mm$.

The specimen was similar to those used for the fatigue experiments so essentially consisted of a rectangular specimen of dimensions 5mm x 40mm x 300mm containing a 2% expanded hole. Using the computer controlled micro manipulators a co-ordinate system describing the specimen was set up describing the xy plane of interest which emanates radially from the hole and is perpendicular to the specimen surface. Using this co-ordinate system one outer edge of the specimen within this plane was defined as having co-ordinates, [0.0,0.0], so that the hole edge on the same surface possessed co-ordinates, [15.5,0.0]. Due to the symmetry of the specimen stresses were measured in only one quarter of this xy plane specifically at fixed y values of 0.5,1.0,1.5 and 2.5 mm.

Relatively long counting periods were needed to measure diffraction peaks of adequate statistical quality. In addition to the relatively poor scattering of aluminium there were two basic reasons for this. The first is the very small sampling volume used. The second involves the crystallographic texture present in the sample. It is usual to use the (311) reflection when measuring stresses in aluminium alloys as the elastic modulus in this direction is close to that of bulk material but the pronounced texture present in this alloy significantly reduced the size of this peak. This meant that ≈2 hours were needed for each peak so that the measurement of the three orthogonal principal strains needed to fully describe the strain tensor involved ≈6 hours of beam time at each point.

Typical strain data obtained for a given depth of the specimen, (in fact y=2.5mm), are presented in figure 4. Appreciation of complex three-dimensional stress distributions from two dimensional sections is extremely difficult and thus we have investigated other ways of presenting this data. The most successful technique employed to date involves the production of 2D-contour stress maps. These are similar to those used to describe elevation in an geographic atlas and are produced by interpolating a grid of derived data to which is fitted a surface of minimum curvature. This surface is then represented by a contour map where all areas between a given stress range are given the same colour.

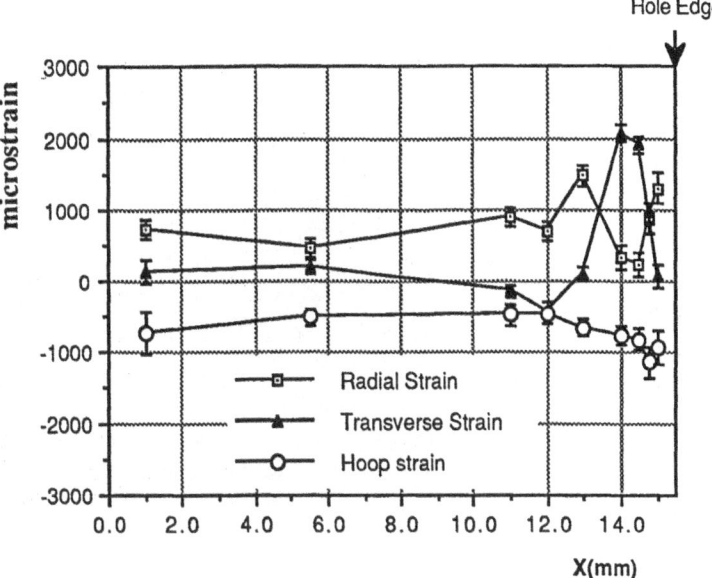

Figure 4 Residual Strain Distribution

The contour map for the hoop stress distribution is reproduced here, (figure 5), but both radial and transverse stresses were, of course, also measured. Due to the curve fitting procedures used, and the difficulty of deriving an accurate stress free lattice parameter, these contour maps are best used to investigate the geometry of the 3D stress distributions and values obtained from the graphs should not be used for any quantitative calculations.

However, figure 5 provides unique evidence for the three dimensional residual stress field around this cold expanded hole and a number of immediate conclusions can be reached on studying the data presented in this fashion. Firstly, stresses are low far from the hole. Indeed the map presented in figure 5 have been deliberately truncated, the outer 10 mm of the specimen not being shown, in order to effectively blow up the area of interest near the hole. However, within 3 mm of the hole the hoop residual stresses near the centre of the specimen are clearly substantially different from those near the surfaces.

Secondly, whilst the hoop stress at the centre of the hole is large and compressive, it is close to zero at the hole edges. This relaxation at the hole corners is probably a result of the extrusion of material sidewards out of the hole. In contrast to the split sleeve method, significant plastic flow of material out of the hole occurs near the workpiece surface, (10). Subsequent deburring and reaming removes this material probably leading to the stress distribution seen in figure 5, which is significantly different from that traditionally thought to exist around split sleeve expanded holes, (11,12).

The reason for this is that the residual stress profile is controlled not only by the residual plastic strain field but also by how that strain field is applied — that is it is strain history dependent. This means that the residual stress distribution produced by roller expansion is not likely to the same as that produced by split-sleeve mandrel expansion. To test this hypothesis we intend to measure the 3D residual stress distribution using neutron diffraction in a specimen expanded to a similar degree by the split sleeve method.

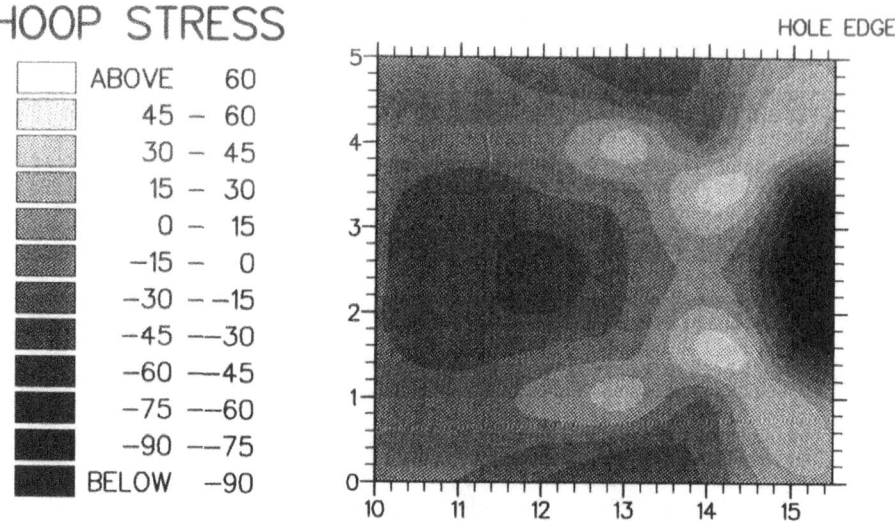

Figure 5   Three Dimensional Residual Stress Distribution

Figure 5 shows that unique three dimensional information on residual stress fields can be obtained using neutron diffraction. However, due to difficulties associated with texture and the acquisition of a stress free lattice spacing, the measurement of absolute values of stress can be difficult. Indeed, in common with many engineering residual stress measurements by neutron diffraction, the stress free zero used to derive the data plotted in figure 5 was obtained from local stress balance considerations.

5. Comparison with Mechanical Stress Measurement, (Sachs)

A destructive method of measuring residual stresses on workpieces of cylindrical symmetry was first proposed by Sachs, (13) and the method has since been further developed, (14). Weiss (15) tried to simplify the Sachs equations but his equations are essentially the differential forms of the conventional Sachs equations indicating that the measurement of residual stresses for this geometry is dependent not on the absolute values of the measured strains but their relative incremental change with hole area.

Essentially, the method involves removing a circular washer of 5mm thick, 40 mm outside diameter symmetrically from around the hole in a specimen similar to those described above and placing strain gauges in both the hoop and transverse directions on its outside edge. The output of these gauges is then monitored as the central hole is sequentially bored out by a stepped copper tool using electrical discharge machining. The basic experimental setup is shown in figure 6. The geometry of the specimen is illustrated by the dummy specimen used for temperature compensation seen to the centre left of the figure. Figure 7 plots the hoop stress as measured by the Sachs method over the same area and on the same specimen as that estimated by neutron diffraction and presented in figure 5. The two measurements give reasonable qualitative agreement but before comparing the absolute values of the residual stresses measured it is worth considering

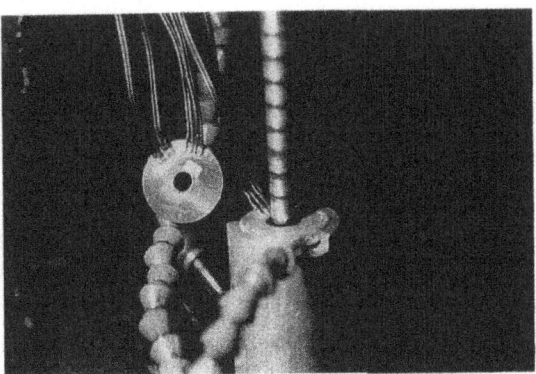

Figure 6 Sachs Experimental Set-up

the relative attributes of the two methods.

The Sachs method possesses a greater spacial accuracy than neutron diffraction in the x-axis but a poorer resolution in the y direction where it is essentially averaging all stresses along the specimen thickness. As the neutron diffraction occurred from a volume of approximately a cubic millimetre, we have the unusual situation where a neutron stress measurement is effectively an average of several Sachs measurements in the x direction whilst the Sachs measurements average areas covered by several neutron measurements.

Under these conditions precise deconvolution of the two sets of values is difficult and it is clear that each method provides you with differing information on the actual residual stress distribution present in the workpiece so that to much extent the methods are complimentary rather than competitive.

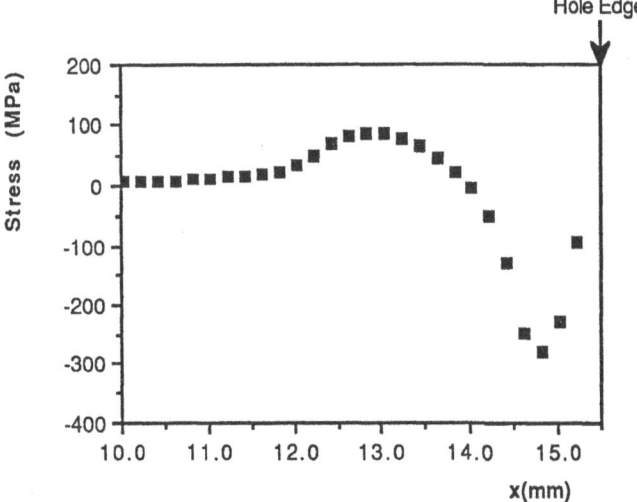

Figure 7 Residual Hoop Stress as measured by the Sachs' Technique

6. Conclusions

1.   A sleeveless method of cold expansion which allows the degree of cold expansion to be infinitely varied has been developed and applied to 5mm thick 7050 plate.

2.   It is shown that this technique can produce significant increases in fatigue life at intermediate stress levels.

3.   The 3D residual stress distribution produced by this method has been studied using neutron diffraction and compared to the residual stress distribution subsequently estimated on the same specimen by a modified Sachs technique..

4.   The relative spacial accuracies of the two methods make direct comparison of the estimated residual stress fields difficult but good qualitative agreement is achieved.

5.   The neutron diffraction results suggest that whilst compressive residual hoop stresses are generated at the bore of the hole substantial relaxation had occurred near the hole edge. Such information cannot be obtained from mechanical methods of stress analysis.

6.   It is clear that neutron diffraction and mechanical residual stress estimation techniques are often complimentary rather than competitive and many situations may require the application of a number of stress estimation techniques if the problem is to be adequately resolved.

7. Acknowledgements

The authors would like to thank the Royal Aerospace Establishment and the SERC for support. The neutron diffraction measurements were made at the Institut Laue Langevin, Grenoble under experiment proposal number 5-23-355 and we are indebted to Dr Peter J Webster and Gordon Mills of Salford University for their invaluable help in their execution. We are also grateful for the technical assistance of Reinhold Hermann, Peter Ledgard, Jim Moffatt and Prof. Nick Reid in developing both the cold expansion tooling and the Sachs experiments.

8. References

1. Champoux, R.L., An Overview of Cold Expansion Methods, in Fatigue Prevention and Design, ed. J.T. Barnby, Chamelon Press, London, 1986, p35

2. Fatigue Technology Inc. Seattle, Washinton, U.S.A.

3. Tool and Manufacturing Engineers Handbook, Vol III, Chapter 16: Mechanical and Abrasive Deburring and Finishing, Society of Manufacturing Engineers, Michigan.

4. Cassatt, G.G. and Tenclay, T.C., A Comparison of Residual Stresses Around Fastener Holes Produced by Three Different Methods of Mechanical Overstraining, in Proc of 1982 Joint Conf. on Exp. Mech., Amer. Soc. for Strain Measurement.

5. Hermann, R. and Reid, C.N., Enhancing the Fatigue Life of Fastener Holes in Al-Li alloy 8090, in Proc Al-Li 5, Eds. E. Starke and J. Saunders Jr., Materials and Component Eng. Publ. Ltd. Warwick, UK, 1989, p1607

6. Hermann, R. and Reid, C.N., Experimental Analysis of the Stress Distribution around Cold-Expanded Holes in Thin Sheets" to be presented at ICRS-3, Tokushima, Japan, July 23-26, 1991, (To be published by Elsevier)

7. Stacey, A., MacGillivary, H.J., Webster, G.A., Webster, P.J. and Ziebeck, K.R.A, Measurement of Residual Stresses by Neutron Diffraction, J.Strain Anal.,1985, 20, p93

8. Allen, A.J., Hutchings, M.T. and Windsor, C.G., Neutron Diffraction Methods for the Study of Residual Stress Fields, Adv. in Physics, 985, 34, p445

9. Webster, P.J., Spacial Resolution and Strain Scanning,this volume.

10. Edwards, L. and Ozdemir, A.T., Measurement of the 3D Residual Stress Distribution Around Cold Expanded Aircraft Fastener Holes, to be presented at ICRS-3, Tokushima, Japan, July 23-26 1991, (To be published by Elsevier)

11. Hsu, Y.C. and Forman, R.G., Elastic-Plastic Analysis of an Infinite Sheet Having a Circular Hole Under Pressure", J. Appl. Mech, 1975, 42, p347

12. Rich, D.L. and Impellizzeri, L.F., Fatigue Analysis of Cold Worked and Interference Fit Fastener Holes, in ASTM STP 637, 1977, p153

13. Sachs, G., Der Nachweis Immerer Spannungen in Stangen und Rohren', Zeitschrift fur Metalkunde, 19, 1927, p352.

14.J. W. Lambert, 'A Method of Deriving Residual Stress Equations', Proc-SESA, 12, 1954, p91.

15.V. Weiss, 'Residual Stresses in Cylinders', Proc-SESA, 15, 1960, p53.

NEUTRON AND X-RAY DIFFRACTION RESIDUAL STRESS MEASUREMENTS ON POWER GENERATION TURBINE BLADES: COMPARISON WITH FINITE ELEMENT ANALYSIS

M. KIJEK, T. R. FINLAYSON and R. L. DAVIS*
Dept. of Physics Monash University
Clayton, Australia 3168

*Australian Institute of Nuclear Science and Engineering
Lucas Heights Research Laboratories, Menai, Australia 2234

ABSTRACT

The residual stresses in used and unused turbine blades from a 380 MW steam power generator were measured by neutron and X-ray diffraction techniques and modelled using finite element analysis computation. One of the main engineering problems with the turbine blades is the growth and propagation of fatigue cracks which may form around the lacing hole under the strain of cyclic rotation and vibration during the service life of the component. Careful analysis of the residual stress distribution is therefore important in assessing the fatigue-limited life of the turbine blades and improving the turbine blade design. The large penetration depth of neutrons compared with that for X-rays allowed us to measure the triaxial residual stress in the vicinity of the lacing hole, averaged over the whole thickness of the blade. Our measurements indicate the existence of triaxial residual stress in both used and unsued turbine blades. The neutron results have been complemented by the data measured, using a laboratory based X-ray facility, from points on the blade surface at locations corresponding to those for the neutron measurements.

Large gradients of the residual stress exist on both sides of the hole in both blades. On one side the stress is compressive whereas on the other it is tensile. This distribution of residual stress is believed to be responsible for the growth and propagation of cracks observed around the hole. We predict that this residual stress distribution was set up during brazing of the metal strip ("Stellite 6") to the leading edge of the blade, to reduce wear due to erosion. The finite elements method employed to simulate the distribution of residual stress in the vicinity of the lacing hole of the blade gave qualitative agreement with experimentally measured values.

M. T. Hutchings and A. D. Krawitz (ed.),
Measurement of Residual and Applied Stress Using Neutron Diffraction, 555.
© *1992 Kluwer Academic Publishers.*

8. COMPARISON OF NEUTRONS WITH X-RAYS AND OTHER STRESS PROBES

CALIBRATION OF PORTABLE NDE TECHNIQUES FOR RESIDUAL STRESS MEASUREMENT

A. J. ALLEN
National NDT Centre
AEA Industrial Technology
B521 Harwell Laboratory
Didcot
Oxon OX11 0RA
United Kingdom

ABSTRACT. Despite the obvious advantages of the neutron diffraction technique for stress measurement, it cannot be used in the field. Hence, one of its most important future engineering applications is likely to be the calibration of portable NDE techniques both for stress measurements and for related structural degradation assessments. A number of such NDE techniques are discussed, their potential assessed, and their need for calibration and validation considered.

1. Introduction

There are a number of emerging advanced NDE methods for the assessment of residual stresses and materials degradation, which require calibration, validation and even interpretation [1,2]. Such techniques may provide partial strain tensor data, path-average data or simply a measure of the plastic damage (plastic strain) present. X-ray diffraction (XRD) provides full strain tensor data but is usually restricted to depths between 20 and 100μm below the surface, depending on the material. Thus there is a need to calibrate the obtainable NDE data against comprehensive stress information. Currently, this is mainly done either using destructive techniques such as hole-drilling and sectioning or by finite element analysis (FEM). Unfortunately, FEM relies heavily on accurate knowledge of the boundary conditions, particularly where there are marked spatial variations in stress. The only technique which nondestructively gives actual comprehensive strain tensor data in the bulk is the neutron diffraction (ND) technique [3], and therefore the potential of many of the portable NDE methods could be greatly enhanced by use of ND for calibration, validation, and interpretation.

In a short paper it is not possible to give a comprehensive review but only to highlight a number of the most promising NDE techniques, as in section 2. In section 3, some examples are given of how ND stress measurement can complement these techniques. General conclusions are drawn in section 4.

M. T. Hutchings and A. D. Krawitz (ed.),
Measurement of Residual and Applied Stress Using Neutron Diffraction, 559–571.
© 1992 UKAEA.

2. Review of Portable NDE methods

The following emergent NDE techniques should all be considered relevant to solving future residual stress problems in industry:

2.1. RADIATION-BASED METHODS

While XRD is the obvious alternative to ND for elastic strain measurement, other nondestructive radiation-based methods have the attraction of being noncontact, and should be considered in the overall NDE assessment of materials degradation. These include positron annihilation, Compton backscattering, infra-red and exoelectron studies [1]. While all are relevant to the assessment of materials degradation, only XRD, which is sensitive to elastic strain, and positron annihilation (PA), which is sensitive to plastic strain, are considered here.

2.1.1. *X-ray Diffraction.* XRD measurement of changes in lattice spacing from Bragg's law, to infer stresses, pre-dates neutron diffraction strain and stress measurement [4]. Indeed, many of the theoretical and experimental aspects are common to both methods. The principal advantages of XRD (compared to ND) are: high incident flux (resulting in 1 - 5 minute measuring times), excellent spatial resolution (3mm spot size down to 40μm), potential portability and much reduced cost compared to neutrons. The principal disadvantages arise from the poor penetrability (10 - 40μm in steel), resulting in considerable sensitivity to the geometry, and a restriction on the data available to near-surface stress tensor information only. The main requirements are a monochromatic X-ray source, a well-defined sample - detector geometry and representative surface conditions.

Some portable XRD equipment [5] has been tested, but the technique is still not widely used on a full NDE plant inspection basis, - probably due to the complications of radiological protection, and also to its relative cost compared to other fully accepted NDE techniques. Nevertheless, the importance of XRD in the lab cannot be overestimated. The power of the method to probe near-surface macrostrain, microstrain, crystallite anisotropy or inhomogeneity and also texture effects has established much of the foundation on which ND can build for measurements within bulk components.

2.1.2. *Positron Annihilation (PA) Lineshape Measurements.* Whereas most stress measurement methods probe elastic strain effects, PA lineshape analysis can provide the complementary plastic strain assessment [6].

Positrons are emitted by some radioisotopes such as germanium-68. In condensed matter, positrons rapidly thermalise and then annihilate with atomic electrons to produce two nearly collinear gamma rays in opposite directions, each of nominal energy 511keV. In practice the distribution of annihilation energies is Doppler broadened by the momentum distribution of the annihilating electrons. In a metal or alloy, annihilations take place with conduction and valence electrons with comparable probabilities. The latter give rise to greater Doppler broadening than the former, and therefore any microstructural effect which changes the ratio of annihilations with the two classes of electron affects the gamma ray Doppler broadening. The introduction of vacancy defects creates sites in the metal matrix which are denuded of valence electrons, but into which unlocalised conduction electrons can spill over from the surrounding matrix. Since this makes the vacancy defect sites negatively charged, thermalised positrons can be trapped there, and annihilations of such trapped positrons are necessarily with conduction electrons. Thus the annihilation gamma ray line narrows with increased vacancy defect damage, and accurate measurement of the gamma ray lineshape can provide a materials degradation probe. The atomic vacancy defect fraction range over which the technique is sensitive is from 10^{-7} to 10^{-4}. For higher concentrations all positrons are trapped at vacancy defects and there is no sensitivity to further damage.

Stress-related vacancy defect damage can be associated with plastic strain. Figure 1 shows schematically the use of positron annihilation lineshape analysis to probe plastic strain. The gamma ray detector is of the high resolution intrinsic germanium type, which can be used for NDE measurements provided that the associated cryostat is sufficiently compact. The positron lineshape measurement parameter, the S-parameter, is determined from the ratio of the counts in the central part of the peak to those over the whole peak. Figure 2 shows a typical plastic strain calibration curve for mild steel. For the assessment of high cycle fatigue damage, one of the most important potential NDE applications of PA, the technique is sensitive between the first quarter to the first half of the fatigue life, before saturation occurs. Using the PA technique, plastic strain can be mapped out over the surface of a component with approximately 1mm spatial resolution. The technique is sensitive to the presence of plastic strain within a small depth below the sample surface: typically 0.25mm in steel or 1.5mm in aluminium. In principle it is possible to make spot measurements in less than 10 minutes per point.

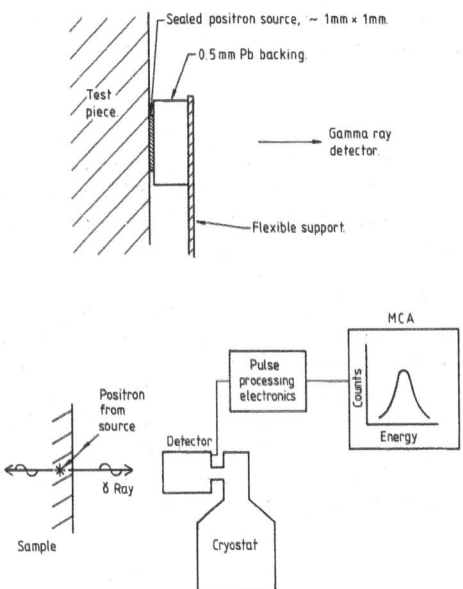

Figure 1. Schematic diagram for typical PA lineshape measurements of plastic damage in metals and alloys.

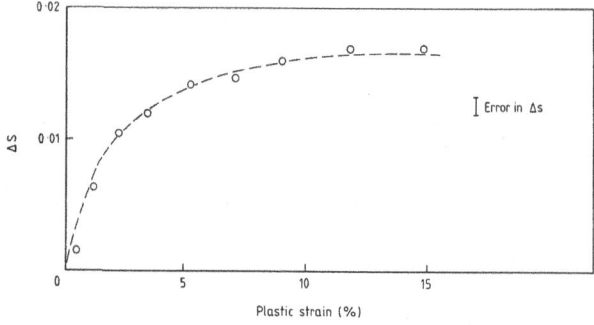

Figure 2.   Typical positron lineshape S-parameter calibration curve.

The main requirements are a radioactive source, a smooth surface, and undamaged reference sample material. With these requirements met and despite a loss of sensitivity at high damage levels, PA can provide useful complementary data to elastic strain measurement when residual stresses are present, as is shown in section 3.

2.2. MAGNETIC-BASED METHODS

While magnetic techniques apply only to ferromagnetic alloys such as ferritic steel, they probably present the best chance of future field-applicable NDE residual stress assessment below the surface of bulk steel components. However, some of the magnetic methods can also provide complementary data on plastic strain and degradation.

2.2.1. *Magneto-Acoustic Emission (MAE).* In a ferromagnetic material such as ferritic steel, a changing magnetic field causes magnetic domain wall motion, and this can lead to the generation of elastic waves known as magnetoacoustic emission (MAE) [7]. MAE arises from the abrupt changes in magnetostrictive strain which occur if the domain walls are of the non-180° type. The presence of stress results in a partial realignment of magnetic domains along the principal stress directions. This results in a significant reduction in MAE, even for small stress levels, because the population and size of non-180° domain walls are reduced. The MAE signal typically decreases by 50-70% for applied stresses of either sign up to 300MPa.

Figure 3 shows schematically a typical MAE measurement configuration. The measurement is of a mean effect along the path between the magnetic pole-pieces and hence the spatial resolution is usually 3 to 20mm. The penetration of MAE is typically 10mm, and can be controlled by varying the oscillation frequency of the applied magnetic field in the range below 100Hz. A typical profile measurement of the MAE signal (in the frequency range 0.1 - 1MHz) as a function of magnetic field is shown in Figure 4 for a quenched and tempered mild steel. Referring to Figure 4, a suitable NDE parameter for measuring MAE is the ratio of the MAE peak signal at positive field (the initial peak in the cycle) or negative field (the final peak) to the MAE signal measured at low (zero) field. The technique is sensitive to the general magnitude of stress levels, positive or negative, ie to the sum of the two principal stresses in the case of a biaxial stress state which lies in the plane of the sample surface.

2.2.2. *Stress-Induced Magnetic Anisotropy (SMA)*. SMA is the orientational variation in magnetic permeability within a ferromagnetic material due to the presence of elastic strain [8]. SMA is sensitive to the magnitude and sign of the difference in principal elastic strains. For example, an uniaxial applied or residual stress results in partial alignment of the magnetic domains along the principal stress axes so as to minimise the stored elastic energy in the material. In general, magnetic anisotropy results in the rotation of any magnetic induction present away from the applied field direction, giving an effective increase in the permeability along the most tensile direction compared to that in the

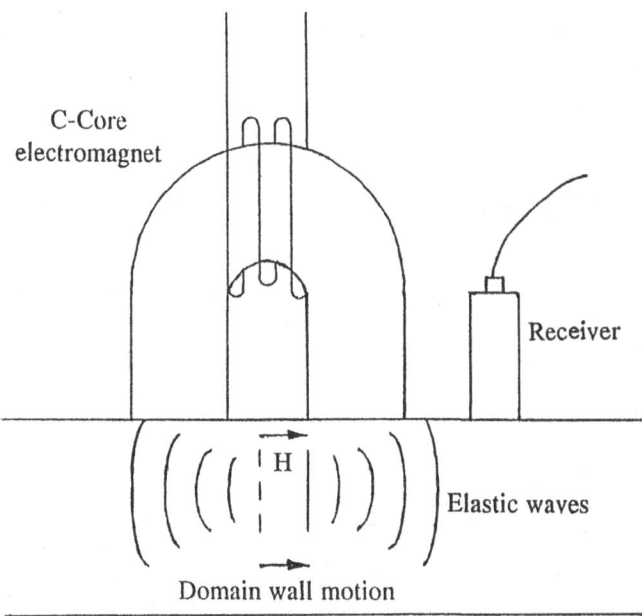

Figure 3.   Measurement configuration for MAE measurements.

564

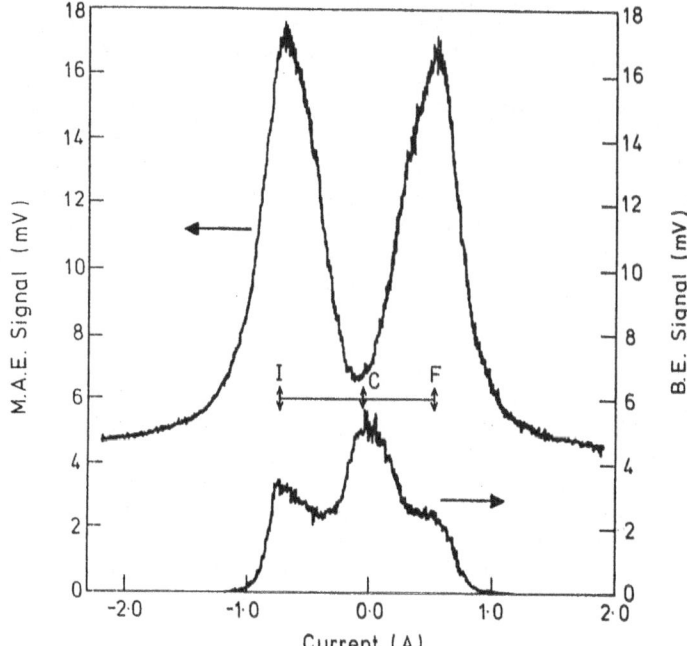

Figure 4.   Typical MAE and BE response curves versus applied magnetic
              field. I, C and F correspond to initial peak, low field and final
              peak parameters.

orthogonal direction. In SMA measurements, an AC magnetic field is applied to the component by a C-core electromagnet and the rotation of the field away from its applied direction is measured by inserting coils or some other flux measuring unit between the poles of the magnet. If coils are used, one set is aligned so as to measure the magnetic flux being applied by the magnet while the other is aligned so as to link any flux in the orthogonal direction (in the plane of the component surface). Thus if the magnetic field is rotated away from its applied direction in the steel the rotation angle in the air just above the steel is simply related to the ratio of the voltages measured across the two sets of coils. The SMA instrument gives a voltage reading which is proportional to this rotation angle. The whole probe is rotated (ie both applied fields and coils) and the output voltage measured as a function of the rotation angle. The maximum penetration in steels is 3 - 4mm. Spatial resolution can be slightly better than for MAE, being determined in part by the size of the search coils.

In the case of biaxial stress, the MAE signal is sensitive to the sum of both principal stress components, while SMA measures the difference in principal stress components. Thus the two techniques can be used together to investigate biaxial stress states in the plane of the surface, including measurement of the principal components and their directions.

The accuracy of these two techniques with calibration is typically ±20MPa for stress level measurement. For principal stress directions in mild steel plate, the accuracies are: ±15° and ±5° for MAE and SMA respectively [9].

2.2.3. *Magnetic Barkhausen Emission (BE)*. BE is the erratic high frequency (5k - 100kHz) magnetic perturbations arising from impeded magnetic domain wall movements under the influence of a smoothly changing (typically 60Hz frequency) applied magnetic field [9]. The cause of these perturbations is domain wall pinning by microstructural inhomogeneities, including dislocation loops. While the underlying cause of BE is the same as for MAE, the BE technique is sensitive to plastic strain effects as well as to elastic strain. As an applied magnetic field is increased so that magnetic domains closely aligned to the applied field direction grow at the expense of other domains, the domain walls repeatedly become pinned and 'break free' from their pinning sites. BE is based on the detection, in a search coil placed close to the specimen, of sharp transient voltage pulses, associated with this discontinuous and thermodynamically irreversible jumping of the domain walls.

Schematically the measurement configuration is close to that shown in Figure 3 for MAE, but the probe is a small search coil placed between the magnet pole-pieces rather than the ultrasonic receiving probe shown for MAE. Unlike for SMA, the plane of the coil is often parallel to the sample surface. The sensitivities and characteristics of BE and MAE are quite different as BE signals can only be detected from near-surface material at depths less than 0.3mm. Spatial resolution is similar to that for SMA. A typical BE profile measurement as a function of magnetic field is included in Figure 4 for a quenched and tempered mild steel; correspondence between features in the BE and MAE profiles emphasises the complementary nature of the two techniques. For BE, a simple NDE parameter is the maximum signal as the applied field swings through zero. BE can be applied to monitor plastic damage (strain) effects and may prove particularly sensitive to high damage (plastic strain) levels. A possible disadvantage is that, unlike PA, BE is highly sensitive to a range of microstructural changes. Also the technique can be sensitive to the surface conditions.

2.2.4. *Other Magnetic Methods*. A number of other magnetic techniques exist which give data on plastic degradation effects, particularly in the case of high cycle fatigue [1]. Principally these are: magnetic flux leakage (a vibrating probe coil is sensitive to flux leakage at microdefects, and hence fatigue damage at the crack initiation stage should be measurable), magnetic energy loss (where magnetomechanical damping is proportional to the stress amplitude in high cycle fatigue), and magnetic induction (B) versus applied field (H) measurements (which utilises the fact that both coercivity and remanence decrease with plastic damage).

All these NDE applicable methods probe surface and near-surface effects (~1mm depth) but the microdefect resolution can be extremely good ($\leq 0.25mm^2$). They are applicable to ferromagnetic components only and require access to rust-free surfaces. Unfortunately, the previous load history of the sample must be well-characterised.

2.3. ACOUSTIC METHODS

While the MAE technique involves the emission of acoustic waves from magnetic perturbations, there are three other acoustic methods relevant to the measurement of residual stresses and their associated plastic strains or degradation. These are: ultrasonic velocity measurements to determine residual stresses, ultrasonic attenuation methods to determine certain kinds of plastic damage, and acoustic emission for in-situ monitoring of plastic damage processes.

2.3.1. *The Ultrasonic Velocity (UV) Combinations Method.* Considerable efforts have been made in recent years to develop ultrasonic NDE methods for the measurement of residual stresses within bulk components [10]. Ultrasonic velocity, v, for an isotropic material is related to the Young's modulus, E, and density,D, of the propagation medium by: v = $(E/D)^{1/2}$, and thus is affected by any overall bulk strain which changes the density. In practice the anisotropic elastic constant variation for different crystallite propagation directions results in more complicated expressions for the longitudinal and shear wave velocities, even in a polycrystalline aggregate. Measurement of the velocities for different propagation modes and directions permits the path-averaged stress tensor components to be determined from algebraic combinations of the velocities. Spatial resolution and penetration are both typically a few millimetres, and each NDE measurement can be made in under one second. The velocities can be measured with one sided sample access using the back-wall echo and a known sample thickness, or using two separate probes to transmit and receive the ultrasonic pulses.

Unfortunately, uncertain texture effects severely limit the usefulness of the technique in materials such as steel where there is considerable crystallite anisotropy in the elastic constants. The variation in the effective elastic constant with direction and mode in textured polycrystalline aggregates can have the primary effect on elastic wave velocity, masking that due to changes in lattice spacing. Nevertheless, the UV technique can provide the best overall NDE stress measurement method for non-magnetic materials if texture effects are either well-characterised or unimportant.

2.3.2. *Ultrasonic Attenuation Methods (UA).* Ultrasonic attenuation methods divide into analysis of the relative attenuation of successive integrated back wall echoes, the frequency dependence of this relative attenuation, the peak frequency in the fast Fourier transform frequency distribution (FFT peak frequency) of the first back-wall echo, and the ultrasonic back-scattered noise characteristics (received at times prior to the first back-wall echo). The UA methods are essentially directed at materials degradation assessment [11]. In the context of stress measurement, they can provide a measure of the plastic damage (strain) present. In particular, creep damage assessments appear to be possible using the noise analysis methods.

The NDE characteristics are similar to those for the UV method. Texture effects do not disturb the results as in the case of UV measurements of stress. However, grain size variation can have a pronounced effect on the attenuation characteristics. Ultrasonic attenuation results from geometric effects (which can be allowed for), true absorption, back-reflection and diffuse scattering. Most of the UA methods have the same spatial resolution and penetration parameters as for UV stress measurement. However, in ultrasonic noise analysis [12], the back-scattered signal can be time-gated and each time-window Fourier transformed separately to give the characteristics associated with a small (~1mm) region at a specified depth below the sample surface. Although at an early stage of development, this technique is likely to prove a powerful field-applicable NDE method for the detection of the degree and extent of deep-seated incipient materials degradation, and prove a valuable complement to residual stress measurements themselves.

2.3.3. *Surface Acoustic Waves (SAW).* Rayleigh waves or SAW [1] can be used to probe surface crack initiation processes, associated with plastic strain and other structural degradation processes. The method detects the generation of harmonics by small defects, and the NDE characteristics are similar to those for UA except that the penetration is limited to ~1mm. SAW is probably better suited to sensing extreme plastic deformation than are the UA methods, but two probes are required to transmit and receive the ultrasonic pulses.

2.3.4. *Acoustic Emission (AE)*. AE involves the detection of elastic waves originating from fracture processes as they happen [1]. Alone of all the techniques described in this paper, AE consists of in-situ monitoring of industrial plant, not service inspection. While the technique neither measures residual stresses nor plastic strains, AE can in principle provide uniquely complementary information on materials degradation processes during service. In principle, spatial resolution of 1mm is possible at 3 metres depth in steel. Major complications are the need to continuously monitor plant in service to determine the accumulated damage accurately and difficulties in data interpretation. Nevertheless such measurements may be useful in determining plastic strain accumulation in inaccessible parts of a component.

2.4. OTHER METHODS AND CONCLUSIONS OF REVIEW

There exist other NDE methods for probing small surface-breaking defects such as may be associated with plastic strain [1]. These include the study of: eddy currents, AC and DC potential drop, optical reflectance, infra-red (thermography), exoelectrons, etching and replication. There is insufficient space to describe these methods here but all are well-developed field-applicable NDE techniques. In general there exist many NDE and partially destructive methods which can probe surface defects. Ultrasonic methods are extremely effective at locating macroscopic cracks and defects in industrial components. Destructive methods such as hole-drilling and sectioning, and theoretical models such as finite element analysis also exist to characterise residual stresses. However, a growing requirement is being identified for genuine NDE assessment of both residual stresses and incipient damage (which may or may not be associated with stress). To this end, the XRD, MAE, SMA and UV methods must be considered the most likely candidates for future NDE stress measurement, and the XRD, PA, BE, UA (noise) and AE methods for incipient damage assessment, including determinations of plastic strain. Whereas destructive methods for validation exist for damage assessment (such as metallography on cut sections), this is not true for residual stress measurement where destructive treatments themselves perturb the residual stress distribution within the remaining bulk material. For validation of NDE stress measurement in the bulk, ND provides the only true experimental means.

3. Calibration and Validation using Neutron Diffraction

3.1. ND AND RESIDUAL STRESS MEASUREMENT

3.1.1. *ND vs Layering*. Figure 5 shows the stress distribution for a section from a mild steel double-vee test weldment, as measured by neutron diffraction and through the use of strain gauged   layering.   The observed yield-point tension at the edges of the weld, and compression in the centre are as predicted from the thermal cycling during the welding process, and agreement between the techniques is generally excellent. These data confirm the validity of ND calibrations of the portable NDE methods, especially where these can only measure outside the Heat Affected Zone (HAZ). By monitoring the width and intensity variations of the diffraction peaks, it is also possible to characterise texture variations within the sample, as well as to infer information on intergranular microstrain effects as an aid to the further interpretation of data from the portable NDE methods.

568

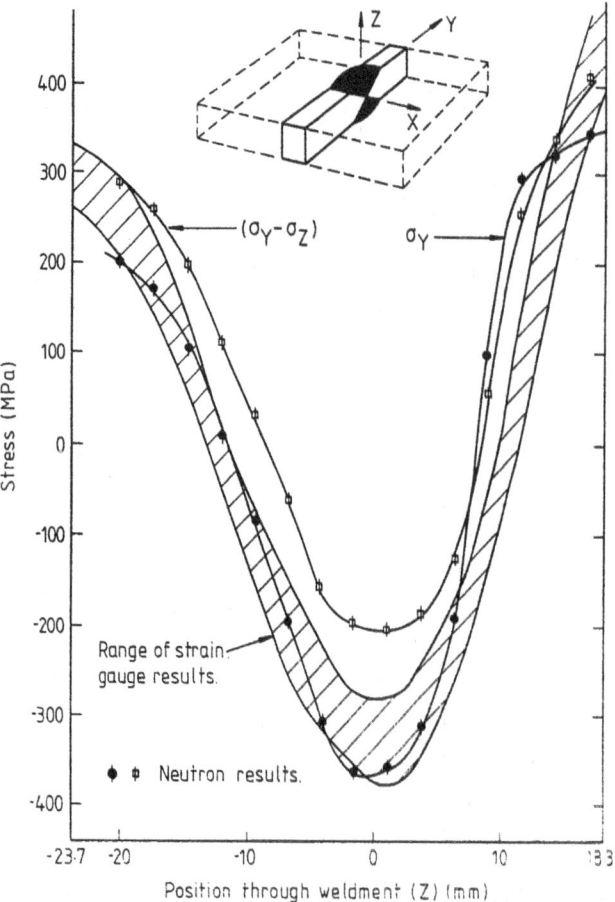

Figure 5.   Comparison of ND and strain-gauged layering   measurements of the
            residual stress across a mild steel double-vee weld section.

3.1.2. *MAE, SMA and UV Measurements of Residual Stress in Simple Welds.* Examples of
MAE and SMA measurements, made across a weldment of A533B steel, are shown in
Figures 6 and 7 respectively. As the weld-line is approached, the SMA increases in a
negative sense, indicating that the stress parallel to the weld-line is more compressive
than the perpendicular component. Over the weld itself, the SMA signal is positive,
indicating a change in sign of the stress difference, so that the stress parallel to the weld is
more tensile. The MAE results show a gradual increase in the sum of the stress
components as the weld-line is approached. Thus, on comparing the results, it can be
deduced that, away from the weld, compressive stresses exist parallel to the weld-line,
while, in the weld itself, both stress components are tensile. These general trends are as
expected, and are consistent with the ND data discussed above for the double-Vee weld and
elsewhere. In a well characterised case, such as this or the cracked fatigue specimen
discussed below, it is also possible to obtain entirely consistent UV data for the stress
distribution [10].

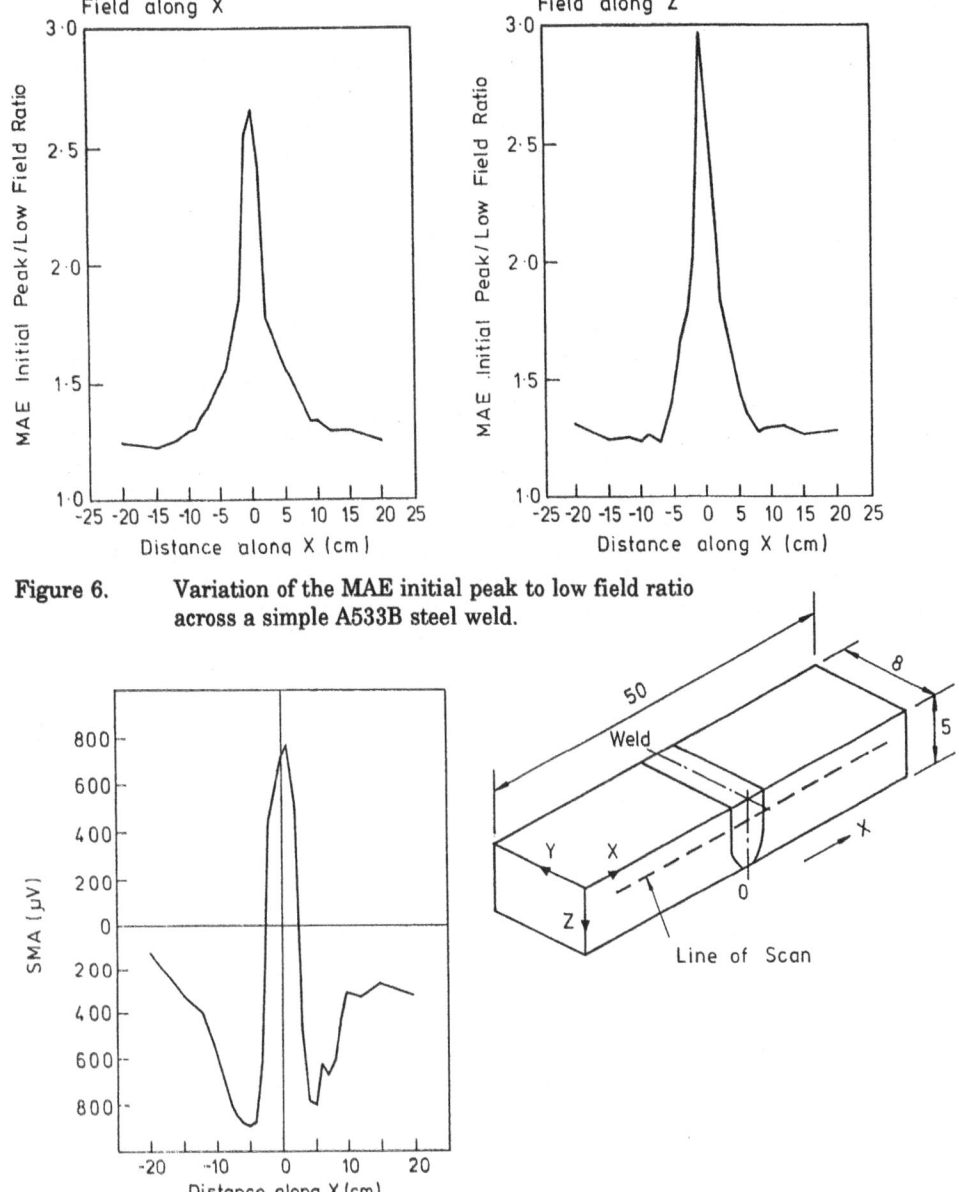

Figure 6.   Variation of the MAE initial peak to low field ratio
             across a simple A533B steel weld.

Figure 7.   Variation of the SMA voltage across a simple A533B steel weld
             and schematic of the weldment itself.

570

3.2. ND AND PLASTIC DAMAGE ASSESSMENT

Plastic damage resulting from static deformation, creep or high cycle fatigue, can usually be associated with the stress or thermo-mechanical history of a component. Thus, its measurement provides important complementary data to those obtainable for residual stress, as well as providing a direct measure of the accumulated incipient structural degradation, which will develop into more apparent damage and ultimately lead to component failure. An example is given below of comparisons between ND residual stress and PA plastic damage measurements in a fatigued compact tension specimen.

3.2.1. *PA Assessment of High Cycle Fatigue Damage* . Figure 8 shows the spatial distribution of incipient plastic damage around and ahead of the crack mapped out with about 1mm spatial resolution over the surface of a fatigue test specimen. The concentration of damage close to the crack tip is as expected, but note the rather extended area of damage, and also the second area of damage towards the far side of the sample, opposite the crack, where the bending moment has been greatest. The stress distribution, as measured by ND (not shown here but see reference [3]), shows the predicted oscillatory stress state ahead of the crack, not wholly coordinated with the plastic damage variation, - a hallmark of high cycle fatigue, since this is virtually the only mechanism which can lead to uncoordinated oscillatory stress and damage distributions. The combined use of residual stress and plastic damage measurement using PA could lead to future improved assessments of fatigue damage accumulation and residual service life. Although the examples quoted involve steels, the PA technique is applicable to most metals and alloys.

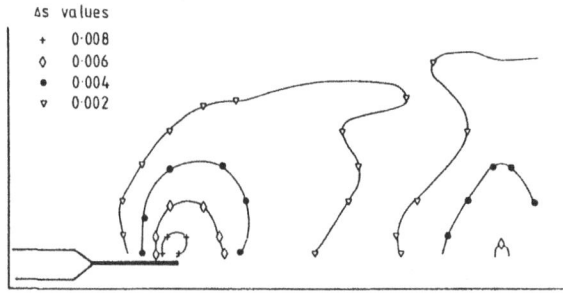

Figure 8.   Spatial distribution of the incipient plastic damage around and ahead of a fatigue crack.

4. Conclusions

A number of emergent portable NDE techniques for residual stress measurement now exist. In relevant generic cases, they all require validation or comparison with comprehensive determinations of the stress field below the surface. High resolution neutron diffraction can provide such validation, as well as investigate generic stress problems directly.

The neutron diffraction technique complements other diagnostic methods in research on the fundamental aspects of microstructure in both existing and new materials. The method is particularly powerful for probing elastic anisotropy and grain boundary interactions, which can influence materials degradation, and which must be characterised and understood before reliable NDE assessments of materials degradation can be made.

This work was carried out as a part of the Corporate Research Programme of AEA Technology.

References

[1] Allen, A.J., Buttle, D.J., Coleman, C.F., Smith, F.A. and Smith, R.L. (1988) "Microstructural examination of fatigue accumulation in critical LWR components", EPRI report NP-5590, Electric Power Research Institute, Palo Alto, California.

[2] Allen, A.J. and Buttle, D.J. (1992) "From microstructural assessment to monitoring component performance - a review relating different nondestructive studies", accepted for publication in Nondestructive Testing and Evaluation.

[3] Allen, A.J., Hutchings, M.T., Windsor, C.G. and Andreani, C. (1985) "Neutron diffraction methods for the study of residual stress fields", Advances in Physics, 34, 445-473.

[4] Macherauch, E. (1976) "Review of European advances in nondestructive XRD-testing", Sagamore Army Materials Research Conference Proceedings, 23, 195-220.

[5] James, M. and Cohen, J.B. (1978) "PARS - a portable X-ray analyser for residual stresses", Journal of Testing and Evaluation, 6, 91-97.

[6] Allen, A.J., Coleman, C.F., Conchie, S.J. and Smith, F.A. (1989) "Applications of positron annihilation to the monitoring of fatigue damage and creep in technological components", in J. Holbrook and J. Bussiere (eds), Nondestructive monitoring of materials properties, MRS Symposium Proceedings, 142, 131-142.

[7] Shibata, M. and Ono, K. (1981) "Magnetomechanical acoustic emission - a new method for nondestructive stress measurement", NDT International, 14, 227-234.

[8] Langman. R.A. (1988) "Measurement of stress by a magnetic method", in J.M. Farley and R.W. Nichols (eds), Proceedings of the 4th European NDT Conference 1987, Pergamon Press, Oxford, 3, 1783 - 1799.

[9] Sundstrom, O. and Torronen, K. (1979) "The use of Barkhausen noise analysis in nondestructive testing", Materials Evaluation, 37.3, 51 - 57.

[10] Allen, D.R. and Sayers, C.M. (1983) "The measurement of residual stress in textured steel using velocity combinations techniques", Harwell report AERE R11115, United Kingdom Atomic Energy Authority, London.

[11] See for example: Krautkramer, J. and Krautkramer, H. (1983) "Ultrasonic testing of materials", 3rd ed, Springer-Verlag, Berlin.

[12] Reynolds, W.N. and Smith, R.L. (1983) "Ultrasonic wave attenuation spectra in steels", Harwell report AERE R10753, United Kingdom Atomic Energy Authority, London.

TRIAXIAL ANALYSIS OF RESIDUAL STRESS FIELDS IN METALLIC PLATES

L. CASTEX and J. BARRALIS
Laboratoire MecaSurf ENSAM
Aix-en-Provence
FRANCE

ABSTRACT

In **X-ray diffraction techniques**, the direction of strain measurements lies in a range of about 45 to 60 degrees from the z-axis normal to the surface of the specimen. The true values of the following terms are directly obtained:

$(\sigma_{11} - \sigma_{33})$, $(\sigma_{22} - \sigma_{33})$, σ_{12}, σ_{13} and σ_{23} as well as the von Mises equivalent stress σ_{eq}. It appears clearly that σ_{33} and hence the hydrostatic pressure are not known.

Three different methods have been proposed and are currently used for this major problem to be dealt with, including experimental determinations and theoretical hypotheses.

1) d_0 is determined by measuring a fine powder representative of the region under investigation.
2) d_0 is determined from the sublayer material which is assumed to represent the stress-free state.
3) σ_{33} is assumed to be zero in the near-surface layer under investigation.
4) Poisson's ratio is assumed to be 0.5.

In **neutron diffraction techniques**, strain measurements are possible along three orthogonal directions respectively taken as x-, y- and z-axes. From these measurements the values of σ_{11}, σ_{22} and σ_{33} can be expressed as functions of the θ_0-angle. The same problem as above must be solved.

To get a reasonable value of the θ_0-angle two main methods are currently available. The first one is based on the fact that the stress field must be self-equilibrated in the cross section of the specimen and the assumption that the θ_0-angle remains constant. The second one uses a reference powder.

Previous results obtained from X-ray diffraction measurements have shown that the component σ_{33} can be taken equal to zero as a valid assumption in the case of metallic plates. In this paper we propose a new method for calculating the value of the θ_0-angle using this hypothesis associated with mechanical considerations on relationships between the hydrostatic pressure and the equivalent stress. This method will be illustrated by using experimental data obtained for a shot peened plate.

M. T. Hutchings and A. D. Krawitz (ed.),
Measurement of Residual and Applied Stress Using Neutron Diffraction, 573.
© 1992 *Kluwer Academic Publishers.*

PITFALLS OF LAYER REMOVAL TECHNIQUES IN X-RAY RESIDUAL STRESS MEASUREMENTS

M. R. JAMES
Rockwell International Science Center
1049 Camino Dos Rios
Thousand Oaks, CA, USA

ABSTRACT

The nondestructive aspect of X-ray residual stress analysis is often heralded as its most important attribute. Unfortunately, errors caused by the presence of steep subsurface stress gradients and difficulties in interpreting surface results often limit the usefulness of surface data. More often than not, subsurface data are required. This necessitates successive layer removal via electropolishing techniques. Correction procedures will be briefly reviewed and examples presented showing common residual stress distributions. It will be shown that, for most cases, the assumptions on which traditional correction procedures are based are usually abrogated by the method of layer removal. Fortunately, the corrections are small and can be neglected except when the stress gradient into the depth is sufficiently steep that the exponentially weighted average of the stress over the volume of irradiation is not representative of the true stress in that region. Other difficulties including coarsening of the grain size and changes in preferred orientation with layer removal will be illustrated.

M. T. Hutchings and A. D. Krawitz (ed.),
Measurement of Residual and Applied Stress Using Neutron Diffraction, 575.
© 1992 *Kluwer Academic Publishers.*

NEUTRONS VERSUS X-RAYS

L. PINTSCHOVIUS
Kernforschungszentrum Karlsruhe, INFP
Postfach 3640, W-7500 Karlsruhe, Germany

ABSTRACT. As X-ray and neutron stress analysis are both diffraction methods, they have many common features. The major advantage of neutrons is their high penetrating power which makes them a very attractive tool for the determination of both macro- and microstresses. However, the comparatively low brightness of neutron sources limits the spatial resolution and / or the throughput. The low speed of the neutron measurements is a particular handicap in view of the very limited beam time available at present day neutron sources. Therefore neutron and X-ray stresses analysis should not be considered as competing, but as complementary techniques.

1. Introduction

Whereas X-ray stress analysis has been a well-established method in materials science for many years, neutron stress analysis is still a somewhat exotic technique, the merits and limitations of which are not well known to many of its potential users. It is the aim of this paper to summarize for which problems neutrons are helpful or even unique and which problems remain the domain of X-rays. To begin with I want to recall those features which are common to X-ray and neutron stress analysis and differ from those of other methods for the determination of residual stresses (RS).

2. Common Features

X-ray and neutron stress analysis are both diffraction methods. The basic information obtained by these two techniques is very much the same: interplanar spacings $d_{\{hkl\}}$ for a sub-set of grains which have properly oriented lattice planes {hkl} to contribute to the diffraction intensity. For X-rays and neutrons well defined reflection lines can be observed only in crystalline solids, hence not in glasses or most plastics. Practically all materials which can be investigated by X-rays can be investigated by neutrons and vice versa. There are a few strongly absorbing (e.g. Cd and B) or weakly diffracting (e.g. V) materials which are not suitable for neutron diffraction but they are of minor technological importance. Only elastic strains are registered, which

M. T. Hutchings and A. D. Krawitz (ed.),
Measurement of Residual and Applied Stress Using Neutron Diffraction, 577–580.
© 1992 *Kluwer Academic Publishers.*

means that diffraction methods inherently discriminate between elastic and plastic deformations. Strong plastic flow may show up only by side effects such as reflection line broadening and/or variation of diffraction intensities due to texture. Residual as well as load stresses can be investigated. As has been explained in a separate contribution [1], the strains measured by diffraction methods are influenced both by macro- and microstresses. In particular, they are phase specific in multi-phase materials. The analogy between X-rays and neutrons in this respect implies that all complications arising from the superposition of macro- and microstresses are completely the same (examples are given in [2]). Hence, the use of neutrons is not automatically a help in such a case. However, it may become a help if the large penetration depth of neutrons is exploited, as explained in the next section.

3. Advantages of Neutrons

In classical diffractometry the main reason to use neutrons is the fact that light elements scatter neutrons on average as well as heavy elements. For stress analysis, this property of the neutron is unimportant. Instead, it is the large penetration depth which makes neutrons a unique tool for probing strains in the interior of bulk solids in a non-destructive way. The use of X-rays for this purpose requires stepwise removal of surface layers. Not only does this destroy, or at least alter, the specimen but it is also a tedious procedure if information is to be obtained over a wide spatial range. Moreover, the reconstruction of the original stress state from the data measured after removal of surface layers is not always unambiguous: Plastic deformation induced by the layer removal procedure or occurring after partial relaxation of RS may influence the stresses observed on the newly created surfaces.

Strains in the thin surface layer probed by X-rays may be strongly influenced by corrosion, decarburization or by machining. Therefore, the specimen is often polished before X-ray values are taken, which is not necessary for neutron stress analysis.

The large penetration depth of neutrons gives them an edge over X-rays for the investigation of composite materials [3]. In order to get strain values which are representative of the bulk, the domains of the different constituents have to be small compared to the penetration depth. If the material is composed of fibres or grains which are several µm thick or even larger, X-ray results will be strongly affected by surface effects, whereas neutron results are not.

For the investigation of grain interaction stresses, the large penetration depth of neutrons may be helpful, because it allows to determine d vs $\sin^2\psi$-distributions up to $\sin^2\psi = 1$ [2]. It is true that modern X-ray equipment allows to cover a wide range in $\sin^2\psi$ as well, that is up to $\sin^2\psi = 0.9$, but in strongly textured specimen the missing region $0.9 < \sin^2\psi \leq 1$ may still be important. Hence are in cold rolled steel the majority of the grains with an 110-axis in the rolling direction and parallel to the surface. Only neutrons can probe the lattice strains of these grains in the rolling direction.

Sometimes it is an elegant way to separate micro and macro RS by illuminating the whole cross-section of the specimen [2]. As macro RS have to be zero when averaged over the whole cross-section, the strains observed with a wide beam reflect directly micro RS. Obviously, this technique can be used only with neutrons.

4. Disadvantages of Neutrons

First I want to point out that what is sometimes suspected by the non-expert to be a drawback of neutrons is not true: irradiation by thermal neutrons does not create defects, they simply have too little energy to do so. The sample may become activated because of neutron absorption, but activation levels remain usually negligible if proper care is taken not to irradiate unnecessarily large areas of the sample. For investigations of microstresses often a wide beam is used, but here exposure times are short which again keeps activation low. If a sample has become activated after prolonged exposure to the neutron beam, it can often be brought back to a normal laboratory after a decay time of a few days or weeks. The real drawbacks of neutrons are the following:

4.1. LOW BRIGHTNESS OF NEUTRON SOURCES

Available neutron sources are much less bright than ordinary X-ray tubes and particularly modern synchrotron radiation sources. The comparatively low brightness of neutron sources means that the gage volume has to be of the order of 10 mm^3 instead of 10^{-1} mm^3 as in the case of X-rays. Hence, the spatial resolution is poorer in the neutron case. In particular, it is impossible to confine the neutron beam to surface layers as thin as the penetration depth of X-rays. Therefore, X-rays are normally the best or even only choice to investigate the RS in the near surface region. Only in case that the near surface region of interest extends to a depth of at least 100 µm stress analysis by neutrons becomes competitive [4].

When the spatial resolution in neutron stress analysis is pushed to its limits (which is often necessary to manage stress gradients), the technique becomes slow. This severely restricts the number of specimen which can be investigated in a reasonable amount of time, much more than for X-rays. Increasing the throughput by optimizing the instrument remains an important task for the people working in this field. One of the aims of the present workshop is to collect ideas for such an optimization (see the corresponding contributions in this volume).

4.2. LIMITED ACCESS TO NEUTRON BEAMS

The speed handicap of neutron stress analysis would be much less significant if there were more neutron diffractometers available. Unfortunately, there are less than 20 sources around the world where neutron stress analysis can be carried out (the number seems to be even decreasing with time), and at

580

these places beam time has normally to be shared with people doing different types of experiments on the same machines. Sharing beam time has also the inconvenience that the delicate system of slits, supports etc. has to be re-installed and calibrated before each experiment. There are very few dedicated instruments so far. Efforts should be made to create new dedicated instruments, e.g. at the HFR Grenoble.

4.3.   HIGH COSTS

Neutron sources are very costly to build and to operate. Consequently, neutron measurements are much more expensive than X-ray measurements. Of course, the ordinary user of a neutron diffractometer does not directly pay for the neutrons and so the high costs are somewhat irrelevant to him. However, for commercial applications the high costs *are* a handicap. The scientific community feels the high costs of neutrons indirectly through the shut-down of existing neutron sources due to budgetary reasons.

4.4.   NO PORTABLE SOURCES

Most X-ray investigations are done in the laboratory, but in-situ investigations are also possible by using portable diffractometers. This possibility does not exist for neutrons and there is very little hope that portable neutron sources will ever become available.

5. Conclusions

Neutron and X-ray stress analysis have both their merits and limitations. The great advantage of neutrons is their penetrating power which is two or even three orders of magnitude larger for most materials than that of X-rays. This makes neutrons highly attractive for a large number of investigations. It has lead to an increasing demand for neutron beam time which cannot be met at existing neutron sources. So it is obvious that the possibilities of X-rays should be fully exploited before neutrons are asked after. There are, of course, some borderline cases, where information on stresses in the interior can be obtained also by X-rays if the destruction of the specimen is acceptable. In such cases, neutron can nevertheless be helpful to check that removal of layers does not induce unpredictable changes of the stress state. In conclusion, neutron and X-ray stress analysis should not be considered as competing, but as complementary, techniques.

6. References

1. Pintschovius, L., 'Macrostrains, Microstrains and Stress Tensors', this volume.
2. Pintschovius, L., 'Grain Interaction Stresses', this volume.
3. Krawitz, A.D., 'Overview of Composites', this volume.
4. Webster, P.J., 'Spatial Resolution an Strain Scanning", this volume

INDEX

AUTHORS INDEX

The manufacturer's authorised representative in the EU is Springer
Nature Customer Service Centre GmbH, Europaplatz 3, 69115 Heidelberg,
Germany. If you have any concerns regarding our products, please
contact ProductSafety@springernature.com

Printed and bound by CPI Group (UK) Ltd, Croydon, CR0 4YY
23/04/2026
02095628-0010